Hypotheses and Perspectives in the History and Philosophy of Science

Raffaele Pisano • Joseph Agassi
Daria Drozdova
Editors

Hypotheses and Perspectives in the History and Philosophy of Science

Homage to Alexandre Koyré 1892-1964

Springer

Editors
Raffaele Pisano
Lille University
Villeneuve d'Ascq, France

Joseph Agassi
Tel Aviv University
Tel Aviv, Israel

Daria Drozdova
National Research University Higher
School of Economics
Moscow, Russian Federation

ISBN 978-3-319-61710-7 ISBN 978-3-319-61712-1 (eBook)
DOI 10.1007/978-3-319-61712-1

Library of Congress Control Number: 2017952599

Printed on acid-free paper

This Springer imprint is published by Springer Nature
The registered company is Springer International Publishing AG
The registered company address is: Gewerbestrasse 11, 6330 Cham, Switzerland

Foreword

A Scholar Without Borders

Like Einstein, Alexandre Koyré could have thought of himself as a citizen of the world.[1] He was one of those scholars who went through the turmoil of two world wars clearly aware of their civic allegiance and deeply concerned about the future of civilization, without ever endorsing nationalistic claims. Born in tsarist Russia in the early years of the oil boom in Baku, Koyré took his secondary education in present-day Georgia and went to Göttingen University in 1908 (Zambelli 2016). Three years later, he moved on to Paris, where he started his PhD research at the *École Pratique des Hautes Études*. In 1914, although he was still a foreigner in France, he first enlisted in the French army and then soon volunteered to fight in a Russian regiment. However, during the October 1917 Revolution, he did not join any of the rival movements and came back to Paris.

Koyré never really settled in France, although he got a number of teaching positions and became a French citizen.[2] He traveled all over Europe, in the Middle East, in Argentina, and in the United States. In 1940, during the German invasion of France, he was lecturing in Cairo. He quickly decided to rally the Free French Troops, and de Gaulle sent him to the United States. After a long journey via India, the Pacific Ocean, and San Francisco, the Koyrés arrived in New York. There, Alexandre joined the New School of Social Research and became dean of the *École Libre des Hautes Études*, a consortium of Belgian and French scholars in exile (Gillispie 2008).

Koyré could feel at ease everywhere, but he really belonged to nowhere. As he was fluent in many languages – at least Russian, German, French, and English – he could study and teach in many countries, including France, Egypt, Lebanon, and the

[1] Einstein is famous for this humorous remark in his address to the French Philosophical Society at the Sorbonne (6 April 1922): "If my theory of relativity is proven successful, Germany will claim me as a German and France will declare that I am a citizen of the world. Should my theory prove untrue, France will say that I am a German and Germany will declare that I am a Jew" (French press clipping, 7 April 1922 [Einstein Archive 36–378], and *Berliner Tageblatt*, 8 April 1922 [Einstein Archive 79–535]).

[2] It seems that he got the French citizenship on November 23, 1922, thanks to a decree of "admission at home" which had to be renewed every five year.

United States. Thanks to his good command of classical Latin, he could also immerse himself in medieval and Renaissance studies. He developed such a remarkable intellectual flexibility that he could read and empathically interpret early modern mystics (Koyré 1929), as well as Descartes' metaphysics, Galileo's physics, and Newton's dynamics. Among his students and collaborators, he was famous for his sharp understanding of subtle nuances in text analysis and translation (Koyré 1943; Cohen 1987).

Across Disciplinary Boundaries

Although at the turn of the twentieth century a disciplinary organization came to prevail in academic research, Koyré never fully embraced a disciplinary identity. For example, he studied philosophy under Edmund Husserl in Göttingen, but, whatever the deep influence the founder of phenomenology had upon him, he was not brainwashed by his first mentor, nor was he converted to the then-booming field of axiomatization of mathematics by David Hilbert's lectures in Göttingen. When he completed his philosophical training in Paris, he attended Henri Bergson's lectures at the *Collège de France*, which attracted crowds of the Parisian intellectual elite (Azouvi 2007). He also attended the Sorbonne courses of André Lalande and Emile Brunschvicg, two of the leading figures of the *Société Française de Philosophie*. Yet Koyré was not really attracted by mainstream and star philosophers. Rather, he became a disciple of a more marginal philosopher of science, Emile Meyerson, who was an immigrant in France, like himself. Meyerson, who was already in his sixties, invited Koyré and his wife Dora to his home and even occasionally supported them financially when Alexandre was jobless. Until Meyerson's death in 1933, Koyré used to call him "*mon cher maître*" and visited him every Thursday in his salon (Meyerson 2009, pp. 226–253).

More surprisingly, Koyré did not dedicate his earlier scholarship to the history or the philosophy of science. He was primarily interested in early modern religious thought and theology. Following his PhD on St. Anselm, he moved on to Descartes' proof of the existence of God (Koyré 1922, 1923). He got a temporary lectureship in history of religious thought at the *École Pratique*. When it was turned into a tenured position of *directeur d'études* in 1932, Koyré lectured about the relations between science and faith. He did not give up the history of philosophy, for that matter. In 1929, as he was appointed lecturer in philosophy at Montpellier University, he tried to introduce his students ("mes gosses" as he referred to them) to Hegel's system. He subsequently wrote a state-of-the-art report and an introduction to Hegel's doctrine. At the same time, he diligently paid tribute to his mentors, editing Husserl's *Recherches Philosophiques* and dedicating reviews and articles to Meyerson in various languages (Koyré 1926, 1927, 1931, 1933). Indeed, this was a characteristic feature of Koyré's career: while he was building up his own philosophical framework, he was keen to spread the views of the thinkers who mattered for him. Sharing knowledge was his priority. He also studied Russian intellectual history for a course at the Institute of Slavic Studies of the University of Paris and published two major volumes on this topic (Koyré 1929, 1950).

Given the variety of domains of expertise that Koyré displayed in his career, how are we to understand that he is mainly considered as a historian of science?

Galileo and Koyré

In 1941, Koyré arrived in the United States with the three volumes of his *Etudes Galiléennes* in his suitcase (Koyré 1939). The book, recently published in Paris, provided the basis of his courses in New York and Cambridge in 1942–1943. The history of science, as it developed in the United States in the interwar period under the influence of George Sarton, the founder of *Isis,* was seen above all as a chronological series of descriptions of discoveries, carefully detached from their social and intellectual contexts. Arnold Thackeray later characterized this early period of institutionalization of the history of science in the United States as the prehistory (Thackray 1980). Against this background, Koyré's lectures on Galileo created a great impression on a dozen scholars. Isaac Bernard Cohen, among them, was immediately attracted by Koyré's approach to the history of science. What Koyré offered was nothing like a boring linear chronology of great discoveries supposedly made due to crucial experiments. Koyré's attempt to analyze Galileo's thinking process and to outline the inner coherence of his system revealed an exciting battle of ideas about natural phenomena. "From Koyré we learned about the inner dynamic quality of scientific thought" (Cohen 1987, p. 56). It is fair to say that Koyré invented a new Galileo, a Platonist Galileo who, in turn, invented Koyré as a post-positivist historian of modern science.

Not all students were convinced by Koyré's portrait of Galileo. Giorgio de Santillana and Leonardo Olschki could not agree that Galileo did not actually perform the experiments that he presented. In addition, they raised objections. Still, the long controversy that ensued with Olschki and later with Stillman Drake enhanced the aura surrounding Koyré in the United States.

Unlike other French refugees, Koyré did not leave his country of exile in the aftermath of World War II. Indeed, he came back to Paris to resume his position at the *École Pratique* and his lectures on religious thought, but he shared his time between the United States and Paris. In 1951–1952, he was elected a fellow at the Institute for Advanced Studies in Princeton, and he spent half of the years there. But he did not stay quietly in Princeton when he was in the United States. Always on the move, Koyré lectured in Chicago, at Johns Hopkins in Baltimore, and at the University of Wisconsin in Madison, thus attracting an increasing number of young scholars at a crucial moment when history of science was being promoted as a profession in the US university system (Aurières 2017).

Koyré's intellectual approach to science, emphasizing the unity of thought through the close interaction between science, religion, and metaphysics, was particularly attractive in the heyday of the Cold War. It potentially provided US historians of science with a robust armor against the increasing influence of the Marxist interpretations of history of science, such as that developed in Britain by John D. Bernal and his followers. Koyré's intellectual history was a good antidote for addressing what was identified as perhaps the most "critical problem" in the field – the intense philosophical debate about the impact of economic and social structures on scientific change. This question pervaded the Conference on the History of Science held in Madison in 1957, the year of Sputnik, and was reflected in the resulting widely read volume edited by Marshall Clagett (Clagett 1959; Fuller 2000).

Through its widespread circulation in the right places at the right moment, the *Etudes Galiléennes* became a landmark book, one of those books which opened up new research pathways for decades. In a minor key, Koyré's *Études Galiléennes* anticipated the later grand impact of Thomas Kuhn's *Structure of Scientific Revolutions.* In this volume, published 2 years before Koyré's death, Kuhn acknowledged Koyré's influence on his vision of scientific change (Kuhn 1962, p. v–vi, 1977, p. 11; Conant and Haugeland, pp. 286).

Ambiguous Heritage in France

The contrast between the impact of Koyré's works in the United States and in France is striking. While he influenced an entire generation of historians of science and thus greatly contributed to professionalizing and dignifying the history of science in American universities, in France, his courses at the *École Pratique des Hautes Études* attracted few students. I.B. Cohen, in keeping with a remark by Gillispie, contrasted the image of Koyré as a historian of science in the United States with his reputation as a philosopher in Europe (Cohen 1987, p. 64). However, this is not exactly the case, since Koyré has never been recognized as a philosopher in France and his sphere of influence there in the history of science remained more limited than in the United States.

In the 1950s, when the *Centre National de la Recherche Scientifique* was established, history of science was integrated into the philosophy section, and Koyré was part of it. The coupling of philosophy and history of science is a major feature of the French system and has long favored the practice of intellectual history focused on scientific concepts and ideas. Yet Koyré did not benefit from this situation. To be sure, the students he recruited at the *École Pratique* gradually assisted him and later became professional historians of science or technology (Taton 1987). But Koyré was rebuffed in 1951, when he applied for a position at the *Collège de France*, just as Meyerson had been turned down 30 years before. Both of them remained at the periphery of the French system, because they did not follow the standard *cursus* of French academics (Chimisso 2008).

Moreover, Koyré's program of a history of scientific ideas was rivaled by the overarching influence of Gaston Bachelard's historical epistemology. Well established at the Centre of History of Science and Technology of the Sorbonne,

Bachelard attracted large audiences and created a research school of philosophers cum historians of science, including Georges Canguilhem. In 1958, an alternative Centre of History of Science and Technology was created for Koyré by the *École Pratique des Hautes Études*, which in 1966 was rechristened the *Centre Alexandre Koyré*. For decades, there were virtually no interactions between the two centers respectively located on the left bank and the right bank of the river Seine. Canguilhem nevertheless later boldly claimed that Koyré was part of the "French tradition of epistemology," on the assumption that Koyré's interpretation of the scientific revolution exemplified Bachelard's "epistemology of rupture" (Canguilhem 1968, p. 14). Although this view distorts Koyré's much more sophisticated history of ideas, it has been widely spread and has become a kind of golden legend among French epistemologists (Bensaude-Vincent 2016). Still, the golden legend about the "French tradition of epistemology" hardly conceals the cleavage between two rival traditions of history of science and the subsequent marginalization of Koyré. Remarkably, it was an Italian scholar, Pietro Redondi, member of the Centre Alexandre Koyré, who edited the volume of Koyré's lectures (Redondi 1986).

The time has come to reconsider Koyré's works in a broad international perspective and the present volume will hopefully pave the way for a reassessment of his influence.

CETCOPRA, Université Paris 1 Bernadette Bensaude-Vincent
Panthéon Sorbonne
Paris, France
bvincent@univ-paris1.fr

References

Aurières E (2017) Alexandre Koyré aux Etats-Unis: un ambassadeur de l'histoire des sciences, PhD. Diss., Université Paris 1 Panthéon-Sorbonne.

Azouvi F (2007) La gloire de Bergson. Essai sur le magistère philosophique. Gallimard, Paris.

Bensaude-Vincent B (2016) Koyré dans la légende dorée de la tradition épistémologique française, Communication delivered at the conference Actualité d'Alexandre Koyré. October 18. Centre Alexandre Koyré, Paris.

Canguilhem G (1968) Introduction : l'objet de l'histoire des sciences in Études d'histoire et de philosophie des sciences concernant les vivants et la vie. Vrin, Paris.

Chimisso C (2008) Writing the History of the Mind: Philosophy and Science in France, 1900 to 1960s. The Open University, Milton Keynes–UK.

Clagett M (1959) (ed) Critical problems in the History of Science. The University of Wisconsin Press, Madison

Cohen IB (1987) Alexandre Koyré in America: some personal reminiscences. History and Technology: An International Journal 4/1–4:55–70.

Conant J, Haugeland J (2000) (eds) The Road Since Structure. Thomas S. Kuhn. Philosophical Essays, 1970-1993, with an Autobiographical Interview. The University of Chicago Press, Chicago.

Fuller S (2000) Thomas Kuhn, A Philosophical History for Our Times. The University of Chicago Press, Chicago.

Gillispie CG (2008) Entry: Alexandre Koyré. Complete Dictionary of Scientific Biography. Retrieved January 2017 http://www.encyclopedia.com/doc/1G2–2830902380.html the weblink does not work

Koyré A (1922) Essai sur l'idée de Dieu et les preuves de son existence chez Descartes. Ernest Leroux, Paris.

Koyré A (1923) L'idée de Dieu dans la philosophie de St. Anselme. Ernest Leroux, Paris.

Koyré A (1926) La tragédie de la raison. La philosophie de Émile Meyerson [in Russian]. Zveno 2–4 and 11–12.

Koyré A (1927) É. Meyerson. La déduction relativiste [in Russian]. Versty 2:269–274.

Koyré A (1929 [1979]) La philosophie de Jacob Boehme. Étude sur les origines de la métaphysique allemande. 3rd édition. Vrin, Paris.

Koyré A (1929 [1976]) La philosophie et le problème national en Russie au début du XIXe s., Librairie Honoré Champion, Paris. 2nd édition. Gallimard, Paris.

Koyré A (1931) Die Philosophie Émile Meyerson. Deutsch–Französische Rundschau 4:197–217.

Koyré A (1933) Du cheminement de la pensée, par É. Meyerson. Journal de psychologie normale et pathologique 27:647–655.

Koyré A (1939) Études galiléennes. Vol. I. À l'aube de la science classique. Vol. II La loi de la chute des corps. Descartes et Galilée, Vol. III Galilée et la loi d'inertie. Hermann, Paris. [Reprint 1966 and 1980]

Koyré A (1943) Traddutore–tradittore, à propos de Copernic et de Galilée. Isis, 34/95 [1966. reprinted in Études d'histoire de la pensée scientifique. Gallimard, Paris, pp. 272–274.

Koyré A (1950) Études sur l'histoire de la pensée philosophique en Russie, Vrin, Paris.

Kuhn T (1962 [1970]) The Structure of Scientific Revolution. The University of Chicago Press, Chicago. 2nd ed. [see also *Id.* (1977) The Essential Tension, The University of Chicago Press, Chicago.

Meyerson E (2009) Lettres françaises, edited by B. Bensaude-Vincent and E. Telkes–Klein. CNRS éditions, Paris.

Redondi P (1986) (ed) Alexandre Koyré. De la mystique à la science. Cours et conférences (1922–1962) [2nd ed. 2016]. EHESS, Paris

Taton R (1987) Alexandre Koyré et l'essor de l'histoire des sciences en France (1933 à 1964). History & Technology 4/1–4:37–53.

Thackray A (1980) The Pre–History of an Academic Discipline: The Study of the History of Science in the United States, 1891–1941. Minerva 18:448–473.

Zambelli P (2016) Alexandre Koyré in incognito. Olschki, Paris.

Acknowledgments

We want to express our appreciation to all contributing authors for their efforts to produce papers of interest and of elevated quality worthy of this Springer volume. The result is excellent.

Therefore, we also express our warm and pleasant gratitude to all our distinguished authors and to Bernadette Bensaude-Vincent (Paris 1 Panthéon Sorbonne University, France) for having kindly written her *foreword* for this book, offering the readers her authority standpoint on the subject.

Finally yet importantly, our acknowledgments one more time are addressed to Lucy Fleet, Springer associate editor, and her staff for doing a good job and the positive reception of our Koyré anniversary project and to the anonymous Springer referees for their excellent suggestions.

March 2017 The Editors

Remarks for the Reader

The papers in this volume have been independently double-blind peer-refereed, and the book followed Springer's double-blind review policy. The authors' contributions are in alphabetical order. The editors have respected individual authors' different ideas and historical, philosophical, and epistemological accounts. Therefore, the editors are not responsible for the contents. Each of the eminent authors is responsible for his/her/their own opinions, which should be regarded as personal scientific and experienced background. The authors are also univocally responsible for images, reprints, quotations, acknowledgments, and all related permissions/approvals displayed/not displayed in their papers.

March 2017 The Editors

Contents

Contributors

Joseph Agassi Tel Aviv University, Tel Aviv, Israel

York University, Toronto, Canada

Bernadette Bensaude-Vincent Paris 1 Panthéon Sorbonne University, Paris, France

Charles Braverman Archives Henri Poincaré–CNRS, Lorraine University, Nancy, France

Paolo Bussotti Department of Human Studies, Udine University, Udine, Italy

Mauro L. Condé Department of History, Federal University of Minas Gerais – UFMG, Belo Horizonte, Brazil

Francesco Crapanzano Department of Cognitive Science, Psychology, Pedagogy and Cultural Studies (COSPECS), Messina University, Messina, Italy

Mario De Caro Dip. di Filosofia, Università Roma Tre, Rome, Italy

Department of Philosophy, Tufts University, Medford, MA, USA

Amelia de Jesus Oliveira Department of Philosophy, IFCH, State University of Campinas, Campinas, SP, Brazil

Jose Carlos Pinto de Oliveira Department of Philosophy, IFCH, State University of Campinas, Campinas, SP, Brazil

Dominique Descotes Institut d'Histoire des Représentations et des Idées dans les Modernités, Clermont Auvergne University, Clermont-Ferrand, France

Antonino Drago Formerly at Physics Department, Napoli Federico II University, Naples, Italy

Daria Drozdova National Research University Higher School of Economics, Moscow, Russian Federation

Massimo Ferrari Department of Philosophy and Science Education, Torino University, Torino, Italy

Stephen Gaukroger Sydney University, Sydney, NSW, Australia

Glenn Allen Hartz Ohio State University, Mansfield, The United States

Gérard Jorland Paris École des Hautes Études en Sciences Sociales (EHESS), Paris, France

Patrick Kuehner Lewtas Department of Philosophy, American University of Beirut, Beirut, Lebanon

Anna Maria Lombardi Dipartimento di Fisica, Università degli Studi di Milano, Milano, Italy

Rodney K.B. Parker Department of Philosophy, The University of Western Ontario, London, ON, Canada

Raffaele Pisano Lille University, Villeneuve d'Ascq, France

Archive Poincaré, Lorraine University, Nancy, France

Unit HPS, Sydney University, Sydney, NSW, Australia

Diederick Raven Utrecht University, Utrecht, Netherlands

Marlon Salomon Faculty of History, Federal University of Goiás, Brazil

John A. Schuster Unit for History and Philosophy of Science & Sydney Centre for the Foundations of Science, Sydney University, Sydney, NSW, Australia

Jean-François Stoffel Haute École Louvain en Hainaut, Montignies-sur-Sambre, Belgium

Anna Yampolskaya National Research University Higher School of Economics, Moscow, Russian Federation

Introduction: Homage to Alexandre Koyré

An Outline

Fifty years have passed since the death of Alexander Koyré (born Taganrog, Russia, August 29, 1892; died Paris, France, April 28, 1964).

This chapter is an homage to Koyré, who was one of the most influential historians of science of the twentieth century and an eminent representative of both the Western European historical-epistemological tradition and (at that time) the tradition of eastern countries. His activities in the United States in the 1940s and 1950s established a bridge between the European tradition of the history of science and American academic environments. Indeed, a whole generation of American historians of science grew up under his direct influence. Very notable was his impact on Thomas Kuhn (1922–1996) and Kuhn's work *The Structure of Scientific Revolutions*. Although Koyré is known primarily as a historian of science, his interest in philosophical questions about the nature of scientific knowledge and the role played by philosophical concepts in the birth and development of modern science is evident. Koyré has compounded historical arguments (e.g., (a) the choice of the infinity in mathematics, "[...] (b) the destruction of the cosmos [...], and (c) the geometrization of space [...]") and philosophical issues (e.g., extra-scientific nature as part of the foundations of scientific theories). Other historians who wrote on the same topic considered artisans' work and inventions to be the main catalyst for the birth of seventeenth century science. Alexandre Koyré, however, proposed an opposite-antipodal thesis: even building upon the work of the most brilliant artisan, the result would inevitably be undermined by the inaccuracy of measurement. In his words:

> The new science, we are told sometimes, is the science of craftsman and engineer, of the working, enterprising and calculating tradesman, in fact, the science of rising bourgeois classes of modern society. There is certainly some truth in these descriptions and explanations [. . .]. I do not see what the *scientia activa* has ever had to do with the development of the calculus, nor the rise of the bourgeoisie with that of the Copernican, or Keplerian, astronomy theories. [. . .] I am convinced that the rise and the growth of experimental science is not the source but, on the contrary, the result of the new *theoretical*, that is, the new *metaphysical* approach to nature that forms the content of the scientific revolution of the

seventeenth century, a content which we have to understand before we can attempt an explanation (whatever this may be) of its historical occurrence. (Koyré 1965, pp. 5–6)

I shall therefore characterize this revolution [the birth of modern science] by two closely connected and even complementary features: (*a*) the destruction of the cosmos and therefore the disappearance from science – at least in principle, if not always in fact – of all considerations based on this concept, and (*b*) the geometrization of space, that is, the substitution of the homogeneous and abstract – however now considered as real – dimension space of the Euclidean geometry for the concrete and differentiated place-continuum of pre-Galilean physics and astronomy. (*ivi*, p. 6).

In the past century Koyré was part of a notable debate concerning the historiography of science (Redondi 1987). As usual the debate identified two types of investigations: (a) internal historiographies,[1] which tend to explain the history of science relying upon the inner development of scientific ideas and on the conceptual views on the scientists, and (b) external historiographies, which tend to underline the decisive role played by social components through the development of science. This classification is clearly suggested by the specific nature of the subject investigated. Throughout the nineteenth century, theories with a different approach, with respect to the Newtonian paradigm, suggested that *subjective history* (e.g., history thought and experienced by scientists) could be different from *effective history* (e.g., history relying on the fundamental choices made by scientists who influenced the interpretation of history by the way in which they constructed the science). Similarly, *subjective history* was also different from the *objective history* presented in textbooks as lists of data concerning mathematical laws, argumentative techniques, and objective concepts (Pisano and Gaudiello 2009a, b; Gillispie and Pisano 2014, chap. 6). In this regard, in 1961, for the celebration of the centenary of the birth of Émile Meyerson, Koyré gave a great contribution:

Indeed we do not know, or at least we know just as we think: the phenomenological description is a difficult thing, and in current thinking the form and the content are inextricably intertwined. So not only it is difficult to give an account of its unconscious assumptions, of the underlying axiomatic that carries and informs it. But it is almost inevitable to confuse the present form, which can be and will be short, with its essential form and structure. And it is much easier to identify it, analyzing and studying a thought that is foreign to us, theories that are no more ours, above all if we continue this analysis throughout long periods of time, so that the variety of contents would let emerge the unity of operations.[2]

Regarding the influence of background/heritage on the integration of the history and philosophy of science, both as methods and disciplines, Koyre wrote:

History of Science without philosophy of Science is blind, I must now undertake to show that philosophy of science without History of Science is empty. (Hanson NR (1962) The Irrelevance of History of Science to Philosophy of Science. The Journal of Philosophy 59:[p. 580] 574–586)

[1] Koyré's notes about two types of historiography (Koyré 1963, 1973) are very interesting, particularly as a reply to Henry Guerlac's (1910–1985) talk (Guerlac 1963). See the proceedings of *Symposium on the History of Science* in 1961, edited by A.C. Crombie (1963).

[2] Koyré 1986, p 138. See also: "Message d'Alexandre Koyré à l'occasion du centenaire de la naissance d'Emile Meyerson" (Koyré 1961b).

On the role played by *history of science* and *philosophy of science* in their mutual researches/disciplines, we also read:

1855

> Paraphrasing of a Kant's passage in *Critique of Pure Reason*: "Thoughts without content are void; intuitions without concepts[ions are] blind". Transcendental Logic. On Logic in general [A51-B75], p. 46 [on the opposite ideas, empiricism and rationalism, etc.]

1971

> "Philosophy of science without history of science is empty; history of science without philosophy of science is blind". (Lakatos I (1971) History of Science and Its Rational Reconstructions In: Buck RC, Cohen RS (eds). PSA Proceedings. The Boston Studies in the Philosophy of Science. VIII. Reidel, Dordrecht, pp. [91] 91–136) [See also debate Lakatos–Kuhn, Notes on Lakatos, etc.].

2012

> Mauskopf S, Schmaltz T (eds) (2012) Integrating History and Philosophy of Science: Problems and Prospects. Boston Studies in the Philosophy and History of Science. Springer, Dordrecht.

The Book

From the perspective of the anniversary of Koyré's death, the book *Hypotheses and Perspectives in the History and Philosophy of Science* aims to discuss:

(a) Koyré's intellectual matrix and heritage, which encourages reflection on the interdisciplinary field of the philosophical history of science.
(b) The history of scientific thought, which cannot be entirely separate from epistemological and philosophical thought.
(c) The concept of the scientific revolution and the birth of modern science.
(d) Historical epistemology and philosophical categories of investigations.

Here the editors and authors bring the history and philosophy of science to a wider audience, synthesizing *methods* and *interpretations*, *events* and *facts* of scholarship on the subject, with a historical philosophical narration and analysis.

The Chapters

Summaries of the arguments dealt with by the authors[3] are provided below. All submitted abstracts and papers were accepted only after having been blind peer-reviewed, for both style and content; consequently, a high level of content, appropriate for a wide international audience, is offered for individual and collective research projects.

---

[3] Listed in alphabetical order.

The Authors

Joseph Agassi (Israel/Canada) offers a distinguished wide-ranging paper: he starts from two considerations regarding two features of Koyré's production: (1) Koyré used the mistakes of his "heroes" to describe important aspects of science's development and (2) although he can be considered an internalist, he did not neglect the importance of metaphysics and theology as far as the construction of scientific theories is concerned. Starting from Koyré, the author develops a series of wide considerations on the professionalization of the history of science after the Second World War and on different tendencies in the history of science (for example: internal or external history of science). Thus, this chapter tells a fascinating story, in which, around the protagonist Koyré, all the most important historians of science of the twentieth century (including the author) play their part.

Charles Braverman (France) offers an interesting and personal perspective on one of the most significant features of Koyré's way of conceiving the history of science: the analysis carried out by Koyré concerning the concept of space. In particular, the author proves that, according to Koyré, the concept of space is the result of a construction involving the whole of human thought and – in any epoch – it cannot be reduced to the result of an experiment or to a series of mathematical theories. As to the concept of space, the entire way of thinking of a civilization is implicated. The author deals, in particular, with the way in which Koyré faced the problem of space in Galileo and he applies Koyré's method to examine the notion of space in Ampère.

Mauro L. Condé (Brazil) deals with a series of early contributions by Koyré concerning philosophy and the foundations of mathematics. The author highlights that Koyré adopted a realist conception of mathematics, but that his interpretation of set theory's paradoxes (and in particular of Russell's paradox) differed from that given by Russell himself. In particular, Koyré did not think this paradox was so remarkable for the coherence of mathematics. The author examines how Koyré's *internalist* approach to the history of science was also influenced by his convictions concerning the nature of mathematics. This chapter is useful for the understanding of Koyré's cultural-mathematical background and its influence on Koyré's production in the history of science.

Francesco Crapanzano (Italy) addresses an aspect of Koyré's works on Galileo: the myth of the leaning tower experiments by which Galileo would have proven that bodies of different weights fall with the same speed. According to Koyré, these experiments were never carried out. Crapanzano offers a profound revisitation of Koyré's argument on this subject, as well as an analysis of Koyré's idea that experiments do not necessarily play a fundamental role in the development of science. The author also presents a vast panorama of opinions held by various scholars with regard to Galileo's supposed leaning tower experiments and their possible importance for the development of Galileo's ideas on the laws of motion.

Mario De Caro (Italy) deals with a classical problem in the history of science: the supposed Platonism of Galileo. Koyré is an important part of this picture. De Caro reminds the reader that, in the first part of the twentieth century, most historians of science believed Galileo to be a Platonist. Later other scholars pointed out the possible influence exerted by Aristotle and Archimedes on Galileo, while further studies stressed Galileo's absolute originality, as well as the difficulty of identifying an ancient tradition that was decisive for him. In substance, nowadays the idea that Galileo is a Platonist is a minority tendency. The author reconstructs this interesting historiographical debate in detail.

Dominique Descotes (France) presents a chapter concerning Koyré's works on Blaise Pascal. The author points out that, in his main publications, Koyré rarely acknowledges the contributions made by Pascal to science. Nevertheless, during a conference held at Royaumont Abbey in 1954, Koyré gave a talk on Pascal. This talk by Koyré was published twice, in two slightly different versions. The author describes the circumstances in which the two versions were published. In the main part of this chapter, Descotes specifies many interesting details of the interpretation given by Koyré of Pascal's scientific work, focusing, in particular, on mathematics, but also highlighting Pascal's views on the Copernican system and other scientific questions.

Antonino Drago (Italy) points out the importance ascribed by Koyré to the metaphysical convictions of scientists in the constructions of their theories. This indicates Koyré's anti-positivist stance. Although Koyré was, in several circumstances, criticized for his historiographical categories, Drago stresses that his concepts are particularly suitable for the correct historical framing of the works of those physicists who did not belong to the – so to say – Newtonian legacy. The author develops a comparison between Koyré's and Kuhn's approaches and shows that, in many cases, Koyré's is preferable. Thorough consideration is also dedicated to a comparison between Koyré and other important historians and philosophers of science, such as Imre Lakatos and Paul Feyerabend.

Daria Drozdova (Russian Federation) presents a stimulating chapter concerning the internal logic of Koyré's principal historiographical assumptions. In particular, she faces the problem concerning the relations among the four *topoi* of Koyré's history of science: (a) *destruction of the Cosmos*; (b) *geometrization of space and mathematization of Nature;* (c) *transition from the world of more-or-less to the universe of precision*; and (d) transition *from the closed world to the open universe.* A careful examination of important statements by Koyré, drawn from his main works, drives us to the general logic of his approach. This offers us useful categories for interpreting Koyré's thought.

Massimo Ferrari (Italy) develops a comparison between Cassirer's and Koyré's approaches to the history of science. The first section of the chapter is dedicated to Koyré. The author presents us with a profound examination of Koyré's philosophical background and shows that his approach to the history of science was deeply influenced by this background. This background is detectable in all the most significant works written by Koyré. The second section concerns Cassirer, and Ferrari describes Cassirer's work as a "a neo-Kantian history of science", and explains how it originated from a neo-Kantian philosophy. In the third section, the author develops a comparison between Cassirer's and Koyré's ideas on the evolution of human knowledge and science.

Stephen Gaukroger (Australia) presents an interesting interpretation of Koyré's influence on the history of science: according to the author, Koyré transformed the history of science. He introduced a wide problematic sphere, including the way in which the concepts of space, time, and matter were transformed, in particularly significant periods of human history, and the way in which mathematics was used in physics, as well as the role of experiments and scientific instruments. Before Koyré, the history of science was, in general, a mere report on how modern science was reached; with Koyré it became a discipline dealing with deep conceptual problems. Koyré also considered the metaphysical and epistemological convictions of scientists. However, all these positive aspects led to a problem: according to the author, Koyré's works were often nearer to epistemological or metaphysical theses rather than historiographical ones. The author attempts to support this idea, based on the way in which Koyré regarded Galileo and Descartes.

Glenn H. Hartz (United States of America) and Patrick K. Lewtas (Lebanon) deal with an internal problem concerning Descartes' thought. In particular, they wonder whether Descartes' voluntarism is coherent with his general philosophical approach and, more specifically, with his natural philosophy. In this context Descartes' *Meditationes* play a significant role. A particularly significant section of this chapter is the sixth one, dedicated to logical voluntarism. This concept was the subject of a dense series of discussions in the seventeenth century. The authors refer to a vast amount of literature on this complex problem, with Koyré playing a role in this picture.

Gérard Jorland (France) presents a dense article dealing with the cesura that Koyré represented for studies on Galileo. Koyré's theses on Galileo's experiments and Galileo's Platonism are an unavoidable reference point, as they are for the scholars who oppose Koyré. The author stresses that the return to pre-Koyrean historians such as Antonio Favaro and Emil Wolhwill is, generally speaking, mediated by the reading of Koyré. The author provides a wide, and, at the same time, detailed, panorama of Galilean studies after Koyré. His aim is to evaluate the limits within which and the form in which some of Koyré's ideas are sustainable. Then there is a detailed examination of authors such as Pierre Costabel, Stillman Drake, David K. Hill, James MacLachlan, Wolfgang Lefèvre, the group of the Max Planck Institute, and others. Thus, the reader can gain a precise idea of Galilean studies and of the possible role that Koyré's interpretations of Galileo can play nowadays.

Anna Lombardi (Italy) deals with the harmony of the world and the music of the heavens in Kepler and with Koyré's fundamental contributions to the full comprehension of the importance of this harmony in Kepler's astronomy. Koyré, in his *La Revolution Astronomique* (1961), dedicated the second and third chapters of the section devoted to Kepler to the concept of harmony in the German astronomer's work. Lombardi provides a picture of the development of the concept of harmony in Kepler, starting from the *Mysterium Cosmographicum*. Then the author focuses on Kepler's third law, its genesis, and its role in Kepler's system. Koyré's analyses play a fundamental role in several parts of this chapter.

Rodney K. B. Parker (Canada) tackles the problem of the influence that Edmund Husserl and his phenomenology exerted on Koyré's approach to the history of science: Husserl's ideas, expounded in his *Logische Untersuchungen* (1900–1901) and, later on, in his writings on phenomenology, were influential, starting from the second decade of the twentieth century. Koyré was one of the students who belonged to the so-called Göttingen circle and he tried to spread Husserl's ideas. Later on, as the author reminds us, Husserl refused to accept Koyré's thesis on the paradoxes of set theory. Nevertheless, the two scholars remained in contact. The author attempts to identify the influence exerted by Husserl on the whole of Koyré's production and he interprets some of Koyré's statements as a possible continuation of and response to Husserl's phenomenology.

Jose Carlos Pinto de Oliveira (Brazil) and Amelia de Jesus Oliveira (Brazil) present a historical and historiographical case study concerning an analysis of Kuhn's epistemology and George Sarton's epistemological concepts. The authors consider this analysis as the first historical investigation of these two distinguished historians and epistemologists. Taking into account Koyré's approach to the history and philosophy of science, Pinto de Oliveira and de Jesus Oliveira discuss the above-mentioned case study, referring to Sarton's treatment of Leonardo da Vinci and related studies of blood circulation in the seventeenth century. In particular, they examine Sarton's concept of the history of science and Kuhn's criticism, and note how different the historiographical approaches of Sarton and Kuhn are.

Raffaele Pisano (France) and Paolo Bussotti (Italy) analyze the concept of force in Kepler's corpus. In particular, the role played by Kepler's physical causes in planetary movements and, consequently, the differences between a kinematic concept of astronomy and a dynamic one are analyzed. This scholarly tradition, of differences between the history and epistemology of science, dates back to Alexandre Koyré (e.g, in *La Révolution astronomique*, 1961) and concerns the scientific (astronomical) revolution in the seventeenth century. Taking into account Koyré's contributions, the authors also discuss the role assigned by the Russian scholar to the beginning phases of physical astronomy in his general concept of the scientific revolution. The influence exerted by Koyré's thought in the second half of the twentieth century, in regard to *Keplers Forschung*, is also outlined.

Diederick Raven (The Netherlands) faces: what role is played by practical and technical activities, and what is their influence, in the development of science? It is well known that Koyré thought that practice was not important for science. The author reminds readers that in 1943 Koyré expressed his opinion in contraposition to the ideas of Leonardo Olschki and Edgar Zilsel. Raven revisits this polemic. He adds details separating the positions of Olschki and those of Zilsel. The author focus on works written on this subject after Koyré. He reaches the conclusion that, although some important contributions do exist, this subject has not yet been studied as profoundly as it deserves.

Marlon Salomon (Brazil) starts from a statement concerning Koyré: in general he is considered the historian of science who placed scientific theories in their time. This was a novelty, because before Koyré ancient science was seen with the lenses of present science. However, the author claims this is only part of the truth: Koyré did not want to construct a wall between past and present. Rather, he was interested in a comparison between old and new science. The history of science is connected to the actuality of science. It is important to detect the nature of this connection in a perspicacious manner, which is not taken for granted. The author claims, hence, that Koyré did not neglect the aspects of past science that could be useful for an epistemic transformation of modern science, while at the same time, he did not neglect those aspects that could be useful for the new perspectives modern science can offer to interpret the past.

John Schuster (Australia) focuses on the early ideas and works of Thomas Kuhn. The author claims that, as a young scholar, Kuhn attempted to identify the shape of the history of physical sciences and to determine the dynamics of scientific traditions. In this work Kuhn was deeply influenced by Alexander Koyré, whom Schuster defines as Kuhn's "historiographical idol." However, Kuhn was certainly not a passive continuer of Koyré's thought. As the author points out, he was creative: he rejected the category of metaphysics as a central one and arrived at the notion of *normal science*. This chapter presents us with an interesting ideal dialectic between Kuhn and Koyré, highlighting the influence the latter exerted on the former, as well as the original approach to ? made by Kuhn.

Jean-François Stoffel (Belgium) offers a picture of traditional readings and debates on the Copernican revolution's anthropological consequences and the geocentric model. The readings and debates date from the seventeenth and eighteenth centuries (Cyrano de Bergerac, Bernard Le Bovier de Fontenelle, Jean-Étienne Montucla, ? Bailly, Pierre-Simon Laplace, and Jean Baptiste Joseph Delambre) through to the nineteenth (Auguste Comte, Darwinism, and Sigmund Freud) and twentieth centuries (Arthur Lovejoy, Hans Blumenberg, Fernand Hallyn, and Remi Brague). Stoffel presents Koyré's studies as an integral part of the cited debates, with Koyré being both a historian and a philosopher of scientific thought. The author discusses to what extent did Koyré remain dependent on the commonplace that the Copernican revolution would have deprecated man? In addition, which are the statements of the traditional reading that he mainly questioned?

Anna Yampolskaya (Russian Federation) deals with a fascinating problem: the influence exerted by Koyré on some authors and concepts of twentieth century philosophy connected to religion and to the idea of God. In particular, the author focuses on Emmanuel Levinas and Michel Henry. The study of infinity characterizes Koyré's works, starting from his books on St. Anselm and Descartes. The author shows how Koyré's ideas on infinity and on the finite-infinite relation influenced Levinas' concept of God. Koyré's interpretation of St. Anselm's ontological proof plays a significant role in this context. Furthermore, the author proves that the way in which Koyré interpreted the thought of Johann Gottlob Böhme was important for Henry's philosophy.

Summary

In this book, scholars from different traditions explain Koyre's various historical, epistemological, and philosophical contributions to the history and philosophy of science, aiming to gather and re-evaluate the current thinking on this subject.

The book brings together contributions from leading experts in the field, and gives much-needed insight into the subject from both historical and philosophical standpoints. It should prove to be absorbing reading for historians, philosophers, epistemologists, and scientists.

Lille University
Villeneuve-d'Ascq, France

Raffaele Pisano

References

Crombie AC (1963) (ed) Scientific change: historical studies in the intellectual, social and technical conditions for scientific discovery and technical Invention, from antiquity to the present – Symposium on the history of science. The University of Oxford, 9–15 July 1961. Heinemann Education Books, London.

Gillispie CC, Pisano R (2014) Lazare and Sadi Carnot. A Scientific and Filial Relationship. 2nd ed. Springer, Dordrecht.

Guerlac H (1963) Some historical assumptions of the history of science. In: Crombie AC (ed) 1963, pp. 797–817.

Hanson NR (1962) The Irrelevance of History of Science to Philosophy of Science. The Journal of Philosophy 59: [p. 580] 574–586.

Koyré A (1957) From the Closed World to the Infinite Universe. Johns Hopkins Press, Baltimore.

Koyré A (1961a) Du monde de « à-peu-près » à l'univers de la précision. M Leclerc et Cie–Armand Colin Librairie, Paris [Les philosophes et la machine. Du monde de l'« à-peu-près » à l'univers de la précision. Études d'histoire de la pensée philosophique].

Koyré A (1961b) Message d'Alexandre Koyré à l'occasion du centenaire de la naissance d'Emile Meyerson. Bulletin de la Société française de philosophie 53:115–s.

Koyré A (1961c) La révolution astronomique: Copernic, Kepler, Borelli. Hermann, Paris.

Koyré A (1963) Scientific Change. In Crombie AC 1963, pp. 847–857.

Koyré A (1965) Newtonian studies. The Harvard University Press, Cambridge, MA.

Koyré A (1966) Études galiléennes. Hermann, Paris.

Koyré A (1971) Études d'Histoire de la pensée philosophique. Gallimard, Paris.

Koyré A (1973) Études d'histoire de la pensée scientifique. Gallimard, Paris.

Koyré A (1986) De la mystique à la science. Cours, conférences et documents 1922–1962. In: Redondi P (ed). Cours, conférences et documents 1922–1962. EHESS Éditions, Paris.

Koyré A (1965) Newtonian Studies. The Harvard University Press, Cambridge, MA.

Kuhn TS (1962) The Structure of Scientific Revolutions. The Chicago University Press, Chicago.
Nagel E (1961) The Structure of Science: Problems in the Logic of Scientific Explanation. Harcourt-Brace & World Inc., N.Y.
Pisano R, Gaudiello I (2009a) Continuity and discontinuity. An epistemological inquiry based on the use of categories in history of science. Organon 41:245–265.
Pisano R, Gaudiello I (2009b) On categories and scientific approach in historical discourse. In Hunger H (ed) Proceedings of ESHS 3rd conference. Austrian Academy of Science, Vienna, pp. 187–197.
Redondi P (1987) (ed) Science: The renaissance of history. Proceedings of the International Conference Alexandre Koyré. Paris Collège de France, 10–14 June 1986. Special Issue History and Technology 4/1–4.

Chapter 1
Alexandre Koyré: His Secret Charm

Joseph Agassi

Abstract There are no classics in the natural sciences akin to those in the humanities; Koyré paved the way to getting some connoisseurs to read some old science texts the way they read old classics in the humanities. Historians of science can help that project by debunking the myth that to be a good historian of science one has to be professional. Indeed, it is hard to say whether as a professional Koyré was a historian of science or of religious thought, or simply of culture. The historiographical innovation of Koyré is the idea that (contrary to the traditional view that science is infallible) historians of science should better not ignore their heroes' errors. Hence, viewing him as a Platonist is highly misleading: whereas Plato deemed science perfect, Koyré used the analysis of scientific mistakes as a tool for understanding the research process in its diverse manifestations. Unfortunately, Koyré did not place sufficient emphasis on scientific controversies, even though, differing from other historians, and opposing Duhem's views, he did not suppress them and he emphasized the role of metaphysics and of theology as their background, thus recognizing the proper place of metaphysics and of theology in research that Duhem had barred from science so as to exclude controversy from science. Thus, Koyré was an internalist not in the positivist sense but in the sense that he studied the problem situations of researchers, for his greatest charm was that he was consistently problem oriented.

Keywords Scientific classics • Historiography of science • Scientific controversies • Koyré • Popper

J. Agassi (✉)
Tel Aviv University, Tel Aviv, Israel
e-mail: agass@post.tau.ac.il

R. Pisano et al. (eds.), *Hypotheses and Perspectives in the History and Philosophy of Science*, DOI 10.1007/978-3-319-61712-1_1

1.1 Are There Scientific Classics?

Ernan McMullin (1994) asked, are there classics in the natural sciences? Some educated people read Shakespeare or even Chasseur, others read Dante, and still others read Cervantes. No doubt, these readers are a very small minority even among the highly educated; their cultural impact on their surroundings is significant nevertheless. This is more so regarding Homer and regarding some of the Platonic or Aristotelian corpus. Are there comparable classic texts in the natural sciences? Michael Faraday's *A Chemical History of a Candle* may count as such, but it is an odd item, being Christmas lectures for the young, namely, popular science, and so comparable not to works of Homer or of Shakespeare but to their simplified versions by Charles Lamb. Is there a corpus of classic texts in the natural sciences proper, then? Consider Hans–Georg Gadamer, the influential twentieth-century obscurantist. He confessed in his autobiography that he found the ancient physics text by Aristotle more interesting than any modern physics text. He ignored Archimedes, who, though ancient, is more akin to Newton than to Aristotle. (This is extremely interesting for students of the scientific revolution, particularly Koyré). Why is that? It is not specific to one (pseudo-)scholar. Thus, the mediocre text of Vitruvius on architecture is more of a living classic than Archimedes on buoyancy, even though its text is superb, judged as scientific or otherwise. What causes the blindness to it that respected classicist Friedrich Solmsen displayed when (1960) he discussed gravity and sided with Aristotle against Plato? Perhaps it is the result of the famous fact that current education compels adolescents to learn to dislike either the natural sciences or the humanities. If so, then the recent (1998) discovery of the Archimedes palimpsest with commentary on Aristotle should make it harder for them to continue to turn a blind eye to Archimedes.

Philosopher John Findlay defended Aristotle's doctrine of the four elements (that Archimedes had superseded). Though such scholars are ignorant obscurantists, facts are sadly on their side: tradition neglects the greatest, old scientific texts, despite the great beauty of some of them—not only *On Floating Bodies of* Archimedes but also *The Elements* of Lavoisier and the *Theory of Natural philosophy derived to the single Law of Forces which exist in Nature* of Roger Joseph Boscovich, not to mention the exceptionally beautiful *Dialogue on the Two World Systems* of Galileo, *The Origin of Species* of Charles Darwin, *Treatise on Thermodynamics* of Max Planck, and *The Meaning of Relativity* of Einstein. Admittedly, some great scientific classics are simply unreadable, including the most important and monumental texts ever: *De Revolutionibus* of Copernicus (1543) and the *Principia* of Newton (1687). Even the once-best-selling and highly readable 1704 *Opticks* of Newton is these days not a classic the way the literary classics are. It is easily available, but as far as I know, it has no readers except for some historians of science. Perhaps *Paradise Lost* by John Milton of 1667 and *Gulliver's Travels* by Jonathan Swift of 1726, the near contemporaries of Newton's *Opticks* are also relatively neglected, but they are read by sufficiently many readers to have their presence felt, however indirectly.

This relative neglect of old science texts may be overcome. As Paul Feyerabend has noted, this requires nurturing open-mindedness about science, an attitude absent (among students of the humanities and much more so among students of the natural sciences) due to prejudice and hostility. The gulf between the humanities and the natural sciences is easy to bridge, he said, since viewing some science (texts) as art proper enriches the understanding of it. One easy way to open up to any piece of art is that of learning about the techniques that go into it.

The scholar who has paved the way in that direction most is Alexandre Koyré, who atypically moved to the history of science after he had made a name for himself in the humanities. He had a pet aversion to the prejudice that the gulf between the humanities and the natural sciences is unbridgeable. He denounced it (together with Gaston Bachelard) as a corollary of positivism (without entering into any discussion of aesthetics, much less the aesthetics of science). Alas, some historians of science clung to positivism even after they admitted the influence of Koyré on them and even after they became some of his best-known followers. Koyré was not the first anti-positivist student of the impact of metaphysics on science; among his august predecessors were Arthur Lovejoy and R. G. Collingwood, not to mention E. A. Burtt. No one before him studied scientific texts with a fine toothcomb, however, bringing the metaphysics and the physics as close together as possible. This he did. "The magisterial influence of Alexandre Koyré" is as obvious as it is generally recognized, writes I. Bernard Cohen in his moving obituary of Koyré. What this influence is or what it amounts to is far from "obvious," however. Cohen characterized it thus:

> A new emphasis is discernible today that was largely absent some twenty or thirty years ago: a stress upon scientific ideas, understood in their own terms and in relation to the living background in which they were embedded—in short, the method of conceptual analysis, based on the model set before us by Alexandre Koyré. (Cohen and Clagett 1966, p. 158)

This is true; yet it is problematic. How did this new emphasis of Koyré signify? What instigated it? Moreover, there is here a more difficult question: how could any history of science ever avoid that "stress upon scientific ideas" that Cohen ascribes to Koyré? In other words, what was the field like before Koyré entered it?

Some leading commentators on current historiography of science speak of the field as if it was born in the twentieth century, as the marriage of the influence of Koyré with the rise of the history of science as a profession (to wit, as a department or subdepartment in some leading universities). This is an insult to the great historians of science of the eighteenth and nineteenth centuries, who were not professional intellectuals in any sense. They were of independent means or else they lived on income from occupations that were scarcely research oriented. Observatories were the first instituted research positions proper in the modern world. Next came those of the Royal Institution of Great Britain, founded in 1799, that became de facto a research institute. The trend of overlooking them began, I am afraid, with the apologetic response to my criticism of the ways current then of writing a history of science my *Towards an Historiography of Science* of 1963 (Agassi 2008). Reviewers of it dismissed old masters as antiquated and claimed that much improvement in the writing of the history of science is due to the professionalization of the field. This assertion is distasteful, especially in history, and this explanation is downright silly: one can be good and amateur and one can be professional and inept. Indeed, some of the greatest scientists ever were amateurs: prior to 1800, very few people were ever paid to do scientific research, and these were pensioners of rich individuals.[1]

My criticism of the admirable old historians of science was very severe, but it would never occur to me to put them down. Indeed, one of my major points in my writing about the history of science in the first place is the claim that we cannot do justice to the great thinkers of the past as long as we are captive to the claim of Francis Bacon that criticism bespeaks contempt (Preface to his *Great Instauration*). The effort to view old scientists as right, which the positivist dogma imposed on historians of science, forced them to select from the old texts small parts and distort them if need be; this made the approach of Koyré quite impossible. This renders his influence much narrower than it should be, for his respectful criticism of Galileo is inconsistent with the Baconian claim that of criticism implies contempt that is still so popular among positivist historians of science.

I criticized the positivist ways of writing the history of science (the Baconian or inductivist-empiricist way as well as the Duhemian or instrumentalist-conventionalist way), adding briefly reference to Koyré and his school as historians of science who have left positivism behind. They do this every time they present science as embedded in the general culture of the day. My criticism of older historians of science was most respectful, and it came to indicate the difference in scientific culture and the superiority of today's fallibilism (Karl Popper-style) applied to science as opposed to the older view of science as perfect. Most of the commentators on my criticism of the older historians of science insulted them by dismissing their works as outdated. It is silly of any historian to dismiss any opinion as outdated, of course, and historians of science have to treat with respect even silly superstitions if they were prevalent, let alone histories of science proper. Yet my critics said, my criticism amounts to very little as the poor works in the field that I had discussed as examples for my target are antiquated, that with the rise of professional history of science things are greatly improved. This is contrary to my intention: my works display greater respect for older historians of science, who expressed the ethos of their days when they wrote of science as perfect; when historians of science display endorsement of the same ethos after the few scientific revolutions, they deserve less attention, and their being professionals is irrelevant to this. Nevertheless, the myth that the history of science is now in a better shape than it was centuries ago due to its professionalization soon became popular. Seymour H. Mauskopf wrote in the Charles Gillispie *Festschrift* (Buchwald 2013), of the great influence that Koyré's work had on the "first generation professional historians of science" (Mauskofp 2013, p. 29). Though meant as a compliment to Koyré and to his followers, it is not.

James B. Stump recognizes the pre-Koyré tradition of writing histories of science, saying (Stump 2001, Abstract), "While history of science as a distinct academic department is relatively new among most universities, writing histories of

[1] For the view of *The Age of Reason* as amateur science, see Agassi 2013.

science is not." True. Prior to World War II, most historians of science, like most researchers in every field, were not academics; most of the rest were academics whose chairs were not in the history of science. Even academic Thomas Kuhn was professionally not a historian of science when his famous *The Structure of Scientific Revolutions* (1962) appeared. (Only in 1964 did he succeed in attaining the coveted position of a professional historian of science proper.) This field was traditionally not professional. (Most fields, incidentally, were not; Ornstein Bronfenbrenner 1913; Stimson 1948). Even medical research in the USA was first amateur, with Benjamin Rush as the most conspicuous physician in the early days of the republic. Professional medical research began early in the twentieth century as described vividly in the 1925 novel *Arrowsmith* by Sinclair Lewis. At the time, there were a few research positions in industry as described in novels of Upton Sinclair.

This topic is too rich for even a brief survey. Let me mention, nevertheless, one of the most important and most neglected historical fact. In the wake of Francis Bacon and in line with his stress on the need for a "just history of nature," it is hard to judge whether The History and Present State of Electricity by Joseph Priestley is science (as intended) or history of science (as he composed it out of the learned press). This explains the paucity of histories of science during the Age of Reason and the fact that every important one then had a specific agenda.

The rapid expansion of the academic professions after World War II created publication pressure. It started already two decades earlier, but then it was still avoidable; it became *de rigueur* only after World War II and under the initiative of Harvard University president James Bryant Conant who also initiated the celebrated department of the history of science there and the series of classic texts in its press. This was the professionalization of the field, and it encouraged the flood of publications of all sorts, including some that were of questionable quality—in the history of science in particular. Conant was greatly indebted to Pierre Duhem; he first failed to mention him in his bibliography. Incidentally, both were amateurs. When Duhem was offered a job in Paris not as the physicist that he was but as a mere historian, he was offended and refused. The flood of publications by professional historians of science, however, is neither here nor there: in any field, run-of-the-mill works naturally tend to mirror the great ideas in it, and this, if anything, is what makes them signify.

Even the great I. Bernard Cohen appreciated professionalism. He noted the paucity of good secondary texts in the history of science (Cohen 1954) and ascribed it to the "relative newness of the history of science as a professional subject." He mentioned two defects that professionalization allegedly cured. First is the absence of adequate training in the field. Second is the "problem of audience which is not encountered, at least on the same scale, by scholars in other disciplines." This suggests that most of the customers for most of the learned publications are co-professionals. Then there can be no classics in the classical sense of the word as Ernan McMullin has used it and as is cited in the opening of the present essay.

Cohen offers an example: the absence of reprints and of studies of Newton's *Principia*, the greatest scientific text of all time. I think it takes little checking to find out that students of the history of physics who wrote extensively of the field recoiled from examining even the most obvious facts about this book. They ignored even the obvious change of the *Hypotheses* in the opening of Book III of the first edition of this masterpiece to *Rules of Philosophizing* in the later editions. This is a point that—amazingly—Koyré is the first to have noted and discussed in print. It is no secret that Newton's views on many matters embarrassed his admirers, as they had endorsed Bacon's maxim that disagreement bespeaks contempt. The public discussion of Newton's less palatable ideas has ceased to be taboo, but it is still not welcome. Leading authorities sabotaged the plan of Richard Popkin to publish the collected works of Newton; now on the Internet, thanks to a few enterprising scholars,[2] allowing for the seamy side of Newton's output, however, is no longer a sufficient innovation for the occupants of the new professional field of the history of science. What else have they to tell us that is new?

This is a tough question. We should backtrack and ask, *can historians innovate*? How? If we find a cogent answer to this question, we may try to apply it to the history of science and then to professional history of science.

One may protest against the last words above, saying that the professional status of a historian matters least. Perhaps. Why then did commentators stress in response to my study that professional history of science is new? They did so, let me contend, because they took for granted the claim that the professional is proficient and that the amateur is a dilettante. This claim is a myth: counterexamples to it abound. Historians of science should be particularly aware of the historical fact that classical modern science was largely amateur. (Even in the twentieth century, some of the greatest discoverers were amateurs, beginning with all the pioneers of computer technologies, not to mention lone individuals like Michael Ventriss, decipherer of Linear B). They are not. For example, despite the popularity of the book of Dorothy Stimson on the Royal Society as traditionally an amateur organization, its official website declares rather emphatically that the nonprofessional members were individuals who financed researchers as their private pensioners. The relevance of all this to my query is this.

Professionalism creates university departments and their associated fields of research. The demand evolves quietly but persistently that a decent professional should not cross professional boundaries (namely, departmental ones) at the peril of being called a dilettante. This demand is called internalism (Fuller 2000). This also invites the demand for the creation of interdisciplinary fields, which demand is important ever since Sputnik and the 1959 call by C. P. Snow (his *The Two Cultures and the Scientific Revolution*, 1961) on historians of science to build a bridge between natural science and the humanities. Whatever innovations one may offer much depends on whether one is an internalist or not and even what kind of internalism one advocates. If one's field is the history of science, then one would do well to describe the rules for that field.

Perhaps all fields of study need specifications of their rules, it is hard to say, since the discussion about internalism itself still lacks the necessary preliminary discussion of what is internal to begin with. (It also has no proper home that should lend it some

---

[2] See http://www.newtonproject.sussex.ac.uk/prism.php?id=22

respectability.) The reason is that internalism is part-and-parcel of positivism and positivists have a clear idea about what they want to keep out: metaphysics (including the theology that Koyré found so important for seventeenth century physics). Arch inductivist Francis Bacon said, nothing stops the growth of science but the pollution of the minds of researchers with non-science—especially theology and metaphysics. This idea, Bacon's doctrine of prejudice, became the received opinion of the scientific revolution and the central idea of the whole scientific tradition. Dr. Johnson's assertion that even sheer fantasy can pollute science sounded then plausible enough to enable him to incorporate it in a novel of his (*The History of Rasselas, Prince of Abissinia*, 1759). The Baconian demand to keep science free of outside influences remained popular until the twentieth century and beyond. Koyré showed the relevance of both theology and metaphysics to the scientific revolution (Elkana 1987). And the arch-instrumentalist Pierre Duhem rejected the prevalent inductivist methodology and with it Bacon's doctrine of prejudice. Yet he too recommended the separation of science from metaphysics. He argued for this from commonsense: this is advisable, he said, in order to maintain the unanimity of science. Koyré criticized this by showing the impact of theology and metaphysics on research, as did (amateur) Émile Meyerson earlier. Koyré dedicated his *Galileo Studies* to the memory of Meyerson; Daria Drozdova (Drozdova 2012) considers him his disciple. Duhem fearlessly dismissed the theological ruminations of scientific researchers, declaring utterly irrelevant what they did in their scientific research proper: he took methodological conflicts among researchers as sufficient evidence that their views on science are unscientific and so better ignored Snow. It may seem Einstein was in agreement with this, as he likewise dismissed all demands from researchers, declaring them opportunists. Yet Duhem fiercely denounced (Einstein as a German and as a Jew as well as) the contributions of Einstein to physics—as instigating a revolution in science: Duhem deemed perverse any revolution in science, since scientists must stand on the shoulders of their predecessors.

Duhem's rejection of Einstein may seem not surprising, as he disagreed with almost everybody. The distressing fact is that he rejected Einstein's output because in his view scientific progress must be in small steps and Einstein took a huge step away from his predecessors. All this seems very clear, except that Duhem never said what step is small and what step is big. This Edmund Whittaker showed when he argued that Einstein's theory was no revolution at all (Whittaker 1953, Chap. 2). This angered Max Born, who was a leading scientist (and a friend of Einstein): he said he had witnessed the revolution as he had battled on the barricades (Born 1954). How big should a step be for us to view it as a revolution? Thomas S. Kuhn, who followed much of what Duhem taught despite his declaration of loyalty to Koyré (justified by his view of science as problem oriented), likewise did not specify; he ended up admitting that revolutions may be small (Kuhn 2000, p. 143). This is a tacit admission of defeat. The reason for the defeat is obvious: Kuhn was a positivist *malgré lui*. Koyré had no difficulty answering the question: a scientific revolution is always external—it is usually metaphysical. But, he added, it could be any change that has influenced research greatly.

It should be obvious that in science as elsewhere some changes are small and few are big (Agassi 1981, Chaps. 9 and 22). Still, the question that Kuhn took as central, namely, what change is big enough to count as revolutionary, is as artificial as his distinction between normal and revolutionary change. Duhem's view that there are no revolutionary changes is challenging, especially for historians of science, as it imposes on them the emergence technique: seek in the literature moves that look like intermediary steps between what looks like a revolutionary move and its known predecessor so as to reduce the gap between them. Should historians of science follow this technique of Duhem? Not if Koyré is right. He has declared the continuity theory inherently defective: any theory has some predecessors in some sense; this is almost trivial; a theory having predecessors, however, does not show that it is not revolutionary. On the contrary, the many researchers who have tried to solve a problem and failed show how very revolutionary is the solution that succeeds and how daring its author was (Koyré 1978, p. 3). Moreover, the revolution is an obvious fact, the fact that the new researchers destroyed the old metaphysics and presented an alternative to it and the rapid growth of science that accompanied this change. These two facts about science Kuhn separated as revolutionary and normal, much contrary to Koyré.

Koyré rejected Duhem's view on continuity by describing the scientific revolution and by characterizing revolutions in general. Duhem's followers may resist this, since it rests on the supposition that research is problem oriented, a supposition that Einstein and Koyré took for granted but that Duhem simply ignored, as his followers still do (with the notable exception of Kuhn), as most positivists still do. Duhem's followers can cling to his view on continuity, and they will do so if they advocate internalism, since revolutions often alter the boundaries that decide what is internal and what is not. Problem-oriented studies then render the demand for Duhem-style internalism impossible to fully abide by; they view as internal the discussions of problems and their solutions and their improvements and replacements by better problems. As science tackles problems, it may refuse to stay within given disciplinary boundaries. In addition, indeed, these keep shifting. This phenomenon is the peak of the opportunism of research Einstein style.

1.2 Problem-Oriented Internalism

Back to our question, then: can historians (of science) innovate? This question depends on the question, do historians (of science) tackle problems? What kind of problems? In my view (1963), most historians of science come to the field armed with views of the proper method of research that they have adopted, often quite uncritically. And then they present an image of the growth of science that fits their views. This they do reasonably, as best they can, and when they find it difficult, their problem is, how best should they confront that difficulty? For, their methodology does not enable them to tackle all such problems the same way or even acknowledge that their research is problem oriented. Moreover, at this stage, the dogmatic and boring historians diverge from those who admit that they have difficulties and acknowledge facts that their theories exclude. Were they more reasonable, they would have given up their views of method. But this is too much to expect of ordinary scholars. The claim that scientific research is logical, inductive, or mathematical reasonably leads to efforts to illustrate the claim by some (historical) case studies and then much depends on what case is reasonable to consider logical or inductive or mathematical. Choices of texts to study are seldom explained, yet they are most significant. They are more rational and more interesting when the chosen texts or case studies are challenging rather than obvious. Discussing (old) texts in disregard for the (unscientific) circumstances in which they appeared is internalist in the sense that it centers on science rather than on its circumstances. If the discussions are problem oriented, then it is internalist—if, and in the sense that, it centers on scientific problems rather than on technological or religious or political surrounds. But this kind of internalism does not respect boundaries of subject matters.

The internalism of the problem oriented (historians of science) is possibly different from the demand of Bacon and his followers that scientific researchers keep theology and metaphysics out of research, depending on whether the demand refers to problems or to doctrines. (I ignore here Bacon's astute observation that the very choice of a problem rests on a prejudice and is thus non-kosher). As long as a discussion of internalism proceeds without stating the doctrine explicitly, the distinction between the different kinds of internalism (scientific and historical) gets blurred and somewhat confused.

Internalist historians of science, Stump suggests, view Koyré as an internalist, and they see this as his contribution. "The standard interpretation of Koyré is that he falls squarely within the internalist camp of historians of science—that he focuses on the history of the ideas themselves, eschewing cultural and sociological interpretations regarding the influence of ideologies and institutions on the development of science." Stump does not advocate internalism. Consequently (in a Baconian stance), he sees no virtue in internalism. He finds Koyré's virtue elsewhere:

> I claim, rather, that a careful reading of Koyré's work suggests that a tension exists between internal and external methodological considerations. The external considerations stem from Koyré's commitment to the unity of human thought and the influence he admits that the 'transscientifiques' (philosophy, metaphysics, religion) have on the development of science. I suggest in conclusion then, that if we are to put a philosophical label on his work, rather than 'Platonist', as has been the custom, 'Hegelian' makes a better fit. (Stump 2001, p. 243)

This quotation is opaque (since it is a part of an abstract); unpacking it requires saying more about what exactly the division to internalism and externalism is and what are Platonism and Hegelianism. Yet even before this unpacking of the text of Stump, we may reject his polemic from the very start. What is utterly objectionable in his polemic is the (regrettably generally admitted) disposition to decide first what one considers the task of a history of science and then express admiration only for texts that illustrate it. Rather, as R. G. Collingwood has suggested (*Autobiography* 1939), researchers should declare what problem they study and what interesting solution to it is available and only then discuss critically the question, which solution if any is satisfactory and perhaps also offer a new one.

1.3 Respectful Dissent

Let me confess: I have an axe to grind. Despite the great success of my 1963 *Historiography*, I have a complaint about most commentators on it. Most of them took it for granted that my criticism was dismissive. Professor Alastair Crombie kindly invited me once to his history of science seminar in Oxford. Opening his discussion with me about my views, he ascribed to me the dismissive attitude. He was genuinely surprised to hear me say in response that I am a great admirer of Sir Francis Bacon. Even placid Koyré dismissed Bacon on the ground that he had no scientific knowledge and no understanding of research. It was Bacon who said that dissent is contemptuous, we remember, in flat contrast to Plato, who repeatedly declared criticism both a favor and a token of appreciation. On this Koyré was a full-fledged Platonist: Plato had recommended putting forward conjectures for critical debate; Bacon distrusted dialectics (as academics were constantly abusing it, he repeatedly and all too rightly stressed), and so he insisted on the avoidance of all conjectures as they are bound to become prejudices. He expected instead slow but sure scientific progress. Alas! Koyré ignored the tremendous influence of Bacon on the research community even while fighting it tooth and nail.

Let me mention a few examples for this influence of Bacon. As the great Luigi Galvani lost a debate with Alessandro Volta (the very invention of the Voltaic pile showed that animal electricity is no different from the electricity that comes from other sources), he left his post in the Academy of Science, first for a pilgrimage to the Holy Land and then to his family estate (Munro 1890, p. 95). A mistake that Sir Humphry Davy made (the advice that he gave to the British Admiralty to electrify the metal plates of battleships made them attract seaweed and reduce speed considerably) drove him to exile (Germany), never to return home (Paris 1831, pp. 270–271). It is thus no surprise that Henry Cavendish had lost priority to James Watt regarding the great discovery of the decomposition of water due to caution (Miller 2004, p. 275; Britannica, 4th ed., Art. Chemistry cited in Agassi 2008, p. 205).

The inability to treat false ideas with respect is the main feature of the positivism of most histories of science, the declared allegiance to Koyré of the latest members of this creed notwithstanding. They often distort information due to the good faith that they hold quite uncritically. For example, instead of describing the Copernican hypothesis as the idea that the sun is the center of the universe, they usually ascribe to him—usually implicitly—the idea that the sun is the center of the solar system; a system that only Giordano Bruno's infinite universe allowed (Koyré 1957, pp. 49–51). They likewise suggest that Newton's theory agreed with that of Copernicus—as Newton himself said—rather than notice that according to Newton's theory the center of the universe does not exist and the center of the solar system is (not the sun but) the center of gravity of the system. As Koyré and Cohen have stressed, that point may reside in empty space. This, incidentally, is an example of the careful reading of texts that Cohen said was characteristic of Koyré and his innovation.

Duhem partly overcame the inability to treat false ideas with respect. In his histories of science and in his writing about them, he denied that scientific theories are descriptions of reality: they are neither true nor false even though it looks that way. Also, he declared them true as implicit definitions (Bonnet and de Calan 2009, pp. 123–125). This idea was novel, and it became standard in mathematics due to the influence of David Hilbert who borrowed it from Poincaré, who borrowed it from Duhem (Jaki 1987, pp. 334–335, p. 354). Furthermore, Duhem declared scientific theories true for the domains of their application: they are true as long as their application is limited to where it is successful. This is like saying that there are no forged passports since a forged passport is not really a passport or that all my conjectures are true because I now withdraw every one of them that in future will turn out to be false.

As a concession to the prevalence of the inability to treat false ideas with respect (common among historians of science), P. P. Wiener established his *Journal of the History of Ideas* in 1940 under the influence of Émile Meyerson, whom Koyré too acknowledged as a major influence, we remember. Here is still Bacon's influence: although the Koyré study of scientific errors makes it unnecessary to separate the history of science from the history of ideas in general, the separation was admitted and is still there. It is odd that this is still hardly noticed, since until today there is too little appreciation of Bacon's ostracism of metaphysics and its value. Koyré had. The title of one of his great books *Metaphysics and Measurement: Essays in the Scientific Revolution* (1968) is a blunt declaration of his view of science as having metaphysical foundations (Koyré 1998). This was bold even though both Meyerson and Burtt had argued so earlier. Unlike Burtt and like Meyerson, however, Koyré was arguing against Duhem. It is when compared with these two that he stands out, as Cohen claimed, as one who read scientific texts closely and analyzed them "in their own terms," that is to say, without attempting to read old texts in the light of new knowledge as most historians of science still do stealthily and as Duhem did majestically. As to Burtt, he did not go into details of the texts he was discussing, and in his pioneering work, he was satisfied to show how imbued with metaphysics the lead scientific texts of the scientific revolution were.

The contrast between Duhem and Koyré is striking. Like Bacon, Duhem wanted science clean of metaphysics. He thus ignored all the superstitions of the medieval authors he was citing. Consequently, they look modern in spirit. If you want to learn more about their own views that place them firmly in their own culture, you may find it useful to look them up in Wiener's *Journal of the History of Ideas* and more so his *Dictionary of the History of Ideas* (1973). Koyré bridged the two by close readings of old texts. This is clearly a first and a marvelous one at that.

This is why Koyré is an acknowledged leading historian of science rather than a leading historian of ideas: he has changed the way the history of science is written, especially on the scientific revolution, where he repeatedly opposed Duhem's denial of the very possibility of scientific revolutions. Duhem argued for his view by showing that every revolutionary idea has close predecessors. This, said Koyré in terrific response, is obvious: every successful revolutionary idea has less successful predecessors. Yet, *pace* Duhem, the more and the closer the predecessors to the revolutionary, the more striking the revolution is. This is precisely what makes for a great revolution. To see this, we should see that the revolutionary had a vision that solved difficult problems. At times a mere hint at the spark that triggers the revolution suffices to impress peers with the whole blaze of it.

In my 1963 discussion of the historiography of science, I singled out Koyré and his followers as those who have systematically avoided the two main positivist pitfalls in the field, the inductivist and the instrumentalist. I took it for granted that it is best to characterize these two trends by their descriptions of scientific method. The older and dominant one, of the inductivist or Baconian school, is that scientific theories are a posteriori valid, inductively based on information. The second, of the instrumentalist or Duhemian School, is that scientific theories are freely invented classification systems that are a priori valid analytic truths. In 1963, I did not discuss the views of the Koyré school and merely claimed that they viewed theories as free inventions but not a priori valid (except for mathematics that is outside the current discussion), in line with Karl Popper's critical or fallibilist view of scientific method. I said this tentatively since historians of science may ignore all views on scientific method and since my evidence was personal, from meetings with Koyré when he came to visit Popper whom I then served as his research assistant. Meanwhile, I. Bernard Cohen, the closest collaborator and the leading follower of Koyré, has published a paper in which he shared this (Cohen 1974, p. 303) as far as philosophy is concerned. He criticizes Popper's history, though. Still, let me mention a difference of opinion between Popper and Koyré. Whether it is central or marginal to their philosophies I am not qualified to judge; it seems to me important. It concerns ways to treat controversies.

The purely intellectual part of any controversy is something that deserve the attention that Koyré and Cohen and their followers were the first to discuss systematically. Unfortunately, controversies seldom appear this way: they often come with rancor. Perhaps the individual most responsible for the perpetuation of this unfortunate aspect of controversy is the great Newton. Augustus De Morgan said he was an unpleasant character. Here the concern is with a question that is quite independent of character. Many details of a response depend on many incidental factors and they may then hardly signify; not so whether a response to disagreement and to criticism is in respect or in scorn; this is very significant. We often consider the attention of a leading thinker a compliment. The notice may be more pleasant if it is critical, as it shows notice of our opinions. The Age of Reason ignored this. A famous slogan then compared controversy to the fire that emits more heat than light. Custom then allowed for public displays of controversies but recommended to block them after one or two rounds. This is an error; if the slogan is true, then effort should be made to change things: to render controversy a generator of cold light—the cold light of reason. This was considered impossible then, literally due to the identification of heat and light as phlogiston or caloric and metaphorically due to the tremendous influence of Francis Bacon, who considered unavoidable the acrimony that often accompanies dispute and that he ascribed to human nature. Respectful and friendly disputes refute his view—such as that between Ricardo and Malthus or between Bohr and Einstein.

Koyré followed the tradition that Robert Boyle has instituted in the scientific revolution of criticizing opinions without mentioning them, much less expressing dissent explicitly. Both Boyle and Koyré made exceptions to this rule, at times criticizing some significant views, but then as briefly as possible, and at times registering dissent with no explanation. They preferred to avoid open controversy and to present as facts criticism of views they scarcely mentioned and then only in brief asides. Although my pleasure in the company of Koyré was brief—we met in the office of Karl Popper whom he visited a few times when I was working as a research assistant—he wrote to me, in great kindness, recommending that I do the same. Unfortunately, before I could answer I learned of his death. I would have thanked him and told him why my preference is to discuss controversies openly and in all the necessary detail. My argument against the method of implicit criticism is from the historiography of science: the avoidance of frank statements of controversies and the suppression of information about them is confusing and thus harmful. Moreover, when it became necessary to mention an adversary explicitly, the unusual character of this conduct led to great acrimony. The case history that I wish to present here is that of Koyré himself. To conclude this point, let me say that Popper's methodology considers the admission of criticism learning, so that it also allows for a better attitude towards controversy. But this is not enough. Popper's ingenious theory of degrees of testability and the reasonable proposal to examine the most testable hypothesis first that he (and before him C. S. Peirce) advocated, come to reduce controversy. This is an error.

1.4 The Secret Charm of Koyré

What was the secret charm of Koyré? What was the source of his influence? Some say, his influence was due to his being an innovator. Descriptions of his innovations are few, vague, and divergent. Some of these refer to his discussion of errors (of Galileo). Now this is no news: historians of science regularly report some errors, the paradigm case of this being the story of the doctrine of phlogiston. This hardly counts, however, since its critics repeatedly declare it pseudo-scientific. Now, the reason for this view is the central thesis (unanimously endorsed since antiquity and until the early twentieth century) that science is the body of proven theories. This is the generally received opinion. Thus, allegedly revolutionary Ludwig Wittgenstein took the certainty of Newtonian mechanics for granted (Wittgenstein 1922, § 6.341), and he never withdrew it. The popularity of Wittgenstein could easily mask the innovations of Koyré. Kuhn confessed having followed both.

By the received doctrine, scientific error is impossible. The consequence of this view is that most views about God's world are not science, including those advocated in antiquity and in the Middle Ages, including most of the ideas about biology produced prior to Harvey (in the early seventeenth century), most of the ideas about chemistry produced prior to Lavoisier (in the end of the eighteenth century), and most of the medical ideas produced prior to Ignaz Semmelweis (in the middle of the nineteenth century). Histories of science that followed this doctrine closely followed either Bacon or Duhem and so they are much too narrow. Koyré could not endorse so narrow a view of science, but he advised me in his last letter to me to ignore the narrow historians and write better history. He did that although he was drenched in a sense of frustration: as he kindly told me, though we were not close friends, his real wish was to study James Clerk Maxwell's harvest the way he studied Newton's, that he saw all his past work as preparatory to that study that he hardly began. He spent his research time studying Galileo and Newton as preparation for the study that he never began. He died in 1964.

Koyré's discussion of the errors of Galileo was new. How so? Galileo's errors that he discussed concern gravity. Now, as is a very well known, Galileo erred about astronomy too, as he deemed all planetary orbits circular. Hence, to appreciate Koyré's discussion of Galileo's error on gravity, we should see its difference from his error about astronomy. "What is served, after all, by dwelling on error?" Koyré asked (Koyré 1978, p. 66). Admitting that discovery is what we are after, and that discovery is success, not failure, he adds, the errors that discoverers make on their way to discovery "sometimes enable us to grasp and understand the hidden process of their thinking." This is lovely as far as it goes, but it does not explain the significance of Koyré: William Whewell had dwelled on the errors of Kepler to that end. For, it was old received opinion that received opinion is often erroneous and that it stands in the way of finding the truth. It is a corollary to Bacon's doctrine of prejudice that says, only the unprejudiced sees facts as they are (Agassi 2013, § 6.3). And so, well before Koyré Whewell tried to explain how researchers corrected their errors thereby opening the door to success. Even Galileo's error regarding gravity that Koyré discussed is rooted in prior doctrines that can pass for a popular prejudice. Koyré offered a detailed study of Galileo's thoughts about gravity to show that it was no prejudice but an intelligent error (that later on Descartes shared with him). The detailed study of series of errors in an effort to see the discoverer's thought-process is not an innovation of Koyré. For, nearly a century earlier, William Whewell took Kepler as an example of this: Kepler told his readers about the series of his errors that led him stepwise to his great discoveries. The study of Koyré differs from that of Whewell. I. Bernard Cohen noted in his obituary of Koyré (Cohen and Clagett 1966) that no one before Koyré took the trouble to read old texts line by line and examine them with all due care. He was thus able, added Cohen, to avoid the schoolmaster's attitude that so many historians of science take as they dismiss old errors as prejudices rather than seek their sources and rationale. This is terrific, except that it reveals an attitude to error that requires more discussion: what was the view of Koyré about science and error? He was clearly a fallibilist, but fallibilism alone does not distinguish between claims for scientific status that are adequate and those that are not. What was his view about the scientific character of scientific theories and what role did it ascribe to error?

Koyré is often declared a Platonist, we remember, and this raised much criticism from commentators averse to the Platonic view of science as a priori valid. Even the careful and erudite and sympathetic scholar Maurice Finocchiaro commits this error (Finocchiaro 2010) citing Koyré to say, "good physics is made *a priori*." He seemingly takes Koyré to have suggested that good physics is a priori valid. These two assertions differ radically: Einstein, for example, advocated the one and strongly opposed the other (as dangerous, no less). Clearly, Platonism is the idea that science is a priori valid and fallibilist Koyré did not endorse it. As to the idea that Koyré rightly ascribes to Galileo, that observation requires clear thinking, his example is mechanical for obvious reasons. Otherwise, a better example is from the observation of the surface of the moon. It is not smooth but rough. The received view that it is smooth rests on the idea that mirrors are brighter than walls. Leonardo expressed this view and van Gogh illustrated it! Galileo disproved it. To use modern parlance, Galileo argued that observations are theory laden. Otherwise, he observed, as you stroll down the street in a moonlit night, you may see the moon jumping like a cat from one rooftop to another.

The report of Cohen that Koyré "was in his heart of hearts a Platonist: he heartily disliked positivism in all its forms and manifestations" (Cohen and Clagett 1966, p. 164) requires some expansion. Koyré, it seems, was not interested in the problem of demarcation of science from metaphysics, but he saw it as central for historians of science, at least for those with a philosophical bent. More specifically, he wanted to demarcate not science but the scientific revolution. In addition, on this he was clear: he endorsed the view that it was the transition from the advocacy of Aristotle's philosophy to the advocacy of Plato's. This, he added, is not so much a metaphysics as a methodology and not so much in general as in two particular points: first, that theories are invented and, second, that good scientific theorizing is mathematical (Koyré 1968, p. 21, n. 3).

There is more to it. Stump has described Koyré as a Hegelian, we remember. This kind of Hegelianism he finds expressed in the interest that Koyré had shown in Hegel. This, however, was not quite an interest in Hegel; it was in his influence and in its diversity, as Elkana has noted (Elkana 1974) and as Gillispie has argued (Gillispie 2008). So says also James M. Edie (Edie 1963): "Koyré properly distinguishes this contemporary interest in Hegel both from the Hegelianism of the historical Hegel himself and the Neo-Hegelianism of the nineteenth and early twentieth centuries in England and Italy." Moreover, Koyré said his interest in Hegel was, "last but not least—the emergence of Soviet Russia as a world power and the victories of the communist armies and ideology [...]. Hegel begot Marx; Marx begot Lenin; Lenin begot Stalin" (Koyré 1961, p. 228; cited in Hyppolite 1974, p. xxxix). Nonetheless, it is hard to deny that in his healthy anti-positivism Koyré tended to ally himself, however tentatively and reluctantly and critically, with parties that could not keep their hold on his interest. This is said in disagreement with Stump, who finds Hegel's influence on Koyré expressed in the essay "Newton and Descartes." He cites Koyré to say, "Human thought is polemic; it thrives on negation. New truths are foes of the ancient ones which they must turn into falsehoods" (Koyré 1939, p. 65). He comments: "This has obvious affinities with Hegel's thought, but Koyré does not mention it by name" (*Ivi*, note 63). Not so. The dialectic that Koyré supported is Platonic, not Hegelian (Popper 1961).

1.5 Conclusion

Many commentators mention the tremendous erudition of Koyré and the breadth of his researches. They also note that he studied the history of scientific thought from antiquity to Newton. This and his having swam against the current made him address mainly expert readers. His output is thus a challenge to his followers who labor at narrowing the gap between the arts and the sciences. What he achieved is the output that may then become the classics that will hopefully open our culture to the possibility that we will treat some scientific texts as all-time classics akin to some great old works of the arts and the humanities. Of course, the aims of art and of science are different, so the two cannot merge. But they can and better be good neighbors.

References

Agassi J (1981) Science and Society: Studies in the Sociology of Science. Boston Studies in the Philosophy of Science, Vol. 65.
Agassi J (2008) Science and Its History. Springer, Dordrecht.
Agassi J (2013) The Very Idea of Modern Science. Springer, Dordrecht.
Bitbol M, Kerszberg P, Petitot J (2009) (eds) Constituting Objectivity: Transcendental Perspectives on Modern Physics. Springer, Dordrecht.
Bonnet Ch, de Calan R (2009) Moritz Schlick: Between Synthetic a Priori Judgment and Conventionalism. In Bitbol, Kerszberg and Petitot 2009, pp. 117–126.
Born M (1954) A History of the Theories of Aether and Electricity. The Modern Theories 1900–1926. By Sir Edmund Whittaker. British Journal for the Philosophy of Science 5:261–263.
Buchwald JZ (2013) (ed) A Master of Science History: Essays in Honor of Charles Coulston Gillispie. Springer, Dordrecht.
Collingwood RG (1939) An Autobiography. Oxford, Oxford University Press.
Cohen IB (1954) Review–Discussion: Some Recent Books on the History of Science. Journal of the History of Ideas 15:163–192.
Cohen IB (1974) Newton's Theory vs. Kepler's Theory and Galileo's Theory. In Elkana 1974, pp. 299–338.
Cohen IB, Clagett M (1966) Alexandre Koyré (1892–1964): Commemoration. Isis 57:157-66.
Drozdova D (2012) Alexandre Koyré, učenik Èmilja Meyersona: neizmennost' i istoričnost' čelovečeskogo razuma [Alexandre Koyré, disciple of Emile Meyerson: on immutability and historicity of human reason]. Epistemologia & Filosofia nauki 31/1:192–206.
Edie JM (1963) Review of Koyré (1961). Philosophy and Phenomenological Research 24:294–295.
Elkana Y (1987) Alexandre Koyré: Between the History of Ideas and Sociology of Disembodied Knowledge. History and Technology 4:115–148.
Elkana Y (ed) (1974) The Interaction between Science and Philosophy. The Humanities Press, Atlantic Heights NJ.
Finocchiaro M (2010) Defending Copernicus and Galileo. The Review of Metaphysics 64:75–103.
Fuller S (2000) Internalism versus Externalism. In Hessenbruch, pp. 380–381.

Gillispie CC (2008) Entry: Alexandre Koyré. The Complete Dictionary of Scientific Biography. Charles Scribner's Sons. Detroit–Michigan, pp. 482–490.
Hessenbruch A (ed) (2000) Reader's Guide to the History of Science. Routledge, London.
Hyppolite J (1974) Genesis and Structure of Hegel's Phenomenology of Spirit. Translated by Cherniak S, Heckman J. The Northwestern University Press, Evanston, IL.
Jaki SL (1987) Uneasy Genius: The Life and Work of Pierre Duhem. Nijhoff, Dordrecht.
Koyré A (1939) Études galiléennes: La Loi de la chute des corps. Hermann, Paris.
Koyré A (1957) From the Closed World to the Infinite Universe. The Johns Hopkins Univerity Press, Baltimore.
Koyré A (1961) Études d'Histoire de la pensée philosophique. Colin, Paris.
Koyré A (1968) Metaphysics and Measurement: Essays in the Scientific Revolution. Chapman, London.
Koyré A (1978) Galilean Studies. European Philosophy and the Human Sciences. The University of Chicago Press, Chicago, IL.
Koyré A (1998) Present Trends of French Philosophical Thought. Journal of the History of Ideas 59:521–548.
Kuhn TS (2000) The Road Since Structure: Philosophical Essays, 1970–1993. The University of Chicago Press, Chicago IL.
Mauskofp SH (2013) A Career in the History of Science as Student of Charles Gillispie. In Buchwald 2013, pp. 25–36.
McMullin E (1994) Scientific Classics and Their Fates. *PSA: Proceedings of the Biennial Meeting of the Philosophy of Science Association* 1994:2, 266–274.
Miller DP (2004) Discovering Water: James Watt, Henry Cavendish, and the Nineteenth Century 'water controversy'. Ashgate, Aldershot–Burlington.
Munro J (1890) Pioneers of Electricity, Or, Short Lives of the Great Electricians. The Religious Tracts Society, London.
Ornstein Bronfenbrenner M (1913) The Role of Scientific Societies in the Seventeenth Century. The University of Chicago Press, Chicago IL.
Paris JA (1831) The Life of Sir Humphry Davy. Colburn and Bentley, London.
Popper KR (1961) What is Dialectic? in his Conjectures and Refutations. Routledge, London.
Snow CP (1961) The Two Cultures and the Scientific Revolution. The Cambridge University Press, New York.
Stimson D (1948) Scientists and Amateurs: A History of the Royal Society. Henry Schuman, New York.
Stump JB (2001) History of Science through Koyré's Lenses. Studies in History and Philosophy of Science 32:243–263.
Whittaker E (1953) A History of the Theories of Aether and Electricity, Volume 2: The Modern Theories, 100–1926. Nelson, London.
Wittgenstein L (1922) Tractatus Logico-Philosophicus. Routledge, London.

Chapter 2
Homage to Koyré: Space as Paradigmatic Example of "The Unity of Human Thought"

Charles Braverman

Abstract My point is to demonstrate that Koyré's work can be synthesised through the notion of space and that Koyré's interest for this notion still has a great value for historians and philosophers of science. Actually, Koyré's analysis shows us that space is perhaps the most efficient concept to use when examining mutations of thought. What is at stake with space? Space is the place where historians can find an expression of the "unity of human thought". Space is then the centre of gravity of the epistemological commitments of many scientists but not only scientific commitments. "The unity of human thought" is to Koyré, a complex network of trans-scientific ideas such as philosophical, metaphysical and also religious ones. Moreover, my homage to Koyré consists of using his methodological principles in order to shed light upon André-Marie Ampère's scientific practices. In my opinion, these practices imply the same kind of unity of trans-scientific ideas as those studied by Koyré.

Keywords Classification • Experiment • Geometrisation • History of science • Indefinite • Infinity • Metaphysics • Mistake • Natural place • Naturphilosophie • Noumena • Phenomena • Qualitative • Quantitative • Revolution • Space • Trans-scientific • Unity of human thought • Unthought

2.1 Introduction

Due to the fact that Alexandre Koyré renewed the practice of the history of science, there is no need to pay homage to him. Actually, the value of his work was revered not only by philosophers and historians of science but also by historians who could agree that the history of science was the black sheep of institutionalised historical studies. This can be seen, for instance, in this statement from a book – written by the renowned French historian Paul Veyne – which studies the principles of historical inquiries:

C. Braverman (✉)
Archives Henri Poincaré–CNRS, Lorraine University, Nancy, France
e-mail: bravermancharles@gmail.com

R. Pisano et al. (eds.), *Hypotheses and Perspectives in the History and Philosophy of Science*, DOI 10.1007/978-3-319-61712-1_2

> The work of A. Koyré consisted, in some ways, in passing the history of science from an axiological history to a pure history, to a history of science "in its time". Before him, the history of science was mainly a history of great discoveries and inventions, a history of established truths and how they were arrived at; Koyré has put in place a history of errors and truths, a history of the too human progression of eternal truths [...] (see below)

> Ceasing to be axiological, the history of science ceases to be a prize-giving and becomes as fascinating as a novel based on fact.[1] (Veyne 1984, 68)

While the history of science was a largely axiological practice which focused on heroes and great discoveries, it might be considered to be the black sheep of the family of historical studies. It is undoubtedly the hermeneutical principle of the "unity of human thought" (Koyré 2007 [1966], p. 11) that enables the history of science to become a part of "pure history" or what Veyne calls also "a history of science in its time". In Koyré's point of view, pointing out the unity of human thought leads one to identify the commitments of scientists – commitments which are not only scientific but also "trans-scientific" (Koyré 2007, p. 11). These commitments include complex networks of ideas that are derived from philosophy, metaphysics and religion. How does unity of human thought express itself and to what elements should historians of science pay attention? What can we learn from Koyré? What specifically in Koyré's research is worthy of homage?

Consequently, the three first sections of my paper will be devoted to a hermeneutical point of view giving a synthesis of Koyré's analysis. Koyré was called, for instance, a "historian of the astronomical revolution" (Taton 1965, p. 147) or a "critic of the mechanical thought" (Costabel 1965, p. 155). Despite these labels, I aim to show that Koyré was fundamentally a historian devoted to the study of space. In doing so, I attempt to shed light upon Koyré's methodological principles regarding the history of science. It is also my hope to unify some of his principal results. Where does the unity of human thought express itself? In the way scientists considered space, I will then show the following: that space is a homonym linked with trans-scientific ideas (see below Sect. 2.2), that mistakes reveal unthoughts about space (which are crucial for scientific development) (cf. Sect. 2.3) and that the evolution of the concept of space is the key to understanding modern scientific criteria, as well as the experiment (see below Sect. 2.4).

However, without showing the actual validity and fecundity of Koyré's methodological principles for research, my homage would be incomplete. Actually, Koyré's analysis shows us that space is perhaps the most usable concept to examine evolutions of scientific thought. Along with that thesis and following a historical approach, I will use (in the conclusive section) Koyré's methodological principles which are first pointed out to show their fecundity for a case study of the early nineteenth century: the scientific practices of André-Marie Ampère (1775–1836).

[1] "L'œuvre d'Alexandre Koyré a consisté, à certains égards, à faire passer l'histoire de la science d'une histoire axiologique à une histoire pure, à une histoire de la science « en son temps ». Avant lui, l'histoire de la science était surtout une histoire des grandes découvertes et inventions, une histoire des vérités établies et de leur acquisition; Koyré a mis en place une histoire des erreurs et des vérités, une histoire du cheminement trop humain des vérités éternelles [...]. Cessant d'être axiologique, l'histoire des sciences cesse d'être une distribution des prix pour devenir passionnante comme un roman vrai" (Veyne 1996, p. 97).

2.2 The Homonymy of the Notion of Space

2.2.1 The Principle of the "Unity of Human Thought"

"A history of science in its time", according to Veyne, is the achievement of Koyré's methodological principles. Veyne opposes "pure history" and "axiological history". What shall we understand by the expression "pure history" and what exactly, in Koyré's historical studies, may be defined as "pure" analysis? The history of science is perhaps a part of history which leans essentially towards axiological judgements because of two philosophical presuppositions: there is a unique truth searched for by scientists; we can retrospectively examine what was right in a doctrine and consequently dismiss what is identified as false results. In the same way that the validity of historical researches are traditionally endangered by moral and political presuppositions, the history of science becomes axiological history when it implies retrospectively choices made by historians upon what is not worth a minute of analysis. Axiological history would have entirely condemned Bellarmine for being narrow minded and would have been labelled as an example of obscurantism and religious insincerity. In the same way, Galileo would be the objective and thoughtful scientist knowing the value of experience and whose triumph was undoubtedly linked with the experiment of the Tower of Pisa. In that opinion, Aristotelian science would be "an incomprehensible monster, ridiculous and misshapen"[2] (Koyré 1961, p. 258). However, that picture of the seventeenth-century scientific revolution is not the one Koyré depicts in his analysis. If Aristotelian science was the monster, it is sometimes said to be how could we explain that its success lasted more than fifteen centuries? In addition, even a caricature of that science cannot totally dismiss its success, for instance, in astronomy. When examining the arguments of Galileo and his opponents, Koyré tries to understand why they used those arguments and the context in which those arguments were made. Instead of an anachronistic distribution of blames and rewards, he tries to explain all the arguments mentioned and presupposed by scientists, insisting on the fact that those arguments were always of different kinds and far from what the *doxa* may expect about the progress of science:

> The history of scientific thought, as I understand it and as I endeavour to do it, aims at knowing the development of that thought during the mere movement of its creative activity. Then, it is essential to replace the studied works in their intellectual and spiritual environment, to interpret them thanks to the mental habits, the preferences and aversions of their authors. [...]. It is equally essential to include in the history of a scientific thought the way it understood itself and its place in comparison with what was previous and contemporary to it. [...]. We finally have to study the mistakes and failures with so much care that the success [...], they reveal difficulties that have to be surpassed and obstacles that have to be got over.[3] (Koyré 2007, p.14)

---

[2] "Un monstre incompréhensible, ridicule et difforme" (Koyré 1961, p. 258).

[3] "L'histoire de la pensée scientifique, telle que je l'entends et m'efforce de la pratiquer, vise à saisir le cheminement de cette pensée dans le mouvement même de son activité créatrice. A cet effet, il est essentiel de replacer les œuvres étudiées dans leur milieu intellectuel et spirituel, de les interpréter en fonction des habitudes mentales, des préférences et des aversions de leurs auteurs. [...].

The meaning of the "unity of human thought" is that scientific developments are not only supported by what we may call scientific commitments (even if for now it is not clear what "scientific" may exactly mean) but also by a complex network of trans-scientific ideas rooted often in philosophy and religion. Consequently, historians should not separate their analysis of scientific evolution from the cultural ground which nourished it. Moreover, an axiological history would irremediably dismiss all mistakes. Koyré insists though on the necessity of carefully examining those mistakes because they may reveal important aspects of the thought process of the scientist. Mistakes are then frequently the expression of the unity of human thought (see below Sect. 2.3).

As a consequence, anachronism leads us to analyse the findings of former scientists with the lens of our own concepts, which are the results of mutations. As a matter of fact, one of Koyré's major aims was the study of those mutations – mutations which demonstrate different kinds of commitments and which express the unity of human thought. Among those concepts – which evolved and whose evolution was determinant of the mutation of science – my point is to show that space is the paradigmatic example of Koyré's method and results.

Before demonstrating that space is the paradigmatic concept of the unity of human thought, I shall illustrate what kind of arguments are examined by Koyré in order to shed light upon that expression "unity of human thought". Actually, an example of trans-scientific arguments is given by Copernicus to support his heliocentricism, and it expresses what one may call his "heliolatry". Koyré mentions it in his book published in French in 1961, *La Révolution astronomique. Copernic, Kepler, Borelli, histoire de la pensée*:

> The old traditions, the tradition of Metaphysics of Light [...] Platonic memories, neo-Platonic and neo-Pythagorean (the visible Sun representing the invisible Sun; the Sun, master and king of the visible world and hence symbolic of God, this is the conception perfectly expressed by Marsilio Ficino in his Hymn to the Sun), these traditions alone are capable of explaining the emotion with which Copernicus speaks of the Sun. He adores it and almost deifies it. Those who, like Digby and Kepler and many others, have associated Copernican astronomy with a kind of Sun-worship, linking it moreover with Christianity were by no means disloyal to the inspiration of the great Polish thinker.[4] (Koyré 2013, p. 65)

It seems to us almost trivial to say that Aristotelian science and modern science do not use the same concepts. Actually, the concepts of "natural place", "natural

---

Il est tout aussi essentiel d'intégrer dans l'histoire d'une pensée scientifique la manière dont elle se comprenait elle-même et se situait par rapport à ce qui la précédait et l'accompagnait. [...]. On doit enfin étudier les erreurs et les échecs avec autant de soin que les réussites [...], ils sont révélateurs des difficultés qu'il a fallu vaincre, des obstacles qu'il a fallu surmonter" (Koyré 2007, p. 14).

[4] "Les vieilles traditions, la tradition de la Métaphysique de la Lumière [...], réminiscence platonicienne et renaissance néo-platonicienne et néo-pythagoricienne (le Soleil visible représentant le soleil invisible, le soleil maître et roi du monde visible et donc un symbole de Dieu, conception dont Marsilo Ficino nous donne une expression si parfaite dans son hymne au soleil) peuvent seules expliquer l'émotion avec laquelle Copernic parle du Soleil. Il l'adore et presque le divinise. Ceux qui, comme Digby et Kepler et bien d'autres encore, ont associé l'astronomie copernicienne à une sorte d'héliolâtrie en la liant d'ailleurs au christianisme, n'étaient nullement infidèles à l'inspiration du grand penseur polonaise" (Koyré 1961, p. 258).

movement" and "violent movement" are progressively replaced by others like "speed", "acceleration", "principle of inertia" or "force of gravity". However, those concepts were unthinkable in Aristotelian tradition. If they were unthinkable, it is not only because they required mathematical tools unknown by the Aristotelian tradition (tools like integration and derivation which took a long time to be precisely defined). Following Koyré's analysis, those concepts were unthinkable because they are founded by a totally different conception of space. The mere exigency of mathematisation of the concepts of modern physics cannot be taken as a pure and obvious exigency of a scientific mind but is linked with a singular "conception of the world"[5] (Koyré 2007, p. 11).

2.2.2 *The Homonymy of the Concept of Space in 1943: The Vanity of Any Experimental Evidence*

Koyré demonstrates the danger of confusing "habits" and "evidence" (Koyré 1943, pp. 335–336) that is rooted in our modern conception of the world inherited from Descartes or Galileo. The example of the principle of inertia is used by Koyré in an article from 1943 entitled *Galileo and the Scientific Revolution of the Seventeenth century* to demonstrate that our conception of the world (which seems obvious to us) is actually the result of habits. The principle of inertia is contradictory with the Aristotelian concepts of natural movements and places because it asserts that without any external forces (or when the sum of these forces is equal to 0), the object will stay at rest if it was at rest and conversely it will continue to move with the same speed and direction it once had. In the Aristotelian point of view, the absence of external power implies rest. Consequently, the object would move either until it reaches its natural place or if it is moved by an external power. Koyré can then assert that what we think to be evident would have seemed "*evidently* wrong and even absurd […] to the Greeks as well to those living in the Middle-Ages" (Koyré 1943, p. 335). Koyré demonstrates the fact that we should not confuse habits and evidence by using the principle of inertia because it is absolutely impossible to experiment an object that is *perfectly* isolated (or for which the sum of the external forces is *perfectly* equal to zero). Koyré concludes:

> The Galilean concept of motion (as well as that of space) seems to us so "natural" that we even believe we have derived it from experience and observation, though, obviously, nobody has ever encountered an inertial motion for the simple reason that such a motion is utterly and absolutely impossible. (Koyré 1943, p. 336)

Koyré refutes the fact that any representation of motion may be grounded obviously in an experiment. Those representations (as well those of the Aristotelian tradition and those issued by Galileo) are the results of habits and presuppose intellectual commitments that historians have to track down. Beyond the principle of inertia,

[5] "Conception du monde" (Koyré 2007, p. 11).

Koyré asserts that the concept of space is equally the result of a construction that expresses the unity of human thought and which is not the mere result of an experiment. Even the debate around the principle of inertia presupposes a modification of the concept of space. Actually, in Aristotelian tradition, to speak about natural or violent movement implies determining natural places and the fact that space is not homogeneous. In order to assume the possibility of the principle of inertia, one has to consider the fact that space is homogeneous and that there is no natural place which would determine the final place where the movement should stop. Space is then conceived in two very different ways, and no matter what experiment one may do, it will never be sufficient to convince others. To change a qualitative space into a homogeneous space cannot be only a matter of experiment, and one has to enter the field of philosophy and, perhaps, religion.

In order to convince his reader of that homonymy of the concept of space, Koyré studies in particular the conjunction between space and mathematics:

> We are equally well accustomed to the mathematical approach to nature, so well that we are not aware of the boldness of Galileo's statement that "the book of nature is written in geometrical characters", no more than we are conscious of the paradoxical daring of his decision to treat mechanics as mathematics, that is, to substitute for the real, experienced world, a world of geometry made real, and to explain the real by the impossible. (Koyré 1943, p. 336)

In a similar fashion, he points out that experiment is not able to find the evidence of the mathematisation of nature. He opposes then the "experienced world" with the "world of geometry" in order to show that the geometrical space is not the same as the daily experienced space. Actually, one may always argue that geometrical forms are abstractions from the experienced world. Those abstractions are then in rupture with experience, and no one can ever experience a pure geometrical form. To say that "the book of nature is written in geometrical characters" is a metaphysical assertion and not an obvious experienced fact. The way Galileo represented space was then an ontological mathematisation of nature as opposed to a qualitative conception of space, which was defended by the Aristotelian tradition.

2.2.3 The Homonymy of the Concept of Space in 1957: The Notion of Infinity

In 1957, Koyré published his book entitled *From the Closed World to the Infinite Universe* (Koyré 1957). In my opinion, it is a development of the thesis already defended in his article from 1943 – and already before in his *Études galiléennes* (Koyré 1939) – which asserted the homonymy of the concept of space and that it expresses the unity of human thought. That development consists then in the "long and thrilling story" which includes "the contemporaries and predecessors of Galileo" (Koyré 1943, p. 333).

The title of Koyré's book from 1957 exemplifies the fact that a mutation of the representation of space is at stake during the scientific revolution he analyses. According to Koyré, the progressive dissolution of the traditional "cosmos" was the key to understanding that revolution. Once again, experiment is not sufficient to understand that change of spatial representation. The "closed world" (the "cosmos") and the "universe" are not at all the same representation of space, and the two names imply, according to Koyré, two very different grounds for science.

Actually, Koyré shed light upon the homonymy of the word "space" which explains profound changes in scientific thought. One of Koyré's greatest credits is avoiding anachronism by showing that space representations did evolve. There is not only one representation of space but many, which involve very different ways of conducting scientific research (for instance, the importance of mathematisation but also, as I will show, the way of understanding experimentation).

This homonymy is then due to the fact that the different representations of space crystallise trans-scientific ideas and illustrate the "unity of human thought". To come back to Koyré's analysis, Aristotle's tradition, Bruno, Galileo or Descartes would illustrate a huge range of arguments linked to their conceptions of space which are not scientific but philosophical, metaphysical or religious.

The book from 1957 is not centred only around Galileo. There are, for instance, long analyses of the physics of Descartes and its philosophical and religious presuppositions. The physics of Descartes is not only important because of its influence but also because of the mathematisation of the nature it implied. The Cartesian mathematisation of nature is correlated to his particular conception of space because its central concept is "extension" which brings with it the possibility of using geometry in order to explain all movements. Actually, that concept of extension and its identification with matter are philosophical theses which are not rooted in experiment and which are not obvious. Thanks to the debate between Descartes and Henry More, Koyré shows how that concept of extension implies theological and philosophical arguments. One must see that More's objections revolve around the concepts of infinity and indefiniteness which are plainly linked to space in two different manners. Actually, Koyré sees in More's letters two main objections:

1. A first objection against the indefinite divisibility of matter
2. A second one against the assertion of the indefiniteness of the world

The objection against the indefinite divisibility of matter aims at asserting the possibility of atoms which may be divisible by God but not by any created being. Geometrisation – due to the Cartesian identification between matter and space – would then not be the principal principle of physics. More tries to give to physics another key principle, which is dynamic and implies a clear distinction between matter and space:

> [...] a special and proper quality or force – impenetrability – by which they [the things] resist each other and exclude each other from their "places". (Koyré 1957, p. 114)

The analysis lies on a philosophical and theological level. More and Descartes think about what God can or cannot do and about the distinction between the creature and the Creator.

They mention the same kind of trans-scientific arguments when it comes to the matter of the indefiniteness of the world. Here, More refutes the Cartesian distinction between indefiniteness and infinity. According to him, there is no alternative between a finite and infinite space. He then assumes that space is infinite because one cannot imagine a real limit of it. Again, that assertion is based upon the distinction between matter and space. More argues that:

> [...] this void space will not be absolutely void, for it will continue to be filled with God's extension. (Koyré 1957, p. 112)

Descartes could not agree with such an assertion because speaking about "God's extension" means nothing to him. His prime philosophical distinction between mind and matter led him to restrict extension to matter and, in no way, may God be reduced to a material being. According to Descartes and his revival of "the famous Anselmian *a priori* proof of the existence of God" (Koyré 1957, p. 124), God may only be defined by the concept of infinity. Also, infinity could only be intensive and never extensive. This is why Descartes introduces the concept of indefiniteness associated with space. Indefiniteness is then due to the extension of matter, of the subtle ether supporting the vortices.

To conclude, I could do nothing better than to quote Koyré's result of his analysis of the debate between Descartes and More where the homonymy of space appears plainly. Here, the reader can see the concept of space being linked with trans-scientific arguments and as the foundation of science:

> Summing up, we can say that we have seen Descartes, under More's pressure, move somewhat from the position he had taken at first: to assert the indefiniteness of the world, or of space, does not mean, negatively, that perhaps it has limits that we are unable to ascertain; it means, quite positively, that it has none because it would be contradictory to posit them. But he cannot go farther. He has to maintain his distinction, as he has to maintain the identification of extension and matter, if he is to maintain his contention that the physical world is an object of pure intellection and, at the same time, of imagination – the precondition of Cartesian science – and that the world, in spite of its lack of limits, refers us to God as its creator and cause. (Koyré 1957, p. 124)

2.3 Mistakes and Unthoughts About the Conception of Space

2.3.1 The Valorisation of the Study of Mistakes

If space expresses the unity of human thought, that does not mean scientists are always aware of all their epistemological commitments. Scientific developments may reveal unthoughts or thoughts, which are never really questioned. Examining the different ways in which scientists represent space affords us the possibility to understand mistakes, erring ways and sometimes the stubbornness of scientists.

I insisted previously on the methodological renewal of the history of science due to Koyré. If he studied science in its time, he also analysed very carefully the mistakes which were traditionally dismissed by other historians. According to Koyré, since mistakes sometimes reveal the commitment of scientists, they are not only a negative production of the mind. In fact, the unity of human thought may be revealed by these mistakes. Consequently, mistakes may also reveal trans-scientific ideas, which have, according to Koyré's assertions, a great weight in scientific research. The way Galileo represents space gives us an example of such mistakes which are highlighted by Koyré. Actually, Koyré shed light upon a double mistake rooted in the unthoughts concerning the notion of space.

But to be honest, Koyré was not the first to highlight the importance of studying mistakes when analysing the history of science. In his article entitled "Meyerson and Koyré: Toward a Dialectic of Scientific Change" (Biagioli 1987), Mario Biagioli emphasises the legacy of Meyerson concerning the necessity of taking mistakes into account. Meyerson himself would have paid attention to the Hegelian dialectic. According to Mario Biagioli, that focus on mistakes is not "a symptom of its biases [the biases of scientific evolution], idols or conceptual shortcomings, but rather a sign of its freedom" (Biagioli 1987, p. 169).

Here, the implicit opposition between Koyré and Bacon is obvious. Bacon's theory of idols was built to explain why our knowledge had made no significant improvements. For instance, the *Idola theatri* ("idols of the Theatre") produce fallacies because every philosophical sect is the result of systematic assertions of which coherence dismisses absolutely the results of experiment (Bacon 1620, Book I, Aphorism XLIV). Indeed, the experiment which is put forward in the *Novum Organum* is far from equivalent to what is called experiment in modern science. Moreover, Bacon's distrust of mathematisation of nature shows also that his condemnation of mistakes is linked to a peculiar conception of science. Mario Biagioli is then right to say that Koyré's rehabilitation of mistakes implies "turning upside down the older views of Bacon, Descartes and Mach" (Biagioli 1987, p. 169).

The case of Descartes' theory of mistakes may also help one understand why Koyré's focus on mistake is considered "as one of his most fundamental contributions to the modern historiography of science" (Biagioli 1987, p. 169). Mistakes are described only as a "disproportion between our finite understanding and our infinite will" (Descartes 1641, fourth meditation). They are the product of a bad use of free will which, in its precipitation, judges something not well examined by understanding. Under this conception of mistakes lies the religious commitment which asserts that human beings are the image of God, that the will is the expression of the absolute perfection of God's one but that the understanding is not so perfect and is the sign of our ontological deficiency. In doing so, Descartes may consider mistakes to be perfectly worthless. Mistakes would only be negative productions of an inattentive mind and would have to be overstepped by a careful method.

According to Koyré, to learn from mistakes does not mean to create a new method for avoiding future mistakes. The history of science should not be:

> [...] a graveyard of forgotten theories, or even a chapter of the history of the idiocy of men.[6]

In Koyré's opinion, the study of mistake is a major heuristic principle which enables historians to understand the unity of human thought constituted by trans-scientific ideas and the obstacles which stood in the way of the development of new conceptions of the world and consequently of science.

James B. Stump – in his article "History of Science through Koyré's Lenses" – focuses on the legacy mentioned by Biagioli in order to suggest "that if we are to put a philosophical label on his work [Koyré's work], rather than 'Platonist'", as has been the custom, "Hegelian" makes a better fit" (Stump 2001, p. 243). That suggestion is based upon the dialectic movement of science implied by the importance of Koyré's analysis of mistakes.

It is clearly legitimate to understand Koyré's methodology through its legacy from Meyerson or Hegel. However, I will complete that approach by showing Koyré's focus on mistakes can be better understood if it is anchored in the notion of space. Indeed, to focus on mistakes linked to space enables historians of science to find a way to analyse the implicit cultural axioms of the studied theories. The mistakes which arouse Koyré's interest are often about space. I will now analyse two such mistakes made by Galileo which were identified by Koyré. Those two mistakes will shed light upon Galileo's trans-scientific ideas about space:

1. The first mistake is his excessive geometrisation of space.
2. The second mistake is an incoherent commitment to maintaining some levels of perfection in a space though defined as homogeneous.

2.3.2 Galileo: A Mistake by Excess Concerning the Geometrisation of Space

One can easily deduce from the second law of Newton that the speed of the fall of a body – in the case of a free fall in a Galilean referential – is proportional to the time.

Instead of asserting that the speed is proportional to the time of the fall, Galileo wanted to calculate the falling motion of a body with an axiom postulating that the speed would be proportional to the distance covered by the body. In retrospect, one may say that it was a huge mistake to assert the proportionality of the speed to the distance, and one may consequently dismiss Galileo's tendency to *explain* the fall-

[6]That quotation is extracted from a conference of Koyré for the American Association for the Advancement of Science at Boston (Koyré 1955). The full text of that conference can be found in a book published in French in 1961 by Koyré himself (and the French translation is then Koyré's own). The following quotation (from which my previous quotation translated to English is extracted) illustrates the way he describes the older conception of history of science: "Un cimetière d'erreurs, ou même une collection de *monstra* justement relégués au cabinet de débris et bons seulement pour un chantier de démolition. *A graveyard of forgotten theories*, ou même un chapitre de la *Geschichte der menschlichen Dummheit*" (Koyré 1981, p. 257).

ing motion. However, Galileo's *description* of that movement was right, he should not have then tried to *explain* that movement in the way that he did.

In order to be more succinct, I may here mention Gerard Jorland's analysis of that mistake in his book named *La science dans la philosophie, les recherches épistémologiques d'Alexandre Koyré* (Jorland 1981, pp. 261–267):

> In 1604, Galileo already knows the law of the falling body, he knows that the distance covered by a free-falling body is proportional to the square of the elapsed time and that the distances covered in equal times are like the series of the odd numbers, but he is not satisfied with that descriptive knowledge. He searches the principle of the falling motion, its essence, of which those descriptive laws may be deduced. Yet, the principle he adopts as an axiom – that the speed of the falling body is proportional to the covered distance – does not enable to deduce those laws.[7]

This mistake reveals that Galileo gave priority to space rather than to time. According to Koyré, it is an "excessive geometrisation"[8] (Koyré 1939, p. 131) that explains Galileo's mistake. Galileo would have given the priority to a quantifiable variable linked to space: the height. By the reification of a pure geometrical space, Galileo would have condemned himself to search for an explanation of the law of the variation of speed by the covered distance and not by a proportionality with time.

Jorland's analysis has the merit to put forward an important implicit notion of Koyré's methodology: the notion of unthought. This is to say that some commitments of scientists may be revealed by their practices and their mistakes, although they do not question them explicitly.[9] Galileo's theory of movement reveals – through the mistake about its explanation of the free fall – that it was sustained by an unthought: an excessive geometrisation of space. It is that excess that would have been unquestioned, which is why that theory left no room for the notion of time.

The example of the mistake made in excess by Galileo is not the only piece of evidence that shows that space is the centre of the unity of human thought and of the revolution of the seventeenth century. As a matter of fact, Koyré's analysis points out another mistake made by Galileo, a mistake implying a deficiency of coherence.

---

[7] "En 1604, Galilée connaît déjà la loi de la chute des corps, il sait que l'espace parcouru par un mobile en chute libre est proportionnel au carré des temps et que des espaces franchis en temps égaux sont entre eux comme la série des nombres impairs, mais il ne se contente pas de ce savoir tout descriptif. Il recherche le principe du mouvement de chute, son essence, dont ces lois descriptives pourraient être déduites. Or, le principe qu'il adopte comme axiome, que la vitesse du mobile est proportionnelle à la distance parcourue, ne permet pas de déduire ces lois" (Jorland 1981, p. 262).

[8] The French expression is "Une geometrisation à outrance".

[9] The same kind of unthought about space could be found in the Cartesian philosophy. Descartes' identification of matter and extension drives him to the valorisation of the geometrical and quantifiable variable of distance. I have already shown – through Koyré's analysis of Descartes' debate with More – that the geometrisation of space tends to exclude the concept of force as modern physics constructed it progressively. However, that may be (and it would have to be developed in a much longer analysis), Koyré's methodological principles lead the way to the question of the importance of the notion of force in the Cartesian thought and its link to his radical geometrisation of space. Cartesian mistakes about the conservation of the quantity of movement may then be an important place of investigation.

2.3.3 *Galileo: A Mistake by Deficiency Linked to the Defence of Some Qualities Relative to Space*

Koyré's analysis points out a mistake made in excess by Galileo's geometrisation of space. He also shows that the notion of space reveals what seem to be incoherent assertions. Such assertion – which forms a mistake by deficiency – reveals that space is really the paradigmatic expression of the unity of human thought because trans-scientific ideas influence its characteristics.

In his article published in 1955 titled "Attitude esthétique et pensée scientifique" (Koyré 2007), Koyré completed his previous analysis of Galileo's representation of space at the occasion of the issue of Panofsky's book, *Galileo as a Critic of the Arts* (Panofsky 1954).

Although Galileo may have known the first two laws articulated by Kepler, how should we explain the absolute silence of Galileo about that important scientific result? One may argue that silence is not a mistake and that would be right. However, Koyré insists on the fact that silence cannot stay unexplained and that an explanation is required because Galileo's silence is associated with some very problematic assertions about the circularity of the fall. Knowing probably Kepler's laws devaluating circular movements and having demonstrated that the trajectory of a thrown body is a parabola, how could he still assert that falls are circular? Koyré explains his own unease:

> Everybody knows them, those assertions [about the circularity of the fall], and nobody can read them without some unease: they seem so anti-Galilean; we cannot admit that Galileo asserted those Aristotelian platitudes, so inadequate to the spirit of the new science, to his science, negating the "natural place" thanks to the geometrisation of space.[10]

According to Koyré, there are two traditional solutions to that kind of mistake due to a deficiency of coherence in Galileo's assertions about space:

- Wohlwill's solution would be based on the hypothesis that Galileo did not believe in the definitive value of Kepler's theory (Wohlwill 1906).
- Koyré's own interpretation takes into account the peculiar context of the edition of the *Dialogo sopra i due massimi sistemi del mondo* (Galileo 1632). Actually, the *Dialogo* was written in Italian, not in Latin, and was made in order to be read by well-educated people to convert them to the Copernican theory (Koyré 2007, p. 281). In that perspective, Kepler's laws would have been a point too complicated and technical. Ignoring Galileo's opinion about Kepler's laws, Koyré points out that it would have then been useless to Galileo's purpose to introduce them in his *Dialogo*.

[10] "Tout le monde les connaît, ces passages, et personne ne peut les lire sans une espèce de malaise: tellement ils nous paraissent anti-galiléens; nous ne pouvons pas admettre que Galilée ait sérieusement professé ces platitudes aristotéliciennes, tellement contraires à l'esprit même de la science nouvelle, de sa science, avec sa négation des « lieux naturels », avec la géométrisation de l'espace" (Koyré 2007, p. 283).

Taking into account the context of the publication of the *Dialogo*, Koyré explains the silence about Kepler's laws in that book. But, this does not explain Galileo's absolute silence about those laws. Could one be satisfied by that explanation while there are so many assertions in which Galileo seems to accept that movements have to be circular?

Koyré speaks then very highly of Panofsky's interpretation in which he shows his disagreement with both Wohlwill's and Koyré's solutions. In fact, Galileo's silence about Kepler's laws forces historians to make conjectures, but all the assertions about the circularity have to be taken very seriously. How many historians interpret Galileo's fascination for circularity while that fascination seems inconsistent with his geometrisation of space leading to a homogeneous space and to the negation of all peculiar qualities?

Panofsky's own interpretation explains that Galileo's work was marked by a highly classical aesthetic which implied the refusal of ellipse. Actually, Koyré takes into account that interpretation showing that it is coherent with his own methodological principle which is the unity of human thought. According to him, Panofsky is right to believe that Galileo's scientific theory was determined by aesthetic ideas. We have previously seen that Koyré identifies philosophical and religious commitments concerning the concept of space as they emerged during the scientific revolution of the seventeenth century. So why not aesthetic commitments? Panofsky sheds light on the fact that Galileo's theory is rooted in an aesthetic context which should not be ignored by historians. Meanwhile, Koyré accepts that aesthetic ideas are a part of the unity of human thought which is again expressed through the concept of space.

In short, Galileo's silence about Kepler's laws could be explained by an "invincible aversion" (Koyré 2007, p. 283) to ellipse which would be only a degraded circle. That classical aesthetic may seem surprising in Galileo's writings, but one should pay attention to the fact that the valorisation of the circle is the result of an ancestral tradition considering circles as the mobile imitation of eternal identity and unity. Galileo may have defended very revolutionary points of view. However, that is not a sufficient reason to presuppose that he was a revolutionary in all his ideas. There could be unthoughts in his scientific practice and especially in his conception of space expressed here by his reverence to the circle.

Indeed, the weight of Aristotelian tradition here is evident. Aristotle's distinction between different natural places allows him to study the movements of bodies thanks to a finalism where all bodies have to reach their natural places. Distinguishing then different elements, Aristotle identifies each element with a natural place that explains the movement of all things. In the *De caelo*, Aristotle demonstrates the existence of a fifth element whose dignity itself expresses a perfectly circular movement (Aristotle 2004, A). Stars made in that fifth element would have a circular movement which would be nothing less than perfect, always identical, due to the definition of circle which implies a geometrical form symmetric in all its parts. This perfection would also be possible because of the great distance between stars and the sublunary world which is characterised by its contingency. Aristotle's space is then determined by different qualities. Circularity is the movement demonstrating perfection of celestial bodies. Though Galileo's geometrisation of space implies a homogeneous space, his valorisation of the circle may then be interpreted as an unthought inherited from Aristotelian tradition and as a rehabilitation of spatial qualities.

In studying Galileo's conception of space, Koyré points out that there is a matter of coherence: Galileo criticises Aristotelian tradition but cannot put an end to the valorisation of the circle which should not be a priori preferred. Consequently, Koyré shows that there is an imperfection, if not a mistake, in Galileo's revolution concerning the notion of space. Finally, this imperfection implying a deficiency of coherence could be linked to a trans-scientific idea – an aesthetic valorisation of the circle which has to be taken into account when analysing the unity of human thought. Galileo's example shows us that the evolution of scientific theories about the movements of bodies implied that the valorisation of circle had to be overstepped (in the same way that the "horror of the vacuum" had to be). Without suppressing that reminiscence of Aristotelian thought about space, Kepler's laws could not be accepted. Koyré mentioned that Kepler himself had some hesitations concerning his own results.

To conclude my analysis of Galileo's mistakes, one may see that mistakes reveal the implicit axioms of Galileo's conception of movement. More specifically, those mistakes give direct access to Galileo's unthoughts about his conception of space: the unthought of an excessive geometrisation of space and the unthought of the valorisation of circle. Consequently, space appears again to be the paradigmatic example of Koyré's methodology, because after being a homonym explaining the rupture at stake during the scientific revolution of the seventeenth century, it now reveals the unthought of scientists.

2.4 Theory and Experiment: The Geometrisation of Space

2.4.1 A Singular Conception of Science Due to the Importance Accorded to the Notion of Space

I have already shown that the unity of human thought is a methodological principle which presupposes to destroy experiment as a definitive, clear and evident criterion of science. If trans-scientific ideas have to be taken into account, it is because experiment is never sufficient to convince and enable scientific progress. My point is to demonstrate that Koyré's thesis – that space implies the expression of the unity of human thought – is the basis of a new understanding of scientific practice, which is the experimental method. Experimentation, quantitative precision and technical achievements must be understood as the outcome of that new way of representing space, which arose thanks to the geometrisation of nature.

My point is that the originality of Koyré's historical studies is clear when one explains his methodology through the notion of space. This notion is the keystone of the conceptions of the world which were those of the Aristotelian tradition and of Copernicus, Galileo or Newton. It enables us to understand the unity of their thoughts – a unity which includes trans-scientific ideas. The heuristic value of that attention to the unity of human thought has already been shown by several commentators.[11] They are right to emphasise that heuristic value. I argue moreover that the value gains a better understanding when it is centred on the notion of space.

However, I would like to go further and show that what is at stake with Koyré's study about space is not only a methodological revolution but correlatively a revolution in the way science is conceived. Indeed, I may point out that methodological assertions about the history of science have to be rooted in an analysis of what science is. Koyré's interest for the notion of space as an expression of the unity of human thought is not only a methodological perspective but the mere result of a "general philosophy of scientific activity" (Costabel 1965, p. 159). In my opinion, the very fact that Koyré's methodology is centred around the analysis of the notion of space reveals that he was opposed to a traditional conception of science, seeing in experiment and technique the engine of its history. On the contrary, Koyré shows that what we consider to be the criteria of modern science are actually the results of profound theoretical mutations of the notion of space.

The idea that the unity of human thought is not only a methodological principle but also a description of what science is clearly appears in Koyré's opposition to some of his contemporaries, Crombie especially. Koyré's review (Koyré 2007, pp. 61–86) of a book written by Crombie about the origin of modern science and its links with the science of the Middle Ages (Crombie 1953) reveals that Koyré's methodological principles are rooted in a singular conception of science. Koyré concedes that "in the realm of history, Mr Crombie built a very beautiful residence".[12] However, Koyré does not hesitate to criticise that "residence", and one should not think that there was only a methodological disagreement between them. Koyré himself asserts that:

> The defenders of a continuous evolution, like those of a revolution, all keep their positions and seem unable to convince one another. In my opinion, not because they disagree about the facts, but because they disagree about the mere essence of modern science and consequently about the importance of some of its fundamental characteristics. Moreover, what seems to some of them a difference of degree is regarded by the others as an opposition of nature.[13]

Upon analysing, on account of Koyré's opposition to Crombie's thesis, one has the ability to understand his conception of science and the weight of the concept of space. Crombie's book entitled *Grosseteste and the Origins of Experimental Science* (Crombie 1953) shows the continuity of scientific evolution and explains that the experimental method was firmly implanted in Middle Age thought. According to

[11] See, for instance, Catherine de Buzon's paper "Alexandre Koyré et l'histoire de la pensée scientifique" (Buzon 1975, p. 203).

[12] This is the French quotation from the paper published in 1956 in *Diogène*: "dans le royaume de l'histoire, M. Crombie a édifié une très belle demeure" (Koyré 2007, p. 86).

[13] "Les partisans d'une évolution continue, tout comme ceux d'une révolution, restent tous sur leurs positions et semblent incapables de se convaincre les uns les autres. Ceci, à mon avis, beaucoup moins parce qu'ils sont en désaccord sur les faits, que parce qu'ils le sont sur l'essence même de la science moderne et, par conséquent, sur l'importance relative de certains caractères fondamentaux de cette dernière. De plus, ce qui semble aux uns une différence de degré, apparaît aux autres comme une opposition de nature" (Koyré 2007, pp. 61–62).

him, far from being a rupture, modern science would be only the result of a progressive process of realisation regarding Middle Age concepts and practices.

Koyré disagrees about the conception of science that lies under Crombie's study. According to Koyré, modern science is not *fundamentally* a revolution based on experiment. Koyré does not deny that Middle Age thinkers may have studied the necessity of experiment. He denies nonetheless that experiment could be understood in the same way during the Middle Ages and during the scientific revolution of the seventeenth century. Moreover, if the nature of experiment changed profoundly, Koyré argues that it was caused by the variation of "the profound reality which underlies those data" (Koyré 2007, p. 86) of experiment. The "profound reality" is undoubtedly the representation of space, and Koyré criticises Crombie's results for not having taken into account the homonymy of that concept of space. Yet, Koyré insists that the development of the modern mechanic and astronomy presupposed a singular way of understanding space. To him, even the modern conception of experiment is only a consequence of the modification of the concept of space. And so, experiment would not be the real criterion to study the history of science.

Consequently, modern science would be neither a *mere* revolution in experimental practices nor the continuous development of Middle Age analysis about experiment. According to Koyré, modern science would be due to a totally new set of axioms allowing one to define scientific researches and goals. Then, in anchoring the definition of science on theoretical and trans-scientific commitments about space, Koyré freed science from the over-evaluated weight of experiment.

2.4.2 A Geometrical Space in Order to Enable Experiment

Koyré tried to emancipate science from the over-evaluated experiment but that does not mean he considered that experiment has nothing to do with modern science. Indeed, Koyré's analysis clearly shows that experiment linked to quantification and mathematisation was a crucial and original gain of the revolution of the seventeenth century. However, that does not mean that experiment should be the criterion of science. Yet, it reveals that profound changes in the unity of human thought were at stake and that the keystone of the history of science is the notion of space.

Although Koyré's works refer frequently to Duhem's analysis of the science of the Middle Ages and not to his book entitled *The Aim and Structure of Physical Theory* (Duhem 1991), one may say that Koyré's critique of experiment is close to Duhem's one. Actually, Duhem's thesis of the underdetermination of theory by experiment linked to his criticism of the induction and of the crucial experiment opened a breach in the confidence too often put in experiment. However that may be, Koyré harshly criticises the importance many historians place on experiment regarding the development of Galilean science.

For instance, a paper written by Koyré – entitled "Galilei and the Experiment of Pisa, About a Legend" (Koyré 2007, pp. 213–223) – shows that it would be very naïve to think that such an experiment could have guaranteed Galileo's victory. To

Koyré, the experiment of the Tower of Pisa cannot be said to be the triumph of experimental modern science over the Aristotelian tradition; it could only be regarded as a myth.

According to Koyré, if that experiment of the tower of Pisa had taken place and if it had been taken seriously by Galileo's contemporaries, it would have turned to Galileo's disadvantage. Aristotelian defenders might have assumed a qualitative point of view and would not have taken into account the nonproportionality of the speed of the fall with the mass of the body. However, Galileo would have been confronted with a serious difficulty because nobody would have seen a simultaneous fall. Due to the resistance of the air, every spectator would have seen a difference between the two falling bodies. Galileo could not ignore the effects of the resistance of the air against such a fall, and he then could not have tested the speed of the fall independent of the mass of the body. Koyré expresses in those words the limits of any experiment:

> There is no perfection in this world; one may come close to it, but never reach it. Between the empirical datum and the theoretical object there is and there will always be an insurmountable gap. It is here that imagination comes on stage.[14]

However, Koyré does not neglect the necessity of experiments. He tries only to show clearly their limits. He believes there is no crucial experiment and that experiment is not the mere motor of the evolution of science. The necessity and the limit of experiment are pointed out in Koyré's paper of 1960 "The *De Motu Gravium* of Galileo. On Imaginary Experiment and its Abuse":

> It is clear that only experiment can give us the numerical data and that without them our understanding of nature would stay incomplete and imperfect. [...] And even when it comes to the fundamental laws of nature – like that of the fall – where the pure reasoning is in principle sufficient – it is only experiment which can guaranty to us that no other unanticipated factors hamper their application and that all happens in the sensible reality, *in hoc vero are* approximatively as it would happen in the Archimedean world of the reified geometry upon which we make our deductions. Furthermore, in a pedagogical point of view, nothing replaces experiment [...]. And the Galilean doctrine of the simultaneous fall of bodies is so new and, at a first look, so contrary to the common sense, that only an experimental confirmation would make it acceptable.[15]

---

[14] "La perfection n'est pas de ce monde; on peut s'en approcher, sans doute, mais on ne peut pas l'atteindre. Entre la donnée empirique et l'objet théorique, il reste, et il restera toujours, une distance impossible à franchir. C'est là que l'imagination entre en scène" (Koyré 2007, p. 225).

[15] "Il est clair que l'expérience seule peut nous fournir les données numériques sans lesquelles notre connaissance de la nature reste incomplète et imparfaite. [...]. Et même lorsqu'il s'agit de lois fondamentales de la nature – telle celle de la chute – où le raisonnement pur suffit en principe, c'est l'expérience seule qui peut nous assurer que d'autres facteurs non prévus par nous, ne viennent pas entraver leur application et que les choses se passent dans la réalité sensible, *in hoc vero are* à peu près comme elles se passeraient dans le monde archimédien de la géométrie réifiée sur lequel portent nos déductions. En outre, d'un point de vue pédagogique, rien ne remplace l'expérience [...]. Et la doctrine galiléenne de la chute simultanée des graves est tellement nouvelle et, à première vue, tellement contraire aux faits et au bon sens, que seule une confirmation expérimentale pourrait la rendre acceptable" (Taton 1965, XIII:197–245; Koyré 2007, p. 258).

It is not the experiment of the Tower of Pisa which interests Koyré. Indeed, he dismisses that experiment for the reasons I previously mentioned. He prefers by far other experiments which try to reduce the gap between empirical data and theoretical objects. The experiments of the inclined plane and of the pendulum reduce the resistance opposed to the fall, but Koyré explains that we have to understand Galileo's experimental mistakes by "turning our attention to the amazing and pitiful poverty of the experimental means at his disposal" (Koyré 1953, p. 224). According to Koyré, what is important in Galileo's experiment is obviously not his results – which are at best approximate – but his way of questioning nature and how he theoretically constructed those questions. Experiment is then not a simple observation but an intellectual construction in order to obtain measurable answers.

However, the necessity of experiment could only be understood in a theoretical context. To Koyré, experiment is essential in the collection of important data for the theoretical construction, in understanding the complexity of the causal determination and in terms of pedagogy and teaching theory to others. It is obvious that Koyré does not want to dismiss experiment but only to understand its links to the primitive theoretical context of its development.

Furthermore, the primitive theoretical context which enables one to understand the necessity of experiment presupposes a profound revolution in the conception of space. If the experimental method is important to modern science – and it is – it does not embody the scientific revolution of the seventeenth century. Under that revolution lies another revolution: the geometrisation of space. Experiment in modern science presupposes the precise measurements of data. Those measurements are needed because the laws of nature are expressed by mathematical variables. The modern idea of experiment was incompatible with Aristotelian tradition because of the attention to qualities, which was the keystone of that form of science. Koyré points out that "quality, indeed, is repugnant to the precision of measure" (Koyré 1953, p. 223):

> Which means that modern science constitutes itself in substituting for the qualitative or, more exactly, for the mixed world of common-sense (and Aristotelian science) an Archimedean world of geometry made real; or – which is exactly the same thing – in substituting for the world of the more–or–less of our daily life a universe of measurement and precision. Indeed, this substitution implies automatically the exclusion from – or the relativisation in – this universe of everything that cannot be subjected to exact measurement. (Koyré 1953, p. 223)

In that context, the importance of the concept of space appears again plainly. The geometrisation of space is the first step in the constitution of "an Archimedean world of geometry made real". It implies the destruction of the old conception of space – the Aristotelian space – with all its natural places and qualities.

The concept of space is a homonym and the place of the expression of the unity of human thought – expression which is often linked to mistakes. One can then deduce that this concept is the key to understanding the evolution of experiment and the development of modern science. There is no doubt that space is a paradigmatic concept of Koyré's historiography, but what is the value of that lesson? Can we use it for scientists other than those studied by Koyré?

2.5 Conclusion

2.5.1 Koyré's Lesson About Space: The Unity of Ampère's Thought

My homage to Koyré is not only founded on a hermeneutical point of view which highlights how strongly space is a paradigmatic example of the unity of the human thought. It is also to show that Koyré's results about space are not worthless and can be used to better understand scientific developments which occurred long after the seventeenth century.

André-Marie Ampère (1775–1836) will serve as a brief case study showing the value of Koyré's methodology. Ampère is not only known for his electrodynamics theory (Hofmann 1996) but also for the range of researches he conducted. Ampère was firstly an institutional renowned mathematician, but he was also very interested in chemistry. And, above all, he was an "encyclopaedist and a metaphysician" (Locqueneux and Scheidecker-Chevallier 2008).

This is why I aim to demonstrate that André-Marie Ampère's conception of scientific research – in chemistry and in electrodynamics – implies trans-scientific ideas. My point is to then shed light on the Catholic and post-Kantian aspects of Ampère's thoughts, which are linked to his scientific practice and are expressed through his concept of space.

2.5.2 Geometry and Chemical Classification

The scientific practice of Ampère expresses a "passion for natural classification" (Locqueneux 2009). To him, classification was undoubtedly the incarnation of the Baconian inductive method, whose legacy could be found in the French *Encyclopédie*. In his first lecture in Bourg-en-Bresse (Ampère 1936, p. 106), but also in his last work entitled *Essay of philosophy of science*, one may see that the aim of science is to seek unity. Beginning with the variation and diversity of the daily sensible world, scientists have to abstract facts, to induce relations between facts which become laws and finally to induce principles which are the highest level of unity and generalisation (Ampère 1834, XIX). Because classification was considered by Ampère as the paradigmatic incarnation of the inductive method, his works concerning electrodynamics (Ampère 1826) can be seen as a classification seeking to unify all electric and magnetic phenomena.

Although science is identified with classification – which expresses induction – one should not be tempted to assert that experiment could be the keystone of Ampère's scientific practice. Koyré's lesson appears here to be fruitful: experiment has to be replaced in its theoretical context, and historians have to seek theoretical – and maybe trans-scientific – commitments of scientists. Classification always implies theoretical choices made by scientists. Ampère is clearly aware of that necessity. Classifications are constructions which imply that the scientist chooses – implicitly or explicitly – some criteria which enable him to order and classify the diversity within certain phenomena. The fact that Ampère was aware of the presence of theoretical choices appears plainly in his claim found in "natural classifications".

According to the legacy of the naturalists, when Ampère claims to make a natural classification, his choices are not arbitrary but correspond to the order of nature itself. There is here a trans-scientific commitment which should not be unseen: classifications are human constructions, but *there is a natural order* that we can discover. Two questions may arise: what is that natural order and how can we pretend to discover it? Consequently, my point is to show that Ampère's presupposition of a natural order is an idea which leads the way to his theoretical commitment in all its scientific practices: the mathematical order of nature and especially of space.

In chemistry, Ampère published his classification of the elements in 1816. The criterion he used was experimental because he focused on the kinds of associations and products to which elements are known to be committed. In their paper on Ampère's studies about chemistry, Robert Locqueneux and Myriam Scheidecker-Chevallier are right to point out that Ampère's classification of chemical elements was preceded by another chemical study but which was not conducted in the same way (Locqueneux and Scheidecker-Chevallier 1994). Actually, those commentators show that Ampère tried, in 1814, to explain all possible chemical reactions by a geometrical analysis inspired from Haüy's crystallography. In that study, Ampère considers that phenomenal experiments of chemical reactions are only an appearance that has to be explained at a microscopic level by a geometric representation. There are spatial relations which interest Ampère in that attempt of giving an explanation of experimented chemical reaction. Even if that attempt was not successful, one may analyse Ampère's theoretical commitments in the same way Koyré studied Galileo's mistake.

Apart from his defence of atomism and his participation in the discovery of fluorine, few historians decide to take Ampère's studies in chemistry seriously. His hypotheses about the geometrical microstructure of elements are then almost ignored. Yet, learning from Koyré, historians should take Ampère's attempt into account because it reveals that, in his opinion, science cannot stay at a phenomenal level but has to be rooted in microscopic mathematical relations. It also shows that mathematics have, to him, an ontological weight and that space is where that ontological mathematisation is manifested.

Actually, in a post-Kantian point of view, Ampère asserts that phenomena are not the same as noumena and that space is a form of our representations linked to our sensitive constitution (Biran 2000, p. 5). He then wants to rectify the Kantian philosophy by showing that a knowledge of noumena is possible. Ampère is ready to accept the fact that our phenomenal representation of space is linked to our sensitive constitution, but he also asserts that there is a noumenal space, a "real extension" which is the field of mathematical relations and which can be expressed by mathematics (Ampère 1834, LXII).

Ampère's attempt to explain chemical reactions thanks to geometric microstructure is then rooted in his philosophical analysis. Consequently, trans-scientific ideas about space support Ampère's scientific practices. Indeed, he defends a structural realism based on a particular conception of space which consists of an ontological mathematisation distinct from daily phenomenal space. According to Ampère, a classification ideally expresses a mathematical structure. But how can we be certain that our mathematical constructions are really objective? How can we get rid of the risk of an arbitrary choice? After all, Kepler's laws are only a certain mathematical construction, and one may imagine an adaptation of the Ptolemaic system in order to save the phenomena. But Ampère believes that "Kepler's beautiful laws" (Ampère and Ampère 1866, p. 12) are right and that they are not just an "explicative hypothesis" (Biran 2000, p. 67). How does Ampère justify his faith in the geometrisation of space and in the necessity to know the reality through mathematical relations? Finally, my point is that Ampère's practices imply a religious commitment, a belief that God "created human beings in order to accomplish the good and to know the truth" (Ampère 1843, p. 39). Thanks to mathematics, we are not condemned to phenomenal representations.

2.5.3 *His Hesitation About Electrodynamics: What Are the Properties Linked to Space?*

I have shown how Ampère's scientific practices implied trans-scientific ideas and that Koyré's paradigmatic interest in the concept of space is valuable in understanding Ampère's thought. Finally, I want to show that Koyré's lesson is not only valuable to understand Ampère's fruitless attempt to study chemistry but also useful to understand the way Ampère considered his results in electrodynamics.

Ampère is renowned for his theoretical description of the electrodynamics phenomena and especially for having written a book called *Theory of Electrodynamic Phenomena, Uniquely Deduced from Experience* (Ampère 1826). Duhem has already showed that we should not to be fooled by Ampère's assumption that he deduces his theory "uniquely from experience" (Duhem 2007, pp. 297–298). In the same way, I have pointed out that Koyré insists repeatedly that experiment – though necessary to science – cannot be considered as the foundation of theoretical developments and that experiment itself must be linked to a geometrisation of space.

Actually, Ampère's research in electrodynamics illustrates Koyré's point of view because empirical data have to be classified in order to reveal the true relations which structure nature and because the choices of Ampère cannot be reduced to empirical experiments. In his *Theory of Electrodynamic Phenomena*, Ampère adopts the standard model of law, which is illustrated by the Newtonian law of gravitation. Action at a distance was then the ground of common descriptions of phenomena, and it was then the criterion chosen by Ampère to classify the data of electromagnetic phenomena.

However, Ampère's correspondence shows us that he hesitated and that he would have preferred a description based upon the assumption of space filled with ether acting like a fluid. What is at stake is different representations of a space characterised by some primary principles, which are beyond the grasp of experimentation. As a matter of fact, a letter to Auguste de La Rive written on July 2, 1824, and another one to Faraday written in 1825 explain Ampère's preference for a space identified with a fluid ether.[16]

Ampère adopted a Newtonian model, but he considered that model as means for calculating the electrodynamics phenomena, and he did not regard it as a true explanation. One may argue that Ampère developed a pragmatic theory that enabled him to make quantitative predictions and also that he was influenced by the institutional way of doing science, specifically Laplacian context and its valorisation of the Newtonian model (Fox 1974). But that explanation must be completed by someone else answering the question why Ampère would have preferred a representation of space implying an imponderable ether. I then follow Kenneth Caneva's interpretation which could perfectly illustrate Koyré's idea of the unity of human thought and its focus on space: Ampère's hesitation would be linked to a post-Kantian influence due to Œrsted's *Naturphilosophie* (Caneva 1980).

For my purposes, I may only retain here that *Naturphilosophie* argues philosophically for a unification of all natural forces in discovering the primitive forces which determine the movements of matter and which should also explain chemical reactions and vital processes (Œrsted 1813, p. 113 and sq.). It postulates a unity and continuity of natural process which begin with matter (Friedman 2006, pp. 7–26, pp. 51–59). Œrsted then considers that space is filled with an ether, that the laws of nature are those of the dynamic of that fluid and that the right way to study them is undulatory dynamics (Œrsted 1813, p. 130).

In that context, Ampère's hesitation about electrodynamics – which is a hesitation about the representations of space and its principle properties – becomes understandable. According to Ampère, one had to use the Newtonian model in order to create a means of calculating phenomena. This is why he made the hypothesis that elementary charges can attract each other at a distance. It was also a way of conforming to the Laplacian programme. But to consider space to be identified with an indefinite fluid, ether was to assert that only action implying mechanical contact is *really* possible and that the model of scientific explanation should be the undulatory dynamics. This point of view has no experimental basis, and if Ampère believed in it, it is because of trans-scientific ideas such as post-Kantian ones. Here, again, one should give homage to Koyré's lesson. He taught us that science cannot be reduced to experiment, that space is homonymous and that mistakes, or hesitations, may reveal that homonymy and the trans-scientific ideas implied in it.

Acknowledgements I would like to thank Kaitlin King, Katharine Linek and anonymous referees for helpful comments and corrections on earlier drafts of this paper.

[16]This letter, number L680, to Faraday is transcribed on the website *Ampère et l'histoire de l'électricité*.

References

Ampère AM (1826) Théorie mathématique des phénomènes électro–dynamiques, uniquement déduite de l'expérience [édition 1883]. A. Hermann, Paris.

Ampère AM (1834) Essai sur la philosophie des sciences, ou Exposition analytique d'une classification naturelle de toutes les connaissances humaines. Première partie (notice de Sainte-Beuve et Littré). Bachelier, Paris.

Ampère AM (1843) Essai sur la philosophie des sciences, ou Exposition analytique d'une classification naturelle de toutes les connaissances humaines. Deuxième partie (notice de Sainte-Beuve et Littré). Bachelier, Paris.

Ampère AM (1936) Correspondance du grand Ampère. Gauthier–Villars, Paris.

Ampère AM, Ampère JJ (1866) Philosophie des deux Ampère. Didier, Paris.

Aristotle (2004) Traité du ciel. Traduction Pierre Pellegrin. Flammarion, Paris.

Bacon F (1620) Novum organum. Traduction Michel Malherbe [édition 2010]. PUF, Paris.

Biagioli M (1987) Meyerson and Koyré: Toward a dialectic of scientific change. History and Technology 4:169–182.

Biran PMD (2000) Correspondance philosophique Maine de Biran-Ampère. Vrin, Paris.

Buzon CD (1975) Alexandre Koyré et l'histoire de la pensée scientifique. L'histoire de la pensée scientifique, pp. 203–205.

Caneva KL (1980) Ampère, the etherians, and the Œrsted connection. British Journal for the History of Science 13:121–138.

Costabel P (1965) Alexandre Koyré, critique de la pensée mécanique. Revue d'histoire des sciences et de leurs applications 18/2:155–159.

Crombie AC (1953) Robert Grosseteste and Origins of Experimental Science, 1100-1700. The Oxford University Press, Oxford.

Descartes R (1641) Meditationes de Prima Philosophia. Traduction par Louis Charles d'Albert de Luynes. Vrin, Paris.

Duhem PM (1991) The Aim and Structure of Physical Theory. The Princeton University Press, Princeton, NJ.

Duhem PM (2007) La théorie physique : Son objet, sa structure. Vrin, Paris.

Fox R (1974) The rise and fall of Laplacian physics. Historical Studies in the Physical Sciences 4:89–136.

Friedman M (2006) The Kantian legacy in nineteenth–century science. MIT Press, Cambridge.

Galileo G (1632) Dialogo sopra i due massimi sistemi del mondo.

Hofmann JR (1996) André-Marie Ampère: Enlightenment and Electrodynamics. The Cambridge University Press, Cambridge.

Jorland G (1981) La Science dans la philosophie : les recherches épistémologiques d'Alexandre Koyré. Gallimard, Paris.

Koyré A (1939) Études galiléennes. Hermann, Paris.

Koyré A (1943) Galileo and the scientific revolution of the seventeenth century. Philosophical Review 52:333–348.

Koyré A (1953) An experiment in measurement. Proceedings of the American Philosophical Society 97/2:222–237.

Koyré A (1955) On the influence of philosophical concepts on the development of scientific theories. The Scientific Monthly 80/2:107–111.

Koyré A (1957) From the Closed World to the Infinite Universe. The Johns Hopkins University Press, Baltimore.

Koyré A (1961) La révolution astronomique. Copernic, Kepler, Borelli, histoire de la pensée. Hermann, Paris.

Koyré A (1981) Études d'histoire de la pensée philosophique. Gallimard, Paris.

Koyré A (2007) Études d'histoire de la pensée scientifique. Gallimard, Paris.

Koyré A (2013) The Astronomical Revolution: Copernicus – Kepler – Borelli. Translation by Maddison. Hermann, Paris.

Loqcueneux R (2009) André-Marie Ampère ou la passion des classifications naturelles [lecture read at the Lille 1 University Workshop entitled: "pourquoi classer ?".

Loqcueneux R, Scheidecker–Chevallier M (1994) La théorie mathématique de la combinaison chimique d'André-Marie Ampère Revue d'histoire des sciences 47/3:309–352.

Loqcueneux R, Scheidecker–Chevallier MM (2008) Ampère, encyclopédiste et métaphysicien. EDP Sciences, Lille.

Œrsted HC (1813) Recherche sur l'identité des forces chimiques et électriques. Dentu, Paris.

Panofsky E (1954) Galileo as a Critic of the Arts. Nijhoff, Leiden.

Stump J (2001). History of science through Koyré's lenses. Studies in the History and Philosophy of Science 32:243–263.

Taton R (1965) Alexandre Koyré, historien de la révolution astronomique. Revue d'histoire des sciences et de leurs applications 18/2:147–154.

Veyne P (1984) Writing History: Essay on Epistemology. Translation by Mina Moore-Rinvolucri. The Manchester University Press, Manchester.

Veyne P (1996) Comment on écrit l'histoire. Seuil, Paris.

Wohlwill E (1906, 1909) Galilei und sein Kampf für die kopernikanische Lehre. Hamburg L. Voss, Leipzig–Verlag.

Chapter 3
"The Philosophers and the Machine": Philosophy of Mathematics and History of Science in Alexandre Koyré

Mauro L. Condé

Abstract This study examines the influence of Koyré's philosophy of mathematics on his later history of science. This chapter focuses on a topic that has been almost unexamined by Koyré's commentators, his first papers on the foundations of mathematics (which present his criticism of Bertrand Russell's paradox), to establish a link between Koyré's mathematical realism – present in his preoccupation with the foundations of mathematics – and his later history of science. Based on this link, I reassess an old controversy: "the philosophers and the machine", as it is called by Koyré, or the problem of *theoria versus praxis*. Finally, I attempt to understand the extent to which Koyré's internalism, which originates in this controversy, continues to raise important issues, such as the autonomy of science, in the face of sociological interpretations of science.

Keywords Koyré • Philosophy of mathematics • History of science • Internalism • *Theoria* • *Praxis*

3.1 Introduction

Alexandre Koyré (1892–1964) began his philosophical career in the early twentieth century with substantial interest in the problem of the foundations of mathematics, antinomies and paradoxes (Koyré 1912, 1922). In that period, he was in contact with Edmund Husserl and was fascinated by mathematical realism. Soon, logic and the foundations of mathematics would no longer be the focus of his research. However, these topics, which he eventually revisited, influenced his future conception of the history of science. Years later, Koyré not only developed his history of science by reaffirming mathematical realism as a key to reading the works of Galileo (Koyré

M.L. Condé (✉)
Department of History, Federal University of Minas Gerais – UFMG,
Av. Antonio Carlos, 6627, 31.270-901 Belo Horizonte, MG, Brazil
e-mail: mauroconde@ufmg.br

R. Pisano et al. (eds.), *Hypotheses and Perspectives in the History and Philosophy of Science*, DOI 10.1007/978-3-319-61712-1_3

1966 [1939]) and Newton (Koyré 1965) but also extended this concept to authors such as Maxwell and Einstein (Koyré 1971 [1961], p. 267).

According to Koyré, "the science of our time, like that of the Greeks, is essentially *theoria*" (Koyré 1973 [1966], p. 399).[1] It is rationality that organises facts and experiences. Nature is written in mathematical language, as stated by Galileo, and the mathematical realism behind this statement was the main element for Koyré in the construction of modern science. Based on this mathematical realism, he affirmed the perspective that would become known as internalism or the idea that science finds its justification in itself independent of influences from social contexts. Consequently, he became extremely critical of the real importance of social and technological aspects (externalism) in the birth of modern science, as advocated at that time by Zilsel, Hessen, Grossmann and others (Koyré 1965, p. 6; 1966 [1939], pp. 12–13; 1973 [1966], p. 167), despite the fact that Koyré himself had worked extensively on the problem of technique and the sociological interpretation of science (Koyré 1971 [1961], pp. 305–362).

This debate on the complex relationship between theory and practice at the beginning of modern science, as synthesised by Koyré, is the problem of "philosophers and the machine" (Koyré 1971 [1961], pp. 305–339), or the difficult task of reconciling *theoria*, on the one hand, and *praxis*, on the other. What Koyré views as *theoria*, mathematical realism and the metaphysical bases of science (or Koyré's internalism, according to tradition), we can also understand as a need for the autonomy of science. Long after this controversy, in a period when the history of science is written primarily from a social perspective, I seek to reassess the issue of *theoria* versus *praxis* to understand the extent to which Koyré's internalism continues to raise important issues, such as the autonomy of science, in the face of these sociological interpretations of science.

With this aim in view, this chapter is structured as follows. First, I review the initial context of Koyré's career, which was dominated by questions concerning the foundations of mathematics and paradoxes. I attempt to show how these issues were present in the thinking of the young philosopher and, to some extent, were still present in his later history of science. Second, I present what became known as Koyré's internalism, particularly through the affirmation of his mathematical realism and his criticism of sociological approaches and techniques as major elements in the foundation of modern science (externalism).[2] Finally, I seek to point out the extent to which Koyré's internalism can be understood as an affirmation of the autonomy of science rather than as a barrier that isolates science from society.

[1] "La science, celle de notre époque, comme celle des Grecs, est essentiellement *theoria*".

[2] The internalism vs. externalism controversy was largely structured with Koyré as an iconic figure of internalism. For an overview of the internalism vs. externalism issue see Shapin 1992. According to Shapin, although this topic has been an important issue in the history and sociology of science, particularly from the end of WWII to the end of the Cold War, it continues to demand new approaches (Shapin 1992, p. 334).

3.2 Foundations of Mathematics, Paradoxes and Rationality

The young Koyré had an enthusiasm for philosophy in a scientific atmosphere marked by significant transformations in the sciences, particularly in the field of logic and mathematics. From the late nineteenth century, the development of mathematics and the birth of mathematical logic created the need to rethink the foundations of mathematics. However, the paradoxes[3] that emerged from this project of foundation threatened, more than mathematics did, the very idea of scientific rationality itself. As Koyré states:

> In the twentieth century, the logico-mathematical paradoxes played an important role, as is well known, in the evolution of mathematical or, more precisely, metamathematical thought. It was their discovery which determined the 'crisis of the foundations' of mathematics; and we owe the rich development of symbolic logic, the intuitionism of Brouwer, the axiomatics of Zermelo and Hilbert to the desire to solve or avoid them. In his brilliant and important paper, A. Fraenckel (1939) says that the discovery has had a 'terrifying' effect. The most secure foundations of science, indeed, of reason itself, seemed to be undermined (Koyré 1946, p. 344)

There were different reactions to this "terrifying" effect of paradoxes: solving, avoiding or seeking new directions for rationality. Paradoxes such as the "liar" or "Achilles and the tortoise" have been known since Greek antiquity, but in this new context created by mathematical logic, these paradoxes acquired decisive importance, not because they had the same logical structure as the traditional paradoxes but rather because the foundations of mathematics and, consequently, the very idea of rationality would be at stake. The works of Frege, Russell and Husserl were central to this reflection on the foundations of mathematics and the search for the solution to the paradoxes.

In the background of the issues of logic and the philosophy of mathematics (which were threatened by paradox), there was the affirmation (or denial) of an essence of formal thought: a mathematical realism from a Platonic perspective. Would mathematics be a "discovery" or an "invention"? For all of these authors, it would be a discovery. Frege and Russell's logicism defended the idea that there was an identity between logic and mathematics: mathematics was derived from certain fundamental logical principles. The tradition of a philosophy of mathematics that originated in Plato by way of Descartes's *mathesis universalis* could finally, with

[3] "Russell's paradox" will be the centre of these discussions. In 1902, it was reported by Russell in a letter to Frege in the following way: "Let *w* be the predicate: to be a predicate that cannot be predicated of itself. Can *w* be predicated of itself?" (Russell 1967 [1902], p. 125). The contradiction that arises here is called Russell's paradox. In fact, Cantor identified the paradox of set theory in 1896 and Zermelo and Russell between 1900 and 1901. However, "Russell's paradox" is best known because it was identified by Russell in Volume 1 of Frege's *Grundgesetze* – in a dramatic episode – shortly before the publication of Volume 2. The drama felt by Frege when he saw his project to outline the foundation of arithmetic shaken by the paradox was reported by him in response to the letter in which Russell announced the discovery of this paradox in Frege's work. Frege said, "Not only the foundations of my arithmetics, but also the sole possible foundations of arithmetic, seem to vanish" (Frege 1967 [1902], p. 128).

the new mathematical logic, not only establish the foundations of mathematics but also carry out the project of a universal language of science, as proposed by Leibniz with his idea of a *characteristica universalis*. Even Husserl, who did not initially advocate mathematical Platonism in his *Philosophy of Arithmetic* (1891), ultimately accepted it in Volume 1 of his *Logical Investigations* (1900). Finally, for these authors, what was at stake in the threat of paradoxes to mathematical realism was not only the foundation of mathematics but also modern scientific rationality itself, as based on the idea of *mathesis universalis*.

When Koyré started his professional preoccupations with these philosophers and mathematicians, he guided his philosophy with this set of issues, in which the problem of paradoxes occupied a central place. In an attempt to preserve mathematical realism, Koyré understood that paradoxes could not shake traditional rationality. Indeed, his entire endeavour concentrated on minimising the "terrifying" effect of the paradoxes, thereby affirming the mathematical realism that would ensure the foundations of mathematics and, as we shall see below, the foundation of his future interpretation of the history of science.

In Germany, the young Russian Alexandre Koyré presented a PhD proposal to Husserl on mathematical issues. Because Husserl did not accept his proposal (Zambeli 1999), Koyré left Göttingen to live in Paris. However, the relationship between the two thinkers was maintained for a long time through letters and visits, demonstrating the considerable influence of Husserl on Koyré. Therefore, Koyré's first philosophical reflections were strongly directed by concerns that came from this logical and mathematical circle. As observed by Zambelli (Zambelli 1999), this technical interest in mathematics would later allow Koyré to analyse the works of Galileo and Newton. In mathematical realism, as supported by this circle, Koyré had found a strong element in his later conception of the history of science.

The first publication by the young Koyré, a short note entitled "Sur les nombres de Bertrand Russell" (1912), commented on Russell's *Principles of Mathematics*, which was published in 1903. With "Russell's paradox" previously identified, in this book, Russell presented his own solution to it by means of the doctrine of logical types, which was used again in the *Principia Mathematica*.[4] In his inaugural text, Koyré had announced what he understood as the epicentre of the issue throughout his long reflection on the paradox: refuting the idea that the paradox could threaten mathematical realism and, consequently, the foundation of mathematics, logic and scientific rationality. Koyré had also already presented his first criticism of the possible solutions to the paradox indicated by mathematical logic. Koyré did not accept Russell's solution to avoid the paradoxes, understanding that the very definition that aimed to solve the problem would itself be paradoxical. Koyré concluded:

> Mr. Russell's definition, being paradoxical, cannot serve as the foundation of mathematics, and his number does not exist, from which it follows that they are not the ordinary numbers

[4] The theory of types appears as Appendix B in *Principles of Mathematics* in 1903. In 1908, Russell published a lengthy paper on this subject, "Mathematical logic as based on the theory of types" (Russell 1967 [1908]).

> of arithmetic. The problem of the definition of number and even the possibility of such a definition must be considered an unsolved problem. (Koyré 1992 [1912], p. 452)

In "Réponse a M. Koyré" (Russell 1912), Russell asserted that the answers to the problems raised by Koyré resided in the adoption of his "theory of types". As demonstrated by "Russell's paradox", some sets lead to paradoxes (sets of all sets) precisely because they do not use the theory of types. Thus, Russell concluded, "We can, and we must, therefore, accept all the detailed arguments of M. Koyré, but the conclusion which must be drawn from them is not that the definition of number must be abandoned, but that logic cannot do without the theory of types" (Russell 1992 [1912], p. 59).

In 1922, Koyré returned to the theme of paradox and published an extensive paper entitled "Remarques sur les paradoxes de Zénon"[5] (Koyré 1971 [1961], pp. 9–35), which focused on the analysis of the classical paradoxes of Zeno. However, this time he did not address the issue from the perspective of mathematical logic but instead analysed how different thinkers conceived of these paradoxes. According to Koyré, there are two main points:

1. Asserting that the problem of paradox is a formal and not a material problem
2. Asserting the supremacy of the Cartesian philosophy of mathematics in the face of the new problems posed by paradoxes and antinomies in mathematics

On the first point, for Koyré, the movement made by Achilles up to the tortoise was not in itself the main problem, that is, it was not a science of motion (phoronomy) that studied a body and the causes of its motion because "paradoxes do not have a meaning and a value that is purely phoronomic" (Koyré 1971 [1961], p. 22).[6] Admittedly, Achilles surpasses the tortoise when walking, but the central aspect addressed by this paradox constituted a formal and not a material question.

> The problem raised by Zeno reveals a deeper level, that of pure mathematics. The level at which the motion arises no longer exists. (Koyré 1971 [1961], p. 29)[7]

Koyré saw the epicentre of rationality in the intrinsic possibilities of the formal system, not in the physical features of the motion per se. The idea behind Koyré's history of science, that is, the idea that science is essentially the theory that guides our understanding of the physical world, is already prefigured here. It is not experience that dictates the rules; rather, it is theory that organises and guides experience. This formal dimension peculiar to mathematics – anchored in mathematical

---

[5] First published in German in the *Jahrbuch für Philosophie und phänomenologische Forschung*, Volume 5, 1922. Interestingly, close to the period in which Koyré wrote about the problem of paradoxes (Koyré 1912, 1922), Husserl also wrote on this topic. Husserl addressed the paradoxes in manuscript AI 35, which has two parts: α, dated 1912 and β, dated 1920. The first part of this manuscript was "concerned with different ways to solve Russell's paradox" (Haddock 2006, p. 218). In the second part, "Husserl tries to show that the way to avoid the paradoxes consists in a constructive axiomatization of set theory" (Haddock 2006, p. 219).

[6] "Les paradoxes n'ont donc pas une signification et une valeur purement phoronomiques".

[7] "Le problème soulevé par Zénon revèle d'un niveau plus profond, celui de la mathématique pure. Au niveau où se pose celui du mouvement, il n' existe plus".

realism – is far from the inaccuracy of experience or the "world of more or less", as Koyré would clarify years later (Koyré 1971 [1961], pp. 341–362). Thus, when Achilles approaches the tortoise, for Koyré, "it is only a virtual divisibility and not a division in action [...] the motion is not just a change of place" (Koyré 1971 [1961], pp. 15–16).[8] Indeed, "if it is accepted, since Descartes, that motion is a *state* of the body similar to the *state* of rest" (Koyré 1971 [1961], p. 19),[9] what allows the assertion of a formal system – in which rationality is founded – is not the existence of the motion but rather the emerging structure based on the contrast between motion and rest. This distinction seems to be an important point from which Koyré interprets classical mechanics when, inspired by Descartes, he understands motion and rest as ontological equivalents, thus making this position one of the pillars of his interpretation (Koyré 1973 [1966], p. 185).

The second important aspect of the paper of 1922 – the defence of Cartesian mathematics in the face of the new problems posed by mathematics – was established with a critique of Cantor that "defined the infinite set as having a property to be equivalent to one of its parts" (Koyré 1922, p. 27).[10] However, in his opposition to Cantor, Koyré did not make a mathematical criticism but rather a philosophical criticism. To Koyré, Cantor's argument was weaker than what had already been presented by Descartes (Koyré 1971 [1961], p. 26) because, according to Descartes, "the infinite is the first and positive notion, and the finite can be understood only by the negation thereof" (Koyré 1971 [1961], p. 28).[11]

In the second half of the 1940s, Koyré, already an important historian of science, revisited the problem of the paradoxes in two extensive papers, "The Liar" (Koyré 1946) and "Manifold and Category" (Koyré 1948). These papers became the basis for his book *Epimenide, le Menteur* (Koyré 1947). Although he addresses the problem from a logical–mathematical perspective,[12] his intention is the same: a philosophical critique of the logical-mathematical approach to minimise the impact of the paradoxes on the foundation of mathematics and classical logic. Paradoxes are not exactly antinomies, even when they generate doubts, sophisms and nonsense. However, for Koyré, "to extend these doubts to a doubt of the validity of the laws of logic is an exaggeration" (Koyré 1946, p. 349). Despite the fact that the contradictions and nonsensical sentences may cause difficulties, with the existence of

[8] "Il ne s'agit que d'une divisibilité virtuelle et non pas d'une division en acte" [...] "Le mouvement n'est pas un simple changement de lieu".

[9] "Si l'on admet, depuis Descartes, que le mouvement est un *état* du corps analogue à *l'état* de repos".

[10] "Cantor définit l'ensemble infini par sa proprieté d'être équivalent à une de ses parties".

[11] "C'est l'infini que est la notion première et positive, et le finit ne se comprend que par la négation de celui–ci".

[12] Koyré's book and papers on logic had a negative reception among logicians (cf. Bar–Hillel 1947) (Cf. Black 1948). However, although Koyré had investigated the field of mathematical logic, his primary goal was not to discuss logical technicalities but to understand in a broader philosophical sense the impacts of the paradoxes on rationality. In fact, although Koyré had radically defended mathematical realism throughout his work in the history of science, he always did so from a philosophical and historical perspective and not from a logical or mathematical point of view per se.

paradoxes, "we have no right to incriminate the rules of classical logic" (Koyré 1946, p. 356). Although the paradoxes present difficulties for logic and mathematics, these difficulties cannot be overcome with types or levels, as Russell intended with his theory of types, meaning that his solution had failed. For Koyré, it would be necessary to understand the paradoxes as structures that lead to counter-sense or nonsense, as argued by Husserl in Volume 1 of his *Logical Investigations* (Koyré 1946, p. 362). A local counter-sense, for Koyré, could not imperil the entire structure of logic. That position was far from one as drastic as Frege's, for example, when he found the paradox in his attempt to establish the foundations of arithmetic.

In "Manifold and Category", Koyré continued his analysis of Russell's paradox. He persevered in his aims:

1. To overcome the theory of types
2. To show that, despite the paradoxes, a reform of logic would not be necessary (Koyré 1948, p. 7, p. 14)

After thoroughly discussing Russell's paradox and the theory of types in a way similar to the previous paper, he again concluded that the paradox was a counter-sense or "something that has no meaning" (Koyré 1948, p. 14). His inspiration for this conclusion was still derived from Husserl's work. Here, Koyré argued that the perplexities raised by paradoxical structures should be considered "empty notions" (Koyré 1948, pp. 19–20), a position Husserl had presented in his *Formale und Transzendentale Logik* (1929). Indeed, there would be no sense in these paradoxical structures; that is, they would be similar to the constants that express thought, being, logic and ontology – as was already established under mediaeval logic. Furthermore, although these are reflexive notions, they do not hinder us from having correct judgments about things; they do not confuse our reason.

In this interpretation of paradox as a "meaningless" structure or an "empty" concept, Koyré perceived the possibility of continuing to assert logical essence and mathematical realism. Therefore, he opposed the vision of paradox as a bankruptcy of rationality. Koyré maintained this view throughout his life.

Having addressed these questions (e.g. paradoxes, the foundation of logic and mathematics, rationality) raised by the logical-mathematical circle and their impact in the early works of Koyré, in the next section, I outline Koyré's internalism, which was inherited, at least in part, from the mathematical realism inherent in this mathematical circle.

3.3 The Philosophers and the Machine, or Koyré's Internalism

Although it is relatively little known that Koyré concerned himself with questions of logic and paradoxes, from the mid–1930s onwards, Koyré's extraordinary contributions to the history of scientific thought transformed him into one of the most prominent historians of science in the twentieth century. Koyré's conception of the history of science is closely related to his belief in the foundations of mathematics. In other words, for Koyré, if the foundation of logic and mathematics resided in mathematical realism, then the history of science should also be conducted based on this assumption. Modern science was not established by the mere accumulation of experiences but rather by means of ideas, the primacy of theory guided by mathematical realism. In this way, Koyré summed up his philosophical conception behind the foundation of modern science:

> [...] the philosophical attitude that, ultimately, is well accepted, it is not that of the positivist or pragmatic empiricism, but rather the mathematical realism. In short, it is not the attitude of Bacon or Comte, but of Descartes, Galileo and Plato. (Koyré 1971 [1961], p. 267)[13]

Throughout his work on the history of science, Koyré passionately defended the view that science is mostly theory. In one of his last articles, written in 1961, Koyré still defended the primacy of theory over practice and assumed the idealism behind this conception (Koyré 1973 [1966], p. 399).

By introducing the analysis of scientific ideas as a main point, Koyré created an alternative to the strong positivism that had prevailed in the history of science. In large measure, the background of this new conception of the history of science was established in mathematical realism or Platonism as advocated by Koyré (1973 [1966], pp. 187–188, p. 191, p. 193). Although the circumstances that led Koyré to his history of science (e.g. religious and philosophical thought) and his possible influences (including Plato, Descartes, Hegel and Husserl) had been very diverse, the defence of mathematical realism was certainly one of the key elements in his conception of the history of science. Moreover, it was an important aspect in the composition of his internalism. Perhaps this mathematical realism or Platonism present in Koyré's work was the link to all of his other influences.

By assigning a decisive role to ideas in the construction of science, Koyré downplayed the social and technological aspects involved in scientific knowledge. Koyré's internalism[14] was characterised by the affirmation of mathematical realism

[13] "L'attitude philosophique qui à la longue s'avère bonne n'est pas celle de l'empirisme positiviste ou pragmatiste, mais, au contraire, celle du réalisme mathématique. En bref, non pas celle de Bacon ou de Comte, mais celle de Descartes, Galilée et Platon". (*Ibidem*).

[14] Although Koyré assumed the idealism of his perspective on the history of science (Koyré 1973 [1966], p. 399), he never attributed to himself the epithet of internalist. The debate over "internalism vs. externalism" and the selection of Koyré as the most representative author of internalism were outcomes of the later historiography of science, particularly the reception of Koyré's ideas in North America. In 1957, based on a selection of papers published in the first 18 volumes of the *Journal of the History of Ideas* that contrasted the two models of history of science (i.e. internalism

as a guide to scientific ideas and an understanding of social and technological aspects as secondary in the production of science. As noted above, contrary to Koyré's view, authors such as Zilsel, Hessen, Grossmann and Merton produced different interpretations of modern science in the same period that were primarily formulated based on the affirmation of these historical, social and technological factors. This position was later characterised as *externalism*. Therefore, unlike Koyré, these authors sought to understand the construction of scientific knowledge as the relationship between society and nature in a specific historical time. To a greater or lesser degree, these authors considered historical and social interactions – within which different scientific practices and technological uses were established – as the main points in the construction of modern science. However, this sociological thesis based on practical knowledge (or technological developments) that was present in the social and economic context of early modernity found it difficult to withstand the critique of Koyré. He stated:

> If the practical interest was the necessary and sufficient condition for the development of experimental science – in our conception of this word – this science would have been created a thousand years before [...] by the engineers of the Roman Empire, if not by those of Roman Republic. (Koyré 1973 [1966], p. 75)[15]

Koyré characterised modern thought as a new "style of thought", a revolution, or a "mutation" in thought (Koyré 1973 [1966], p. 18). For him, science was at the heart of this shift. The scientific revolution was an extraordinary event; however, science was still essentially theory (*theoria*). This theoretical interpretation of the world was not based on a collection of facts but rather on the understanding of universal laws, which were strictly mathematical. The use of mathematical language to understand nature revealed the rational world of precision against the "world of more or less" experience (Koyré 1971 [1961], pp. 341–362). The experimental method mainly involved the use of reason over experience. Thus, it was characterised by the primacy of theory over facts, practices and techniques. Indeed, even the law of inertia would be formulated from the development of ideas and not from the accumulation of experiences. The Greeks characterised the rest as the primary condition of being. Starting from this point, the modern age improved this idea by placing motion on the same ontological level (Koyré 1973 [1966], p. 185). This ontological equivalence between rest and motion helped to create an understanding of the world machine or classical mechanics. Thus, for Koyré, this process occurred because there were connections between the ideas of Plato and Galileo (Koyré 1973 [1966], pp. 167–195) rather than because Galileo was attempting to solve problems of ballistics in the arsenal.

Koyré did not establish an individual critique of each of these historians of the social perspective, but at different times, he criticised conceptions that assigned the

---

and externalism), Wiener and Noland proposed Koyré as an iconic figure of internalism and Edgar Zilsel as a symbol of externalism (Wiener and Noland, 1957).

[15] "Si l'intérêt pratique était la condition nécessaire et suffisante du développement de la science expérimentale – dans notre acception de ce mot – cette science aurait été créée un millier d'année [...] par les ingénieurs de l'Empire romain, sinon par ceux de la République romaine". (*Ibidem*).

core elements of the construction of modern science to engineers and craftsmen, to cities, and to the critical tradition. I reproduce Koyré's citation that, although long, summarised his critique of these authors and simultaneously consolidated his own position:

> Some people stress the role of experience and experiment in the new science, the fight against bookish learning, the new belief of modern man in himself, in his ability to discover truth by his own powers, by exercising his senses and his intelligence, so forcefully expressed by Bacon and by Descartes in contradistinction to the formerly prevailing belief in the supreme and overwhelming value of tradition and consecrated authority. Some others stress the practical attitude of modern man, who turns away from the *vita contemplativa*, in which the medieval and antique mind allegedly saw the very acme of human life, to the *vita activa*, who therefore is no longer able to content himself with pure speculation and theory, and who wants a knowledge that can be put to use: a *scientia activa*, *operativa*, as Bacon called it, or, as Descartes has said, a science that would make man master and possessor of nature. The new science, we are told sometimes, is the science of the craftsman and engineer, of the working, enterprising, and calculating tradesman, in fact, the science of the rising bourgeois classes of modern society. (Koyré 1965, p. 5)

When he presented his synthesis of the sociological theses concerning the advent of modern science, Koyré recognised that they were partially correct but unsatisfactory. In other words, he recognised that they were "necessary conditions" but not "sufficient conditions" (Koyré 1973 [1966], p. 75) for the construction of modern science.

> There is certainly some truth in these descriptions and explanations: it is clear that the growth of modern science presupposes that of the cities, it is obvious that the development of firearms, especially of artillery, drew attention to problems of ballistics; that navigation, especially that to America and India, furthered the building of clocks, and so forth – yet I must confess that I am not satisfied with them. I do not see what the *scientia activa* has ever had to do with the development of calculus, nor the rise of the bourgeoisie with that of Copernican or Keplerian astronomy. And as for experience and experiment – two things which we must not only distinguish but even oppose to each other – I am convinced that the rise and growth of experimental science is not the source but, on the contrary, the result of the new *theoretical*, that is, the new *metaphysical* approach to nature that forms the content of the scientific revolution of the seventeenth century, a content which we have to understand before we can attempt an explanation (whatever this may be) of its historical occurrence. (Koyré 1965, pp. 5–6)

Ultimately, for Koyré, the scientific revolution was not derived from facts, techniques, experiments or historical and social aspects. Although these may be necessary conditions, the advent of modern science ultimately resulted first from a change in thinking, a metaphysical attitude. However, even when stating his perspective, Koyré did not neglect to thoroughly examine the question of technology and even the central aspects of the sociological theses. In particular, in his 1948 articles "Les Philosophes et la Machine" and "Du Monde de l' "à-peu-près" à l'univers de la précision", Koyré analysed in detail the problem of technology and the psycho-sociological conception, as he called it, to consolidate his claim that science was essentially theory. Science was the result of "a shift of metaphysical attitude" (Koyré

1966 [1939], p. 13)[16] and not the fruit of social and technological development. Thus, he stated that

> [...] it is undeniable, even impossible, as I believe, to give a sociological explanation of the birth of scientific thought, or the appearance of great geniuses who revolutionised the development of science – Syracuse does not explain Archimedes, no more than Padua or Florence explains Galileo. (Koyré 1971 [1961], pp. 323–324)[17]

To combat the sociological theses that claimed the leading role of social and technological factors in the construction of modern science, Koyré noted a number of arguments while acknowledging that this issue seemed to have no satisfactory solution (Koyré 1971 [1961], p. 341). Always asserting the primacy of scientific ideas, Koyré argued that a solution (for convenience) showed that technique became technology only after the advent of science. Thus, the advent of science occurred before the advent of technology. The advent of technology would have been impossible without the development of science, although there was, of course, a connection between the history of technique and intellectual thought (Koyré 1971 [1961], p. 345). In other words, the history of technique was related to science, and science certainly has taken advantage of technique, but science was not created by technicians and engineers (Koyré 1966 [1939], pp. 11, p. 13). In Greece, technique was rudimentary, and physical science did not exist. After the advent of modern physics, the conditions for the construction of technology were opened up, as were those of the technological impacts that stemmed from this process.

At first glance, according to Koyré, the clock, for example, seemed to be a merely technological object; behind it, however, was a scientific (and even cultural) conception that was created before the clock itself. Therefore, the creation of the clock owed much more to Galileo, Huygens and Hooke than to technicians, even though the watchmakers had been proficient makers of this artefact used in everyday life. This example shows that "a theoretical object can become a practical object" (Koyré 1961 [1971], p. 357).[18] The shift from natural time to the artificial time of the clock would require a scientific change before there could be a technical event (Koyré 1961 [1971], p. 354). Likewise, the telescope maker was a craftsman rather than an optical scientist. Thus, Koyré concluded that:

> [...] the proper function of the instrument, which, in itself, is not an extension of the sense, but rather, in the stronger and more literal meaning of the term, is the incarnation of the spirit, the materialisation of thought. (Koyré 1961 [1971], p. 352)[19]

---

[16] "Changement d'attitude métaphysique". (*Ibidem*).

[17] "Il est incontestable que, même s'il est impossible, comme je le crois, de donner une explication sociologique à la naissance de la pensée scientifique, ou à l'apparition de grands génies qui en révolutionnèrent le développement – Syracuse n'explique pas Archimède, pas plus que Padoue ou Florence n'expliquent Galilée". (*Ibidem*).

[18] "Un objet théorique peut devenir objet pratique". (*Ibidem*).

[19] "La fonction propre de l'instrument qui, lui, n'est pas un prolongement du sens mais, dans l'acception la plus forte et la plus littérale du terme, incarnation de l'esprit, matérialisation de la pensée". (*Ibidem*).

To summarise, although modern science had been revolutionary, for Koyré, the driving force of this revolution was not the social and technological aspects; they derived from theoretical changes influenced by philosophy (Koyré 1971 [1961], pp. 255–256) and religion.

By asserting the important role of theory in the history of science inserted into the broader framework of metaphysical conceptions, Koyré understood that there was an intertwining of science, religion (theology) and philosophy (metaphysics) that formed the "unity of thought" of mankind (Koyré 1973 [1966], p. 11). Techniques and social behaviours were not the main elements that modelled the "thought of style" in the early modern period; instead, the main elements were metaphysical beliefs, that is, religious, philosophical and scientific ideas. However, Koyré characterised his internalism by asserting his mathematical realism, metaphysical attitude and the primacy of theory over experience, practice and technology. Both experimentation and technology would be the fruits of this theoretical condition, but not vice versa. Nevertheless, by sustaining the unity of thought between science, philosophy and religion – that is, by affirming that science has interfaces with other areas – and even by analysing in detail, the technological and social aspects involved in the production of science, Koyré showed himself to be someone who could not be characterised as a mere internalist.[20]

3.4 Internalism and Autonomy of Science

Science was not destined to begin during the early modern period; however, it eventually found there – rather than in another time or context – the "necessary conditions" to begin. From these initial conditions, it created the "sufficient conditions" to establish itself effectively. Here, we can address the question initially posed by Koyré: "Why did the Roman engineers not produce science"? However, before seeking to answer such a question, we need to distinguish between the different aspects implied in this type of question. When Koyré searches for a "why" in the birth of science and technology, he is interested in a theoretical or metaphysical "why", not a sociological one. This is the reason for his critique of sociological

[20] Based on these assumptions, Elkana (1987) has gone so far as to assert that by showing many of the social aspects that provide contour to the history of science, Koyré was one of the founding fathers of the "historical sociology of scientific knowledge" despite all of his strong epistemological claims in defence of mathematical realism and the prominence he ascribed to theories in the construction of modern science. In contrast, Stump believes that although Elkana's thesis seems unconvincing (Stump 2001, 256), to understand Koyré's epistemology based only on his Platonism would be a mistake. Insofar as the author of *Newtonian Studies* is concerned with the unity of thought between science, philosophy and religion, perhaps it is better to understand him as a Hegelian (Stump 2001, 257). However, although it is possible to accept Stump's thesis, certainly Koyré's internalism would still not be safe; we would still have Koyré's critique of the role of technology and social influences as determinant elements in the construction of modern science. Finally, we would continue to have the distinction between the "philosopher and the machine".

approaches. For Koyré, a sociological "why" presents the necessary conditions, but only a metaphysical "why" may provide the sufficient conditions.

As we know, some societies have devoted their knowledge to magic, religion or politics. They had no interest in science, as Koyré emphasises (Koyré 1971 [1961], p. 324). To produce science is not an aim to be a priori followed by a society – as intended by some positivist conceptions – but rather a practice established based on the needs, interests and circumstances of a given society in a certain historical context. In this sense, the reason science begins in the early modern period is, in some respects, a contingency presented by the constraints of that historical context. Once a historical event is established in its various necessary conditions or motivating circumstances, we can analyse it as:

1. A sociological phenomenon by attempting to understand "why" it happened in that time and space, or
2. A theoretical or metaphysical phenomenon in which we seek to understand how this society has established the production of its own knowledge

According to Koyré, the sociological "why" explains these "necessary conditions", and the theoretical or metaphysical "why" explains the "sufficient conditions". Thus, when analysing the construction of modern science and technology, we must seek to understand not only the social "why" but also the metaphysical "why" this phenomenon was established from the interweaving of science, philosophy and religion. In fact, from Koyré's perspective, the interaction among these fields in a specific context during the early modern period enabled the production of ideas, theories and language and, consequently, techniques, practices, artefacts and products.

However, one must observe this distinction between these different sociological and metaphysical aspects of the question to understand what is at stake in several of the important issues raised by Koyré concerning the birth of modern science and technology. Many of the issues he raised are, in fact, questions of a metaphysical "why". Thus, he downplays the sociological "why" that provides answers concerning the political, social and economic circumstances of the context in which the events occurred. With this metaphysical perspective in mind, Koyré asks, "Why did *technical thought* not develop in antiquity?"; "Why did the inventors of *episteme* not apply it to *praxis*?"; "Why […] has Greek science not developed a technology, if it has formulated this idea?" (Koyré 1971 [1961], p. 338).[21] Concerning the sociological "why", the psycho-sociological approach – criticised by Koyré – offers plausible answers to these questions, although these answers are not epistemological answers. Certainly, there are historical factors that converged to explain the birth of science and technique. The psycho-sociological explanations proposed by Meyerson and Schuhl (Koyré 1971 [1961], p. 318) may provide good reasons for arguing that

[21] "Pourquoi la *pensée technique* de l'Antiquité n'a-t-elle pas progressé?"; "Pourquoi les inventeurs de l' ἐπιστήμη ne l'ont-ils pas appliqué à la πράξις?"; "Pourquoi […] la science grecque n'a pas développé une technologie dont elle avait cependant formulé l'idée?" (*Ibidem*).

the lack of technology in the ancient world was due to the presence of slavery, the prejudice against manual labour and aristocratic society. According to Koyré:

> [...] the psycho-sociological explanation tells us that, due to certain historical and social reasons, the Greek sage despised work and "mechanical" issues; in other words, because Greek science did not constitute a technology, ancient technique has not passed a certain relatively primitive and undeveloped level over the course of centuries. (Koyré 1971 [1961], p. 337)[22]

In short, the psycho-sociological thesis, refuted by Koyré, claims that antiquity did not need machines (Koyré 1971 [1961], p. 308). Interested in the metaphysical "why", Koyré refuses all of these explanations. For him, they would be neither appropriate nor satisfactory. Establishing a list of historical reasons by noting why antiquity did not create technology and experimental science may provide us with plausible answers. However, those answers would always be the answers to a sociological "why", not a metaphysical one. When Koyré disqualifies the psycho-sociological answers, he concludes that Greece did not use technology, only technique, because to create technology, it was first necessary to create science, and this is something that only modernity has created. Certainly, technology could not be created before science. We cannot understand science and technology as isolated processes because the meaning of both is specific to the context of modern science in which they jointly developed. Modern science is a type of "emergent property" that arises from the sum of these two forms of knowledge: practical knowledge and theoretical knowledge. However, according to Koyré, there is a primacy of theory over praxis.

When the psycho-sociological explanation outlines a possible answer to "why" social relations contributed to the birth of science and technology, Koyré also does not accept this response. For him, from a mere sociological perspective, Syracuse could never conceive an Archimedes nor could Florence conceive a Galileo (Koyré 1971 [1961], pp. 323–324) because even if the institutionally "necessary conditions" to produce science were present in these cities, according to Koyré, they still would not be the "sufficient conditions" required to produce science.

Admittedly, as noted by Koyré, because some civilisations preferred magical or religious knowledge (Koyré 1971 [1961], p. 324), they did not create a mechanism that enabled the advent of critical, philosophical or scientific thinking; that is, they did not create the necessary institutions to allow some type of scientific thought. In these civilisations, science certainly could not arise. However, despite the fact that the ideal of youth was not to attain knowledge but rather to gain power and joy (Koyré 1971 [1961], p. 330), it was in Greek society, as opposed to other societies, that the conditions for the production of rational knowledge were created. From Koyré's metaphysical perspective, it was this Greek society that invented the idea of

---

[22] "L'explication psychosociologique nous affirme que c'est parce que, pour des raison historiques et sociales determinées, le savant grec a méprisé le travail et les questions 'méchaniques', en d'autres termes, parce que la science grecque n'a pas constitué de technologie, que la technique antique n'a pas depassé un certain niveau, relativement primitif, et s'est si peu développée au cours des siècle". (*Ibidem*).

episteme, although it was not widespread knowledge among the entire population of Greece. At this point, Koyré introduces into the discussion certain ideas from Lucien Febvre that can help us clarify these issues. The eminent historian of the *Annales* School understands that we have different perspectives on the generation and use of knowledge in society. For example, one can live in a society that does not make use of mathematics, and one who does not know mathematics can live in a society that makes extensive use of it (Koyré 1971 [1961], p. 349). Insofar as a society presents complexity in its institutions, it is possible that one of these institutions is precisely the institution in which science is conducted, even though not all of the members of a society can do science.

It is true that science needs these socially necessary conditions for its development (e.g. only in certain social conditions can one attribute importance to knowledge or create scientific schools), but from where do the "sufficient conditions" for the construction of science derive? As previously noted, from Koyré's perspective, social conditions are necessary but not sufficient. To explain why the Roman engineers could not develop science, Koyré remarks that Baconian empiricism establishes the foundations of science in the records of actions and the classification of facts. However, "Descartes, in his turn, draws exactly the opposite conclusion, namely, that of instilling theory into action, that is, the possibility of the *conversion* of theoretical intelligence into the real, and thus the possibility of a *technology* and a *physics*" (Koyré 1971 [1961], p. 346).[23]

However, a practical activity does not spontaneously turn into a theoretical activity. Thus, Koyré is right when he says that "it is not from the spontaneous development of the industrial arts, by those who exercise them, but it is from the conversion of theory into practice that Descartes expects the progress that will make man 'master and possessor of nature'" (Koyré 1971 [1961], p. 346).[24] Koyré recognises that there is a direct relationship between technique and language.

> It is possible, moreover, that technique, properly speaking, does not have an origin, just like language: man has always possessed tools the same way he has always had language. It actually seems like man has always been able to produce. Is that why the definition of man through language (*parole*) may be opposed to labor: man as man would essentially be *faber*, maker of things, maker of tools [...] One could ask, however, if this opposition is legitimate, and if language (*parole*) and the tool do not necessarily go together. (Koyré 1971 [1961], pp. 316–317)[25]

---

[23] "Descartes, lui, en tire une conclusion exactement opposée, à savoir celle de la possibilité de faire pénétrer la théorie dans l'action, c'est-à-dire, de la possibilité de la *conversion* de l'intelligence théorique au réel, de la possibilité à la fois d'une *technologie* et d'une *physique*".

[24] "Ce ne pas du développement spontané des arts industriels par ceux qui les exercent, c'est de la conversion de la théorie à la pratique que Descartes attend les progrès qui rendront l'homme 'maître et possesseur de la nature'". (*Ibidem*).

[25] "Il est possible, d'ailleurs, que la technique, à proprement parler, n'ait pas plus d'origine que le langage: l'homme a toujours possédé des outils, de même qu'il a toujours possédé le language. Il semble même avoir toujours été capable d'en fabriquer. C'est bien pour cela qu'à la définition de l'homme par parole on a pu opposer celle par le travail: l'homme en tant qu'homme serait essentiellement *faber*, fabricateur de choses, fabricateur d'outils [...] On pourrait se demander, toutefois, si cette opposition est legitime, et si la parole et l'outil ne vont pas, nécessairement, ensemble". (*Ibidem*).

By analysing these issues, Koyré remains in dialogue with Febvre. The historian of the *Annales* School supports the idea that "the history of technique is inextricably linked to intellectual history, and they cannot be separated" (Koyré 1971 [1961], p. 345).[26] Moreover, according to Febvre, each context has its own "material and mental tool" (Koyré 1971 [1961], p. 349).[27] Mental and material tools are forged together in a context. Accordingly, not only practices but also the theory that guides these practices tend to be built in parallel. In this sense, Koyré is right to affirm that a society that does not develop technologies also does not develop the languages (theories) that guide the uses of these technologies. "It is also true that it is not only the measuring instruments that are missing but also the language that could serve to express the results" (Koyré 1971 [1961], p. 349).[28] Therefore, Koyré understands that there is a parallel between theory and technique, although, for him, the main point is the primacy of theory over technique. Thus, to explain a technique that has failed to develop, Koyré says, "It is not a technical failure; it is the lack of the idea that gives us the explanation" (Koyré 1971 [1961], p. 351).[29] Without ideas, one certainly cannot have explanations.

Regarding the question of the autonomy of science, although science constitutes a "unity of thought" along with philosophy and religion, according to Koyré, science has its own reasons. Science is in a constant relationship with these other domains, which often provide ideas that help constitute science itself. Ultimately, however, these influences are not the main elements in the justification of the final discourse and practice of science. As Koyré argues:

> [...] whatever the para-scientific or ultra-scientific ideas are that have guided a Kepler, a Descartes, a Newton or even a Maxwell toward their discoveries, such ideas, in the end, have little – or no – importance. What counts is the effective discovery, the established law, the law of planetary motion and not the harmony of the world, the conservation of motion and not divine immutability. (Koyré 1971 [1961], p. 255)[30]

Certainly, Koyré seeks here to emphasise the "unity of thought" among science and other fields of knowledge by demonstrating the presence of these non-scientific factors in science, which are often deliberately ignored. However, although Koyré attempts to assert the role of the unity of thought, he ultimately shows this distinction between scientific and non-scientific knowledge. Science is certainly permeated by other spheres of knowledge, and in this sense it is part of a type of "unity of

[26] "L'histoire de la technique est inséparablement liée à l'histoire intellectuelle et ne peut pas en être séparée".

[27] "Outillage matériel et mental". (*Ibidem*).

[28] "Il est vrai également que ce ne sont pas seulement les instruments de mesure qui manquent, mais le langage qui aurait pu servir à en exprimer les résultats". (*Ibidem*).

[29] "C'est ne pas l'insuffisance technique, c'est l'absence de l'idée qui nous fournit l'éxplication". (*Ibidem*).

[30] "Quelles que soient les idées para- ou ultra-scientifiques que aient guidé un Kepler, un Descartes, un Newton, ou même un Maxwell vers leurs découvertes, elles n'ont, en fin de compte que peu – ou pas – d'importance. Ce qui compte, c'est la découverte effective, la loi établi, la loi des mouvements planétaire et non l'Harmonie du Monde, la conservation du mouvement et non l'immutabilité divine". (*Ibidem*).

thought", as intended by Koyré. However, to the extent that the rules that constitute science also establish their own criteria of evaluation, science is autonomous. In a sense, science "internally" establishes its criteria; thus, it is internalist.

If science justifies its sufficient conditions in itself through its development as an institution that has its own rules with theories and practices legitimised in their own actions (although it is influenced by other knowledge, such as philosophy and religion), then, according to Koyré, we may conclude that in the Roman context of knowledge production, scientific ideas were not "sufficiently" – theoretically and metaphysically – established to enable the Roman engineers to develop science. Nevertheless, in that context, despite the existence of many favourable situations that offered the "necessary conditions", the level of social and technological institutionalisation by itself did not reach the theoretical and metaphysical threshold necessary to produce the "sufficient conditions" for the advent of science. In fact, for Koyré, science could never have begun with the Roman engineers.

Koyré's internalism has many critics as well as many enthusiastic supporters. Certainly, the large number of published works focusing on the social aspects of science – especially after the publication of Kuhn's *The Structure of Scientific Revolutions* – has contributed greatly to the critique of the epistemological perspective advocated by Koyré, despite the excellence of his historical approaches. Even when the history of science is written primarily from a social perspective, Koyré continues to be read for his contribution to history of science. Those who defend the idea that scientific theories are autonomous in their dialogue with nature – and that social and technological aspects would be secondary in this process – continue to regard Koyré's internalism as a strong reference. Particularly among scientists, whether they are willing or unwilling – consciously or otherwise – to accept epistemology, science is understood as the knowledge that holds the position of supremacy in the production of objectivity – or truth, as Koyré prefers to say (Koyré 1973 [1966], p. 399) – and, that is, the main defence of his internalism.

Consequently, scientists, who continue to refer to internalism in their scientific practice, demand, if not an independence of science in the face of social factors, then at least the autonomy of science from these factors. Even today, in this scenario, with a strong sociological approach to science, it is impossible to deny internalism. That is, it becomes difficult to understand how science goes its own way, with its own ideas and theories autonomous from (but not independent of) these social factors. Thus, it seems that by naively and simplistically refuting internalism, we run the risk of throwing the baby out with the bath water. Perhaps the internalism present in the scientific imagination during a period in which science is highly marked by social factors could be translated as follows: how does one understand that science, even considering the social aspect – which is subject to external influences – has its own rules of behaviour that grant it autonomy with respect to such social factors? That is, although internalism has been criticised, this criticism does not mean that the need to understand science in its "internal" logic no longer exists.

Koyré's internalism seems to continue to be relevant when it alerts us to the need to think about it as a type of autonomy of science in relation to these social practices. Indeed, even if it is abandoned, as Shapin has noted, the issue of internalism vs.

externalism does not seem to be an entirely resolved question (Shapin 1992). My purpose here was not to address this issue in all of its complexity but to understand in what sense Koyré's internalism continues to make important points that are not sufficiently clarified in this criticism. In particular, internalism seems to reveal a need for the autonomy of science.

3.5 Conclusion

In this chapter, I attempted to show that Koyré began his career by defending mathematical realism as the foundation of logic and mathematics. Later, he adopted this assumption of mathematical realism as a guide for understanding the transformation of ideas and scientific theories that culminated in the construction of modern science. However, science resulted much more from a change of "metaphysical attitude" than from a true process supported by social and technological transformations. By adopting this perspective, Koyré established the primacy of theory over practice in the controversy, which he termed "the philosophers and the machine". According to him, the sociological and technological aspects present in modernity were "necessary conditions" but not "sufficient conditions" to create modern science. Such major conditions would be the mathematical realism of Plato, Galileo and Descartes – finally, a theoretical or metaphysical attitude. In this sense, according to Koyré, the primacy of theory over practice was the major element in the scientific revolution. This gap between science, technique and social practices configured Koyré's internalism. Although apparently overcome by the sociological interpretation of science that developed following Kuhn's *The Structure of Scientific Revolutions*, internalism continues to raise important issues, such as the need for an effective autonomy of science with respect to social practices. There are many opportunities to address the issue of the autonomy of science raised by Koyré's internalism, an issue that remains pertinent today. Although Koyré's internalism has remained in the background given the current social approaches to science, it is undeniable that it points to the important issue of the autonomy of science. This issue continues to be an open question for the philosophy of science.

References

Bar–Hillel J (1947) The revival of the liar. Philosophy and Phenomenological Research 8:245–253.
Black M (1948) Review of Koyré, Epimenide le menteur. The Journal of Symbolic Logic 8:146–147.
Elkana Y (1987) Alexandre Koyré: between the history of ideas and sociology of disembodied knowledge. History and Technology 4:115-148.
Frege G ([1967] 1902) Letter to Russell. In Heijenoort 1967, p.128.
Haddock G (2006) Husserl's philosophy of mathematics: its origin and relevance. Husserl Studies 22:193-222.

Heijenoort J (1967) From Frege to Gödel: a source book in mathematical logic, 1879–1931. Harvard University Press, Cambridge.
Koyré A (1912) Sur les nombres de M. Russell. Revue de Métaphysique et de Morale 20:722–4. English Translation in Russell 1992, pp. 451–452.
Koyré A (1922) Bemerkungen zu den Zenonischen Paradoxen. Jahrbuch für Philosophie und Phänomenologische Forschung 5:603–628. French Translation in Koyré, 1971, pp. 09–35.
Koyré A (1946) The Liar. Philosophy and Phenomenological Research 6:344–362.
Koyré A (1947) Epimenide le menteur (Ensemble et catégories). Hermann, Paris.
Koyré A (1948) Manifold and categories. Philosophy and Phenomenological Research 9:1–20.
Koyré A (1965) Newtonian studies. Harvard University Press, Cambridge.
Koyré A ([1966] 1939) Études galileennes. Hermann, Paris.
Koyré A ([1971] 1961) Études d'histoire de la pensée philosophique. Gallimard, Paris.
Koyré A ([1973] 1966) Étude d'histoire de la pensée scientifique. Gallimard, Paris.
Russell, B ([1967] 1902) Letter to Frege. In Heijenoort 1967, p. 125.
Russell, B ([1967] 1908) Mathematical logic as based on the theory of types. In Heijenoort pp. 150–182.
Russell, B (1912) Réponse à M. Koyré. Revue de Métaphysique et de Morale 20:725–6. English Translation in Russell 1992, pp. 57–59.
Russell, B (1992) Logical and philosophical papers, 1909–1913. Routledge, London.
Shapin S (1992) Discipline and bounding: the history and sociology of science as seen through the externalism-internalism debate. History of Science 30: 333-369.
Stump J (2001) History of science through Koyré's lenses. Studies in History and Philosophy of Science 32: 243-263.
Wiener P, Noland, A (1957) Roots of scientific thought: a cultural perspective. Basic Book Publish, New York.
Zambelli P (1999) Alexandre Koyré im der mekka der mathematik. NTM International Journal of History and Ethics of Natural Sciences, Technology and Medicine 7:208-230.

Chapter 4
Koyré and Galileo: The Myth of the Leaning Tower's Scientific Experiment

Francesco Crapanzano

Abstract Koyré's belief about the apriorism of scientific discoveries should not be regarded as a fantasy or a bias: as early as 1937, that is, a few years before the publication of his *Études galiléennes*, he addressed the issue in an article entitled "Galilée et l'expérience de Pise." Investigating on how to value the experiment in the context of the foundation of mechanics, therefore, is not confined to the level of theoretical statements, but translates into considering the genesis of some Galilean theories whose conceptual aspect Koyré stresses to the detriment of their experimental one. One of the best known examples is that of the experiments on falling bodies made around 1589 at the Tower of Pisa, reported by Galileo's first biographer and disciple, Vincenzio Viviani. Koyré comes to the conclusion that they were never carried out; the Vulgate, of course, was and remains the one handed down to us by Viviani, but can we relegate Koyré's observations to the margins? Can the "history of effects" (*Wirkungsgeschichte*) wipe the well-grounded doubt that this tale is celebratory or a *iocularis quaedam audacia*?

Keywords Galileo • Koyré • Experiment • Tower of Pisa • Mechanics

4.1 Introduction

There is in Koyré an intimate logical-argumentative connection between the understanding of Platonic philosophy as aprioristic and therefore as an enemy of experimentalism, primarily of Aristotelian experimentalism, the attribution of this philosophy to Galileo and the exaltation of the virtual (or thought) experiment as the foundation of classical mechanics. The Galilean apriorism and mathematicism remain constant in his analysis,[1] and in this paper we will consider a specific case

[1] My intention is to proceed along the traditional interpretative line – rich in insight – that has seen in Koyré a neat contrast between theory and experimentation. On the Galilean Platonism and for full reference, see Crapanzano 2014, pp. 41–78.

F. Crapanzano (✉)
Department of Cognitive Science, Psychology, Pedagogy and Cultural Studies (COSPECS), Messina University, Via Concezione, 10, 98122 Messina, Italy
e-mail: fcrapanzano@unime.it

R. Pisano et al. (eds.), *Hypotheses and Perspectives in the History and Philosophy of Science*, DOI 10.1007/978-3-319-61712-1_4

that I believe telling: the experiment on the free fall of bodies made by Galileo at the leaning tower of Pisa in the years he spent teaching at the local University (1589–1592).

The first source referring the story is the *Racconto istorico della vita di Galileo* ('Historical Tale of Galileo's Life'), written by his youngest disciple, Vincenzio Viviani, a good 12 years after his master's death (1654), and published only in 1717. The passage in question reads that, at the time when Galileo was in Pisa

> [...] as he seemed to learn that the investigation of natural effects necessarily demanded a knowledge of the nature of motion, granting the philosophic and familiar axiom, 'Ignorance of motion spells ignorance of Nature', he gave himself wholly to the contemplation of this. And then, to the dismay of all the philosophers, very many conclusions of Aristotle were by him [Galileo] proved false through experiments and solid demonstrations and discourses, conclusions which up to then had been held for absolutely clear and indubitable; as, among others, that the velocity of moving bodies of the same material, of unequal weight, moving through the same medium, did not mutually preserve the proportion of their weight as taught by Aristotle, but all moved at the same speed; demonstrating this with repeated experiments from the height of the Campanile of Pisa in the presence of the other teachers and philosophers, and the whole assembly of students; and also that the velocity of a given body through different media kept the reciprocal proportion of the resistance or density of the said media, a point which he deduced from the very obvious absurdities which would [otherwise] follow as a consequence and against reason. He upheld the dignity of this professorial chair with so great fame and reputation, before judges well–disposed and sincere, that many philosophasters, his rivals, stirred with envy, were aroused against him. (Viviani 1907 [1654], p. 606. Translation from: Cooper 1935, pp. 26–27)[2]

This is an important disavowal of Aristotelian physics with respect to the motion of bodies in free fall: Aristotle thought that these bodies would acquire different speeds according to their weights,[3] while Galileo would give decisive proof that this could not happen and, on the contrary, that they fall to the ground at identical speeds. The

[2]At that time, "parendogli d'apprendere ch'all'investigazione delli effetti naturali necessariamente si richiedesse una vera cognizione della natura del moto, stante quel filosofico e vulgato assioma *Ignorato motu ignoratur natura*, tutto si diede alla contemplazione di quello: et allora, con gran sconcerto di tutti i filosofi, furono da esso convinte di falsità, per mezzo d'esperienze e con salde dimostrazioni e discorsi, moltissime conclusioni dell'istesso Aristotele intorno alla materia del moto, sin a quel tempo state tenute per chiarissime et indubitabili; come, tra l'altre, che le velocità de' mobili dell'istessa materia, disegualmente gravi, movendosi per un istesso mezzo, non conservano altrimenti la proporzione delle gravità loro, assegnatagli da Aristotele, anzi che si muovon tutti con pari velocità, dimostrando ciò con replicate esperienze, fatte dall'altezza del Campanile di Pisa con l'intervento delli altri lettori e filosofi e di tutta la scolaresca; e che né meno le velocità di un istesso mobile per diversi mezzi ritengono la proporzion reciproca delle resistenze o densità de' medesimi mezzi, inferendolo da manifestissimi assurdi ch'in conseguenza ne seguirebbero contro al senso medesimo. Sostenne perciò questa cattedra con tanta fama e reputazione appresso gl'intendenti di mente ben affetta e sincera, che molti filosofastri suoi emuli, fomentati da invidia, se gli eccitarono contro" (Viviani [1654] 1907, p. 606).

[3]Cf. Aristotle 1991b, book III, 2, § 301a–301b, pp. 328 f. "[...] Thus the weightless body will move the same distance as the heavy in the same time. But this is impossible. Hence, since the motion of weightless body will cover a greater distance than any that is suggested, it will continue infinitely" (Aristotle 1991b, book III, 2, § 301b 13–17, pp. 328 f.). The same idea is confirmed in Aristotle 1991a, book IV, 8, § 215a–216a, pp. 92–94.

story, though brief, is quite clear and the circumstances well-defined – what on earth would Koyré want to add to it? Meanwhile, Viviani was not the only one who had a say on the matter: Galileo himself declared, Sagredo speaking on his behalf, to have experienced it.[4] In any case, the problem of bodies in free fall was not new: in 1611, Giovanni Battista Baliani (1582–1666), an eminent political figure of Savona – then part of the Republic of Genoa – and a scholar in mechanics and astronomy, would have carried out experiments where he observes that bodies of different materials and weights dropped from great heights take up the same speed and touch the ground at the same time[5]; Niccolò Cabeo (1586–1650), a Jesuit mathematician, expresses himself in a similar way and indirectly spurs the monk and friend of Galileo's and Viviani's, Vincenzio Renieri (1606–1647), to test – and disprove – his statements.[6] Galileo knew very well Baliani's treatise *De motu naturali gravium*,

[4] Cf. Galileo [1638] 1898, VIII, pp. 106 ff. Unfortunately, the Galilean sentence reported in quotes is not in the "Dialogo on p. 222" (Koyré [1937] 1973, p. 220 n) as Koyré reports; more likely, it is the passage in the *Dialogo* reading: "Questo non importa niente, perché palle di una, di dieci, di cento, di mille libbre, misureranno sempre le medesime cento braccia nell'istesso tempo" (Galileo [1632]1897, VII, p. 249); or rather, it is the reworking of another passage taken from the *postille* to *Antonio Rocco's Philosophical Exercises*, where Galileo writes: "Son dunque, Sig. Rocco, d'oppinione, che pigliando qualsivoglia mobile grave [...] venendo da qualsivoglia altezza, si moverebbono con in medesimi gradi di velocità per appunto; talché, partite dalla quiete nell'istesso tempo, si troverebbono sempre di conserva negli stessi momenti, tanto nella distanza di 10 braccia dal primo termine, quanto nella distanza delle 100 o 1000, e così in tutte le altre" (Galileo [1634]1897, VII, p. 737). In the *Discorsi e dimostrazioni intorno a due nuove scienze*, on the other hand, we can read: "Ma io, Sig. Simplicio, che n'ho fatto la prova, vi assicuro che una palla d'artiglieria, che pesi cento, dugento e anco più libbre, non anticiperà di un palmo solamente l'arrivo in terra della palla d'un moschetto, che ne pesi una mezza, venendo anco dall'altezza di dugento braccia" (Galileo [1638]1898, VIII, pp. 106–107; Cf. Koyré [1937] 1973, p. 220 n).

[5] Koyré refers to the preface to the *De motu gravium* where Baliani states: "Inter alia dum anno millesimo sexcentesimo undecimo, per paucos menses, ex patriae legis praescripto, Praefectum Arcis Savonae agerem, ex militari bus observationibus quae occurrebant, illud maxime deprae-hendi, ferreos, et lapideos tormentorum bellicorum globos, et sic corpora gravia, seu eiusdem, seu diversae speciei, in inaequali satis Mole, et gravitate, per idem spatium, aequali tempore, et motu, naturaliter descendere, idque ita uniformiter, ut repetitis experimentis mihi plane constiterit, duos ex praedictis globis, vel ferreos ambos, vel alterum lapideum alterum plumbeum, eodem plane momento temporis dimissos sibi, per spatium quincaginta pedum, etiam si unus esset librae unius tantum, alter quincaginta, in indivisibili temporis momento, subjectum solum ferire, ut unus tantum amborum ictus sensu perciperetur" (Baliani 1638, p. 3). See also Baliani 1638, p. 4, p. 17, p. 22.

[6] Cf. Koyré [1937] 1973, p. 220. Niccolò Cabeo is called into cause by Koyré as the inspirer of Vincenzio Renieri, who, on reading him, would decide to verify his assertions about the fall of bodies. This can be inferred, as indicated, from the Letter of Vincenzio Renieri to Galileo, 13 March 1641, where we can read: "Habbiamo qui havuto occasione di far un'esperienza di due gravi cadenti da alto, di diversa materia, cioè uno di legno et uno di piombo, ma dell'istessa gran-dezza; perché un tal Gesuita [Cabeo] scrive che scendono nello stesso tempo, e con pari velocità arrivano a terra [...]. Ma finalmente habbiamo trovato il fatto contrario, poiché dalla cima del Campanile del Duomo tra la palla di piombo e quella di legno vi corrono tre braccia almeno di differenza. Si fecero anche esperienze di due palle di piombo, una della grandezza uguale a un'ordinaria d'artiglieria e l'altra da moschetto, e si vedeva tra la più grossa e la più piccola, dal'altezza dello stesso campanile, esservi un buon palmo di differenza, del quale la più grossa anticipava la più piccola" (Galileo 1906, XVIII, p. 305).

since the author himself took care to send it to him as soon as he had it printed (receiving several observations and starting off an interesting exchange of letters) and many, among whom Bonaventura Cavalieri (1598–1647), Famiano Michelini (1604–1665), Antonio Santini (1577–1662), and Daniele Spinola (? – 1662?), informed him of their opinions by letter, sometimes in positive terms and sometimes expressing some reserves.[7] Renieri, for his part, cites Father Cabeo, but the problem must have probably arisen with Baliani's text as well (Cabeo and Baliani were friends), except that, having Baliani such a cordial relationship with Galileo and also with him, Renieri would not like to contradict him directly.[8]

So much so that Cabeo, as Koyré points out, did not give up to the disavowal and reaffirmed of having repeatedly experienced that bodies of different weights (made of lead and wood), in free fall, do reach the ground with the same speed and at the same time.[9] To those who reproached him for not considering air resistance, he said he did not understand the question, air not possessing any power.[10] In his utterly different view, the Jesuit Giovanni Battista Riccioli in his *Almagestum novum* declares it impossible to positively verify what Cabeo claimed: the latter's measurements lacked in precision and, above all, he observed what had actually happened in the experiments he himself conducted in 1634 at the bell tower of the Church of Jesus in Ferrara (together with his fellow brother Cabeo) and especially in those at the Asinelli Tower in Bologna (1640, 1645, 1648, and 1650).[11] Notably, the bodies

---

[7] Indeed, on the 7th of January 1639 (3829), Galileo writes to Baliani: "La graditissima lettera di V.S. Ill.ma mi fu resa hieri, insieme col. suo libro Del Moto" (Galileo 1906, XVIII, p. 11; Cf. Letter [3824] of Giovanni Battista Baliani to Galileo, 17 December 1638 in Galileo 1906, XVII, pp. 413–414; Letters [3839, 3868] of Bonaventura Cavalieri to Galileo, 25 January and 18 April 1639 in Galileo (1906, XVIII, pp. 21, 43; Letter [3842] of Famiano Michelini to Galileo, 8 February 1639 in Galileo 1906, XVIII, p. 24; Letter [3854] of Antonio Santini to Galileo, 23 March 1639 in Galileo 1906, XVIII, p. 34; Letters [3855, 3898, 3937] of Daniele Spinola to Galileo, 25 March, 3 August and 29 October 1639 in Galileo 1906, XVIII, pp. 35, p. 79, p. 118).

[8] Cf. Letters of Vincenzio Renieri to Galileo, 10 February, 13 and 28 April 1640, 28 May [3965, 3991, 3999] and 15 June 1641 [4142, 4145] in Galileo 1906, XVIII, pp. 145, 177–178, pp. 184–185, p. 330, pp. 332–333.

[9] Cabeo dedicates a page of his commentary to Book I of Aristotle's Meteorology to reaffirming that: "Neque tamen haec ratio, si non esset experimentu, mihi persuaderet", i.e., that "omnia gravia aequaliter cadunt" (Cabeo 1646, 1, pp. 97–98).

[10] Indeed: "Aerem nihil efficere in isto motu nec pro nec contra velocitatem" (Koyré [1937] 1973, p. 221n). This sentence happens to be not in the place Koyré indicates, that is, Cabeo (1646, 1, p. 68), but below and in another form: "Neque tamen hæc ratio, si non esset experimentu, mihi persuaderet. Qui enim aero adducunt, sine pro, sine contra motum, ut inducant velocitatem, vel tarditatem. [Nor is there anything that] nobis persuadet aere nihil efficere in isto motu nec pro nec contra velocitate" (Cabeo 1646, 1, p. 97). This *lapsus calami*, not an exception in Koyré's writings, is of little relevance but allowed to find the text on the matter that Koyré followed quite precisely, that is, Caverni (Caverni 1895 p. 281), where the same (wrong) page is suggested.

[11] In this case, Koyré explicitly refers to Caverni (Caverni 1895); however, as soon as he wants to cross-refer directly to Riccioli, problems arise again: "Riccioli explique qu'il est à peu près impossible de mesurer directement des différences de temps aussi petites et suppose que Cabeo avait observé des chutes trop courtes pour pouvoir noter quoi que ce fût. V. Almagestum novum, Bononiae, 1651, vol. II, p. 392" (Koyré [1937] 1973, p. 221n). Sadly, on the referenced page, we

dropped from the Bologna tower with all experimental accuracy showed to get to the ground always at different times: two spheres of clay – then of the same material – of different weights (10 and 20 ounces), hit the ground with a gap of about 15 feet from each other, with the heavier always preceding the lighter (Cf. Koyré [1937] 1973, pp. 221–222; Riccioli 1651, 2, pp. 387–389).

Koyré's thesis begins to manifest itself with an increasing force of expression, though unaccompanied by an equally important effort in clarifying all references in its support. We can accept, in fact, that Galileo had no need to be acquainted with the experiences of Renieri or Riccioli in order to know that two bodies of the same material but with different weights, dropped from a certain height, would not move at the same speed nor would get to the ground at the same instant. However, Koyré only reminds that Galileo spoke of falling speeds that are equal in the void – which Baliani, Cabeo, and Renieri would not understand – then of an "abstract and fundamental case" (Koyré [1937] 1973, p. 222) and not of what we could experiment and observe. Perhaps because of the brevity of the article, or of the fact that his Galilean interests were relatively recent, he does not reference Galileo's text correctly.[12] But by invoking its inspirational reading, Caverni's *Storia del metodo sperimentale in Italia* ("A History of the Experimental Method in Italy"), we can integrate and in some way support the argumentations. As regards the resistance of the medium, the first reference is to the first day of the *Two new sciences* (*Discorsi e dimostrazioni matematiche intorno a due nuove scienze*), when Salviati states:

> We are trying to investigate what would happen to moveables very diverse in weight, in a medium quite devoid of resistance [...]. Hence just one space entirely void of air – and of every other body, however thin and yielding – would be suitable for showing us sensibly that which we seek. (Galileo [1638] 1898, VIII, p. 117; English translation [by Drake] from Galileo [1974] 1989, p. 76)[13]

The second is in the *postille* ("annotations") to the *Esercitazioni filosofiche* ("Philosophical exercises") by Antonio Rocco, and runs:

> We see, however, that even when the medium is the air, which is corporeal and therefore resistant, if we mitigate and lighten this same medium we can see two bodies greatly different in weight in motion, for a small amount of space, moving at speeds naught or very slightly different, of which we are certain that they eventually become different not because of heaviness, being this always the same, but because of impediments and obstacles of the

---

cannot find what promised, and the right reference is in Caverni's text, precisely on "p. 382" (Cf. Caverni 1895, p. 281; Riccioli 1651, 2, pp. 382 f.). About the experiments held by Riccioli, see Riccioli 1651, 2, pp. 383 ff., esp. pp. 387–89; and what we can find in Caverni 1895, pp. 281 ff., 391 ff.

[12] Koyré actually speaks of a Galileo who would express himself "là-dessus avec toute la clarté desirable" (Koyré [1937] 1973, 222), in the *Discorsi* but also in the *Dialogo*, without giving any reference.

[13] "Noi siamo su 'l volere investigare quello che accadrebbe ai mobili differentissimi di peso in un mezzo dove la resistenza sua fusse nulla [...] perché solo uno spazio del tutto voto d'aria e di ogni altro corpo, ancor che tenue e cadente, sarebbe atto a sensatamente mostrarci quello che ricerchiamo" (Galileo [1638] 1898, VIII, p. 117).

> medium, which constantly accrue. Why should not we hold fast to this, that once heaviness, density and all the other impediments of a solid medium are removed completely, all metals, stones, woods and in sum all bodies do move in the void at the same speed? (Galileo [1634] 1897, VII, p. 744)[14]

To these and other passages suggested by Caverni,[15] I take the liberty of adding a couple: the first, again taken from the *postille*, stresses that the fall of bodies from on high always records the same speeds for those bodies and their simultaneous arrival on the ground, provided that "one could remove all impediments of medium" (Galileo [1634] 1897, VII, p. 737); and the other from the *Discorsi*, when Salviati explains Simplicio that Aristotle was right when he observed that the bodies would fall at speeds proportional to their weights, but had not taken any account of the "great changes [they receive] from the medium, which alter the simple effect of heaviness alone"[16] (Galileo [1638] 1898, VIII, p. 109, Drake's translation in Galileo [1974] 1989, p. 69).

The problem of the medium's resistance, which was more than clear to Galileo, was then not correctly understood by Renieri, who wrote to the Pisan on 20 March 1641 that he had not yet carefully read the last dialogue, that he did not take Niccolò Cabeo's experiences seriously, and that he vaguely remembered having read or heard that his interlocutor deemed it impossible for two bodies of the same material and of different weights to hit the ground at different speeds and times.[17]

The stories briefly reported by Koyré get somewhat complicated as soon as we delve into them: as we have seen, Galileo receives a letter from Renieri in which the latter announces that he had carried out experiments on the fall that belie Cabeo as well as – so he believes – Galileo himself. The Pisan, however, does not make an issue of this; rather, he invites him to read carefully his last dialogue, i.e., the *Discorsi*. Caverni, who Koyré took his cue from, speaks more widely of the matter and reconstructs it in a different way: after Renieri had communicated the results of

---

[14] "Tuttavolta che noi veggiamo che, con l'attenuare e alleggierire il mezzo, anco nel mezzo dell'aria, che pure è corporeo e però resistente, arriviamo a vedere due mobili sommamente differenti di peso, per un breve spazio, muoversi di velocità niente o pochissimo differenti, le quali poi siamo certi farsi diverse non per le gravità, che sempre sono l'istesse, ma per gl'impedimenti e ostacoli del mezzo, che sempre si agumentano; perché non deviamo tener per fermo, che rimosso del tutto la gravità la crassizie e tutti gli altri impedimenti del mezzo pieno, nel vacuo i metalli tutti, le pietre, i legni ed in somma tutti i gravi, si movesser con l'istessa velocità?" (Galileo [1634] 1897, VII, p. 744).

[15] Caverni 1895, pp. 283 ff. See also recently Crapanzano 2017, pp. 17–35.

[16] The "grande alterazione [che] ricevono dal mezzo, che altera il semplice effetto della sola gravità" (Galileo [1638] 1898, VIII, p. 109).

[17] Cf. Letter [4121] of Vincenzio Renieri to Galileo, 20 March 1641 in Galileo 1906, XVIII, p. 310. This thing was on the contrary understood by the Czech doctor and erudite, the Jesuit Jan Marek Marci (Johannes Marcus Marci von Kronland, 1595–1667), who in his *De proportione motus* had written: "Motum quatenus a gravitate procedit eiusdem speciei seu gradus, eadem celeritate ferri in omnibus, quantumvis mole, figura, pondera a se different" (Johannes Marcus Marci 1639, folio 113; Cf. Koyré [1937] 1973, p. 222 n). Caverni considers Marcus Marci the man who, at the same time and independently from Galileo, demonstrates the laws of the fall of bodies and lays the foundations for the science of motion (Caverni 1895, pp. 310–311, p. 417, pp. 582–584).

his experiments, conflicting with what Galileo wrote in the *Dialogo* and the *Discorsi*, he recommended him reading his "first dialogue *On motion*" again (part of the so-called *De motu*) where he explained how much the medium (air) influenced the motion of bodies in free fall;[18] the theory was then safe from danger. But Caverni goes on to quote the letter of Vincenzio Renieri to Galileo, 20 March 1641 (n. 4121), where Renieri said he had not been able to read "the last dialogue of Your Most Excellent Lordship" (Galileo 1906, XVIII, p. 310),[19] leaving ambiguously untold whether he was referring to the *Dialogue Concerning the Two Chief World Systems* (*Dialogo sopra i due massimi sistemi del mondo*) or rather to the *On motion* (*De motu*).[20] Actually, Renieri did not mean that, and I think he is clearly referring here to the *Discorsi*, and thus Caverni's subsequent expression could be explained, according to which

> Renieri [...] naively replied that *in two years* [my emphasis] he had not had the time to read the book with the attention that the therein mathematically demonstrated propositions required.[21] (Caverni 1895, p. 280)

And 2 years are just a bit less than the time elapsed between the letter and the publication of the *Discorsi* (1638). It is likely that in this case Koyré, on reading Caverni, had some problems in giving his discourse coherence by opting for a different hermeneutical choice, easier and closer to the facts. Following this thread, Renieri would not have understood or read the remarks on the effects of the medium (air) contained in the *Discorsi*, and Galileo had to be very sure – maybe even since his Pisan times and in any case well before 1638 – that air resistance was roughly proportional to surface (in the case of a sphere, to the square of its radius) and weight to mass (then to the cube of the radius); therefore, it could be greater for a musket ball than for a cannon ball.[22] As to the argument, it was not new: Giambattista

---

[18] Thus Caverni: "Galileo allora dichiarò meglio in qual senso si dovesse interpretare quel luogo [del Dialogo], in cui intendevasi formular la legge assolutamente, astraendo dalle accidentalità prodotte dall'impedimento del mezzo, gli effetti del quale, da che solo potevano dipendere le differenze nelle varie cadute sperimentate, diceva di aver minutamente considerati e discorsi nel primo dialogo Dei moti, alla lettura del quale, se voleva avere intera scienza di quelle cose, rimandava il Renieri " (Caverni 1895, p. 280).

[19] "L'ultimo Dialogo di V. S. Ecc.ma non è stato da me letto se non in qua e il la [...]. Leggerò per tanto questi pochi giorni di vacanza l'ultimo suo Dialogo, benché la total lettura me la riserbi a far questa futura estate con più comodo: in tanto torneremo a far l'esperienza delle palle, e vedere se ci fossimo ingannati la prima volta nella osservatione che quando s'avvicinano a terra pieghino e non vadino a perpendicolo" (Galileo 1906, XVIII, p. 310).

[20] See *De motu* dialogue in Galileo [1589–1592] 1890, I, pp. 367–408.

[21] "Il Renieri [...] rispondeva ingenuamente di non aver avuto tempo *in due anni* [my emphasis] di leggere il libro con quell'attenzione, che richiedevan le proposizioni matematicamente ivi dimostrate" (Caverni, 1895, p. 280).

[22] Cf. Koyré [1937] 1973, pp. 222–223. In the case in point, Koyré is quite sparing in his explanations, so we need to clarify. The resistance that air opposes to a body in free fall depends on its surface, that is, for a sphere, on the square of its radius. Mass, on the contrary, grows in proportion to volume, then according to the cube of the same dimension. As a consequence, in bodies of the same nature, the "smallest" will have less mass and more surface, the "biggest" can have more

Benedetti (1530–1590) had dealt with it at least two decades before in his *Diversarum speculationum liber;*[23] if anything, what was different were the self-confidence acquired by Galileo and, above all, the wider background of the foundation of dynamics in which he consciously moved.

Koyré's conclusions about the Pisa experience should not be surprising: if Galileo had always favored theory over experimentation (showing his adherence to a "mathematical and ontological Platonism" that left a number of questions open), and if he had so cleverly supported and defended his model of falling bodies, then he could not help but contradict Aristotle where he says that bodies were to fall at speeds proportional to their weights. In fact, despite the medium's resistance, lighter bodies reached the ground before and with speeds higher than expected, never simultaneously, though. This is the underlying reason "for which Galileo did not do the Pisa experience; and he did not even imagined it"[24] (Koyré [1937] 1973, p. 223).

The outcome of what Koyré calls "little investigation" is disruptive, and, to make matters worse, his gloss wants to elicit further investigation because it leaves readers with the task of deriving a moral or a meaning from it,[25] which has duly happened, as we will see.

4.2 On the Legend of the Leaning Tower

The meaning or the "moral" of Koyré's little essay could not be unique: it was certainly conceived and written in a time when Koyré was committed to becoming the Koyré we all know, the historian of scientific thought. He was actually working on the *Études Galiléennes*, a study that more than others would back the interpretation of a "Platonist" Galileo.[26] It is in this direction that, I think, we must place the survey on the experience of falling bodies, together, of course, with what he had written so

mass and less surface. Of course, everything depends on density; but it is possible that a large body is less sensible than a small one to air resistance, so bodies with different densities may offer less mass over an identical surface or more surface over the same mass. From this, it results that a body's lower specific resistance will cause an increase of air resistance.

[23] Indeed, Benedetti dedicates to the topic one of his Disputationes on some of Aristotle's theories, and in particular to the twelfth, entitled: "Maior hic demonstratur esse proportio ponderis corporis densioris ad pondus minus densis in mediis densioribus, quam sit eorundem corporum in medio minus denso, nec corporum pondera servare proportionem densitatis mediorum" (Benedetti 1585, p. 175; see also pp. 168–195).

[24] "Galilée n'a pas fait l'expérience de Pise; et ne l'a même pas imaginée" (Koyré [1937] 1973, p. 223).

[25] "Quant à sa morale […] nous en voudra-t-on de laisser aux lecteurs le soin de la tirer eux-mêmes?" (Koyré [1937] 1973, p. 223).

[26] I wish to remind that Koyré gave anticipations of the topics that would eventually shape his *Études* in the years before their publication, particularly of the first (see Koyré 1935, 1936) and of the second (see Koyré 1937). The third, the wide-scoping *Galilée et la loi d'inertie* was left unpublished (evidently conceived after the first two) up until the moment the *Études* saw light in 1939–1940.

extensively here and there on the Pisan's various experimentations. Indeed, if Galileo was a Platonist in the meaning that Koyré wanted the adjective to mean, i.e., that he believed that reality had a mathematical structure and that, as a consequence, the scientific method would present a priority of theory over experimentation, then we can understand his interest in showing, whenever he had the chance, how Galilean experiments were much less "sensible" than what the Pisan meant them to be; up to the extreme case of the bell tower of Pisa, that is, of pure invention.

Whatever the opinions around the fact, it is worth suggesting some readings that in the course of time some prominent historians, physicists, and even humanists gave of it; we will then make a partial reconstruction of what the Germans call *Wirkungsgeschichte* (the 'history of effects'). Koyré himself is of help by reminding us how the image of Galileo dropping objects of different weights from the Leaning Tower was so widely known that it soon became an icon of the birth of the new mechanics.[27] So it is, for example, in the pages of the prolific Orientalist and scholar, Angelo De Gubernatis (1840–1913): having taken a university course on Galileo, he immediately declared that in Pisa he began

> [...] his scientific campaign against Aristotelians, with great scandal among many of his colleagues in the Studium, especially because [...] he was determined to make public experiments on the fall and descent of bodies, which many a time he reiterated at the bell tower of Pisa in the presence of Pisan Lecturers and Learners. (De Gubernatis 1909, p. 10)[28]

In an even more celebratory way, the skilled engineer and science historian John Fahie paints Galileo in his presentation as a sort of godsend, who came at the right moment

> [...] and, above all, he came armed with a weapon of convincing force experiment. He was not content, like his precursors, with merely giving an opinion, supported or not by wordy metaphysical arguments, but what he asserted as well as what he denied he proved to ocular demonstration. (Fahie 1903, p. 23)

---

[27] In fact, "les historiens de Galilée – et les historiens de la science en général – attribuent aux expériences de Pise une grande importance; ils y voient habituellement un moment décisif de la vie de Galilée: le moment où celui-ci se prononce ouvertement contre l'aristotélisme et commence son attaque publique de la scolastique; ils y voient également un moment décisif de l'histoire de la pensée scientifique: celui où, grâce justement à ses expériences sur la chute des corps effectuées du sommet de la Tour penchée, Galilée porte un coup mortel à la physique aristotélicienne et pose les fondements de la dynamique nouvelle" (Koyré [1937] 1973, p. 213).

[28] He began "la sua campagna scientifica contro gli Aristotelici, con grande scandalo di molti suoi colleghi dello Studio, specialmente perché [...] egli determinò di fare pubblicamente dell'esperienze sulla caduta e discesa de' gravi, che più volte reiterò sul campanile di Pisa alla presenza dei Lettori e della Scolaresca pisana" (De Gubernatis 1909, p. 10). Koyré indicates "p. 9" in De Gubernatis's text, instead of 10 (Koyré [1937] 1973, p. 213n; De Gubernatis 1909, p. 10). De Gubernatis, in turn, cites Giovan Battista Clemente Nèlli (1725–1793) – and not "Nessi," as Koyré writes – a noble Florentine erudite, precisely the page that runs: "Queste ingegnose, ed egualmente vere di lui speculazioni è credibile, che le palesasse pubblicamente in Cattedra. Onde contradicendole i Professori Aristotelici dello Studio, ad effetto di convincerli, si determinò il Galileo di fare pubblicamente dell'esperienze sulla caduta, e discesa de' gravi, che più volte reiterò sul Campanile di Pisa alla presenza dei Lettori, e della Scolaresca Pisana con gran meraviglia, e dispiacere de' medesimi, perché col mezzo di esse restavano per lo contrario convinte false, ed erronee le proposizioni della Peripatetica Filosofia" (Nèlli 1793, I, p. 44).

Beyond the Pisan's meritorious publications, he does not want to underestimate the importance of his experiments as they would represent the true epistemological revolution that originated classical physics. And among the experiments, in the first place, he includes the very ones at the Leaning Tower:

> We must, however, say something here of his celebrated experiments on falling bodies, on account of their associations with the Leaning Tower of Pisa – one of Italy's many curious monuments. Nearly two thousand years before, Aristotle had asserted that if two different weights of the same material were let fall from the same height, the heavier would reach the ground sooner than the lighter in the proportion of their weights. The experiment is certainly not a difficult one, but nobody thought of that method of argument, and consequently this assertion was received upon Aristotle's *ipse dixit* among the axioms of the science of motion. Galileo, however, now appealed from the authority of Aristotle to that of his own senses, and maintained that, with the exception of an inconsiderable difference due to the disproportionate resistance of the air, they would fall in the same time. The Aristotelians ridiculed and refused to listen to such an idea. But Galileo was not to be repressed, and determined to make his adversaries see the fact as he saw it himself. So one morning, before the assembled University, professors, and students, he ascended the leaning tower, taking with him a 10-lb. shot and a 1-lb. shot. He balanced them on the over-hanging edge and let them go together. Together they fell, and together they struck the ground. (Fahie 1903, pp. 24–25)

Fahie believes that Galileo founded dynamics thanks to his experimental genius, an idea that was the exact opposite of what Koyré believed and tried to show; consistently, he assumes that the experience of Pisa had occurred.[29]

Émile Namer, a professor of Italian philosophy at the Sorbonne, painted Galileo in a somewhat "amateurish" way,[30] believing that he "had no intention of adopting the traditional methods which he had so strongly disapproved of in his student days" (Namer 1931, p. 27). He estimates that:

> If Aristotle had lived in our days, he said, he would have revised his doctrines himself and subjected them the new experiments. Disciples who thought themselves faithful to him were really traitors, because they did not seek to complete his work, or try to enrich human knowledge with their own thoughts and deductions. With unbelievable audacity Galileo relegated Aristotle to dusty library shelves. He proposed to open the great Book of Nature and read its laws with fresh eyes. (Namer 1931, p. 28)

But, most importantly, he accounts for the Pisa incident with a joyful tone and abundance of details (mostly out of sheer invention):

---

[29] The discourse on Galilean experimentation and, in particular, on the dropping of weights made from the Tower of Pisa is taken almost literally by Fahie years later in an essay: "From professorial denunciation he proceeded to those public experiments with which the leaning tower of Pisa has become for ever associated. [...] One morning, before the assembled professors and students, he ascended the leaning tower, taking with him a 10 lb. shot and a 1 lb. shot. Balancing them on the overhanging edge, he let them go together. Together they fell, and together they struck the ground. Neglecting the resistance of the air, he now boldly announced the law that all bodies fall from the same height in equal times" (Fahie 1921, p. 216; Cf. Koyré [1937] 1973, pp. 214–215).

[30] Indeed, Namer concludes without any delay (and documentation): "It was not his way to take texts of Aristotle, turn them about in all directions, define and redefine them, and be satisfied to admit without discussion, in spite of all proof to the contrary, that Aristotle had said the last word about the nature of things" (Namer 1931, p. 27; Cf. Koyré [1937] 1973, pp. 215–216).

> When Galileo heard that all the other professors were expressing their doubts as to the conclusions of this insolent innovator, he took up to the challenge. He solemnly invited those grave doctors and all the student body – in other words, the entire University – to assist at one of his experiments. But not in the customary setting. No, that wasn't big enough for him. Out in the open, under the sky, in the vast Cathedral Piazza! And the academic chair clearly indicated for these experiments was the Campanile, the famous Leaning Tower. The professors at Pisa, like those in all other towns, had always maintained, according to the teachings of Aristotle, that the speed of the fall of a given object was in proportion to its weight [...]. Galileo, on the contrary, claimed that weight had nothing to do with it, and that they would both reach the earth at the same moment. (Namer 1931, pp. 28–29)

So Galileo, confident in his theory and therefore heedless of the risk he was about to run, challenged the Aristotelians donned "in long velvet robes" (Namer 1931, p. 29),

> [...] climbed the steps of the Leaning Tower, holding himself well in hand in spite of the laughter and the booing of the crowd. He realized the solemnity of the hour [...]. The moment came. Galileo released the two balls of iron. All eyes were raised. A silence. The two balls were seen to leave together, stay together at the foot of the tower. Amazement! Indignation! A confusion of sounds rose from this mixed and impatient crowd that had gathered together as if at a pilgrimage. Some of them had been confident that the miracle would be performed by the new prophet; others had been equally certain that he would be proven an impostor. (Namer 1931, p. 30)

It is fair to remind how, many years later, Namer would use a more circumspect phrasing, signaling the impending doubts on the episode, perhaps after reading Koyré.[31]

Among those who dealt with the issue of the experiments carried out at the leaning tower of Pisa, the first to show skepticism was Emil Wohlwill (1835–1912). A German chemist and science historian, Wohlwill argued convincingly – at least for Koyré – that Viviani had exaggerated in telling in a legendary fashion episodes in the life of Galileo, including the "controversial" one.[32] Maybe in order not to weigh down his text, Koyré does not provide other examples of readings for the episode that on the other hand would all show the same stereotype: an attack to Aristotelianism, a successful public demonstration of Aristotle's error from the top of the Leaning Tower, dismay of the Aristotelians who, in spite of evidence, would continue to follow their tradition. Then, Viviani's story "is based on nothing, the Pisa experiments are a myth"[33] (Koyré 1973 [1937], p. 217).

After following Koyré's brief exposition, we can further deepen his contentions, update the "history of effects," and draw some reflections. Firstly, it is remarkable

[31] It is in 1975, in fact, that in a text dedicated to the Galileo affair (then focusing on another Galilean issue) he comments favorably on Koyré's production. About Galileo's Pisan period, he limits himself to reminding that at that time "se situeraient les expériences sur la chute des corps effectuées du haut de la Tour de Pise. Les dates ne sont pas établies, les faits son fort contestés, mais il reste que ces années préparatoires furent décisives" (Namer 1975, p. 38; see also Namer 1975, pp. 253 ff.).

[32] Cf. Wohlwill 1926, pp. 260 ff; Koyré [1937] 1973, pp. 217–218.

[33] "Le récit de Viviani sur l'expérience de Pise n'est lui-même fondé sur rien. Les expériences de Pise sont un mythe" (Koyré 1973 [1937], p. 217).

that Koyré, albeit in a quick overview of the literature on the issue, did not take into account the essay by Lane Cooper (1875–1959), professor of English language and literature at Cornell University, entirely devoted to the problem (*Aristotle, Galileo, and the Tower of Pisa*); after its publication (Cooper 1935), this would eventually become a reference in supporting the theory that Koyré exposes in 1937, that is, that Galileo did not perform the experiment[34]. It is remarkable because it was published 2 years before his article and, above all, dealt with the question carefully by focusing on documentation. It is not possible here to give a comprehensive account of Cooper's work, aimed at disproving Viviani's tale; however, I would like to point out the presence in it of fairly strong arguments in support of the thesis: the performance of similar experiments by Simon Stevin (1548–1620) and Jacopo Mazzoni (1548–1598) well before Viviani's reconstruction; the missing reference to contemporaries to the fact, even by an insider who was working on the same topic; the significant downsizing of the episode itself made by Renieri; the importance that the opposition to Aristotelianism should hold, and so on.[35] The conclusion is that the experiments from the leaning tower of Pisa are probably a fantasy, like other assumptions that should be distinct from historical facts:

> His general habit of contemplation argues against Viviani's story about the repeated experiments from the tower of Pisa. Galileo was perhaps more likely to watch a pendulum that was already swinging, and to climb a tower only for the sake of his telescope, though the story of the pendulum seems also to have gone the way of Newton's apple, while his use of the telescope in the tower at Venice remains historical fact. (Cooper 1935, p. 42)

Several are the findings and the argumentations of Stillman Drake, who, in 1989, devotes a section of his *History of Free Fall* to the scientific context in which Galileo was operating during his Pisan period, perhaps in order to refute Koyré;[36] it is there – and not in the following chapter on the "Discovery of the Law of Fall," where his point draws upon the *De motu* papers – that we find the reference to the leaning tower experiment:

> Written about the time of Galileo's famous demonstration of equal speeds in fall from the Leaning Tower of Pisa, it demolishes fanciful reconstructions of that event in which it is

[34] Koyré mentions Cooper in the third part of his *Galileo Studies* (*Galileo and the law of inertia*), in a note where he expounds a very interesting distinction regarding the law of the fall of bodies: "Historians of Galileo, and of physics, usually mix two entirely different propositions: (i) that which Galileo is alleged to have established at Pisa by experiments which in fact he never performed, and had no need to perform, (cf. L. Cooper, Aristotle, Galileo and the Tower of Pisa [...] and the present author's article, 'Galilée et l'expérience de Pise' [...]); this proposition, to the effect that bodies of the same nature fall with the same speed, was actually already established by Benedetti [...]; (ii) the second proposition, of which a proof is provided for the first time in the *Discorsi*, says that all bodies regardless of their natures fall with the same speed" (Koyré [1939] 1978, p. 228).

[35] Cf. especially Cooper 1935, pp. 14, 26–33, 48 ff. What is more, Cooper presents references to Wohlwill, Fahie, and Olschki. See name index in Cooper 1935, pp. 100–101. Olschki's work on Galileo is Olschki (1927).

[36] I wish to remind that most of Drake's research was characterized by the acknowledgement of Galileo as an experimental physicist and, of his importance as such, the exact opposite of what Koyré did. Even if Drake tries as much as possible not to mention him, it is easy to notice how often he disagrees with him. See, among others, Drake 1970; 2001[1980], pp. 11–12; [1989] 2000.

> frequently said that one falling weight was leaden and the other wooden. The original account, written by Vincenzio Viviani in 1657 [1654], who had it from Galileo in 1640, stated that both weights were of the same material, which in 1590 was the only case that had been considered by Benedetti, Stevin, or Galileo has been mathematically established or verified by observation. From the above it is evident that Galileo might expect the lead weight to move much more swiftly than that of wood after the initial motion. The proof for equal speed fall held only for bodies of the same material, and it would have been foolish for him to risk his reputation with his students by using weights that he would not expect to move with the same speeds. (Drake [1989] 2000, pp. 31–32)

Drake is not interested in a reading of the experiment going beyond what could be its meaning, not believing that it had a role in the repudiation of Aristotelianism or that it was instrumental in the formulation of the law of fall. He imagines that Galileo had done before, or roughly at the same time (there is no agreement on the date), the same experiment as Stevin, Benedetti, and others did. So Drake, who in the broader interpretation of Galileo is at odds with Koyré, on this single fact shows a possible consonance, limited to the part where the latter shows doubts about Viviani's tale and the fanciful readings that were given of it, never on the incident itself;[37] in any case, Drake mostly avoided stressing the incident understanding its problematic status.[38] Thomas Settle, who in the course of years has repeatedly, scrupulously, and successfully carried out several of the experiments described by Galileo, stated that "Galileo then took the balls to relatively high towers, perhaps even to the Leaning Tower. There seems to be enough evidence for us to conclude that Galileo was carrying out experiments with falling bodies at this time" (Settle 1967, p. 326).

[37] Years before, Drake had showed a greater receptivity to a "persuasive" value of the experiment in an anti-Aristotelian sense: "Historians generally have doubted the story about Galileo and the Leaning Tower of Pisa, first told after Galileo's death by a protégé who was not born until long after the incident. According to this story, the demonstration was performed in the presence of Galileo's students and some professors. It is probable that his students, who had been taught Aristotle's rules by their professors of philosophy, would argue against him that weight must affect speed of fall. Galileo's Leaning Tower demonstration would then have been not just to show the students, but to convince the professors that Aristotle's physics must be revised, as he was already arguing performed in the presence of Galileo's students and some professors" (Drake [1980] 2001, pp. 21–22; Cf. Drake1978, pp. 19–21, pp. 413–415).

[38] This is what happens in his *Galileo Studies* where, though putting up a pressing and sometimes harsh confrontation with Koyré, he does not consider Galileo's Pisan period as crucial for the elaboration of the law of motion of falling bodies (Cf. Drake 1970, pp. 240–278). On the other hand, he does not miss the chance for marking his position about Galilean experimentations and observations: "I hasten to add that I do not mean by elaborate experiments such as those he later described as having performed in order to corroborate the rules. It would be grossly anachronistic, both with respect to the history of experimental physics and with respect to the know procedures of Galileo, to assume that he reached the mathematical law of fall by carefully controlled measurements of falling bodies. That he confirmed the law in that way is virtually certain. But he could have arrived at it indirectly in a much simpler and more plausible way […]. Thus we need not assume that Galileo found the law of falling bodies by reasoning from the conclusions of previous philosophical or physical writers; he might have found it by rough observation, or by mathematical reasoning of his own; perhaps by a combination of both, or in some other way that has not occurred to me" (Drake 1970, pp. 218–219).

Since 1980 the science historian, Maurice Finocchiaro, has opted for an explicit criticism of Koyré, embodied in the negative evaluation of his *Galileo Studies* stating that there are not simply a number of mistakes or *lapsus calami*[39] but conclusions not supported by evidence. In particular, as to the law of falling bodies, Koyré would not separate the context of discovery from that of justification,[40] inserting elements of the first – namely, Platonic apriorism – in the field of the second and forgetting that:

> Galileo does not need the epistemological principle to justify his physical argument, but rather his physical argument (the fact that such an argument can be given) justifies the epistemological principle. In short, if Galileo can answer the [experimental] argument in the a priori way that Koyré thinks, then his answer is supporting, not assuming, apriorism. (Finocchiaro 1980, p. 214)

Galileo is indeed rationalist, but he is not apriorist in the meaning Koyré intends: he started from the observation of facts to arrive to a theory, not the other way round[41]; it is understandable then that Finocchiaro does not mention the tower experiment; it would be very risky to labor about a fact of which there is no certainty that cannot be ascribed *sic et simpliciter* to either of the two contexts (discovery or justification) and that certainly had a persuasive and not a verification intent for the man who performed it.

Michael Segre, in a 1989 article, made it clear from the outset that the value of the experiments from the Leaning Tower is low for the history of science as well as for that of physics; in fact:

> […] the experiment certainly had no impact on Galileo's thought [and] after all, it was a social event, not an organic part of his scientific work. (Segre 1989, p. 435)

Nevertheless, taking its veracity for granted, the fact has always been charged with meanings that range, as we have seen, from demonstrating the superiority of the empirical method over the aprioristic one, to repudiating Aristotelian physics, even to the exaltation of it as a symbolic gesture that claimed intellectual freedom for science.

Without dwelling on the interesting account he offers of the feedback and discussion that have sparkled in the course of time,[42] we can say that Segre too wants to

---

[39] These mistakes due to Koyré's "poor erudition" are explicitly taken from Garin (Cf. Garin 1957, pp. 407–408; Finocchiaro 1980, pp. 203–205).

[40] For Hans Reichenbach's classic – and in some respects outdated – distinction between context of discovery and context of justification, see Giordano 2011.

[41] Thus, Finocchiaro corroborates the concept: "Galileo is not a rationalist in the sense of an apriorist, but in the sense that he likes to use reasoning and arguments as much as possible, we might say a rationalist in the sense of a logician, a logician–in–action. The difference is that the logician will not limit himself to a priori reasoning, reasoning from a theory; some reasoning is reasoning based on facts, where by starting from facts one attempts to justify a conclusion" (Finocchiaro 1980, p. 216).

[42] Here as well we tried to offer an outline of it, but Segre gives an undoubtedly wider one by presenting the opinions of Caverni, Wohlwill, Favaro, Cooper, Mieli, and the same Koyré (Cf. Segre 1989, pp. 437–445). I only add that Favaro, editor of the *Edizione Nazionale delle Opere di Galileo* ("National edition of Galileo's works"), assumed an articulated standpoint: on the one hand, he could not conceal some inaccuracies in Viviani's Galilean biography, while on the other, he deemed

understand the reasons why Viviani invented the story. His contentions appeal to the literary canon that biographies had to follow at the time (the model was Vasari) and lead to consider how:

> [...] thus Viviani's story of the leaning tower experiment is to be understood as belonging to a literary style in which truth had less importance than it has in modern biographies. What was important then was to embellish Galileo's image, even by means of invented stories. But even this embellishment was subject to certain rules, dictated by the taste of Viviani's audience. (Segre 1989, p. 446)

And Viviani's audience was not only made up of philosophers, of specialists, but also of educated people who poorly digested the abstract formulas of mathematical demonstrations, though would certainly understand the theory's – and Galileo's – value through the account of an experience. It is clear, at this point, what Segre could conclude: in the biography by Viviani, there are true elements mingled with fancy ones, the latter instrumental in embellishing Galileo's image; other scholars accomplished, at the same time or after, similar experiments at the Tower of Pisa (Giorgio Coresio), and Galileo himself did them from the St Mark's Campanile in Venice. Viviani then,

> [...] on this occasion, was not writing as a scientist or a historian of science, but as a man of letters, addressing an audience interested in literature. (Segre 1989, p. 450; see also pp. 449–451)

The issue has lost its clout for some time, and this is not at all bad, if we think of the several controversies that it sparked in the past up to becoming a battleground between historiographical schools or even between individual historians, with no holds barred.[43] However, the hints for discussion are certainly not exhausted, and it is still worth to reflect on the alleged experiments from the Tower of Pisa; let us see why.

---

them trivial and reckoned that he was in general a careful historian, implicitly supporting the account of the experiments at the leaning tower (Cf. Favaro 1915; Segre 1989, pp. 438–441). Concerning Koyré, Segre considers his *Galileo Studies* where he finds that the Russian-born historian of the scientific thought "was writing about the role of experiment in modern science in general, and did not consider the historiographical question of whether or not Viviani had reported the truth" (Segre 1989, p. 443). This can be accepted, within limits, even if Segre overlooks the little essay on the topic written 2 years before the publication of the Galileo Studies, on which we deployed our analyses (Cf. Koyré [1937] 1973; see also Segre 1980). For a comprehensive overview on the Galileo biographies from the sixteenth to the nineteenth century, see Micheli 1988, where, among other things, we can read that Favaro assumed a fragmentary approach not devoid of consequences in the "valutazione della vita di Viviani, di cui curiosamente sostiene la sostanziale veridicità, pur avendo egli stesso contribuito a rilevarne le numerose inesattezze. Egli difende la vita di Viviani, soprattutto contro E. Wohlwill, l'acuto storico tedesco che aveva per primo rilevato l'inconsistenza di uno dei miti galileiani più consolidati, le famose esperienze pisane" (Micheli 1988, p. 181).

[43] Beyond his conclusions on the events, Segre closes his article by noticing how "somehow it is disquieting when important scholars such as Antonio Favaro or Aldo Mieli make use of their prestige and of rhetorical arguments – rather than rational ones – to *refute* the views of their opponents" (Segre 1989, p. 451).

4.3 Again with the Galilean Experiment of the Tower of Pisa?

Although valuable from a literary standpoint, Viviani's presentation indirectly offered generations of scientists, students, teachers, physicists, and many others the chance for a simple and direct introduction to the law of falling bodies. I also think that the teaching effectiveness of the story is a fact and explains the reasons for its success. Regardless of its veracity, many physics books introduce the subject of motion by citing the episode, and it would be useless to pretend that they deepen the issue or avoid it in the wait – which might be endless, I would add – of knowing the factual truth.[44]

We can overcome the doubts about the responsibility of the account, i.e., whether it was Galileo himself who suggested the circumstances, when still alive, or Viviani who reconstructed the incident: the latter wrote the biography 12 years after Galileo's death and did it as it was customary at the time, that is, in a celebratory way, even making a mistake on his master's date of birth.[45]

There remain, however, the ones on the reasons for the hoax – if it was such: why would Viviani incorporate into his story an incident that never occurred? Koyré has no opinion on the matter, while Cooper suggests that the young disciple had confused Galileo's reference to demonstrations and examples on falling bodies (made against Aristotle) with a real experimentation.[46] Alistair Crombie, taking a more nuanced position towards Galileo's pupil, says he was convinced that it was

---

[44] In the opposite view, there are two American physicists who pointed out that some texts do bad information by presenting a Galileo who discovered the law of falling bodies with his tower experiment, instead of his experiences with the inclined plane: "Some texts, such as the one quoted at the beginning of this article also misrepresent what Galileo would have been trying to prove if he actually performed the experiment. However, an account of the legend which presented in proper context and probable results would be so heavily qualified that it would be of doubtful pedagogical value in teaching the modern theory of falling bodies at an elementary level" (Adler and Coulter 1978, p. 201; Cf. also Highsmith and Howard 1972, p. 74). For a reappraisal under a "sociological" outlook of the legendary experiment at the Tower of Pisa, see Crease 2003.

[45] As Favaro found out more than a century ago, Viviani postponed Galileo's birthdate by 4 days (Cf. Favaro 1887).

[46] Cf. Cooper 1935, pp. 49 ff. In fact, "there are two passages in which Galileo mentions experiment or experience as a test of Aristotle on our point; in neither does he say that he performed the experiment; one of these is early (1590), and contain the expression 'a high tower', while the other is late (1638), and contains no such expression" (Cooper 1935, p. 49). Favaro, on the other hand, writes: "Il fatto delle esperienze sulla caduta dei gravi eseguite dall'alto della torre di Pisa, per dimostrare le nuove verità alle quali era pervenuto, è dal Viviani, il quale deve averlo raccolto dalle labbra istesse di Galileo, affermato in modo così sicuro ed esplicito da non potersi revocarlo in dubbio, e tanto meno recisamente negare perché non se ne trova conferma in altri documenti contemporanei" (Favaro [1910] 1988, p. 232). In addition, while Favaro was inclined to resolutely believe the episode as really told by Galileo, Cooper observed: "Why 'must' have had it, from the 'very lips' of Galileo? Viviani nowhere says that he so learn it, but we have seen that he could have got the notion of 'experiment', 'demonstration', and 'discourses', from the treatise and the dialogue *De Motu*. Nor is he writing in our day of scholarly exactitude, when a serious author does tend to exclude marvels from biography" (Cooper 1935, p. 54).

> [Galileo's] reputation as the founder of the experimental method […] that encouraged the strange elaboration in the nineteenth century of the story of Galileo dropping two different weights from the Leaning Tower of Pisa, in order to prove, as his law of falling bodies stated, that all bodies fall with the same acceleration, and to disprove the Aristotelian teaching that the speed would be proportional to the weight. (Crombie [1956] 1996, pp. 258–259)

Drake, on his part, is inclined to believe that Galileo rose to the top of the tower to make the famous experiment, and Viviani added the episode when he learned of it many years later, after receiving a not better specified letter "from the professor of mathematics at Pisa" (Drake [1989] 2000, p. 31n) on the topic of the fall of bodies that he read and discussed with a Galileo now blind.[47] Segre, as we have seen, suggests that Viviani's description is "simply a classic anecdote in a classic biography of those times" (Segre 1989, p. 450), laying the reasons on esthetic-literary grounds. Recently, John Heilbron, in his *Galileo*, called "Iconoclasts" those who "have thrown doubt on this vignette [of the leaning tower] although the tower's tilt made it a perfect platform for the experiment" (Heilbron 2010, p. 59), and added, referring to Koyré and Segre on the one hand and to Settle and Camerota[48] on the other, that

> […] owing the discrepancies in Viviani's account of his master's adventures with the tower and the lamp, historians take care when he is the only witness. Some have been so bold as to assert that he made things up and that Galileo experimented only in his head. This was to go too far and, in recent times, Galileo has become an exemplary, pioneering experimentalist. (Heilbron 2010, p. 60)

Heilbron gives the impression of being unwilling to join the fray, but I think that he is not sufficiently detached on the topic, attributing as he does radical positions to researches and academics who, to be honest, are not. If it is indeed true that Koyré and Settle committed themselves, respectively, to show how Galileo could not or, on the contrary, could carry out the experiment, we are not dealing either with an iconoclast (Koyré) or with a fanatic of experimentalism (Settle). Camerota, for his part, is inclined to believe Viviani's story, but his usual attention to the Galilean context, combined with a balance in his judgments, lead him to notice how "Viviani's narration includes doubtful details" (Camerota [2004], 2006, I, p. 72).[49]

---

[47] Drake writes that in his last years, "Galileo was blind and Viviani took his letter in dictation. A letter from the professor of mathematics at Pisa recounted recent experiment from the Leaning Tower, unaware of Galileo's much earlier demonstration there. That is why Viviani had the details right, though he was not born until long after the event" (Drake [1989] 2000, p. 31n). Even if he does not give any reference, I think he refers to the letter of Vincenzio Renieri to Galileo, 20 March 1641, where we read that "due gravi ineguali di peso, ma della stessa materia, cadendo dall'istessa altezza a perpendicolo, habbiano ad arrivar con diversa velocità et in diverso tempo al centro, mi pareva d'haver da lei udito o letto, chè ben non mi ricordo, non poter essere" (Galileo 1906, XVIII, p. 310).

[48] Cf. Heilbron 2010, p. 400, notes 110 and 111; see also Koyré 1943, Segre 1989, Settle 1961, Settle 1967, Camerota and Helbing 2000.

[49] Indeed, Camerota writes: "Il resoconto di Viviani costituisce la fonte di un episodio celeberrimo e leggendario della vita di Galileo: l'esperimento della Torre di Pisa. L'aneddoto […] è stato considerato largamente inattendibile da gran parte della critica moderna. Gli storici della scienza contemporanei hanno, infatti, teso – alla luce, soprattutto, della sua posteriore fortuna – a considerare

Paolo Galluzzi believes that the figure of the great scientist is quite different from his stereotypical portrait, "that, starting from the *Vita* by his faithful pupil, Viviani, proposed and is still proposing a historiography motivated by the will to glorify, rather than by the need to historically understand"[50] (Galluzzi 1988, p. 54).

Of course, we could go on with this list and find different articulations but identical conclusions: either the disciple of Galileo reported (more or less accurately) the truth or he wanted to alter a story that happened in other times, ways, and circumstances or even totally made the episode up with a celebratory intent. From a scientific point of view, the question has little relevance because knowing if Viviani did or did not adhere to the truth does not affect either mechanics or its history of effects, let alone touch its teaching effectiveness. As to the historiography of science, on the other hand, I think that things are not much different today, since persisting to investigate into an incident whose truth will be difficult to establish and which is quite limited and marginal to the establishing of Galilean science would not impact significantly on the field of research, all the more since the topic has become the occasion for ideological (and sometimes personal) fight among historians for a while.

Even so, I will take the liberty to add a thought on the matter and quote Evangelista Torricelli (1608–1647), a pupil of the Jesuit Benedetto Castelli (1577–1644), as well as an illustrious disciple of Galileo's. In fact, he not only happened to live with the great scientist in Arcetri, near Florence, during the last months of his life, but was a young and very talented man of science who fully understood his master's works and theories. Now, in a letter to Cardinal Michelangelo Ricci (1619–1682) dated 10 February 1646, Torricelli defended Galileo's arguments on motion in a way that better than others can interpret the same Torricelli's – and maybe even Viviani's – scientific spirit:

> Whether the principles of the *de motu* doctrine be true or false, I care really a little. Because if they are not true, we have supposed and should pretend they are compliant to truth, and then take all other speculations derived from those principles, not only as mixed as they are, but also in a Geometrical way. I feign or suppose that some body or point moves downwards or upwards at the known proportion and horizontally at an equal speed. When this happens to be, I'll say it follows all that Galileo said, and I as well. If lead, iron, or stone balls do not eventually follow that assumed proposition, to its detriment, we will say we are not going to talk about them.[51] (Torricelli 1919–1944, p. 357)

---

il ragguaglio del Viviani nei termini di un importante tassello del mosaico celebrativo ed esaltatorio volto a perpetuare l'immagine di un Galileo primo e provetto sperimentatore. Per quanto la narrazione del Viviani includa dettagli dubbi (la sede stessa dell'esperimento, la Torre di Pisa, e la presenza alla prova di tutto il corpo docente e degli scolari dello Studio appaiono, per molti versi, implausibili), nondimeno va osservato che la descrizione dell'evento data *dall'ultimo discepolo* trova indiretti riscontri nella marcata coloritura empirica e para-sperimentale che caratterizzò il dibattito pisano sulle problematiche *de motu*, nella seconda metà del Cinquecento" (Camerota [2004] 2006, I, pp. 71–72).

[50] The stereotypical portrait "che, a partire dalla *Vita* del fido allievo Viviani, ha proposto e viene proponendo una storiografia piuttosto motivata dalla volontà di esaltare che non dal bisogno di comprendere storicamente" (Galluzzi 1988, p. 54).

[51] "Che i princìpi della dottrina *de motu* siano veri o falsi a me importa pochissimo. Poiché se non son veri, fingasi che sian veri conforme abbiamo supposto, e poi prendansi tutte le altre specolazioni derivate da essi princìpi, non come così miste, ma pure Geometriche. Io fingo o suppongo

This consideration – methodologically unconventional – was juxtaposed[52] to what Galileo wrote to the French mathematician and future illustrious royal librarian, Pierre de Carcavy (?–1684), in a letter dated 5 June 1637 where he declares of researching *ex suppositione*, that is, by figuring

> [...] a motion towards a point which, starting from stillness, goes on accelerating, his speed growing proportionally to the growth of time; and of this motion I demonstrate conclusively many facts: I then add that, if experience should show that such facts happen to verify in the motion of bodies naturally falling, we could unmistakingly affirm this to be the same motion that was assumed and defined by me; in case it shouldn't, my demonstrations built on my assumption would lose nothing of their strength and conclusiveness; so much so nothing would compromise the conclusions reached by Archimedes about the spiral, given in nature no body that in such a spiral fashion moves.[53] (Galileo 1906, XVII, pp. 90–91)

Both quotes show the "Platonism" that Koyré placed at the center of his reading of Galileo; however, we need to agree on what this "Plato's philosophy" is: it is not Platonic metaphysics, nor a mystical mathematicism or even "political science"; there come into play an ontology and an epistemology that Plato could not produce as they are, but that Koyré, so to speak, "derives" and attributes to him (See Crapanzano 2014). At the core, we find the primacy of ideas or of theory over practice, the reality ontologically written in mathematical signs and the suitable method (*ex suppositione* and geometric) to read it. Experimentation, to Koyré, is there and remains

che qualche corpo o punto si muova all'ingiù o all'insù con la nota proporzione ed orizzontalmente con moto equabile. Quando questo sia io dico che seguirà tutto quello che ha detto il Galileo, ed io ancora. Se poi le palle di piombo, di ferro, di pietra non osservano quella supposta proporzione, suo danno, noi diremo che non parliamo di esse" (Torricelli 1919–1944, p. 357).

[52] Cf. Rossi [1962] 2002, p. 124. For Rossi, the ideas expressed in the two excerpts of Galileo and Torricelli would unequivocally demonstrate the Galilean Platonism that Koyré underlined. Lanfranco Belloni, commenting on these very two quotations, announces he does not want to go into the question of Galileo's Platonism, but notes: "Torricelli non riprende semplicemente la posizione di Galileo, rafforzandola o rendendola più esplicita [e risulta] che egli ponga l'accento sul valore 'matematico' della dottrina galileiana, e che resti invece indifferente alla questione della sua rispondenza alla realtà fisica. [...] A mio parere, la posizione del Torricelli non andrebbe genericamente etichettata come 'platonismo'. Parlerei, semmai, di un certo *Bellarminismo* relativo alla dottrina dei moti, nella misura in cui egli non fa altro che estendere a questa le cautele che il Bellarmino consigliava nella trattazione dell'ipotesi copernicana. Per cui tutta la scienza galileiana, non solo l'eliocentrismo, doveva essere prudentemente considerata una pura ipotesi matematica, un mero discorso *ex suppositione*" (Belloni 1975, pp. 31–32). In my opinion, the misconception remains the one related to the significance of Galilean "Platonism." Belloni understands it in an ontological sense and is not mistaken; however, there are also other aspects, such as, for example, Platonism as apriorism, i.e., for Galileo, a priority of theory over practical experience.

[53] Galileo declares that he researched *ex suppositione*, that is, by figuring "un moto verso un punto il quale partendosi dalla quiete vada accelerandosi, crescendo la sua velocità con la medesima proporzione con la quale cresce il tempo; e di questo moto io dimostro concludentemente molti accidenti: soggiungo poi, che se l'esperienza mostrasse che tali accidenti si ritrovassero verificarsi nel moto dei gravi naturalmente discendenti, potremmo senza errore affermare questo essere il moto medesimo che da me fu definito e supposto; quando che no, le mie dimostrazioni, fabbricate sopra la mia supposizione, niente perderanno della loro forza e concludenza; sì come niente pregiudica alle conclusioni dimostrate da Archimede circa la spirale, il non trovarsi in natura mobile che in quella maniera spiralmente si muova" (Galileo 1906, XVII, pp. 90–91).

fundamental for the physical knowledge but does not have any priority in the genesis of scientific thought as it does in Galileo's theories.[54] Then, it is perhaps appropriate to consider the issue as a whole by recognizing how Galileo, Torricelli, and Viviani are at least epistemologically "in tune": when Galileo states that the deviation of experimental results from what expected does not affect much a well-structured theory; when Torricelli confirms that we can omit the reference to observations contradicting a theoretical framework, we should not be surprised that Viviani had pushed a little further by making up a fact (without much of an effort, given the abundance of similar experiments in the context of the scientific culture of the period, which he was acquainted with) to celebrate the grand master in the best way possible, that is, by corroborating his stance with an experiment that publicly disavowed Aristotelianism.

4.4 Conclusion

At this point, it should not be difficult to understand one last thing: if the alleged Galilean experiments from the Tower of Pisa are no longer of great importance for the physicists nor, to some extent, for the historians of science, different is the story for those who have, like us, tried to address the issue from Koyré's point of view. In the evolution of his thought, in effect, the although brief analysis of the experiment at the leaning tower is one of the first signs of the prosperous and controversial thesis on Galileo's Platonism that will emerge in full force in his *Galileo Studies*. For Koyré, disproving the experiments reported by Viviani is indeed functional to the affirming of a perspective on the genesis of scientific theories and thus is relevant both to an "internal" perspective (in support to Galilean Platonism) and to an "external" one, that is, the reconstruction and understanding of his thought.

[54] In this we can find a partial identity of views with what Einstein stated in the foreword to the *Dialogue Concerning the Two Chief World Systems*: "It has often been maintained that Galileo became the father of modern science by replacing the speculative, deductive method with the empirical, experimental method. I believe, however, that this interpretation would not stand close scrutiny. There is no empirical method without speculative concepts and systems; and there is no speculative thinking whose concepts do not reveal, on closer investigation, the empirical material from which they stem. To put into sharp contrast the empirical and the deductive attitude is misleading, and was entirely foreign to Galileo " (Einstein [1953] 2001, p. xxviii). Indeed, as Gérard Jorland stressed: "Alexandre Koyré n'a nullement nié le rôle de l'expérience dans la science classique, il s'est. attaché tout au contraire à en comprendre le statut ontologique et épistémologique. Parce que l'expérimentation a pour fonction indispensable de montrer comment le modèle théorique, mathématique, pouvait trouver à se réaliser dans le cours des choses, il en a conclu que les principes sur lesquels repose ce modèle ne pouvaient pas être tirés de l'expérience et que c'était elle, à l'inverse, qui les présuppose […]. Ces principes devaient être posés a priori et c'est en ce sens que la science galiléenne fut une science archimédienne" (Jorland 1981, p. 308). To better appreciate this reading of Koyrean epistemology, see Jorland 1981, pp. 307–310.

References

Adler CG, Coulter BL (1978) Galileo and the tower of Pisa experiment. American Journal Physics 46/3:199–201.
Aristotle (1991a) Fisica. Opere. Vol. 3. Translated by Russo A. Laterza, Roma–Bari, pp. 3–238.
Aristotle (1991b) De Caelo. Opere. Vol. 3. Translated by Longo O. Laterza, Roma-Bari, pp. 241–363.
Baliani GB (1638) De Motu Naturali Gravium Solidorum. Typographia Mariae Farroni, Genuae.
Belloni L (1975) Foreword. In Torricelli 1919–1944, pp. 9–42.
Benedetti G (1585) Diversarum Speculationum Mathematicarum et Physicarum. Haeredem Nicolai Bevilaquae, Torino.
Cabeo N (1646) In Quatuor Libros Meteorologicorum Aristotelis Commentaria, et Quaestiones Quatuor Tomis Compræhensa. Two tomes. Typis Hæredum Francisci Corbelletti, Roma.
Camerota M ([2004] 2006) Galileo Galilei e la cultura scientifica nell'età della Controriforma. 2 vols. Il Giornale, Milano.
Camerota M, Helbing MO (2000) Galileo and Pisan Aristotelianism: Galileo's 'De motu antiquiora' and the 'Quaestiones de motu elementorum' of the Pisan professors. Early Science and Medicine V/4:319–366.
Caverni R (1895) Storia del metodo sperimentale in Italia. Vol. 4. Civelli, Firenze.
Cooper L (1935) Aristotle, Galileo, and the Tower of Pisa. The Cornell University Press, Ithaca–New York.
Crapanzano F (2014) Koyré, Galileo e il 'vecchio sogno' di Platone. Olschki, Firenze.
Crapanzano F (2017) Per quelle confuse carte… The Galilean De motu in Raffaello Caverni's Reading. In Pisano and Bussotti (eds), Philosophia Scientiae 21/1:17–34.
Crease RP (2003) The Legend of the Leaning Tower. Physics World 16/2:15
Crombie AC ([1956] 1996) Galileo Galilei: A Philosophical Symbol. In Crombie AC, Science, Art and Nature in Medieval and Modern Thought. Hambledon Press, London–Rio Grande, pp. 257–262.
De Gubernatis A (1909) Galileo Galilei. Le Monnier, Firenze.
Drake S ([1980] 2001) Galileo. A very short introduction. Oxford University Press, Oxford.
Drake S ([1989] 2001) History of Free Fall: Aristotle to Galileo. Wall & Emerson, Toronto.
Drake S (1970) Galileo studies: Personality, Tradition and Revolution. The University of Michigan Press, Ann Arbor.
Einstein A ([1953] 2001) Foreword. Galileo, Dialog Concerning the Two Chief World Systems: Ptolemaic and Copernican. Translated by Drake S. Introduction by Heilbron JL. Modern Library, New York, pp. xxiii–xxix.
Fahie JJ (1903) Galileo, his Life and Works. Murray, London.
Fahie JJ (1921) The Scientific Works of Galileo. History and Method of Science. Vol. 2. Singer C (ed). The Clarendon Press, Oxford, pp. 204–284.
Favaro A ([1910] 1988) Galileo Galilei [only the first 32 pages]. In Galluzzi P (ed). Galileo, La sensata esperienza. Amilcare Pizzi, Milano, pp. 230–235.
Favaro A (1887) Sul giorno della nascita di Galileo. Miscellanea galileiana inedita. Antonelli, Venezia, pp. 9–17.
Favaro A (1915) Sulla veridicità del "Racconto istorico della Vita di Galileo" dettato da Vincenzio Viviani. Archivio Storico Italiano 278/1:323–380.
Finocchiaro MA (1980) Galileo and the Art of Reasoning. Rhetorical Foundation of Logic and Scientific Method. Springer, Dordrecht.
Galilei G (1890–1909) Le opere di Galileo Galileo: Edizione nazionale sotto gli auspici di sua maestà il re d'Italia. 20 vols. Favaro A (ed), Barbèra, Firenze
Galilei G ([1974] 1989 Two New Sciences. Drake S (ed). Wall & Emerson, Toronto.
Galluzzi P (1988) La fondazione galileiana della moderna scienza del movimento. Una vicenda esemplare. In Galluzzi P (ed). Galileo, La sensata esperienza. Amilcare Pizzi, Milano, pp. 15–54.

Garin E (1957) Chi legga di A. Koyré … Giornale Critico della Filosofia Italiana 46/3:406–408.
Giordano G (2011) Sul contesto della giustificazione e il contesto della scoperta. In Maldonato M (ed). Fenomenologia della scoperta. Mondadori, Milano, pp. 195–204.
Heilbron JL (2010) Galileo. The Oxford University Press, Oxford–New York.
Highsmith PE, Howard AS (1972) Adventures in physics. Saunders, Philadelphia.
Jorland G (1981) La science dans la philosophie: les recherches épistémologiques d'Alexandre Koyré. Gallimard, Paris.
Koyré A ([1937] 1973) Galilée et l'expérience de Pise. A propos d'une légende. Études d'histoire de la pensée scientifique. Gallimard, Paris, pp. 213–223.
Koyré A ([1939] 1978) Galileo Studies. Harvester Press, Hassocks.
Koyré A (1935) À l'aube de la science classique. Annales de l'Université de Paris 10:540–551.
Koyré A (1936) À l'Aurore de la Science moderne. La jeunesse de Galilée. Annales de l'Université de Paris 11:32–56.
Koyré A (1937) La loi de la chute des corps. Revue philosophique de la France et de l'Étranger 62:149–204.
Koyré A (1943) Galileo and Plato. Journal of the History of Ideas 4:400–428.
Marci von Kronland JM (1639) De proportione motus seu regula sphygmica ad celeritatem et tarditatem pulsuum ex illius motu ponderibus geometricis librato absq. errore metiendam. Typis Ioannis Bilinæ, Pragæ.
Micheli G (1988) L'idea di Galileo nella cultura italiana dal XVI al XIX secolo. In Galluzzi P 1988 (ed). Galileo, La sensata esperienza. Amilcare Pizzi, Milano, pp. 163–186.
Namer É (1931) Galileo Searcher of The Heavens. Translated by Harris S. McBride & Company, New York.
Namer É (1975) L'affaire Galilée. Gallimard–Julliard, Paris.
Nèlli GBC (1793) Vita e commercio letterario di Galileo Galilei. 2 vols. S.N. [but Moucke], Losanna [but Florence].
Olschki L (1927) Galileo und seine Zeit. Niemeyer, Halle.
Riccioli GB (1651) Almagestum novum […] 2 tomes. Typographia Hæredis Victorij Benatij, Bologna.
Rossi P ([1962] 2002) I filosofi e le macchine. 1400–1700. Feltrinelli, Milano.
Segre M (1980) The Role of Experiment in Galileo's Physics. Archive for History of Exact Sciences 23/3:227–252.
Segre M (1989) Galileo, Viviani and the tower of Pisa. Studies in History and Philosophy of Science 20/4:435–451.
Settle TB (1961) An experiment in the history of science. Science 133:19–23.
Settle TB (1967) Galileo's use of experiment as a tool of investigation. In McMullin E (ed). Galileo Man of Science. Basic Books, New York, pp. 315–337.
Torricelli E (1919–1944) Opere di Evangelista Torricelli. Loria G, Vassura G. (eds). 4 Vols. Typ. Montanari G, Faenza.
Viviani V ([1654] 1907) Racconto istorico della vita di Galileo. In Galilei 1890–1909, 19, pp. 597–632.
Wohlwill E (1926) Galilei und sein Kampf für die copernicanische Lehre. Bd. 2. Leopold Voss, Leipzig.

Chapter 5
On Galileo's Platonism, Again

Mario De Caro

Abstract Several decades ago Alexandre Koyré's interpretation of Galileo as a Platonist of a specific sort was the dominant view, but today it is largely out of fashion. In this paper I argue that, if wrong regarding the experimental side of Galilean science, Koyré's interpretation was substantially correct as to its crucial ontological and epistemological components. In this light I defend the view that Galileo should be seen as an advocate of a physico-mathematical version of Platonism.

Keywords Galileo • Platonism • Scientific Revolution

5.1 The Philosophical Matrix of the Scientific Revolution: An Obsolete Question?

Until a few decades ago, it was commonly believed that the Scientific Revolution was accomplished under the aegis of Plato. Cassirer (1906, 1946), Whitehead (1925), Olschki (1927), Burtt (1932), Koyré (1939, 1943), Banfi (1949), and Crombie (1959), among others, held that view. Those authors also had no doubt that Galileo was the greatest advocate of Platonism in early modern science. In this perspective, Alexander Koyré wrote:

> I have [...] called Galileo a Platonist. And I believe that nobody will doubt that he is one. (Koyré 1943, p. 425)

In the last decades, however, very few defenses of the Platonic interpretation of Galileo have appeared in press (Shea 1972; Galluzzi 1973; De Caro 1993, 1996, 2012; Hankins 2000), since the consensus seems to be that that view has been proven false. First, some authors have proposed to bring Galileo's scientific experience back, in an exclusive way, to other philosophical or scientific traditions, like Aristotelianism (Geymonat 1957; Randall 1961; Girill 1970; Crombie 1975;

M. De Caro (✉)
Dip. di Filosofia, Università Roma Tre, via Ostiense 234, 00146 Rome, Italy

Department of Philosophy, Tufts University, Medford, MA, USA
e-mail: mario.decaro@uniroma3.it

R. Pisano et al. (eds.), *Hypotheses and Perspectives in the History and Philosophy of Science*, DOI 10.1007/978-3-319-61712-1_5

Wallace 1981, 1984, 1992a, b, 1998; Valleriani 2010) or Archimedism (Høyrup 1990; Machamer 1998b; Dollo 2003). Second, and more fundamentally, the very assumption that the Scientific Revolution, and in particular Galileo's scientific experience, can be linked to any of the classical traditions has come under attack.

A very eloquent example of this attitude is offered by Rivka Feldhay in the *Cambridge Companion to Galileo*, where she criticizes "the futility of any attempt to reduce Galileo's options to the dichotomy of a Platonic or Aristotelian discourse" (Feldhay 1998, p. 121; see also Finocchiaro 1994, 1997, pp. 335–356; Hatfield 2004). In the same *Companion* one can find an even more significant comment by Peter Machamer who is also the editor of the volume:

> From the 1930s through the 1960s much of the debate about the nature of early modern science revolved around attributing the labels of Platonism and Aristotelianism to various practitioners. By and large these Plato-Aristotle debates were centered on the concept of the proper scientific method. The general lines were that Aristotelians went back to the Posterior Analytics and experience and, therefore, experiment, whereas the Platonists made use of mathematics. So Alexandre Koyré characterized the Renaissance debate between Aristotle and Plato by claiming that if a thinker believed in the descriptive power of mathematics, he was a Platonist. (Machamer 1998a, p. 55)

This passage is interesting for various reasons. First, it alludes to the alleged obsolescence of the debate on the prevailing philosophical underpinnings of the Scientific Revolution. In this spirit, in the same article Machamer (*Ivi*, p. 56) notes that "the categories depicted by the names 'Plato' and 'Aristotle' by themselves as historical personages do not seem today to be of much help in understanding the precursorship of modern science." According to Machamer, the reason for this is twofold: on the one hand, during the late Renaissance, there were so many different forms of Platonism and Aristotelianism that it would be very problematic to derive useful historiographical categories from them and, on the other hand, even if those were useful categories from the historical point of view, they would still not be adequate to characterize Galileo's scientific experience.

The first thing to notice here is that Machamer neglects that, as noted above, both the Aristotelian interpretation and (even more) the Platonic one continue to be represented in the contemporary historiographic debate about the Scientific Revolution. Then, more interestingly, he claims that the twentieth-century debate on the philosophical matrix of the Scientific Revolution was centered on the Renaissance doctrines of the method. It is indeed true that methodological issues played a crucial role in that discussion; not less important, however, were some epistemological issues (what are the forms of scientific knowledge? What are its limitations? In what areas can we achieve certainty?) and some ontological ones (what is the structure of the natural world? What properties and things really exist?). Thus, in evaluating Koyré's characterization of the Galileo's Platonism, one should adequately consider the latter issues as well.

According to Koyré – as quoted by Machamer himself (Machamer 1998a, p. 55) – the Platonists gave a realist interpretation of mathematized science, i.e., an anti-instrumentalist one (in Machamer's words: "If a thinker believed in the descriptive power of mathematics, he was a Platonist" (*Ivi*)). Evidently, however, this is not a mere methodological characterization, because it presupposes an ontological one: if mathematics correctly describes reality, the Platonists would argue that this is because reality is inherently mathematical. And this ontological characterization immediately generates an epistemological issue: how, in general, can the human mind grasp mathematical truths and, more specifically, the mathematical truths that concern the natural world? Indeed, one of the main reasons of the devaluation of the Platonic interpretation of Galileo is the disproportionate relevance attached to the methodological level to the detriment of the ontological and epistemological ones, which often were conceptually prior (a lack which is already found in Geymonat 1957 and Girill 1970).

It should also be noted that the critics of the Platonic interpretation of Galileo's science keep referring, uncharitably, to the letter of the pioneering proposal advanced by Koyré, which is now outdated in at least one very important respect (I shall go back to this below). The second reason is that most of the scholars who today work on the Scientific Revolution have a purely historical (and sometimes even philological) approach and devalue the attempts at understanding the overall philosophical significance of that great intellectual phenomenon. Gary Hatfield for example, writes:

> Galileo's distinctive philosophical contribution to the rise of the new science was to show how one can seek to establish the appropriateness of one type of approach to natural science over its competitors without first establishing a metaphysical framework and support. (Hatfield 2004, p. 118)

This way of putting things is unconvincing. First of all, it is not true that before Galileo scientists would undertake their research only after establishing an adequate metaphysical framework (it is not the case that, say, Archimedes or Copernicus or Vesalius did science only after "establishing" the "underlying metaphysical framework and support" of their researches); consequently, this cannot have been the great contribution of Galileo; nor is it true that Galileo can take credit for understanding that, in doing science, it is much better to proceed by simply ignoring the alleged metaphysical preliminaries. In fact, as shown in all the best philosophical and scientific reflection of the last decades, the idea that we can do science without some metaphysical background is the result of a naive paleo-positivist view, and yet typically such metaphysical background is not the result of a prior rational inquiry. Some kind of underlying metaphysics, in fact, is present in the mind of all scientists and crucially influences their ways of doing science by helping to determine the scope, purpose, method, and epistemic values of reference of their investigations. So – to make an example not too distant in time – the continued adherence to a realist metaphysics in the evaluation of quantum mechanics brought Einstein to conclusions very different from those of an adamant positivist such as Niels Bohr (Jammer 1974, p. 109; Beller and Fine 1994).

This does not mean, of course, that the metaphysical background cannot be renegotiated, both for internal reasons and because the concrete scientific research can push one into entirely new directions. Actually, it was precisely in this way that some of the greatest scientific revolutions were accomplished (think, e.g., of when

Darwin realized that the metaphysical-religious background he had grown up in was keeping him from appreciating the significance of the findings he had gathered in his journey around the world).

Furthermore, the current devaluation of the Platonic interpretation of Galileo can also be explained in a distinctly philosophical way. In fact, because of their philosophical education, not a few interpreters (especially in the Anglo-Saxon world) sympathize with anti-realist epistemological conceptions such as instrumentalism, positivism, or empiricism and tend to project them onto Galilean science (see, as a meaningful example of this attitude, Drake 1978, p. 199). Then there is a further political-cultural reason: some authors are motivated by the attempt to interpret the experience of Galileo in the context of a strongly ideological reinterpretation of the Scientific Revolution, which is actually meant for their contemporaries. Thus Feyerabend (1975), founder of the "anything goes" epistemology, portrayed Galileo as a "methodological opportunist" for whom public rhetoric mattered more than experimentation, while the Dominican William Wallace (1984, 1991) advocated a rereading of Galileo's intellectual agenda such that, in the end, it would owe very much to the teachings of the Collegio Romano (an extremely influential Catholic institution). Finally, and more interestingly, the Platonic interpretation of Galileo's science faces a problem of conceptual determination. In fact, throughout the history of philosophy, the category of "Platonism" has been used all too commonly, and, as noted, also by Machamer, the late Renaissance was no exception in this respect: Cassirer (1927), for example, identified 16 different types of Platonism represented in that period. Merely defining Galileo as a Platonist would therefore be of very little interest, if one were not able to specify precisely what precise sense should be attributed to the label "Platonism" in that context.

In the last decades, many innovative historiographical and philological analyses have clarified, enriched, and sometimes even profoundly changed our ideas on the practical modalities and the cultural context in which Galileo's scientific experience developed. Now, however, is time to go back, in a much more informed way, to the old debate about the philosophical influences of the great Pisan scientist, in which Koyré played such a pivotal role.

In the following, I will focus on some of the main reasons to favor a specific Platonic interpretation of Galileo, which can help to update Koyré's groundbreaking proposal. Then, in the final paragraph, I will outline and criticize an alternative interpretation proposed by William A. Wallace, according to whom the experience of Galileo should instead be traced back to an Aristotelian-Jesuit matrix.

5.2 The Mathematization of Physics

Historiographical reconstructions may be anachronistic for two reasons. The first is unavoidable: past events are to be described in a selective way, and obviously the selection is determined by the point of view of the historian who is doing the reconstruction. The second reason is more interesting and depends on the fact that often historians make use of interpretive categories that are not contemporary with the events they aim at explaining (this happens, e.g., with Weberian "ideal types" such as "feudalism" or "Protestant ethics"). In these kinds of cases, it is necessary, but also difficult, to assess the adequacy of the anachronistic categories. In the field of history of philosophy, this is frequently the case: to argue that Plato was an ontological realist or that St. Thomas defended a theory of truth as correspondence is to use anachronistic categories, whose legitimacy has to be accurately established.

The application of the categories of "Aristotelianism" and "Platonism" to the period of the Scientific Revolution is anachronistic only in the first sense – the more harmless. Already in the late Renaissance, in fact, these two categories were used to identify two opposed theoretical options regarding the applicability of mathematics to physics. On one side stood the Platonists, who believed that such application was legitimate or rather indispensable; on the other stood the Aristotelians, who assumed the opposite position. The philosopher Jacopo Mazzoni explained very clearly the terms of that discussion in his 1597 volume *In universam Platonis et Aristotelis philosophiam praeludia*:

> There is no other question which has given place to more noble and beautiful speculations [...] than the question whether the use of mathematics in physical science as an instrument of proof and a middle term of demonstration, is opportune or not; in other words, whether it brings us some profit, or on the contrary is dangerous and harmful... It is well known that Plato believed that mathematics was quite particularly appropriate for physical investigations, which was the reason why he himself had many times recourse to it for the explanation of physical mysteries. But Aristotle held a quite different view and he explained the errors of Plato by his too great attachment to mathematics. (Mazzoni 1597; English Translation from Koyré 1943, p. 421)[1]

As said before, underlying the late Renaissance *Methodenstreit*, there were issues such as the ontological superiority of the categories of quantity and relation over those of quality and modality, the epistemic priority of mathematical deduction over syllogistic reasoning, the definition of certain knowledge, and the possibility to consider nature a "mathematical multiplicity," to use Husserl's terminology (Husserl 1929, pp. 28–36, 1936, § 9). About each of these questions, in Galileo's age it was very clear what answers a Platonist and an Aristotelian would respectively give.

Mazzoni initially wished for an irenic solution of this discussion (he held chairs both in Aristotelian and Platonic philosophy), but later he explicitly joined the Platonic party. He taught at Pisa, a university that also included some of the leading members of the Aristotelian party, such as Francesco Buonamici and Girolamo Borro, whose physical theories entirely devoid of mathematics perfectly fit

[1] "Non est enim inter Platonem et Aristotelem quaestio, seu differentia, quae tot pulchris, et nobilissimis speculationibus scateat, ut cum ista, ne in minima quidem parte comparari possit. Est autem differentia, utrum usus mathematicarum in scientia Physica tanquam ratio probandi et medius terminus demonstrationum sit opportunus, vel inopportunus, id est., an utilitatem aliquam afferat, vel potius detrimentum et damnum. Credidit Plato Mathematicas ad speculationes physicas apprime esse accommodatas. Quapropter passim eas adhibet in reserandis mysteriis physicis. At Aristoteles omnino secus sentire videtur, erroresque Platonis adscribet amori Mathematicarum" (Mazzoni 1597, p. 188).

Mazzoni's account. After having been his teacher (together with Buonamici and Borro), Mazzoni became a colleague and a good friend of Galileo. This clearly suggests that at the beginning of his scientific career the Pisan scientist was already well aware of the controversy over the mathematization of physics. Nor can we think that Galileo had doubts about which party to choose in that controversy. As a matter of fact, in a letter written on May 30, 1597, Galileo congratulates Mazzoni for having finally embraced the party of Plato – "[t]hat great Master under whose command appear to serve, and should, all those who undertake to investigate the truth" (Galilei 1597; Engl. Transl. from Finocchiaro 2010, p. 48).[2] So Galileo writes in relation to Mazzoni's book *In Universam Platonis et Aristotelis Philosophiam Praeludia*:

> It has given me the greatest satisfaction and consolation to see that Your Most Excellent Lordship has written about some of those questions concerning which in the first year of our friendship we disputed together with so much merriment, and you are now inclined to take the side which I regarded as true and you as the opposite. (Galilei 1597; English Translation from Finocchiaro 2010, p. 48)[3]

To Galileo's disappointment, however, in the *Praeludia*, Mazzoni still defends the Aristotelian-Ptolemaic astronomical system. Thus, in order to counter an argument offered by Mazzoni in favor of the geocentric model, Galileo presents a mathematical proof in defense of "the opinion of the Pythagoreans and of Copernicus" (ibidem).[4] In this regard, it has to be noted that in Galileo's writings, the term "Pythagoreans," while being used as a synonymous of "Copernicans," also has three semantic nuances that strongly suggest the natural alliance between heliocentric astronomy and Platonism in the sense defined by Mazzoni. First, in Galileo's usage the term "Pythagoreans" does not have any of the numerological and mystic connotations that were typical of the Neoplatonic tradition.[5] Second, as Galileo certainly knew, late Renaissance mathematicians (such as Francesco Barozzi) followed Proclus's *Commentary on the First Book of Euclid's Elements* in using the term "Pythagorism" as denoting the view that "mathematical entities belonged to things connected with matter" (Feldhay 1998, p. 96; see also De Pace 1993). Finally, the use of that term in the context of a discussion of astronomy clearly indicates Galileo's predilection for the realist interpretation of the Copernican theory against

[2] "Quel gran Maestro, sotto la cui disciplina pare che militino, e che così far debbano, quelli che si danno ad investigare il vero" (Galilei 1597, p. 197).

[3] "Ha egli in me in particolare arrecata grandissima soddisfazione e consolazione, nel vedere V.S. Eccellentissima, in alcune di quelle questioni che nei primi anni della nostra amicizia disputavamo con tanta giocondità insieme, inclinare in quella parte, che da me era stimata vera ed il contrario da lei" (ibidem).

[4] "La opinione dei Pitagorici e del Copernico" (Galilei 1597, p. 198).

[5] Galileo explicitly refuses as "sciocchezze" ("follies") the numerological and hermetic connotations of Pythagorism as it was interpreted by the Neoplatonists (on that, see Galileo 1632, p. 32; English Translation [by Drake] from Galilei 1953, p. 11). This is made clear also by Galileo's now well-documented aversion to astrology, often a structural component of the Neoplatonic views (Bucciantini and Camerota 2005). Moreover, Galileo was strongly influenced by his father Vincenzo's harsh criticism of Gioseffo Zarlino's numerological musical Pythagorism (Walker 1973–1974; Camerota 2004, pp. 27–30).

its instrumentalist interpretations, which famously began with Osiander's preface to the *De revolutionibus orbium coelestium*. According to Galileo, heliocentrism, instead of merely "saving the phenomena," offers us the true astronomical account; then the Earth does not have any special status in the universe – and therefore there is no reason for thinking that terrestrial physics should be in principle different from celestial physics (i.e., cosmology) regarding the relationship to mathematics. And all this clearly shows that in his early career, Galileo was already firmly aligned with the Platonic party (as it was described in Mazzoni's *Praeludia*).

More generally, if one takes the many philo-Platonic remarks by Galileo throughout his career and reads them in the light of the late Renaissance discussion on the role that mathematics should play in physics, they clearly testify to an unambiguous Platonist stance to which Galileo stayed faithful until the end. Consider, for example, the following passage from the *Dialogue Concerning the Two Chief World Systems*, in which Galileo gets the peripatetic Simplicio – the Aristotelian ridiculed throughout the book – to present the debate on the mathematicity of physical entities in terms that are very similar to those used by Mazzoni 35 years earlier. Simplicio of course defends in his usual pompous way the position of the Aristotelian party:

> I have known some very great Peripatetic philosophers, and heard them advise their pupils against the study of mathematics as something which makes the intellect sophistical and inept for true philosophizing; a doctrine diametrically opposed to that of Plato, who would admit no one into philosophy who had not first mastered geometry. (Galilei 1632; English Translation [by Drake] from Galilei 1953, p. 397)[6]

To this, Salviati, Galileo's spokesperson, responds by making fun of the Aristotelians:

> I endorse the policy of these Peripatetics of yours in dissuading their disciples from the study of geometry, since there is no art better suited for the disclosure of their fallacies. (Galilei 1632; English Translation [by Drake] from Galilei 1953, p. 397)[7]

One may wonder whether Galileo's stance on the *querelle* concerning Platonism and Aristotelianism changed at the end of his career. However, in his late masterpiece *Two New Sciences*, he asked, "Was not Plato perfectly right when he wished that his pupils should be first of all well grounded in mathematics?" (Galileo 1638, p. 137).[8] But this was only a rhetorical question to which Simplicio, the hyper-Aristotelian, had already responded few pages earlier:

[6] "Io ho conosciuti e sentiti grandissimi filosofi peripatetici sconsigliar suoi discepoli dallo studio delle matematiche, come quelle che rendono l'intelletto cavilloso ed inabile al ben filosofare; istituto diametralmente contra a quello di Platone, che non ammetteva alla filosofia se non chi prima fusse impossessato della geometria" (Galilei 1632, p. 423).

[7] "Applaudo al consiglio di questi vostri Peripatetici, di distorre i loro scolari dallo studio della geometria, perché non ci è arte alcuna piú accomodata per scoprir le fallacie loro" (Galilei 1632, p. 423).

[8] "Con che gran ragione voleva Platone i suoi scolari prima ben fondati nelle matematiche?" (Galilei 1638, p. 175; English Translation from Galilei 1914).

> If I were again beginning my studies, I should follow the advice of Plato and start with mathematics, a science which proceeds very cautiously and admits nothing as established until it has been rigidly demonstrated. (1638, pp. 90–91)[9]

It may be useful to emphasize that Galileo's physico-mathematical Platonism was quite different from the mathematical Platonism of contemporary analytic philosophy, which is exemplified by the conceptions of Gottlob Frege, Kurt Gödel, and W.V. Quine. Present-day mathematical Platonism concerns mathematical entities, such as numbers or sets, and attributes three properties to them: abstractness, existence, and ontological independence from the minds that know them (Linnebo 2013). Galileo's physico-mathematical Platonism concerns physical entities, conceived of as inherently mathematical – more precisely, geometrical – and besides attributing to them the properties of existence and independence, it adds that of concreteness. It is the essential mathematicity of the concrete entities studied by physics that, according to Galileo, ensures the applicability of mathematics to that science.[10]

According to an objection by Girill (1970), which is worth mentioning here, conceiving physical entities as intrinsically mathematical should not be considered a form of Platonism. This is, claimed Girill, because Plato did not defend at all such a position: for him, in fact, mathematical entities have a preternatural character and are therefore entirely separate from the physical world – which is what, instead, matters for Galileo. This objection sounds plausible, but it is legalistic. Even if it were true, as Girill believes, that Plato never directly defended the mathematization of physics, this would not prove that considering Galileo's view as a form of Platonism is historically illegitimate. Analogously, one can legitimately claim that Aquinas held an Aristotelian conception of the individual soul, even if (differently from Aristotle) he believed in its immortality. Platonism and Aristotelianism – as much as, say, Epicureanism, Kantianism, and Marxism – are philosophical categories, not philological ones. They emphasize some crucial aspects of the thought of the respective eponymous philosophers while neglecting other aspects (which, of course, may still be important in themselves). In judging the appropriateness of these categories, the parameter to be considered is not their exegetic correctness but their explicative fertility.[11]

---

[9] "Se io avessi a ricominciare i miei studii, vorrei seguire il consiglio di Platone e cominciarmi dalle matematiche, le quali veggo che procedono molto scrupolosamente, né vogliono ammetter per sicuro fuor che quello che concludentemente dimostrano" (Galilei 1638, p. 49).

[10] In this regard it is interesting to look at the few Jesuits of the Collegio Romano who defended mathematical Platonism (Baldini 1992): in their case, the analogy with today's mathematical Platonists seems more appropriate than for Galileo. Giuseppe Biancani, for example, asserted the existence of mathematical entities, as archetypes of all things, both in the divine and in the human mind (Feldhay 1998, p. 99). Biancani's Platonism was different from Galileo's physico-mathematical Platonism, because it did not concern physical entities. At any rate, the fact that some of the mathematicians of the Collegio Romano may have sympathized for the Platonic party does not call into question the legitimacy of dividing late Renaissance scientists into the two opposite parties of Platonism and Aristotelianism – what should be questioned, instead, is the idea that all Renaissance Jesuits were Aristotelians.

[11] In this perspective, establishing how much direct acquaintance Galileo had with Plato's *Dialogues* is interesting, but not decisive. However, Hankins (2000) argued convincingly that an important mediation was played by Ficino, who Galileo certainly read.

5.3 Plato and Archimedes

As said, when he aligned himself with the Platonic party, Galileo was referring to the physico-mathematical version of Platonism defined above, which had methodological, ontological, and epistemological components. As to the methodological point of view, one should notice that the famous interpretation given by Koyré, according to which Galilean science was essentially a priori, without any serious experimental component, has been definitively refuted. Nowadays, historians have no doubts that Galileo performed experiments and very accurate measurements (Drake 1978; Machamer 1998b; Palmieri 2008) and dealt with lots of engineering problems (Valleriani 2010). And he did so because he thought that, while any mathematical theory that does not "save the appearances" – one that does not offer correct predictions and consistent explanations in a determinate field – is empirically false, at most one of the theories that save the appearances can be empirically true (Palmerino 2005, 44).[12] Thus, the physicist has to devise experiments to see which among the adequate theories is the true one.

Moreover, even if one believes (as I think one should) that Galileo's scientific career developed in a Platonic framework, one should not ignore that in his training he was influenced, to varying degrees, also by other traditions of ancient science and philosophy.[13] In particular, in the last decades, many interpreters have insisted on the role played in the Scientific Revolution, and specifically in Galilean science, by a specific tradition, i.e., Archimedism. Some of Archimedes's scientific texts had been available as early as in the twelfth century, but they could offer a decisive stimulus to the development of science only when appropriated by mathematicians as illustrious as Francesco Maurolico, Federico Commandino, Guidobaldo del Monte, and Girolamo Cardano, who were able to appreciate their extraordinary scientific achievements and to develop the Archimedian intuitions regarding mechanics, hydrostatics, and the method of exhaustion (Clagett 1964–1984; Laird 1991). Therefore, late Renaissance thinkers looked back to Archimedes with very different eyes from both medieval scholars (according to whom he had represented primarily the embodiment of moral virtue) and humanists (for whom, instead, Archimedes was the most brilliant and engineer architect; Dijksterhuis 1987). Galileo, in particular, made Archimedes the tutelary deity of his own science,

[12] On the notion of "saving the appearances" and for an interesting defense of the idea that the Greek astronomers' epistemological and ontological views were different from those of the Renaissance instrumentalists, such as Osiander and Ursus, see Lloyd 1987.

[13] Galluzzi (2011, p. 30) rightly notes how the Galilean project was based on the strategic convergence, in a Platonic and anti-Aristotelian perspective, of Archimedes, Democritus, the atomists, and Copernicus. But probably the ancient influences on Galileo were even wider, including also Euclid, Galen, and Pappus (Shea 1972; Wisan 1978; Redondi 1983; Camerota 2008). Finally, some interpreters have noted the role played by the Hellenistic and medieval anti-Aristotelian theory of *impetus* (Clagett 1964–1984; Shapere 1974) (but, in my view, the relevance of this theory for Galileo's mature science should not be overstated). All this notwithstanding, it was the physico-mathematical Platonism discussed above that offered Galileo the general framework that made it possible to synthetize all these disparate contributions in a successful way.

calling him “the greatest and superhuman intellect” (Galilei 1637, p. 250), and was constantly inspired by his writings.

In the light of these facts, some interpreters have argued that Galileo should be labeled as an Archimedean, not as a Platonist (Høyrup 1990; Machamer 1998b; Dollo 2003). However, if superficially plausible, this proposal loses sight of a fundamental aspect of the matter. Archimedes provided modern physicists with a new method, but for the modernization of science to take place, that method had to be contextualized in an appropriate general ontological and epistemological context. In fact, the Archimedean methodology would, in itself, be compatible with an instrumentalistic interpretation of science – which Galileo and the other protagonists of the Scientific Revolution strongly refused. In order to get a realistic interpretation of the new science, the Archimedean methodology had to be framed in the epistemological and ontological framework of physico-mathematical Platonism: only in that way modern science could overthrow the bastions of the premodern conception (the distinction between the superlunary and the sublunary worlds, the primacy of quality over quantity, the rejection of atomism, the inconceivability of the universal laws of mathematical nature).[14]

In Galileo’s works one can find several explicit defenses of Platonic epistemology. The most famous is an explicit reference to the cognitive model based on reminiscence presented in the *Meno*, which is summarized in the formula “*Nostrum scire sit quoddam reminisci*” (Galilei 1632, p. 244). Although some critics have interpreted Galileo’s mention of that formula as a merely rhetorical device (Feldhay 1998, pp. 120–121),[15] this is too explicitly a profession of Platonism to be ignored or underestimated. Not surprisingly, in fact, in Galileo the hypothesis of “reminiscence” assumes (in line with the *Meno*) a proper mathematical significance:

> That the Pythagoreans held the science of the human understanding and believed it to partake of divinity simply because it understood the nature of numbers, I know very well; nor am I far from being of the same opinion. (Galilei 1632; English Translation [by Drake] from Galilei 1953, p. 11)[16]

Galileo is also explicit in his reliance on a Platonic ontology – and more precisely on the physico-mathematical ontology of the tradition that began with the *Timaeus*

---

[14] Noticeably, the affinity of Archimedean science with Platonic philosophy (especially that derived from Plato’s *Meno* and *Timaeus*) was already clear to the ancients, as it is shown by the fact that the doxographic tradition considered Archimedes a “Philosophus Platonicus” (Koyré 1968, p. 38, n. 3).

[15] According to Feldhay, in his personal theory of reminiscence, Galileo – unlike Plato – attached great importance to prior sensory experience and imagination of everyday objects and situations. But, again, in this way the main point of contention gets lost. The point, of course, is not whether Plato was a proto-Galilean but that, according to Galileo, in order to produce scientific knowledge, sensory experience and imagination have to be examined with mathematical rigor. And this view could be developed only within the physico-mathematical tradition derived from Plato and Archimedes.

[16] “Che i Pittagorici avessero in somma stima la scienza de’ i numeri, e che Platone stesso ammirasse l’intelletto umano e lo stimasse partecipe di divinità solo per l’intender egli la natura de’ numeri, io benissimo lo so, né sarei lontano dal farne l’istesso giudizio” (Galilei 1632, p. 35).

(deprived of its vitalistic components), according to which the natural world is made up only of geometric principles and entities (Palmerino 2005). This is evident in Galileo's most famous quote:

> Philosophy is written in this grand book, the universe, which stands continually open to our gaze. But the book cannot be understood unless one first learns to comprehend the language and read the letters in which it is composed. It is written in the language of mathematics, and its characters are triangles, circles, and other geometric figures without which it is humanly impossible to understand a single word of it; without these, one wanders about in a dark labyrinth. (Galilei 1623; English Translation from Drake 1957, pp. 237–238)[17]

This ontological characterization – which also made it possible to incorporate the atomistic doctrine in Galilean physics, as shown by Hankins (2000, p. 213), Camerota (2008) and Galluzzi (2011) – lead Galileo to support a profoundly anti-Aristotelian form of scientific realism, in which not only is mathematics real as an object of thought but it structures reality through the ubiquitous and inescapable laws of nature (Stabile 2003).

In Galileo's science, the correlation between the ontological, the epistemological, and the methodological levels is inextricable. The world has an inherently mathematical structure that we can penetrate with certainty because our intellect, applied to mathematical questions, may "intensively" equal the divine one.[18] Only in this ontological and epistemological framework (which is unequivocally Platonic in spirit) could the Archimedean method – centered on experiments made rigorous by mathematization – flourish.

5.4 Galileo and the *Ex Suppositione* Method

Among the various attempts to refute the Platonic interpretation of Galilean science, one of the most significant and ingenious was certainly that operated by William A. Wallace (1984, 1992a, b, 1998). According to Wallace, Galileo was deeply influenced by the study of the texts of the Jesuits of the Collegio Romano, which were marked by Aristotelianism, and – at least in a specific sense (a crucial one, though) – he always remained an Aristotelian. Wallace proposes many

[17] "La filosofia è scritta in questo grandissimo libro che continuamente ci sta aperto innanzi agli occhi (io dico l'universo), ma non si può intendere se prima non si impara a intendere la lingua, conoscere i caratteri ne' quali è scritto. Egli è scritto in lingua matematica, e i caratteri son triangoli, cerchi ed altre figure geometriche, senza i quali mezzi è impossibile intendere umanamente parola; senza questi è un aggirarsi vanamente per un oscuro labirinto" (Galilei 1623, p. 232).

[18] Galilei (1632, pp. 128–130; English Translation from [by S Drake] Galilei 1967, pp. 103–105). In this sense, it is crucial to consider the answer that Galileo gives to an important Aristotelian objection: how can one apply the principles of geometry to physical reality, considering that in the natural world, one cannot find the perfect solids and figures one finds in geometry? According to Galileo, it is up to the scientist "to make a true accounting" ("fare i calcoli giusti"), i.e., to measure precisely the difference between the physical bodies and the ideal ones, since "whatever form" a natural body has, it has that "perfectly" (Galilei 1632, pp. 234–240; English Translation [by Drake] Galilei 1967, pp. 207–212).

arguments to substantiate his thesis, but here I will analyze the four more interesting ones: (1) an epistemological argument, (2) an ontological argument, (3) a methodological argument, and (4) a textual argument based on a statement in which Galileo allegedly admits his adherence to Aristotelianism.

According to the epistemological argument, inspired by Crombie (1975), Galileo, with the crucial mediation of the Jesuits of the Collegio Romano, appropriated the ideal of science as true and certain knowledge of the natural world developed in Aristotle's *Posterior Analytics* and always remained faithful to that ideal. However, differently from Crombie, Wallace claims that Galileo was an Aristotelian also regarding his ontology and his methodology. More specifically, Wallace's ontological argument is that Galileo defended a "common sense realism" inspired by the Aristotelianism of the Jesuits:

> [Galileo] thought that the world exists independently of his thinking about it, that objects in nature are real, that as presented in sense experience they can be known, and that the natural light of intellect is adequate to the task of knowing them as they are. (Wallace 1992a, p. 302)

Then Wallace presents a methodological argument, which is based on an extensive analysis of an early work by Galileo, the "Manuscript number 27," which includes two youthful treatises (one devoted to pre-knowledge, the other to demonstration), largely inspired by the logical and methodological conceptions of the Jesuits of the Collegio Romano. Those two treatises were not included in the National Edition of Galileo's works, because its curator, Antonio Favaro, considered them mere scholastic exercises. According to Wallace, instead, they date back to a period between 1589 and 1591 (when Galileo was between 25 and 27 years old and was already teaching mathematics in Pisa) and are of considerable historical interest since they include a programmatic exposition of Galileo's scientific method. Consequently, Wallace claims, the other interpreters, having neglected those precious early intellectual experiences, have lost a crucial key to understand Galileo's thought: "Parvus error in initio magnus in fine" ("A little error at the beginning, a big one at the end"; Wallace 1998, p. 27). In particular, the crucial methodological element that Wallace sees in Manuscript 27 is its reference to the so-called *ex suppositione* method, which the Jesuits had taken from the Paduan Aristotelianism. That method, Wallace claims, marked Galileo's entire scientific career.

In the Aristotelians' view, the object of physical sciences, being subject to constant change – i.e., being contingent – could always be different from how it is. But there may be science (i.e., certain knowledge) only of what does not change – that is, of what is necessary. Then a major methodological problem arises: is there a method that can guarantee us the possibility to know nature scientifically? The *ex suppositione* method, according to its proponents, has this prerogative: since it can go back up to the true causes of natural phenomena, it makes it possible to identify the necessary connections between the causes and their effects.

In the philosophical literature of the Middle Ages, recalls Wallace (Wallace 1974), a typical example of demonstration *propter quid* (i.e., from effects to causes) based on the *ex suppositione* method concerned the appearance of the rainbow. This method does not allow one to predict that at a given moment, in a certain part of the sky, a rainbow will appear: however, it allows one to know with certainty that *if* a rainbow appears (*suppositio*), *then* this will necessarily be due to the reflection and refraction of light rays. Every time we have to give an account of the appearance of a rainbow, we can refer to this explanation, which shows the real causes of that phenomenon. The outline of this argument is evidently that of the hypothetical syllogism known as modus ponens: as a form of reasoning, then, it is certainly valid. The first premise of the syllogism is a law that goes back from effects to causes ("if there is a rainbow, then there is the refraction of light rays"); the second is the assumption that the antecedent of this law is actually true ("there is the rainbow"). Finally, in the conclusion of the syllogism, which follows necessarily from the premises, we get the cause of the phenomenon, which is therefore knowable.

According to Wallace, Galileo remained faithful to a distinctly Aristotelian method such as the *ex suppositione* reasoning since his youth up to the *Discourses and Mathematical Demonstrations Relating to Two New Sciences* (1638), his scientific masterpiece, and even beyond. However, Wallace grants that the Galilean method had at least one original trait, namely, that the first premise of the hypothetical syllogism that is generated by the application of the *ex suppositione* method, i.e., the physical law, should generally be demonstrated mathematically.

Finally, in support of his Aristotelian interpretation of Galileo's science, Wallace presents a textual argument by referring to some passages from a letter that in 1640 Galileo, who then was 77 years old, wrote to Fortunio Liceti. In that letter, Galileo allegedly discloses his fully conscious and lifelong adherence to the principles and methodology of Aristotelianism. "I am glad to hear," Galileo writes, "that Your Excellency, like many others, believes that I am adverse to the peripatetic philosophy, since this gives me the opportunity... to claim that I am an admirer of the great man that Aristotle was."[19]

Summarizing, according to Wallace, Galileo's entire intellectual journey (from when he was studying the texts of the mathematicians of the Collegio Romano until his late maturity, when he wrote the letter to Liceti) was internal to the Aristotelian paradigm. Wallace arguments, however, are much weaker than they may look at first sight. First, with respect to the issue of full scientific certainty, one may mention again Galileo's view according to which, within the field of mathematics, humans can come to know with "absolute certainty" and "necessity," to the point of matching (albeit only *intensive*) divine knowledge:

> Extensive, that is in respect of the multiplicity of things to be known, which is infinite, the human mind is as nothing (even if it understood a thousand propositions, because a thousand compared with infinity is like zero): but taking the understanding, in so far as this term means to grasp intensely, that is, perfectly a given proposition, I say that the human mind understands some propositions as perfectly and has of them as absolute certainty as Nature herself can have; and of that kind are the pure mathematical sciences, that is, geometry and arithmetic, of which the divine intellect knows of course infinitely more propositions, for

[19] "Mi giunge grato il sentire che V.S. Eccel.ma insieme con molti altri, sì come ella dice, mi tenga per avverso alla peripatetica filosofia, perché questo mi dà occasione… di mostrare quale io internamente sono ammiratore di un tanto uomo, quale è Aristotele" (Galilei 1640, p. 247).

> the simple reason that it knows them all; but as for those few understood by the human intellect, I believe that its knowledge equals the divine in objective certainty, because it succeeds in understanding their necessity, beyond which it does not seem that there can exist a greater certainty. (Galilei 1632; English Translation [by Drake] from Galilei 1953, p. 114)[20]

Therefore, Galileo's model of the certainty lies in mathematics, and particularly in Euclidean geometry, not in the Aristotelian science of *Posterior Analytics*.

Also Wallace's claim that Galileo accepted the ontological and epistemological views of Jesuit Aristotelianism throughout the course of his scientific career lends itself to a very convincing objection. Indeed, with very few exceptions (such as Drake 1978), most interpreters agree that Galileo had a realist, non-positivistic view of science and consequently believed that science describes the natural world as it is. However, the kind of realism to which Galileo adhered was quite different from Aristotle's realism. More specifically, Galileo did not recognize any fundamental epistemological value to the testimony of common sense as such, as the Aristotelians instead did. He rather defended a rigorous mathematical realism that relegated secondary properties in the context of subjectivity and credited as real only primary properties. Since, according to Galileo secondary qualities do not have any ontological consistency, there cannot be any scientific knowledge of them. This is evident, for example, if we read two famous passages from *The Assayer*:

> I say that whenever I conceive any material or corporeal substance, I immediately feel the need to think of it as bounded, and as having this or that shape; as being large or small in relation to other things, and in some specific place at any given time; as being in motion or at rest; as touching or not touching some other body; and as being one in number, or few, or many. From these conditions I cannot separate such a substance by any stretch of my imagination. But that it must be white or red, bitter or sweet, noisy or silent, and of sweet or foul odor, my mind does not feel compelled to bring in as necessary accompaniments. Without the senses as our guides, reason or imagination unaided would probably never arrive at qualities like these. (Galilei 1623; English Translation from Drake 1957, p. 274)[21]

---

[20] "L'intendere si può pigliare in due modi: cioè intensive, o vero estensive: e che estensive, cioè quanto alla moltitudine degli intelligibili, che sono infiniti, l'intendere umano è come nulla, quando bene egli intendesse mille proposizioni, perché mille rispetto all'infinità è come uno zero; ma pigliando l'intendere intensive, in quanto cotal termine importa intensivamente, cioè perfettamente, alcuna proposizione, dico che l'intelletto umano ne intende alcune così perfettamente, e ne ha così assoluta certezza, quanto se n'abbia l'istessa natura; e tali sono le scienze matematiche pure, cioè la geometria e l'aritmetica, delle quali l'intelletto divino ne sa bene infinite proposizioni di più, perché le sa tutte, ma di quelle poche intese dall'intelletto umano credo che la convinzione agguagli la divina nella certezza obiettiva, poiché arriva a comprenderne la necessità sopra la quale non par che possa essere sicurezza maggiore" (Galilei 1632, pp. 128–129).

[21] "Io dico che ben sento tirarmi dalla necessità, subito che concepisco una materia o sostanza corporea, a concepire insieme che ella è terminata e figurata di questa o quella figura, ch'ella si muove o sta ferma; ch'ella tocca o non tocca un altro corpo, ch'ella è una, poche o molte, né per veruna immaginazione posso separarla da queste condizioni; ma ch'ella debba essere bianca o rossa, amara o dolce, sonora o muta, di grato o ingrato odore, non sento farmi forza la mente di doverla apprendere da cotali condizioni necessariamente accompagnata: anzi, se i sensi non ci fussero scorta, forse il discorso o l'immaginazione per sé stessa non v'arriverebbe già mai" (Galilei 1623, pp. 347–348).

> I think that tastes, odors, colors, and so on are no more than mere names so far as the object in which we place them is concerned, and that they reside only in he consciousness. Hence if the living creature were removed, all these qualities would be wiped away and annihilated. (Galilei 1623; English Translation [by Drake] from Galilei 1957, p. 274)[22]

In general, Galileo advocated a complete overturning of the Aristotelian categories, by giving priority to quantity and relation to the detriment of quality and modality. In this view, *pace* Wallace, the "*sensate esperienze*" ("sensible experiences") that Galileo mentions as crucial in his scientific method, are not the experiences of everyday life, as it was for the Aristotelians, but the observations, experimentations, and thought experiments (Massimi 2010, 176) carried out with mathematical rigor: only the latter experiences can lead one to understand the real properties of things, i.e., the mathematical ones. By appealing to those experiences, the scientist can determine which of the theories that have been proven true in abstract, through certain mathematical proofs ("*certe dimostrazioni*"), are true also of the natural world.[23]

As to the alleged relevance of manuscript 27 for understanding Galileo's methodology, on which Wallace insists so much, one cannot fail to observe that (beyond the controversial issues on the dating of the text) it does not incorporate the most original acquisitions of Galileo's later views. In the two treatises it contains, for example, there is no mention of the mathematization of physics, while the doctrine of the four causes, which Galileo would later abhor, is taken for granted. If, then, Favaro may have been mistaken in excluding that work from the National Edition, nevertheless, he was not wrong in considering it as a scholastic exercise in which Galileo was not ready to present the basic rules of his science yet.

Moreover, going into the details of the alleged extraordinary potential of the *ex suppositione* method, it is not difficult to see that, *pace* Wallace, this method is not able to go with certainty from effects to causes. Once one has forsaken the metaphysical guarantees on which premodern science was based, there is no way to prove that a specific phenomenon is the cause of another specific phenomenon – that is, there is no way to formulate a law that, ascending from effects to causes, could hold as the major premise of the hypothetical syllogism that constitutes the *ex suppositione* method. To return to Wallace's example, in reality we can never be certain that a specific rainbow was not caused by a phenomenon unknown to us, and this proves that we do not have certain knowledge of rainbows (and indeed, unlike Galileo, today we can even concretely imagine a non-standard cause of rainbows, e.g., a beam of light cast by powerful projectors). The *ex suppositione* method,

---

[22] "Per lo che io vo pensando che questi sapori, odori, colori, etc., per la parte soggetto nel quale ci par riseggano, non sino altro che puri nomi, ma tengano solamente lor residenza nel corpo sensitivo, sì che rimosso l'animale, sieno levate ed annichilite tutte queste qualità" (Galilei 1623, p. 348).

[23] On the interplay between *sensate esperienze* and *certe dimostrazioni*, see Galileo 1613.

therefore, cannot give us any certainty, differently from mathematical deduction (which, as said, was the only true paradigm of certainty, according to Galileo).[24]

Finally, there is another difficulty, candidly admitted by Wallace himself. He grants that, according to the dictates of Galileo, in order to know the law that should serve as the major premise of the hypothetical syllogism underlying the *ex suppositione* method, the contribution of mathematics is essential. But this is no small thing, because this is exactly where the physico-mathematical Platonism plays a crucial role. In the light of these considerations, Galilean science can safely be interpreted as a sharp break from peripatetic science.

Finally, let's consider Wallace's textual argument. According to it, in the early days of his career, Galileo noted in a brief, private manuscript the principles that would have guided his entire scientific career, until he finally decided to declare his uninterrupted adhesion to Aristotelianism in a personal letter written in the very last years of his life. This reconstruction wobbles in many respects. First, it is very implausible from a psychological standpoint: if it were true that Galileo remained faithful to the Aristotelian method disclosed by the Jesuits at the Collegio Romano, it would be quite mysterious why he did not make that clear earlier, especially in the period in which he was being fiercely attacked by the Aristotelians and the Jesuits – to the point of losing his freedom and perhaps even risking his life. Quite the contrary, Galileo was very explicit in saying that he did not recognize himself at all in the syllogistic Aristotelian methodology. In the *Dialogue*, for example, he makes Sagredo say: "I remember that when I studied logic, I was never entirely convinced by this much-vaunted powerful demonstration of Aristotle" (Galilei 1632; English Translation [by Drake] from Galilei 1967, p. 276).[25]

Moreover, if one reads carefully the rest of the letter to Liceti on which the whole interpretation by Wallace rests, one clearly sees that the reference to Aristotle is nothing more than a *captatio benevolentiae*, through which Galileo (old, blind, sick, and held under house arrest for years) was trying to get the sympathy of his enemies, who were present in large numbers in the Aristotelian party. Galileo writes:

> I consider (and I believe you do too) that to be truly a peripatetic – that, an Aristotelian philosopher – consists principally in philosophizing according to Aristotelian teachings, proceeding from those methods and those true suppositions and principles on which scientific discourse is founded, supposing the kind of general knowledge from which one cannot deviate without great disadvantage. Among these suppositions is everything that Aristotle teaches us in logic, pertaining to care in avoiding fallacies in discourse, using reason well so as to syllogize properly and deduce from the conceded premises the necessary conclusion, and all this teaching relating to the form of arguing correctly. As to that one, I believe

---

[24] The method that infers a physical effect from its cause in accordance with the laws of nature (which is the true Galilean method) is much more solid than the *ex suppositione* method. Of course, also in this way we may err in explaining or predicting phenomena, but (if determinism, or at least an approximation to it, is true) this only happens because we are mistaken in the formulation of the laws of nature or in the description of the events.

[25] "Mi ricordo che quando studiavo logica, mai non potetti restar capace di quella tanto predicata dimostrazion potentissima di Aristotile" (Galilei 1632, p. 217). In the *Dialogue* Sagredo represent a nonacademic intellectual, so his language can be more direct – still he generally expresses ideas with which Galileo substantially agrees.

> that I have learned sureness of demonstration from the innumerable advances made by pure mathematicians, never fallacious, for if not never, then at least very rarely, have I fallen into mistakes in my argumentation. In this matter, therefore, I am a peripatetic. (Galilei 1640, p. 248)[26]

From this passage it is evident that Galileo is actually using the term "Aristotelian" in a very broad sense, to include everyone who is able to correctly draw the conclusions of a proof from its premises. After that, as if to dispel any doubt about his own view, Galileo immediately makes it clear that the kinds of reasoning by which he learned to "righteously argue" were the "pure mathematical progresses" (the "*certe dimostrazioni*"). Galileo's "scientific discourse" is made up of mathematical reasoning, not of the syllogisms of Peripatetic science.[27]

At any rate, Galileo had already made himself perfectly clear on this issue in the *Dialogue*:

> The logic [...] is the organ with which we philosophize. But [...] one might be a great logician and still be inexpert in making use of logic [...]. Painting is acquired by continual painting and designing; the art of proof, by reading of books filled with demonstrations – and these are exclusively mathematical works, not logical ones. (1632; English Translation [by Drake] from Galilei 1967, p. 277)[28]

5.5 Conclusion

Contrary to what claimed by Wallace, Galileo was not a member of the Aristotelian party. Hanging on to his insignificant youthful exercises to reverse this obvious truth, as Wallace did, was *magnus error in initio, maximus in fine*.

The times are mature for reconsidering Koyré's suggestion that Galileo's scientific experience was accomplished under the aegis of Plato. Of course, today there

---

[26] "Io stimo (e credo che essa ancora stimi) che l'esser veramente Peripatetico, cioè filosofo Aristotelico, consista principalissimamente nel filosofare conforme alli Aristotelici insegnamenti procedendo con quei metodi e con quelle vere supposizioni e principii sopra i quali si fonda lo scientifico discorso, supponendo quelle generali notizie, il deviar dalle quali sarebbe grandissimo difetto. Tra queste supposizioni è tutto quello che Aristotele ci insegna nella sua Dialettica, attenente al farci cauti nello sfuggire le fallacie del discorso, indirizzandolo e addestrandolo a bene silogizzare e dedurre dalle premesse concessioni la necessaria conclusione; e tal dottrina riguarda alla forma del dirittamente argumentare. In quanto a questa parte, credo di avere appreso dalli innumerabili progressi matematici puri, non mai fallaci, tal sicurezza nel dimostrare, che, se non mai, almeno rarissime volte io sia nel mio argumentare cascato in equivoci. Sin qui dunque io sono Peripatetico" (Galilei 1640, p. 248).

[27] It may also be remembered that in his magnum opus, published in 1638 (only two years before he sent his letter to Liceti), Galileo had written "Was not Plato perfectly right when he wished that his pupils should be first of all well grounded in mathematics?" ("E con che gran ragione voleva Platone i suoi scolari prima ben fondati nelle matematiche?") (Galilei 1638, VIII, p. 175).

[28] "La logica [...] è l'organo col quale si filosofa; ma [...] può esser un gran logico, ma poco esperto nel sapersi servir della logica. [...]. [I]l dipignere s'apprende col continuo disegnare e dipegnere; il dimostrare, dalla lettura dei libri pieni di dimostrazioni, che sono i matematici soli, e non i logici" (Galilei 1632, p. 60).

may be no doubt that Koyré was wrong in claiming that Galileo was not an experimentalist, and in this respect his debt to Archimedes's method was very high. However, as Koyré taught us, that method could fully flourish only if one assumed, as Galileo did, that the natural world has an inherently mathematical structure that we are endowed to grasp when we reason mathematically.

Acknowledgments I am grateful to Antonio Clericuzio, Stefano Gattei, Merry White, and an anonymous referee for several useful comments on a previous version of this article. My thanks also go to Michele Camerota, Marco Romani Mistretta, George Smith, and to the audiences of Tufts University, University of Notre Dame, Harvard University, and Università Roma Tre for some very useful discussions on the issues dealt with in this article.

References

Baldini U (1992) Legem impone subactis. Studi su filosofia e scienza dei Gesuiti in Italia, 1540–1632. Bulzoni, Roma.
Banfi A (1949) Galileo Galilei. Ambrosiana, Milano.
Beller M, Fine A (1994) Bohr's Response to EPR. In Faye J, Folse H (eds). Niels Bohr and Contemporary Philosophy, Kluwer, Dordrecht, pp. 1–31.
Bucciantini M, Camerota M (2005) Once More on Galileo and Astrology: A Neglected Testimony. Galileana: Journal of Galilean Studies 2:229–232.
Burtt EA (1932) The Metaphysical Foundations of Modern Physical Science: A Historical and Critical Essay. Routledge–Kegan Paul, London.
Butts RE, Pitt JC (1978) (ed) New Perspectives on Galileo. Reidel, Dordrecht.
Camerota M (2004) Galileo e la cultura scientifica della Controriforma. Salerno editrice, Roma.
Camerota M (2008) Galileo, Lucrezio e l'atomismo. In Beretta M, Citti F (eds). Lucrezio. La natura e la scienza. Olschki, Firenze, pp. 141–175.
Cassirer E (1906) Das Erkenntnisproblem in der Philosophie und Wissenschaft der neueren Zeit. I. Verlag von Bruno Cassirer, Berlin, pp. 289–324.
Cassirer E (1927) Individuum und Kosmos in der Philosophie der Renaissance. Teubner, Leipzig.
Cassirer E (1946) Galileo's Platonism. In Ashley Montagu MF (ed). Studies and Essays in the History of Science and Learning Offered in Homage to George Sarton in the Occasion of His Sixtieth Birthday. Schuman, New York, pp. 277–297.
Clagett M (1964–1984) Archimedes in the Middle Ages. Madison-Philadelphia, Memoirs of the American Philosophical Society. 5 Vols., 10 Tomes. The Clarendon University Press, Oxford.
Crombie A (1959) Medieval and Early Modern Science. Doubleday Anchor, Garden City.
Crombie A (1975) Sources of Galileo's Early Natural Philosophy. In Rignini Bonelli ML, Shea WR (eds). Reason, Experiment, and Mysticism in the Scientific Revolution. Science History Publication, New York, pp. 157–175.
De Caro M (1993) Galileo's Mathematical Platonism. In Czermak G (ed). Philosophy of Mathematics. Holder–Pichler–Tempsky, Wien, pp. 1–9.
De Caro M (1996) Sul platonismo di Galileo. Rivista di filosofia 82:25–40.
De Caro M (2012) Galileo e il platonismo fisico-matematico. In Chiaradonna R. (ed). Il platonismo e le scienze. Carocci, Roma, pp. 119–138.
De Pace A (1993) Le matematiche e il mondo. Franco Angeli, Milano.
Dijksterhuis EJ (1987) Archimedes. The Princeton University Press, Princeton, NJ.
Dollo C (2003) Galileo Galilei e la cultura della tradizione. Rubettino, Soveria Mannelli.
Drake S (1957) Discoveries and Opinions of Galileo. Doubleday & Co., New York.
Drake S (1978) Galileo at Work: His Scientific Biography. The University of Chicago Press, Chicago.

Favaro A (ed) (1890–1909) Le opere di Galileo Galileo: Edizione nazionale sotto gli auspici di sua maestà il re d'Italia. Favaro A (ed), Barbèra, Firenze [1968: reprinted by Giunti Barbera, Firenze].
Feldhay R (1998) The Use and Abuse of Mathematical Entities. In Machamer 1998a, pp. 80–145.
Feyerabend PK (1975) Against Method: Outline of an Anarchistic Theory of Knowledge. Verso, London.
Finocchiaro M (1994) Galileo and the Art of Reasoning: Rhetorical Foundations of Logic and Scientific Method. Reidel, Dordrecht.
Finocchiaro M (1997) Galileo on the World System: A New Abridged Translation and Guide. The University of California Press, Berkeley.
Finocchiaro M (2010) Defending Copernicus and Galileo. Critical Reasoning in the Two Affairs. Springer, Dordrecht.
Galilei G (1597) Lettera a Jacopo Mazzoni. In Favaro 1890–1909, II, pp. 197–202.
Galilei G (1613), Lettera a Benedetto Castelli. In Favaro 1890–1909, V, pp. 279–288.
Galilei G (1623) Il Saggiatore. In Favaro 1890–1909, VI, pp. 197–372. [English (partial) Translation: Drake 1957, pp. 231–280].
Galilei G (1632) Dialogo sopra i due massimi sistemi del mondo. In Favaro 1890–1909, VII, pp. 21–520. [English Translation: Galilei 1967]
Galilei G (1637) Lettera a Pietro de Carcavy del 5 giugno. In Favaro 1890–1909, XVII, pp. 88–93.
Galilei G (1638), Discorsi e dimostrazioni matematiche intorno a due nuove scienze. In Favaro 1890–1909, VIII, pp. 39–318. [English Translation: Galilei 1914].
Galilei G (1640) Lettera a Fortunio Liceti del 15 settembre 1640. In Favaro 1890–1909, XVIII, pp. 247–251.
Galilei G (1914) Dialogues concerning Two New Sciences by Galileo Galilei. Translated by de Salvio H and S. Dover, New York.
Galilei G (1953) Dialogue Concerning the Two Chief World Systems: Ptolemaic and Copernican. Translated by Drake S. The University of California Press, Berkeley–Los Angeles–London.
Galluzzi P (1973) Il Platonismo del tardo Cinquecento e la filosofia di Galileo. In Zambelli P (ed). Ricerche sulla cultura dell'Italia moderna. Laterza, Roma–Bari, pp. 37–95.
Galluzzi P (2011) Tra atomi e indivisibili: la materia ambigua di Galileo. Olschki, Firenze.
Geymonat L (1957) Galileo Galilei. Einaudi, Torino.
Girill TR (1970) Galileo and Platonistic Methodology. Journal of the History of Ideas 31:501–520.
Hankins J (2000) Galileo, Ficino and Renaissance Platonism. In Kraye J, Stone MWF (eds). Humanism and Early Modern Philosophy. Routledge, London.
Hatfield G (2004) Metaphysics and the New Science. The Cambridge University Press, Cambridge, pp. 93–166.
Høyrup J (1990) Archimedism not Platonism. In Filosofi og videnskabsteori på Roskilde universitetscenter 3, Preprints og reprints. Roskilde Universitet, Roskilde.
Husserl E (1929) Formal and Trascendental Logic. Versuch einer Kritik der logischen Vernunft. In Jahrbuch für Philosophie und phänomenologische Forschung 10:77–166. [English Translation: *Id.* (1969) *Formal and Transcendental Logic*. Nijhoff, The Hague].
Husserl E (1936) Die Krisis der europäischen Wissenschaften und die transzendentale Phänomenologie: Eine Einleitung in die phänomenologische Philosophie. Philosophia. Belgrad. 1:77–176. [English Translation: *Id.* (1970) The Crisis of European Sciences and Transcendental Phenomenology: An Introduction to Phenomenological Philosophy. Northwestern University Press, Evanston].
Jammer M (1974) The Philosophy of Quantum Mechanics. Wiley, New York.
Koyré A (1939) Études galiléennes. 3 vols. Hermann, Paris.
Koyré A (1943) Galileo and Plato. Journal of the History of Ideas 5:400–428.
Koyré A (1968) Metaphysics and Measurement: Essays in Scientific Revolution. The Harvard University Press, Cambridge, MA.
Laird W (1991) Archimedes among the Humanists. Isis 82:628–638.

Linnebo Ø (2013) Platonism in the Philosophy of Mathematics. In E.N. Zalta (ed.), The Stanford Encyclopedia of Philosophy. Via https://plato.stanford.edu/archives/win2013/entries/platonism-mathematics
Lloyd GER (1987) Saving the Appearances. The Classical Quarterly 28/1:202–222.
Machamer P (1998a) (ed) The Cambridge Companion to Galileo. The Cambridge University Press, Cambridge.
Machamer P (1998b) Galileo's Machines, His Mathematics, and His Experiments. In Machamer 1998a, pp. 53–79.
Massimi M (2010) Galileo's Mathematization of Nature at the Crossroad between the Empiricist and the Kantian Tradition. Perspectives on Science 18/2:152–188.
Mazzoni J (1597) In Universam Platonis et Aristotelis Philosophiam Praeludia, sive de comparatione Platonis et Aristotelis. Venezia.
Olschki L (1927) Galileo und seine Zeit. Max Nyemeyer, Halle.
Palmerino CR (2005) The Mathematical Character of Galileo's Book of Nature. In van Berkel K, Vanderfjagt AJ (eds). The Book of Nature in Modern Times. Peeters Publishers, Leuven, pp. 27–45.
Palmieri P (2008) Reenacting Galileo's Experiments: Rediscovering the Techniques of Seventeenth–Century Science. Edwin Mellen Press, Lewiston, NY.
Randall JH (1961) The School of Padua and the Emergence of the Modern Science. Antenore, Padova.
Redondi P (1983) Galileo eretico. Einaudi, Torino.
Shapere D (1974) Galileo: A Philosophical Study. The University of Chicago Press, Chicago.
Shea WR (1972) Galileo's Intellectual Revolution. Science History Publications, New York.
Stabile G (2003) Lo statuto di 'Inesorabile' in Galileo Galilei. In Hamesse J, Fattori M (eds). Lexiques et glossaires philosophiques de la Renaissance. Fédération Internationale des Institute d'Études Médiévales, Louvain–La Neuve, pp. 269–275.
Valleriani M (2010) Galileo Engineer. Springer, Dordrecht.
Walker, DP (1973–1974) Some Aspects of the Musical Theory of Vincenzo Galilei and Galileo Galilei. Proceedings of the Royal Musical Association 100:33–47.
Wallace WA (1974) Galileo and the Reasoning *Ex suppositione*. The Methodology of the Two New Sciences. In Cohen S (ed). PSA 1974: Proceedings of the 1974 Biennial Meeting of the Philosophy of Science Association. Reidel, Dordrecht, pp. 79–104.
Wallace WA (1981) Prelude to Galileo: Essays on Medieval and Sixteenth–Century Sources of Galileo's Thought. Boston Studies in the Philosophy of Science, 62. Reidel, Dordrecht.
Wallace WA (1984) Galileo and His Sources: The Heritage of the Collegio Romano in Galileo's Science. The Princeton University Press, Princeton, NJ.
Wallace WA (1991) Galileo, the Jesuits and the Medieval Aristotle. Collected Studies Series, CS346. Variorum Publishing, Aldershot.
Wallace WA (1992a) Galileo's Logic of Discovery and Proof: The Background, Content, and Use of His Appropriated Treatises on Aristotle's *Posterior Analytics*. Boston Studies in the Philosophy of Science. Vol. 137. Kluwer, Dordrecht.
Wallace WA (1992b) Galileo's Logical Treatises. A Translation, with Notes and Commentary, of His Appropriated Latin Questions on Aristotle's *Posterior Analytics*. Boston Studies in the Philosophy of Science. Vol. 138. Kluwer, Dordrecht.
Wallace WA (1998) Galileo's Pisan Studies in Science and Philosophy. In Machamer 1998a, pp. 27–52.
Whitehead AN (1925) Science and the Modern World. McMillan, New York.
Wisan WL (1978) Galileo's Scientific Method: A Reexamination. In Butts RE, Pitt JC (1978) (ed) New Perspectives on Galileo. Dordrecht, Reidel, pp. 1–58.

Chapter 6
Alexandre Koyré and Blaise Pascal

Dominique Descotes

Abstract Alexandre Koyré's views on Pascal have been published twice, in the proceedings of the "colloque de Royaumont", éd. de Minuit, 1956, and in the *Études d'histoire de la pensée scientifique*. His speech in Royaumont is still famous, as he disturbed the French audience by questioning the genius of Pascal as a mathematician and the reality of the celebrated Rouen experiments, two largely admitted ideas in France. However, by this speech, he opened a new field, which had never clearly appeared before: the important part of the "art de persuader" in Pascal's scientific works.

Keywords Geometry • Koyré-Pascal • Physics • Royaumont Abbey

6.1 Introduction

Pascal seems to be of secondary importance in Alexandre Koyré's works. Only a few short references to the author of the *Pensées* are to be found in his books on the astronomical revolution. The French reader, who remembers the passage in the *Pensées* about the double infinity of the greatness and smallness of the universe, may wonder at this.[1] But, as we will see, Pascal only devoted a few pages of the *Pensées*, of the *Provincial letters* and of his *Letter to Father Noël* to the astronomical revolution. However, during the conference that took place in Royaumont Abbey in 1954 for the tercentenary of the *Memorial*, Alexandre Koyré gave a paper on Pascal's scientific works, which left its mark on his audience and played a crucial part in the history of Pascalian studies: it still has repercussions today.

[1] Fragment 199 in Louis Lafuma's edition (Pascal 1951) and Sellier 230 (Pascal 1991). In the rest of this paper, I will present the references to the *Pensées* as follows: Laf. 199, Sel. 230.

D. Descotes (✉)
Institut d'Histoire des Représentations et des Idées dans les Modernités,
Clermont Auvergne University, 49, Bd François-Mitterrand – CS 60032,
63001 Clermont-Ferrand, France
e-mail: dominique.descotes@uca.fr

R. Pisano et al. (eds.), *Hypotheses and Perspectives in the History and Philosophy of Science*, DOI 10.1007/978-3-319-61712-1_6

6.2 Alexandre Koyré and Pascal in the Royaumont Abbey (1954)

This "masterful"[2] paper was published twice. The first time was in the first issue of the *Cahiers de Royaumont*, published in 1956 by the *Éditions de Minuit* under the title *Blaise Pascal, L'homme et l'œuvre* (Koyré 1956). The second time was in the *Études d'histoire de la pensée scientifique*, in the *Bibliothèque des idées* of the NRF of the Presses Universitaires de France (Koyré 1966). Since we cannot know whether the written version is identical to the spoken paper, we may wonder in which version it is better to read the text. At first glance, it seems reasonable to turn to the second one; the text went through a few minor modifications, for instance, the words *combinaisons* and *siphon* replaced *combinations* and *syphon*. These corrections were incidentally not always fortunate: in the letter from Petit to Chanut (*OC* II, p. 353), the word *sarbatane* was replaced by *sabartane* in 1956 and then became *sarbacane* in 1966. And the publishers from Gallimard wanted to stick to the original text so much that they kept the mistake which says the tubes used for the Rouen experiment were 50 m long, even though it should obviously be corrected to 50 ft (which brings their length back to about 16 metres) (Koyré 1956, p. 275; Koyré 1966, p. 382). But quite naturally, the *Études d'histoire de la pensée scientifique* is of particular interest as they make possible to situate Pascal in the rest of A. Koyré's research and contextualize the paper on Pascal in the studies Koyré devoted to the other scientists of the classical period: Galileo Galilei, Bonaventura Cavalieri, et al. Comparing Pascal's work on indivisibles with Cavalieri's *Geometry of Continuous Indivisibles*,[3] or comparing Pascal's vacuum experiments with Galileo Galilei's, brings a rich background to the paper on "Pascal, scientist".

However, the Royaumont conference proceedings teach us more about Alexandre Koyré's thinking and methods. Indeed, on this occasion, he found himself confronted to the best experts on Pascal: historians of philosophy such as Maurice de Gandillac, Henri Gouhier and Yvon Belaval; historians of religion such as Louis Cognet and Jean Orcibal; specialists of text edition such as Jean Mesnard and Louis Lafuma; historians of sciences such as René Taton, Robert Lenoble, Pierre Costabel and Bernard Rochot; and not to mention Marxist ideologists Henri Lefebvre and Lucien Goldmann. The conference proceedings give us the complete text of the debates that followed the talks: they show us a living image of A. Koyré in the presentation and the discussion of his ideas and of the impact, both enthusiastic and critical, they met among the specialists of Pascal.[4]

Koyré's talk assumed great significance as it took place in a transitional time for Pascalian studies. The first part of the twentieth century remained marked by the

[2] J. Mesnard in Pascal 1964–1992, II, p. 494. Hereafter, I will refer to this edition by the acronym *OC*, followed by the volume number.

[3] The paper on Cavalieri was published a year before the conference on Pascal.

[4] Even the silences which marked the discussion were significant. In the discussion that followed the talk about "Pascal, scientist", Marxist L. Goldmann, who usually specialized in interminable interventions, did not say a word.

debates on Pascal's *Jansenism* which shook the Catholic circles and which tried to determine what could be salvaged in the works of an author both dangerous and appealing in order to serve the religious orthodoxy. Pascal was dangerous, because the *Lettres provinciales* and the *Écrits sur la grâce* were proof enough of his relationship with Port Royal, which was still suspected of heresy. But on the other hand, it was difficult to ostracize the author of the *Mystère de Jésus*, of the *Memorial* and of the *Pensées*: Henri Brémond and Maurice Blondel suggested to solve the problem by dividing Pascal into two parts, by rejecting the *Lettres provinciales* and keeping the *Pensées* as completely Catholic. But others attempted to salvage parts of his works, especially L. Goldmann and Henri Lefebvre, who tried to pass the dialectical aspects of the *Pensées* as precursors of the Marxist doctrine. As a result, the papers on the *Pensées* were largely dominant during the conference. The *Lettres provinciales*, the *Écrits sur la grâce* and *a fortiori*, Pascal's scientific works, remained largely undiscussed. But an evolution was underway. The work of Louis Lafuma, Jean Mesnard and Yoichi Maeda was beginning to bring to light the interest of the technical and paleographic studies of the genesis of the *Pensées*, as revealed by the manuscript and the Copies. As for the history of science, the great editions of Descartes, Fermat and Desargues made it possible to replace the lyrical eulogies of *The Genius of Christianity* with a more precise and more moderate assessment of Pascal's position in his century. Koyré's lecture supplied an important contribution to the Royaumont symposium by unveiling technical aspects or Pascal's scientific works.

Alexandre Koyré skilfully made sure of adapting his speech to the audience, for instance, in the choice of the editions on which he based himself: in a note, he referred to the easily available 1954 Pléiade edition of the *Œuvres complètes* made by Jacques Chevalier, who devoted a few illuminating passages to Pascal's scientific texts. But he actually based himself on Brunschvicg, Boutroux and Gazier's edition of the *Grands Écrivains de la France*, which reunited a large wealth of data and of documents relating to Pascal's scientific works.[5] Moreover, the same desire of adapting his words to his audience is perceptible in the way he carefully avoided any too much technical development: as we will see later, he too refrained from translating the *Treatises* on the cycloid into formulas of integral calculus, which would certainly have been misunderstood by a non-mathematician's assistance. In these proceedings, the reader discovers truly interdisciplinary debates, which show how the subtlety and complexity of Alexandre Koyré's ideas could point out new trends in Pascalian studies.

For instance, in the discussion during which, as a specialist of the astronomical revolution, he spoke about the few pages Pascal devoted to this subject. Pascal does not believe that Copernicus's cosmological system was fully demonstrated. In the

[5] Both these editions are unsatisfactory as far as the *Pensées* are concerned. The Brunschvicg edition (Pascal 1904–1914) is based on a principle of thematic classification, which cancels out Pascal's apologetic project. As for the Chevalier (Pascal 1954) edition, it is an attempt to reconstruct Pascal's plan, mostly owed to the editor's imagination. The Lafuma edition was published in 1951. The Sellier edition (Pascal 1991) and Le Guern edition (Pascal 1977) give us a more satisfactory version today. But they are based on completely different principles. Today, the only really satisfactory edition of Pascal's *Œuvres complètes* is Jean Mesnard's, by Desclée de Brouwer (Pascal 1964–1992).

Provincial letters and his *Letter to the P. Noël*, Pascal clearly states that since the planets' movements can be explained with equal exactness by Ptolemy's, Tycho's and Galileo's hypothesis, it is still impossible, for lack of experiments, to know if Galileo is right or wrong about the movement of the Earth. "This is how" he says:

> […] when one talks in a human way of the motion or of the stability of the Earth, all the phenomenons of planet motions and retrogressions perfectly follow the theories put forward by Ptolemy, Tycho Brahe, Copernicus, and other theories that can be made, of which only one can be true. But who will dare make such a great choice between those theories, and who can, without risking a mistake, support one against the other […] without making himself ridiculous? (Lettre au P. Noël, *OC* II, p. 524)

In fact, Pascal's *pensée de derrière la tête* seems to be that, assuming the universe is infinite, one can just as well choose the Earth, or the Sun, or any place, for a point of reference, which makes the problem pointless. Anyway, only a short sentence of *La révolution astronomique* allows us to know what Koyré thought about the *Pensées:* when it became clear, says he, that the geocentrical and anthropocentrical cosmology was definitively dismissed, the "old world" tried two different ways to defend itself – the Galileo trial in order to destroy the new vision of the universe and Pascal's *Pensées*, aiming to provide that revolution with an answer consistent with the principles of Christianism (Koyré 1961, p. 17).[6] Unfortunately, Koyré said no more about it. But this gives us an insight into the place he attributed to Pascal in the history of cosmological ideas.

Nevertheless, during the Royaumont proceedings, he targeted the incoherence of the cosmology in the fragment *Disproportion de l'homme* (Laf. 199, Sel. 230. See Koyré 1956, pp. 383–384), with the Earth at the centre of the universe, the sun rotating around it to light the Earth and the planets and the stars which turn in the firmament around the sun, all this in an infinite universe. Koyré pointed out that it is not easy to know if this description ought to be understood in a Ptolemean or a Tychonian way, and he added it was incompatible with an infinite universe. But this critical intervention prompted in Koyré's audience a fruitful collective reflection on the complexity of Pascal's rhetoric, which combines cosmological elements with a biblical and religious outlook, in keeping with the viewpoint of the earthly observer, in order to provoke the reader into a religious meditation on his situation in nature, rather than a cosmological investigation.[7] This interdisciplinary cooperation was

[6] Koyré mainly thinks of Giordano Bruno's writings.

[7] An echo of this discussion can be found in later studies, for example see Lanavère 1971, p. 82. See also Mesnard 1993, p. 89. The place of this cosmological vision in the spiritual progress and conversion is stated in Pascal's *Écrit sur la conversion du pécheur*, *OC* IV, pp. 40–44. I must add that the transcript of the debates is also interesting because it highlights the confusion which often creeps in the heat of the debates. It happened when it came to the argument of the wager, which the interlocutors agreed to nullify, on the pretext that the infinity of God is of a different nature from the mathematical infinite. Actually, the infinity of God only intervenes in the first part of the text in order to establish that the finite mind of men cannot comprehend God's immensity. But from the moment when Pascal sets to determine what can reasonably be wagered, the only infinite which comes into his calculations is that of supernatural eternity, whose duration and worth can be expressed mathematically. Such a confusion is not rare during conferences.

quite new in Pascalian studies. We will see later this was not the only opportunity Koyré seized to show new trails for Pascalian research.

Before he even started to speak, Alexandre Koyré was visibly conscious of the impact his study was about to create. This impact was all the more remarkable as he was preceded with a reputation that M. de Gandillac summarized humorously by awarding him with the title of "official iconoclast" and as it was precisely in that light that the paper on "Pascal, scientist" was received. Not that Koyré wished to depreciate Pascal, whom he considered as one of the few authors who truly measured the importance of the cosmological revolution which followed Copernicus' works. Furthermore, Koyré was free with his praise of Pascal's literary genius, his skill of reducing to "ten" crystal-clear "lines" "three (muddled) pages written by Marin Mersenne, or one page written by Roberval", and of the "magic" of his "style". But he took the precaution of pointing out that the Pascalian hagiography made it very difficult to "form a precise" and "unbiased opinion" of the "character and of the scientific works". In addition, the praise given to the "magic" of Pascal's "style" comes before this warning: "I, however, fear that this magic of his style may deprive us, to an extent, of our critical faculty, and may prevent us from examining the content of Pascal's tales" (Koyré 1956, pp. 270–271). It is true that echoes of the anti-Pascalian trend, which dates back to the seventeenth century, can be found on several occasions in Koyré's talk. And when he says that Pascal generally stood out rather for his talent for expressing ideas formulated before him, we do feel like we recognize an echo of the old tune which makes Pascal a mere "secretary of Port Royal", who gave brilliant structure to the theories dictated by the Jansenist doctors: a popularizer often admired for things he did not create. In short, we recognize the voice of the Jesuit priests Father Nouët or Father Morel during the argument of the *Lettres provinciales* or Voltaire's voice in the *Letters on the English.*[8]

Koyré repeats several times in his talk his warning against "the magic of Pascal's language", "a dangerous thing: it is very difficult to resist, and yet it is vital to do so". He concludes: "that is, however, what I will attempt to do today, at the risk of seeming against Pascal" (Koyré 1956, p. 259).

Koyré paid little attention to the epistemological side of Pascal's works (Vincent 2014), while he mentioned its more controversial side, the *Letter to Father Noël* and to Le Pailleur, the absence of the *De l'esprit géométrique*, of the *Preface to the Treatise on the Vacuum* and of the *Conclusion* of the great 1663 treatises appears clearly. But his main idea was that as far as actual scientific creation is concerned, Pascal must be placed after the true trailblazers of the seventeenth century: Descartes, Desargues and Fermat. He maintains that Pascal did not stand out for discovering new principles or new methods, as Descartes or Desargues did. According to Koyré, while Descartes was the man of one brand new algebra, which could supposedly be

[8] It is not completely certain that Koyré may not be somewhat bad-mouthing Pascal. Indeed, nothing allows us to suppose that by publishing his Lettre à M. ADDS (Arnauld, docteur de Sorbonne) on the equality of the parabolic arc and of the spiral arc, Pascal meant to "deal a bad blow to Torricelli", to please Roberval, his "master and friend", who was Torricelli's rival in the invention of indivisibles.

applied to all problems, Pascal was more able to take over ideas invented by others than to create his own method, clearly express and get in order ideas formulated before him, and find several particular and specific methods, each of which was suitable for treating individual cases. Incidentally, Koyré easily listed Pascal's sources: Desargues for the conic sections, Mersenne and Stevin for hydrostatics, and Roberval for the indivisibles. His theory immediately received a moderate endorsement from R. Taton:

> Pascal presented his works with more method than his contemporaries, but his original contribution was often overvalued. (Koyré 1956, p. 287)

Incidentally, on this important topic, Royaumont's version is more telling than later editions: after a talk followed with "passionate interest" (said R. Lenoble), the discussion brought about the necessary corrections, by reminding everybody of two parts of Pascal's works which bore witness to a true creative genius and were overlooked by Koyré, partly because of the time limits imposed by the conference – the invention of Pascal's calculator[9] and the "geometry of chance", which was at the origin of the computation of probabilities. This was an opportunity for Koyré to make more than corrections to his own paper and to show he had rich suggestions to produce: he gladly allowed that Pascal's "share of originality" was unquestionable in the foundation of the computation of probabilities. And in a few words, he opened a completely new perspective on Pascal's legal inspiration in this area: "In the problem of the points intervenes the problem of the player's right to the stakes. Fermat was a legal practitioner and Pascal lived among legal practitioners, and I think this made them more sensitive to this side of the problem, where Galileo Galilei only saw combinatorics". In these words, we recognize the origin of Ernest Coumet's works on the legal origins of the *theory of chance* (Koyré 1956, p. 291; Coumet 1970, pp. 574–598).

6.3 Pascal the Mathematician According to Koyré

The way Koyré presented the mathematician in Pascal is very coherent. He avoided the traps of recurrent history. Contrary to the nineteenth century historians, Koyré refrained from looking for the beginnings of the subsequent development of science in Pascal's works and from making Pascal the inventor of mathematical theories he could not know.

Even though in 1883, Maximilien Marie transcripted the *Lettres de A. Dettonville* on the cycloid into formulas of integral calculus (Marie 1883), Koyré uses none of them: there are absolutely no equations in his presentation of *Dettonville's Letters*

[9] Koyré reckons it is "unusable". In the 1950s, the restoration of the remaining machines, which showed that was not the case, had not yet taken place. Mourlevat (Mourlevat 1988) also showed that the secrets of the functioning of the *Pascalines* (especially the repartition of the weights of the inner parts) had not all been grasped at the end of the twentieth century.

and of the *Arithmetical Triangle.*[10] There were no equations either in Koyré's presentation of the treatises of physics, although P. Duhem had devoted famous studies to them. When he discussed Pascal's *Traités des trilignes rectangles, des sinus du quart de cercle* and *des arcs de cercle*, he carefully avoided translating their propositions into integral calculus formulas, for even if we grant that it is very easy to express "Pascal's reasonings" in "the language of infinitesimal calculus", a translation of his skillful geometrical demonstrations into algebraic formulas cannot give an adequate idea of his genuine spirit, which generally bans abstraction (Koyré 1956, p. 269). His own genius was marked by a natural hostility to algebra, which made him miss the Cartesian revolution. Pascal, Koyré said, was, like Desargues, a geometrician who knew how to use his imagination to see in space and to trace a multitude of lines without getting their connections and relations mixed up (Koyré 1956, pp. 260–261).

For instance, it is true that one can see that the BACK solid, which Pascal calls an *onglet*, made up by the sum of the (OD.DF) rectangles (See Fig. 6.1), is half the solid made by the sum of the triangles each equal to EG^2. Pascal then states that *the sum of the squares of the EG ordinates is double of the sum made by the rectangles* (DF.DO).[11] The geometrical nature of this theorem is essential for Pascal's method, because the BACK *onglet* is meant to give a concrete expression to a centre of gravity problem he dealt with in Dettonville's *Letters* (Fig. 6.2).

But though the translation into the formula

$$\int_a^0 y^2 dx = 2\int_b^0 xy\,dy$$

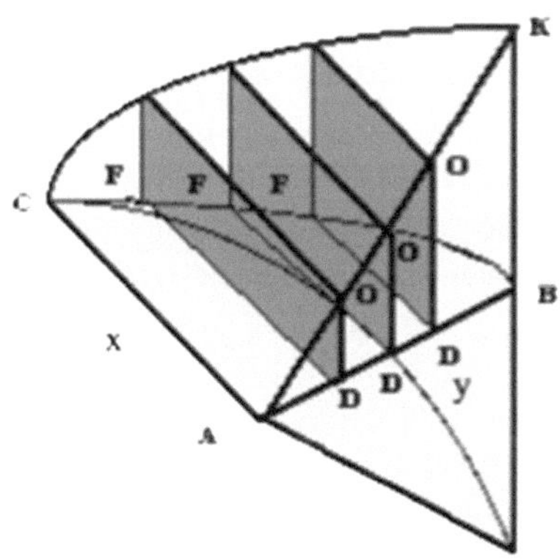

Fig. 6.1 Double onglet (Source: Author's figure)

[10] *Dettonville* was the pseudonym Pascal used for his works on the cycloid.

[11] Traité des Trilignes, Proposition II: "La somme des carrés des ordonnées à la base est double des rectangles compris de chaque ordonnée à l'axe et de sa distance à la base." (*OC* IV, pp. 444–445).

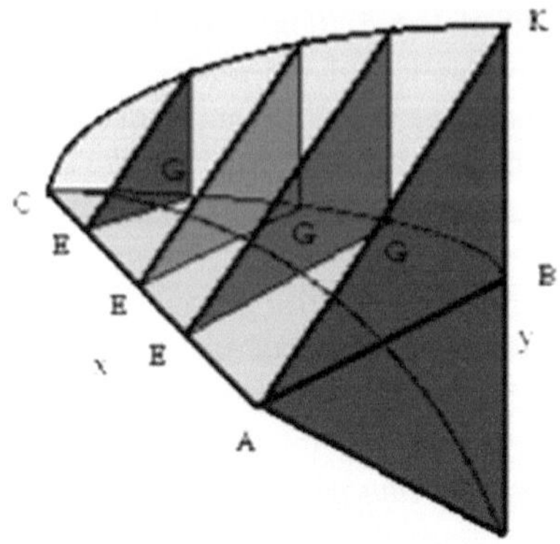

Fig. 6.2 Double onglet (Source: Author's figure)

is very easy, it loses the geometrical signification in Pascal's mind. And such an algebraic translation could mislead the audience to believe Pascal had already discovered the Leibnizian or the Newtonian calculus, which would be a complete nonsense. Koyré's theory on Pascal's talent for organizing other people's works is especially strong concerning Pascal's works on conic sections. The research led by Poncelet, by Poudra and by R. Taton – the latter's work were recent then – showed how much Pascal owed to Desargues's *Brouillon projet d'une atteinte aux événements des rencontres du cône avec un plan.* Pascal himself recognized his debt in his *Essay on Conic Sections*. And indeed, in the *Generatio conisectionum*, which is all that is left of his Latin treatise, he managed to present the method of unification of curves of degree 2 (parabola, hyperbola and ellipse) as optical projections of a circle placed on a cone much more clearly than Desargues did. As for the consequences, Desargues himself paid tribute to his student for the invention of the proposition he called *la Pascale*. "Pascal's work", Koyré concluded, "is Desargues's work, clearer and more systematised", but the fact remains that "Desargues, not Pascal, was the true creative genius, the inventor of a new form of geometry".[12]

The same theory is less convincing when it comes to the arithmetical triangle. It remains true that in this work too, just like in the conics section, Pascal was able to make clearer and more systematic matters, which others, like Mersenne and Gassendi, dealt in great confusion. It is also true that the arithmetical triangle is a very old thing, much more ancient than Pascal's *Traité du triangle arithmétique*. It was developed by many texts in the Arabic language and even from China. The triangle is also discussed by Stifel, Tartaglia and Stevin's work. However, none of these sources seem to be essential. In my opinion, Hérigone's *Cursus mathematicus*, to which Pascal referred in his short *Usage du triangle arithmétique pour trouver les puissances des binômes et apotomes*, is not truly significant: this *Cursus* is not a

[12] Like many others before him, Koyré assimilated *l'hexagramme mystique* to the *théorème de l'hexagone*, which says the intersections of the opposite sides of a hexagon inscribed in a conic section cross on lined-up points. Actually, René Taton (1964) and Jean Mesnard (1993) contested this identification: *hexagon* means that the figure has six *sides*, while *hexagram* means that it is comprised of six *letters*, which is not the same thing.

Tabella pulcherrima & vtilißima Combinationis duodecim Cantilenarum.

I.	II.	III.	IV.	V.	VI.	VII.	VIII.	IX.	X.	XI.	XII.
1	1	1	1	1	1	1	1	1	1	1	1
2	3	4	5	6	7	8	9	10	11	12	13
3	6	10	15	21	28	36	45	55	66	78	91
4	10	20	35	56	84	120	165	220	286	364	455
[illegible]	15	35	70	126	210	330	495	715	1001	1365	1820
6	21	56	126	252	462	792	1287	2002	3003	4368	6188
7	28	84	210	462	924	1716	3003	5005	8008	12376	18564
8	36	120	330	792	1716	3432	6435	11440	19448	31824	50388
9	45	165	495	1287	3003	6435	12870	24310	43758	75582	125970
10	55	220	715	2002	5005	11440	24310	48620	92378	167960	293930
11	66	286	1001	3003	8008	19448	43758	92378	184756	352716	646646
12	78	364	1365	4368	12376	31824	75582	167960	352716	705432	1352078
13	91	455	1820	6188	18564	50388	125970	293930	646646	1352078	2704156
14	105	560	2380	8568	27132	77520	203490	497420	1144066	2496144	5200300
15	120	680	3060	11628	38760	116280	319770	817190	1961256	4457400	9657700
16	136	816	3876	15504	54264	170544	490314	1307504	326876	7726160	17383860
17	153	969	4845	20349	74613	245157	735471	2042975	5311735	13037895	30421755
18	171	1140	5985	26334	100947	346104	1081575	3124550	8436285	21474180	51895935
19	190	1330	7315	33649	134596	480700	1562275	4686825	13123110	34597290	86493225
20	210	1540	8855	42504	177100	657800	2220075	6906900	20030010	54627300	141120525
21	231	1771	10626	53130	230230	888030	3108105	10015005	30045015	84672315	225792840
22	253	2024	12650	[illegible]5780	296010	1184040	4292145	14307150	44352165	129024480	354817320
23	276	2300	14950	30730	376740	1560780	5852925	20160075	64512290	193536720	548354040
24	300	2600	17550	[illegible]8280	475020	2035800	7888725	28048800	92561040	286097760	834451800
25	325	2925	2047[illegible]	[illegible]8755	593775	2629575	10518300	38567100	131128140	417225900	1251677700

Fig. 6.3 Mersenne's *Table des variétés* [Mersenne 1636a, De Cantibus, Book VII, Proposition XII, p. 136) (Source: gallica.bnf.fr / BnF [French version: Mersenne 1636b, De Chants, Livre second, Proposition XVI, Corollaire II, p. 145])

first-rank mathematical treatise but a mere textbook. In fact, Koyré did not mention the most probable source, even though it would strengthen his theory, Marin Mersenne's *Traité des chants* in *L'harmonie universelle*, which is largely devoted to combinations. In this big book, Mersenne gives a *Tabella pulcherrima et utilissima Combinationis duodecim Cantilenarum*, which shows in how many ways one can take 12 notes in a lot of 36 (Mersenne 1636a, p. 136, b, p. 145) (Fig. 6.3).

Had Mersenne added a vertical column of units on the left, his numerical table would have been identical to Pascal's arithmetical triangle. So Mersenne's *Harmonie universelle* announces one of the ideas Pascal wanted to highlight: the connexion between figurate numbers and combinations. But this particular member of the *Order of Minims* can hardly be considered as Pascal's intellectual guide: in the same *Traité*, Pascal showed that the arithmetical triangle was also able to compute the "partis" and establish the principles of the "géométrie du hasard", which is the ancestor or the probability calculus. In spite of his skill when it comes to combinations, Mersenne never thought of such an invention. What is really new in the *Traité du triangle arithmétique* is precisely Pascal's own creation.[13] But, in any event,

[13] Of course, we know that Fermat too had his own method for computing "partis", which he compared with Pascal's. But their correspondence shows that their principles were different.

Koyré was perfectly right to insist on the fact that Pascal was mainly interested in the *Uses of the Arithmetical Triangle*, as an ingenious and powerful instrument of solving combinations and probabilities problems, as well in the establishment of a parallel between numbers and geometry's continuous magnitudes he took from in the *Potestatum numericarum summa*.

Concerning *Dettonville's letters* on the cycloid, the demonstration is just as questionable. With regard to the consequences, Koyré summarized impeccably the main results established by Pascal, the way in which he measured solid and curved surfaces and determined centres of gravity by using geometrical constructions with four, five or even six dimensions. He correctly underlined the deep difference between Cavalieri's method for geometrical quadratures and Pascal's. In his lecture about Cavalieri, Koyré clearly explained the way, in his *Geometria indivisibilibus continuorum* (Cavalieri 1635), he calculated geometrical objects by the mean of indivisible elements having one dimension less: Cavalieri calculated areas by summing lines and three-dimensional solids by summing two-dimensional surfaces. We do not know if Pascal knew Cavalieri's method or if he has been able to read his book. But Pascal had another idea of indivisibles than Cavalieri: the opuscule *De l'Esprit géométrique* makes obvious he thought that correct geometry could not proceed by violating the rule of homogeneity – a line adds nothing to a surface; it is, as Pascal says, a *naught of surface*. It is therefore impossible to generate an area by adding lines, because an area has length and breadth, but a line has only one dimension. It is also impossible to achieve a three-dimensional solid by adding two-dimensional surfaces. So one has to keep homogeneity by adding lines to lines or surfaces to surfaces (Fig. 6.4).

In fact, Pascal thought it was also possible to say that the semicircle CMFZ, for instance, could be regarded as the sum or rectangles (ZM.ZZ), built up on the ordinates ZM and the segments ZZ. If these ZZ become very small, the difference between the semicircle and the rectangles can be made less than any chosen area. And as Pascal makes all the ZZ segments equal to another, $\sum$ (ZM.ZZ) can be expressed in an abbreviated way, as the sum of all the ZM ordinates. In fact, what Pascal calls *indivisibles* really are divisible, because the ZZ segments can always be bisected.

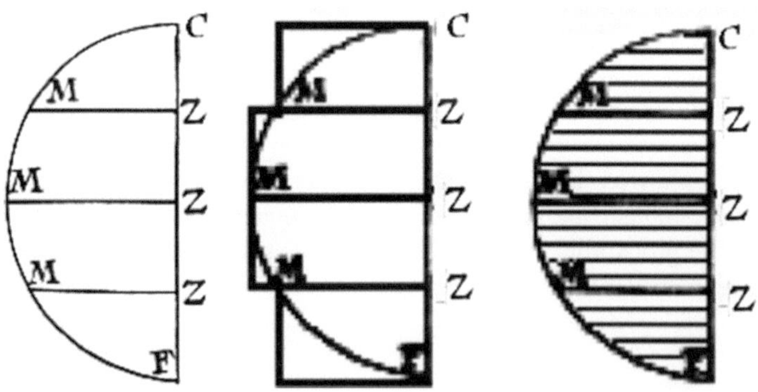

Fig. 6.4 Sum of lines (Source: Author's figure)

Once again, Koyré is right when he says Pascal was not first to use this method. It is certain that this conception was shared by the scholar group which appeared around Mersenne and especially by Roberval. He is right too when he adds that Pascal gave the first clear and rigorous explanation of this second indivisibles method. And he is right again when he says that Pascal was able to give a great extension to this method: we know that in the *Lettres de A. Dettonville*, he did not stop at two nor three geometrical dimensions; he went on by adding three-dimensional figures and generating four-, five- and six-dimensional figures, far away from the classical Euclidean geometry, arousing father Antoine de Lalouvère's great wrath. In his *Lettre à Carcavy*, Pascal supplied important explanations about the special way he spoke about sums of lines or areas, but some scientists thought he went too far in what they called *Arabisms*.

However, Koyré concluded from this that Pascal, who was lucky enough to be Desargues's student in his youth, was also unlucky enough to become Roberval's later in life, which blinded him to Cavalieri's worth. It may be regrettable that Pascal did not follow Cavalieri's bold ideas, although nothing proves that he actually read his works. But the idea that he was Roberval's *student* is not admissible. Roberval's ideas about indivisibles were certainly older than Pascal's, but they are also less clear and less meticulous. Moreover, it is not clear that Pascal had a real knowledge of Roberval's mathematical discoveries. On this subject, recent conclusions tend not to overestimate Roberval's influence, whose obsessive secretiveness makes it difficult to believe that he made Pascal his disciple. In fact, Roberval had the bad habit to hide his inventions, in order to disclose them only when he had to show he deserved more than anyone Ramus' chair in the *College Royal de France*. The best proof that Pascal learned very little from Roberval is that his ignorance of his discoveries created a very awkward situation for him when he arranged his competition about the cycloid, discovering that a part of the problems he proposed had been secretly resolved.[14] This conclusion is made even more imperative, as in Pascal's mind, the main point of his method did not lie in the use of indivisibles but rather in his original method using *triangular sums* to determine centres of gravity, for which he has no precursors.

Pascal calls a *simple sum* of magnitudes A, B, C and D, the sum (A + B + C + D).

Suppose now A, B, C and D are four weights hanging at regular intervals on the balance OD fixed at point O

[14] Koyré yielded here to the same tendency as Jesuit Henri Bosmans's, who, for similar reasons, presented Jesuit André Tacquet as Pascal's master when it came to indivisibles, arguing that Pascal did not know the geometrical homogenity law until he read Tacquet's *Cylindrica et annularia* book, which shows that every demonstration by means of heterogeous indivisibles should be reduced to classical proofs. See Bosmans 1924, pp. 130–161, pp. 424–451. Father Tacquet's treatise was printed twice: *Cylindricorum et annularium libri IV* (1651) were comprised of four books; in 1659, the Jesuit added a fifth one. The date is significant: the Society of Jesus was looking for a geometrician capable of winning recognition in the competition arranged by Pascal, as Father Lalouvère was being ridiculed. Book V focused on the problems of the centres of gravity, the most difficult problems posed by Pascal.

$$\begin{matrix} O & A & B & C & D \\ & 2 & 7 & 3 & 2 \end{matrix}$$

Their simple sum is: 2 + 7 + 3 + 2 = 14.

Pascal calls the *triangular sum* of the same magnitudes beginning by A this new sum:

$$\begin{matrix} A & B & C & D & \\ & B & C & D & \\ & & C & D & \\ & & & D & = A+2B+3C+4D \end{matrix}$$

The triangular sum of the previous weights beginning with A will be

$$[(1\times 2)+(2\times 7)+(3\times 3)+(4\times 2)]=33.$$

It will express the *momentum* of the weights on balance OD. Translating boldly this new type of sum into geometrical magnitudes, Pascal is able to find the gravity centres of various geometrical figures, which allowed him to resolve the most difficult problems related to the cycloid.[15] Pascal thought this method was his own true discovery. The great editor of Pascal's Works, J. Mesnard, concluded that on this matter

> Pascal appears to be the disciple of none of his predecessors. We can observe no dependence comparable to that which was established, for example, towards Desargues regarding conic sections. (*OC* IV, p. 400)

6.4 Pascal and the Void According to Koyré

However, it was not Pascal's mathematical works that provoked most of the reactions during the Royaumont conference but his work on physics, on the vacuum and on atmospheric pressure.

Once again, it seems that Koyré is perfectly right when he says the main quality of Pascal's work is the art of creating an order in matters which other scholars only handled with some confusion. He was not the first to say Pascal borrowed some important ideas from previous scientists: in a celebrated paper about *Pascal's principle* (Duhem 1905, pp. 599–610), Pierre Duhem showed some important ideas of the *Traités de l'équilibre des liqueurs et de la pesanteur de la masse de l'air* (hereafter *Traités*. Pascal 1663) had been borrowed from Mersenne, Stevin and Benedetti. But one only has to read the first chapters of the *Traités* to understand how truly

[15] The onglet we mentioned can be considered as the simple sum or rectangles (FD.DO), but it is also the triangular sum of the FD ordinates (Costabel 1964, pp. 154–168; Merker 2001; Descotes 2001).

Koyré says Pascal produced a very clear and well-ordered explanation of the principles of hydrostatics and found the proper experiments to demonstrate them.

In his first book, Pascal begins by stating the basic *fact*: liquors weigh in proportion to their height. The second and third chapters explain the *reason* why it is so, illustrating the hydrostatical paradox by a number of experiments with communicating vessels and hydraulic press. Then Pascal explains how a solid body behaves floating in a liquid, how this body behaves when it is completely submerged and how animals behave underwater.

The *Traités* then draws the consequences of the previous: since air is a liquid, atmosphere can be regarded as an ocean. Therefore, Pascal can state this new principle, the air mass has weight and presses everything within it, and the following consequences:

1. Since every part of the air has weight, the whole atmosphere has weight, and the weight of the whole air mass is not infinite.
2. The air mass presses every part of the Earth.
3. The air mass presses more the lowlands and less the highlands.
4. The air mass presses in every direction.
5. The lower parts of the atmosphere are more compressed than the higher.

The following chapters show by a lot of original experiments each point Pascal announced.

The images added to the *Traités* visually show the analogy between what happens in the open air and what happens underwater (Fig. 6.5).

But the main point of Koyré's analysis focused on the *Expériences nouvelles touchant le vide* and Pascal's Rouen experiments. He assumed that because he wanted to avoid cutting corners, Pascal settled for proving the existence of an *apparent* vacuum in 1647 and that he meant to show later that this vacuum was real but that he had actually already written a whole treatise on the vacuum designed to establish that the effects attributed to *horror vacui* are owed to the weight of air. After quoting the description of the most spectacular experiences and pointing out that 46 ft-long tubes and scalene siphons were "very difficult to make, even nowadays",[16] Koyré says that the manipulation of such instruments presented considerable difficulties, and that with no diagrams and no concrete description, Pascal left the reader without any way of knowing the tricks he used to manipulate these tubes; he concluded again that only Pascal's "magic of style" explains why nobody doubted these extraordinary assemblies actually took place. If not for the prestige of Pascal's name, one would have to wonder if these experiments really happened and, if they did, if they were truly faithfully described. Koyré felt how iconoclastic this supposition was and added:

> Let us be clear: I do not wish to imply that Pascal did not carry out the experiments he described. [...] However, I think I can safely affirm that he did not describe them as they

[16] The 46 ft of the tube corresponded to 15,088 m then. The 32 ft of the water column represented 10,496 m.

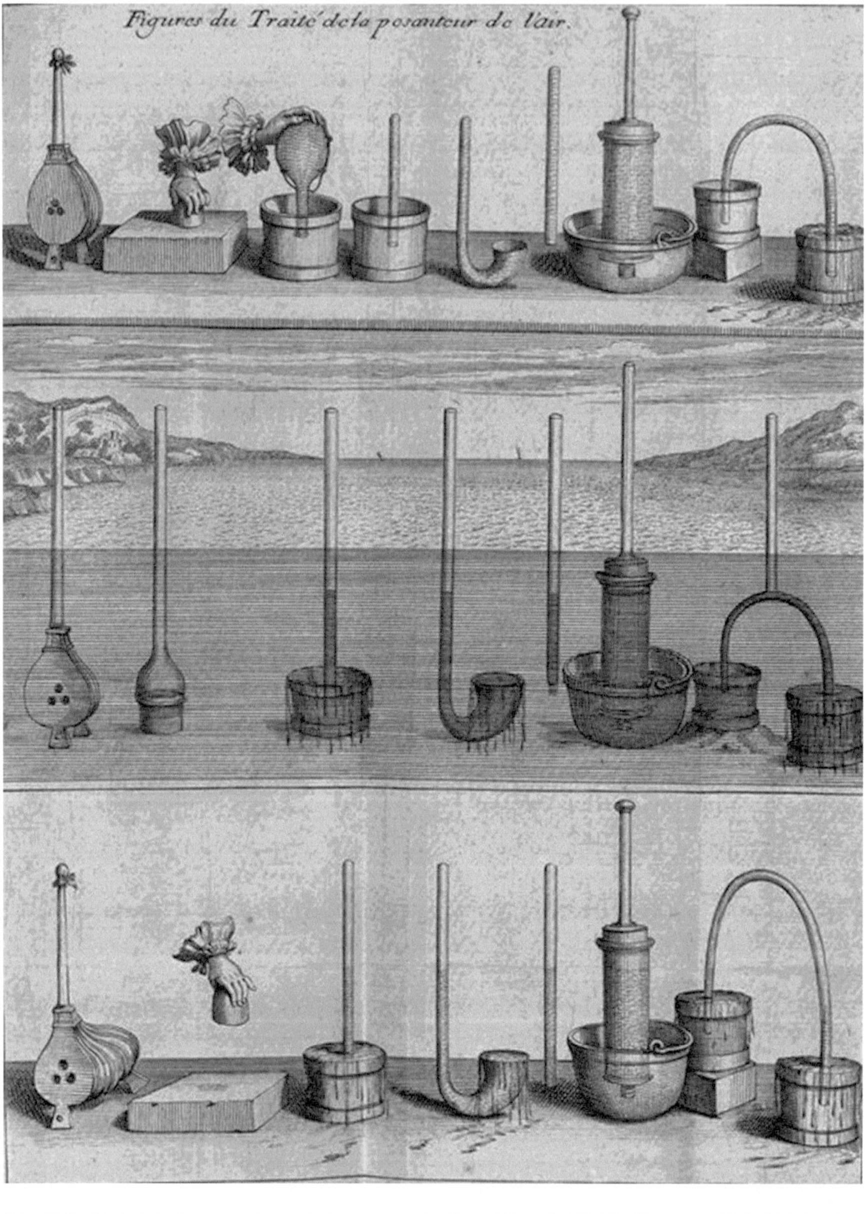

Fig. 6.5 *Traité de la pesanteur de la masse de l'air* (Pascal 1663). (Source: Bibl. Patrimoine Clermont).

> happened, and that he did not present the results as they unfolded before his eyes. He most certainly concealed something from us. (Koyré 1956, pp. 275–276)

To tell the truth, his *pensée de derrière la tête* might very well be that Pascal did not actually carry out the experiments he described. For the argument of how difficult these experiments were to carry out, technically speaking, serves no purpose if we only maintain that Pascal did not describe the experiments he carried out as he observed them. It is only relevant if we suppose that these experiments were not, or could not be, carried out. This is actually the way in which his talk was understood: during the discussion that followed it, epistemologist M. Ullmo observed that not only were imaginary experiments common in the seventeenth century, they were still common in the twentieth century; as a proof, he put forward Heisenberg's works. Besides, it was not long before some critics took the plunge: K. Koyanagi consistently maintains that most of Pascal's experiments were fictitious (Koyanagi 2001, pp. 137–157). However, if we are to contest the carrying out of the Rouen experiments, we have to put the reports of several authors in doubt, especially Guiffart and Pierius' works.[17] The public nature of the experiments which took place in the Saint Sever glass factory rules out the possibility of such a clumsy deception.[18]

However, by supposing that Pascal "actually carried out the experiments he speaks of" but contesting that he "described them exactly and completely", Koyré takes on a very strong position.

Indeed, he had a very precise idea of what Pascal neglected to mention: the bubbling produced in the water inside the tube, which scholars contemporary to Pascal were able to observe, as Roberval's reports are evidence of. Koyré himself was able to confirm it in 1950 in the *Palais de la Découverte*.

> This water starts to bubble. It was very natural: the air dissolved in the water escaped from it, forming bubbles. [...]. Could this phenomenon have eluded Pascal? I don't believe so [...]: all the more so as this bubbling is not the only remarkable phenomenon happening in the tube: because of the atmospheric pressure (and because of the steam) the water column decreases, and this decrease reaches 1.5 m in 24 hours.
>
> [The necessary conclusion was that this phenomenon] was quite embarrassing for the supporters of the vacuum: those who negated it could, with some reason, affirm that the space above the water only appeared empty, but that it was really full of air and steam.
>
> [For] it became clear, in any case, that it could not be accepted as true that apparent vacuum was identical to actual vacuum. [And, then what?] There are air bubbles in this water, and even in the mercury? Big deal! It was of no importance to Pascal. He *imagined* the experiments he did–or did not do–so well, so clearly, that he deeply grasped the crux of the matter, to wit the interaction of the liquids which balance one another.
>
> [For] Pascal did not want to lay his cards out on the table too early: indeed, he kept to hand an entire *Treatise* which would provide the required demonstration, and explain at the

---

[17] Roberval apparently did not witness for himself the Rouen experiments.

[18] Koyré reminds us that Robert Boyle already maintained in the seventeenth century that some of the experiments of the *Treatise on the Equilibrium of Liquids* were only mental experiments. He also doubted that it was possible to send a man 20 ft underwater. Indeed, it is even less likely that this same man should be sent in a cistern full of oil, as Pascal imagined in *The Weight of the Air* (*OC* II, p. 1078). However, the characteristics of the experiments of *The Equilibrium of Liquids* are not the same as those of the Rouen experiments.

> same time by the theory of the balance of fluids the reason why the vacuum appeared in the pipes. In the meantime, he did not wish to give weapons to his opponents or to cast doubts in the minds of the simple–minded, who on the contrary must be prepared to accept future evidence.
>
> [Koyré's conclusion hit the bull's eye, especially as it was grounded in a fact that was difficult to challenge] Pascal was not a faithful disciple of Bacon, a first version of Boyle. (Koyré 1956, pp. 276–278)[19]

In fact, the problem is not completely solved yet. For instance, we know by Roberval's *Relation* of the Rouen experiments that to show the possibility of creating a visible void, Pascal used two great tubes, the first one filled with water and the other one filled with wine. It was generally believed that Roberval did not report accurately this experiment, writing that the wine column in the second was higher than the water column in the first. But recently, two scientists of Orsay University, Armand Le Noxaïc and Pierre Lauginie, showed that by using a seventeenth century-type wine, it was really so (Le Noxaïc and Lauginie 2010, pp. 48–55. See *OC* II, pp. 468–472). Following the track Alexandre Koyré suggested, the discussion about the technical realisation of Pascal's experiments still goes on.

The way Koyré presented Pascal made an impression. On this essential matter, he questioned a largely admitted idea and opened a new field, which had never clearly appeared before: the part of rhetoric and of the art of persuasion which comes into the proceedings of the experiments and which prepares the reader to admit the physical theories. This perspective did not elude the audience then, especially not Jean Duvignaud, who underlined its interest. This broad question could not be seriously dealt within the discussion of a paper, Koyré answered. But a whole trend of research developed in that direction. From there came the idea that a distinction must be made between the kind of physics which cultivates the project of a complete report, rounding up all the details of the realization of the experiment, including those which seem the most useless, and a kind which, on the contrary, favours a simplified description which does not show what the observer sees but what he *should see*. Nowadays, the rhetoric of science is a discipline which gave rise to very active researches.

6.5 Conclusion

Koyré's talk on Pascal opened more perspectives and more directions in research than the importance of an ordinary talk made it possible to solve. He made a lasting impression. In a time when a paper is often forgotten before it is even published, the example of a master who managed to outline in a few pages the directions the research would take for the next 40 years or so deserves to be contemplated. Although he repeats at the end of his talk his warning against "the magic of Pascal's language", Koyré did Pascal the considerable favour of contributing to answering the question Sainte-Beuve asked in the nineteenth century, namely, whether the

[19] Actually, Pascal did not have a treatise up his sleeve in 1647.

same style can be found in Pascal's scientific works as in the *Pensées* and in the *Lettres provinciales*, and to paving the way to an all-encompassing research in which the geometrician, the philosopher, the writer and the Christian make up one single person.

References

Bosmans H (1924) Sur l'œuvre mathématique de Blaise Pascal. Revue des Questions scientifiques 85:130–161, 424–451.

Cavalieri B (1635) Geometria indivisibilibus continuorum: nova quadam ratione promota. Duciis, Bononiae.

Costabel P (1964) Essais sur les secrets des Traités de la Roulette. In L'œuvre scientifique de Pascal. Presses Universitaires de France, Paris.

Coumet E (1970) La théorie du hasard est.-elle née par hasard? Annales 3:574–598.

Desargues G (1639) Brouillon projet d'une atteinte aux événements des rencontres du cône avec un plan.

Descotes D (2001) Blaise Pascal: littérature et géométrie. Presses Universitaires de France, Clermont–Ferrand.

Duhem P (1905) Le principe de Pascal. Essai historique. Revue générale des sciences 2:599–610.

Duhem P (1905–1906) Les Origines de la Statique. Hermann, Paris.

Guiffart P (1647) Discours du vide, sur les expériences de Monsieur Pascal et le traité de M. Pierius. Auquel sont rendues les raisons des mouvements des eaux, de la génération du feu et des tonnerres, de la violence et des effets de la poudre à canon, de la vitesse et du poids augmenté par la chute des corps graves. Jacques Besongne, Rouen.

Hérigone P (1634–1642) Cursus mathematicus, nova, brevi et clara Methodo demonstratus, per Notas reales et universales, citra usum cujuscumque Idiomatis Intellectu faciles. Henry Le Gras, Paris.

Koyanagi K (2001) Cet effrayant petit livret... Expériences nouvelles touchant le vide de Blaise Pascal. In Cléro J-P (ed). Les Pascal à Rouen. Presses universitaires de Rouen, Rouen, pp. 137–157.

Koyré A (1956) Pascal savant. In Blaise Pascal. L'homme et l'œuvre. Éditions de Minuit, Paris, pp. 259–285.

Koyré A (1961) La révolution astronomique. Copernic, Kepler, Borelli. Hermann, Paris.

Koyré A (1966) Pascal savant. In Études d'histoire de la pensée scientifique. Gallimard, Paris, pp. 362–389.

Koyré A (1973) Du monde clos à l'univers infini. Gallimard, Paris.

Lanavère A (1971) L'argument des deux infinis chez Pascal et chez La Bruyère. In Mesnard J (ed). Les Pensées de Pascal ont trois cents ans. De Bussac, Clermont, pp. 79–103.

Le Noxaïc A, Lauginie P (2010) Reconstitution de l'expérience des liqueurs de Blaise Pascal. Courrier du Centre International Blaise Pascal 32:48–55.

Marie M (1883) Histoire des sciences mathématiques et physiques. Vol 4. Gauthier–Villars, Paris.

Merker C (2001) Le chant du cygne des indivisibles. Le calcul intégral dans la dernière œuvre scientifique de Pascal. Presses Universitaires de France, Comtoises, Besançon.

Mersenne M (1636a) Harmonicorum libri. Baudry, Paris.

Mersenne M (1636b) Harmonie universelle contenant la théorie et la pratique de la musique, où il est traité de la nature des sons et des mouvements, des consonances, des dissonances, des genres, des modes, de la composition, de la voix, des chants et de toutes sortes d'instruments harmoniques. Cramoisy, Paris.

Mesnard J (1993) Les Pensées de Pascal. 2nd ed. SEDES–CDU, Paris.

Mourlevat G (1988) Les machines arithmétiques de Blaise Pascal. La Française d'édition et d'imprimerie, Clermont–Ferrand.
Pascal B (1904–1914) Œuvres complètes. Brunschvicg L et al. (ed). Hachette, Paris.
Pascal B (1951) Pensées. Lafuma L (ed). Éditions du Luxembourg, Paris.
Pascal B (1954) Œuvres complètes. Chevalier J (ed). Gallimard, Paris.
Pascal B (1964–1992) Œuvres complètes. Mesnard J (ed). Desclée de Brouwer, Paris.
Pascal B (1977) Œuvres complètes. Le Guern M (ed). Pléiade, Paris.
Pascal B (1991) Pensées. Sellier P (ed). Bordas, Paris.
Pascal B (1663, 1698) Traités de l'équilibre des liqueurs et de la pesanteur de la masse de l'air. Desprez, Paris.
Pierius J (1646) An detur vacuum in rerum natura.
Tacquet A (1651) Cylindricorum et annularium libri IV, item de circulorum Volutione per planum Dissertatio physico-mathematica. Jacobum Meursium, Antverpiae.
Tacquet A (1659) Cylindrica et Annularia quinque libris comprehensa. Jacobum Meursium, Antverpiae.
Taton R (1951) L'œuvre mathématique de Girard Desargues. Presses Universitaires de France, Paris.
Taton R (1964) L'œuvre de Pascal en géométrie projective. In L'œuvre scientifique de Pascal. Presses Universitaires de France, Paris, pp. 17–72.
Vincent J (2014) Pascal via Duhem and van Fraassen. Courrier du Centre International Blaise Pascal 36:5–12.

Chapter 7
Koyré's Revolutionary Role in the Historiography of Science

Antonino Drago

Abstract Koyré's analysis of the birth of modern science has surely changed the old historiography of science because he recognized as an essential component of the science of this case study a metaphysics which allowed him to improve considerably the interpretative power of his accounts. He summarized this metaphysics by means of two "characteristic features". My previous works suggested that two basic dichotomies formed the foundations of science – one on the kind of infinity and the other one on the kind of the theoretical organization. In the light of them, Koyré's characteristic features correspond to the choices, concerning the two dichotomies, on which Newton implicitly based his mechanics. This result explains the adequacy of Koyré's interpretation in the examined historical events before and during Newton's time. On the other hand, the historiography of the other major representative historian of science, Kuhn, is less related to these foundations of science. However, from a comparison of the categories of both historians, a general scheme for characterizing as a whole the "new historiography" of last decades is obtained. In addition, as a verification of the previous interpretation by means of the two dichotomies of Koyré's categories, by opposition to them, new categories are obtained for interpreting the nineteenth-century scientific theories which do not belong to the Newtonian paradigm – e.g. chemistry and thermodynamics. They apply also to the first two theories of modern physics. By quickly examining the other historiographies of the last decades, I conclude that (1) the period of the new historiographies has ended; (2) of these, Koyré's was the best one because it approached more closely than all others the foundations of science; (3) very remarkably, the works on the history of science, Koyré's in particular, suggested great advancements in philosophy of science.

Keywords New historiography • Categories • Metaphysical content • Two dichotomies • Foundations of science

---

A. Drago (✉)
Formerly at Physics Department, Napoli Federico II University, Naples, Italy
e-mail: drago@unina.it

R. Pisano et al. (eds.), *Hypotheses and Perspectives in the History and Philosophy of Science*, DOI 10.1007/978-3-319-61712-1_7

7.1 The Birth of the New Historiography: Koyré's

Without doubt, Koyré's studies on above all Galileo, Descartes and Newton (Koyré 1939, 1957, 1965) have the merit of having significantly renewed the history of science and of giving birth to a "new historiography".

The previous dominant philosophy was positivism, which conceived the academic discipline of historiography of science according to specific rules. Positivist historiography dealt with hard historical facts, first of all, dates, in particular the date of a scientific discovery together with the question of which scientist, among the several concurrent ones, was the actual discoverer; moreover, by taking as yardstick the viewpoint of present-day science, this historiography considered the historical development of science as determined by the final result; as a consequence, of the historical development of science, it disregarded both errors and potential alternatives and only took into account the accumulation over time of new scientific discoveries.

Koyré' made radical criticisms of the positivist view. His studies represented a new historiography of science for several reasons. Much more than subsequent historians – e.g. Jammer, who analysed the historical evolution of some basic physical notions, such as space (1954), or Kuhn, who analysed some specific historical events, such as the Copernican Revolution (1957) – he decisively overcame the traditional history. Above all, Koyré offered an authoritative interpretation of the birth of modern science. A common evaluation is that it is still the best one on the science of this period of time.

Koyré introduced as essential elements of a historiography of science new issues, i.e. the historical context of the given case study, the particular culture shared of the time, the scientists' debate and even the philosophical assumptions, i.e. the kind of metaphysics inherent in the science of the time; in particular, he stressed that Galileo's mathematics was informed by a Platonist philosophy. By doing so, he inaugurated "a new historiography", in which the categories for an analysis of the history of science brought together scientific and extra-scientific notions.

In the following, I will suggest that in order to advance the present method of the history of science, one has as a:

First step to revisit, after half a century, the starting point of the new historiography; this programme involves a patient work of achieving a deeper comprehension of Koyré's writings.

A *second step* is to discover how Koyré's method can be applied to science after Newton's time, when the finite cosmos of ancient Greeks had been definitely abandoned and mathematics had advanced well beyond the Cartesian geometrization of space.

These two steps will be performed in Sects. 7.2–7.3 and 7.5–7.8, respectively, by means of a new conception of the foundations of science which will be presented in Sect. 7.4.

A general conclusion will be drawn about the glorious new historiography, which was capable, mainly through Koyré's studies, of coming much closer to the foundations of science than the philosophers' studies.

7.2 Koyré's Historiographic Innovations: Linking Science with Metaphysics

In a more detailed way, one may stress that:

1. Koyré changed the horizon of a historian of science in that he suggested abandoning the history of hindsight, in favour of comparing the authors and their texts in order to illustrate how true they were with respect to their times and not to ours (Cohen 1966, p. 161).
2. Firstly, he masterly illustrated both the method (Galilei) and the technical aspects (i.e. Galilei, Descartes and Newton) of a crucial period in the history of modern science, its birth.
3. He was the first to declare which foundational notions he made use of in interpreting this case study.
4. Koyré was the first to suggest exploiting the notion of infinity as a basic interpretative category of science history, as the title of his celebrated book (Koyré 1957) claims. He showed that this notion constituted the logical thread connecting the achievements of all the founders of modern science, so that its persistence in several scientists determined the birth of science as a stable intellectual structure. Hence, he shared the suspicion advanced by Burtt's lucid words (Burtt 1924, p. 303); after the great effort made by the scientists to purge science of metaphysics, the small remaining part played an essential role in the constitution of science; hence, it has to be included in the interpretation of the history of science.
5. His analysis stressed that a mathematics is not at all neutral with respect to metaphysics, rather it includes a metaphysics which has pervaded the new scientific notions and conceptions. Hence, his analysis has added evidence to what both Enriques (1919) and Husserl (1962) had already suggested, i.e. modern science hid a metaphysics in at least mathematics.
6. He summarized the metaphysics of that historical period through his celebrated formula "Dissolution of finite Cosmos and geometrization of the space", although he never explained it conclusively.

Later, both Kuhn and Feyerabend (Feyerabend 1965) also stressed a crucial influence of non-scientific issues on scientists producing new science. Yet, neither dealt with the subject of infinity. Lakatos (1976) discussed it in philosophical terms in the case study of Cauchy, yet without producing certain results (Feferman 1998, Chapter 3). Even the philosophers of science were very late in recognizing this novelty of attributing a metaphysical import to mathematics.

In the subsequent decades, the scientific and philosophical milieus, which for a long time were dominated by the positivism, rejected Koyré's historiography owing to its overcoming of the traditional separation between science and metaphysics. Koyré's studies were accused of being "idealist" (maybe, antiscience?); at last, their evaluation remained sub judice. To Koyré's interpretation of the birth of modern science an interpretation (Crombie 1953) was opposed which surely stands far from any metaphysics because it attributes to the advancements of artisans' techniques as the decisive role for this birth.

7.3 Kuhn's New Historiography

Rather than arousing opposition (as Koyré had done), Kuhn's account of the history of classical physics (1969) fired a strong interest in historiography of science. Kuhn's book was a great success, not only in terms of the number of books sold (Naughton 1982) but also in terms of audience; it suggested a new picture of scientific development that was accepted by both philosophers and laymen. In particular, Kuhn's basic word "paradigm" entered common language (Naughton 2012).

Kuhn's account became the most relevant interpretation of the history of science, although he ambitiously generalized from his historical studies of a few cases to the entire history of classical physics, which he surprisingly illustrated in a relatively short book of no more than 200 pages, using unprecedented categories lacking any verification. In addition, Kuhn claimed to have discovered both a new historical dynamics of classical physics and a new methodology for the historical interpretation of the science of all times.

Let us inspect Kuhn's interpretative method. First of all, Kuhn's studies made Koyré's historiography less objectionable. Kuhn (1969, p. 3, pp. 16–17) declared himself in agreement with both Koyré's historical method and *grosso modo* with Koyré's historical analysis of the birth of modern science (in fact, no subsequent historian has examined by means of different categories the contributions of all the scientists taken in account by Koyré). Moreover, also Kuhn's illustration of the history of science made use of categories which transcend objective scientific facts and theories. However, with respect to the old positivist tradition, these new categories represented a less radical innovation than that of Koyré, who introduced categories that were also philosophical and metaphysical. Kuhn's historiography (Kuhn 1969) introduced notions, which apparently do not pertain to metaphysics, but rather to either sociology or psychology. In addition, he left aside mathematics, considered by him – in agreement on this point with positivism – to be a mere instrument of the theories. For these reasons, Kuhn's book was much more acceptable to both scientists and historians than Koyré's.

Although his sociological and psychological notions give a pseudoscientific dress to his account, his introduction of non-scientific categories gave to his "new historiography" a non-rigorous method, in contrast to that of the hard sciences. Shortly after the publication of his celebrated book, precisely his main category, the "paradigm" made apparent the weakness of his historiography. A well-known analysis of the contexts of the occurrences of this word in Kuhn's book identified more than 20 different meanings (Mastermann 1970). In reply to Mastemann's analysis, Kuhn defended the relevance of at least two meanings of his main notion but did not remove its ambiguity (Kuhn 1970).

Moreover, other basic notions proved to be either ill-defined in their corresponding scientific disciplines – e.g. the notion of "scientific community" in the sociology of knowledge – or ill-linked to historical events, e.g. the notion of Gestalt.

All of this leads to the suspicion that there was a metaphysics hidden under some Kuhn's imprecise and polysemantic categories. Indeed, these notions leave the following questions without answers: Why did the notion of a paradigm persist as an ill-defined notion in Kuhn's mind, if not because it included an unknown metaphysics? What is the category "scientists community" if not a sociological notion implicitly covering the dominant metaphysics shared by contemporary scientists? What is a Gestalt switch if not an abrupt change of an unexplained metaphysics?

In actual fact, some scholars have expressed a similar dissatisfaction with Koyré's categories, which he so often repeated and which are the basis of his historiography. They have received a full explanation neither from Koyré nor from subsequent scholars (Coumet 1987); rather, one historian (Panza 2001) rejected them.

To worsen the situation of the new historiographies, several of its notions – e.g. Platonist mathematics in theoretical physics, incommensurability, non-cumulativity, etc. – suffered the charge of "irrationalism" because they deny that the historical development of science represents an autonomous application of human reason to reality.

Owing to the above deficiencies of its main authors, the ill-defined method of the entire new historiography was accused of being at the least dubious. Thus, it is not surprising that, at the present time, the "new historiography of science" is positively appreciated mainly because it suggests a subjective, phantastic picture of scientific development (Naughton 1982) rather than a first step towards further advancements in the history of science, all of which makes it apparent that the historical change introduced by the "new historiography" has in its turn to be interpreted.

7.4 Two Foundational Dichotomies

My first interpretative hypothesis concerning the "new historiography" takes seriously what may be considered its representative motto:

> Philosophy of science without history of science is useless. History of science without philosophy of science is blind. (Lakatos 1978 , vol. 1, p. 102)

In the light of this strong link between the two disciplines, I suggest, as a second hypothesis, that the development of the "new historiography" represented a great, inductive effort to achieve an appraisal on what constitutes the real foundations of science. Certainly, the scholars of the new historiography considered this appraisal a lateral or subordinate target of their research, which focused much more on a faithful interpretation of the historical case study at issue. However, all were aware that the public considered their implicit effort to identify the foundations of science to be the most appealing feature of their historical accounts.

Owing to the above-mentioned inaccuracy of the categories introduced by the more representative authors of the "new historiography", this effort was inconclusive; however, I will show later that this effort has come close to achieving the goal. On the other hand, I would point out that in the meantime, no better results have been obtained from the philosophers of science.

My third interpretative hypothesis concerns the foundations of science, which, as I have stressed, were what the new historiography was trying to identify. My work on the history of science suggested that they are constituted by both mathematics and logic (Drago 1991). This assertion is no longer trivial if we add the two following qualifications. Some technical analyses of the foundations of mathematics recognized in them a basic dichotomy between a mathematics including actual infinity (AI) and a mathematics that limits itself to using potential infinity (PI) only (Markov 1962; Bishop 1967, pp. 1–10). One dichotomy holds true also for logic (Dummett 1977; Prawitz 1977; Drago 1996) between classical and intuitionist logic – the latter dichotomy being equivalent to a dichotomy regarding the kind of the organization of a scientific theory, between a pyramidal organization of all deductions from few axioms (AO), as suggested by Aristotle, and an organization aimed at discovering a new scientific method for solving a universal problem (PO; Drago 1991, 1996).

Let us remark that these two dichotomies pertain at the same time to the realm of philosophy – in that no future experiment will be able to decide if one choice or another of a dichotomy is valid – and to the realm of science, in that no scientist constructing a theory can avoid choosing between them. Owing to their double nature, these dichotomies – amounting to four philosophical notions of their choices on them plus the corresponding four formal scientific theories – offer greater explicative power than all the previous philosophical notions, viz. determinism, which have been employed for interpreting the foundations of science.

7.5 Interpretation of Koyré's Categories by Means of the Choices Between the Two Dichotomies

According to the two dichotomies, the birth of modern science was a true revolution since a change occurred in the basic choices, from those of the ancient scientific attitude, PI (only the finite) and PO (apart from Euclidean geometry), to the two choices AI and AO, the first of which is represented by infinitesimal analysis and the second by Newton's deductive theory from which the new laws describing motions in the entire universe are drawn. Newton translated and emphasized the idealist nature of these choices by regarding the metaphysical notions of absolute space and absolute time as basic and, moreover, by seeing force as a metaphysical cause and in particular gravitational force as representing God's continuous intervention in the world.

In his writings examining the scientific revolution of the birth of modern science, Koyré declared his categories (with little variations each time) no less than 13 times: "Dissolution of the finite cosmos and geometrization of space". Whenever he tried to explain their meanings, he unfortunately added no more than what the above words already make intuitively apparent (Drago 1995, pp. 660–662). However, they sound well because Koyré included in his categories the notions which have traditionally been recognized as the basic ones – i.e. space and geometry.

Let us scrutinize each word of his formula in the light of the two dichotomies. The words "finite cosmos", characterizing the old theoretical world, constitute an apparent metaphor for the choice PI.

Furthermore, the word "geometrization" represents Descartes' basic tenet and indeed Descarte's invention of analytical geometry contributed greatly to the foundation of modern science. Yet, in the history of mathematics, this invention was merely a first step with respect to a more powerful advance, i.e. infinitesimal analysis, which Newton, in a historically important decision, introduced into theoretical physics. Through the notion of an infinitesimal, this calculus relied upon the manifest choice of AI; however, Koyré could not synthesize Newton's novelty through a single word ("infinitesimalization"?); hence, he was justified in merely alluding to it by means of the first important process of mathematization of nature, whose summarizing word, "geometrization", sounds intuitive to a wide audience. In sum, we may consider the word "geometrization" as a metaphor for the choice AI on which it is founded that infinitesimal analysis which for two centuries governed theoretical physics and was considered by most philosophers to be an indispensable tool for theoretical physics.

Furthermore, in Koyré's categories, the notion of "space" – above all astronomical space – may be understood as the container of all physical events, prior to their occurrence. For both Descartes and Newton, this container may have been a metaphor for an a priori organization in a systematic form of all the laws of a scientific theory, i.e. AO. In fact, both chose it for their theories.

The first word of Koyré's characteristic formula, "dissolution", alludes to a process of destruction of an organization; he refers to the organization of the cosmos, the theory of which, in the ancient times, was constructed as an answer to the most fundamental question of that time: what the world is as a whole. Hence, the organization of this theory is PO, a choice that was dismissed in modern times. In sum, Koyré's category is a metaphor for Newton's rejection of PO, which represents the alternative to the choice AO.

In conclusion, Koyré's categories represent an implicit attempt to provide a version of the two above-mentioned choices of the foundations of science by means of some intuitive notions. His categories allude to both the two positive choices and the two rejections underlying both the mathematics and the organization of Newton's mechanics, i.e. scientific theory which triumphantly concluded a process of at least three centuries of accumulation of partial contributions.

Conversely, it is no surprise if Koyré's historical account firmly relies on the choices AO and AI:

1. Disregarded much of Galilei's uncertainty, which is manifested in his last two books, on both the kind of infinity (III Journey in *Discourses*) and the kind of organization (in both books he alternated formal theorems and investigative dialogues)
2. Ignored Newton's unsuccessful effort to include optics in an AO (Shapiro 1984) as well as Newton's divergence with Huygens' optics, a theory clearly based on a problem – how a front point of a light wave may be composed of the contributions of the wavelets emitted from all the points previously attained by the front wave
3. Underrated Leibniz as an alternative to Newton in metaphysics and science

Finally:

Koyré's categories result from a translation of Newton's positive choices – AI and AO – as well as his rejections, PI and PO, into intuitive historico-scientific notions: space, dissolution, finite, etc.

Koyré was right to call the birth of modern science a "scientific revolution", because, as we have already remarked, this birth changed both basic choices; it was based on the opposite choices to the preceding ones. Moreover, Koyré's words of his two characteristic formulas came very close to the fundamental change represented by this revolution; they translated very well the four choices involved and therefore their role (positive or negative) in the revolution.

7.6 Interpretation of Kuhn's Categories by Means of the Two Dichotomies

Kuhn's account of the history of the classical physics as a whole relies upon several basic notions – i.e. normal science, scientific community, paradigm, anomaly, crisis, incommensurability, revolution, etc. A moment of reflection suggests that two of these notions – i.e. "scientific community" and "Gestalt" – constitute two pillars, in that they justify all the other basic notions that Kuhn made use of. Indeed, the former notion represents the intellectual context in which Kuhn's account is developed, and the latter notion represents a characteristic change that Kuhn suggested had occurred in the historical development of science.

Let us inspect these two notions. A "scientific community" means, first of all, an organization, i.e. the organization of all the most representative actors of the historical development of science. It is the top that makes the rules for conducting the production of new scientific laws at the bottom, since it instructs new scientists through recognized textbooks and pronounces through authoritative verdicts on the validity of the results of scientific research. Hence, through the sociological notion of "scientific community", Kuhn is referring to the pyramidal organization of a scientific theory that alludes to AO.

According to Kuhn, a Gestalt phenomenon occurs in scientists' minds in a very short period of time, i.e. it constitutes a metaphor for a change in an infinitesimal span of time, which alludes to the choice of AI. In conclusion, Kuhn's two basic notions represent metaphors for the two positive choices of Newton's mechanics, AO and AI.

In actual fact, Kuhn presented as his main category the notion of a "revolution". Yet, when we read his book and focus our attention on this notion, we notice that what the title of his book claims – i.e. as science developed some revolutionary changes of paradigm occurred – was not treated by him. In fact, the book offers a considerable amount of historical evidence for the thesis that the theories of classical physics may be all subsumed under a single paradigm, the Newtonian; even Lavoisier's well-known revolution of the birth of chemistry is attributed to some – unexplained by Kuhn – "supramechanical aspects" (Kuhn 1969, 105) which, being of a theoretical nature, link the birth of this science to Newtonian mechanics. Thus, according to Kuhn's account, no change of paradigm occurred before the twentieth century.[1]

One more fact proves that this interpretation of Kuhn's categories in his celebrated book relies on AI and AO. After the success of his book, Kuhn wanted to apply his historiography to the decisive case study of a well-known revolutionary event, i.e. the birth of black-body theory. Yet his attempt was unsuccessful, and therefore he (cleverly) investigated an old historiographical kind of problem, i.e. who has priority, Planck or Einstein. The above-mentioned foundations of science suggest a plain explanation of his failure. Since Kuhn's mind was informed by the Newtonian paradigm and its choices AI and AO, his attempt was doomed to failure, since the black-body theory is surely extraneous to such a theoretical framework but rather pertains to an opposite theoretical framework, which is based on the choices PO (see the problem of how to join thermodynamics with electromagnetism or, more specifically, the problem of the theoretical justification of Planck's formula) and PI (see quanta).

All of which suggests a strong link between Kuhn's historiography and the Newtonian theory. In general, a link between the historiography of science and Newtonian mechanics is supported by a revealing remark of a celebrated historian of science, Brush. He overtly declares that a common feature of all historians of science is to take the Newtonian paradigm as an essential reference theory.

> Newtonian mechanics was a paradigm for most scientists during the greater part of the last two centuries and is therefore the paradigm of the paradigms for historians of science… one may recall that until around 1900 it was assumed that any problem in physics could be solved, at least "in principle", by applying Newton's law of mechanics; it was only necessary determine forces and the mechanical properties of the parts of the system and then compute a solution for the appropriate set of differential equations. Since this was the most

[1] Otherwise, we obtain a paradoxical result; if Lavoisier's advances constituted a Kuhnian revolution, the Newtonian paradigm would then have to be superseded by a chemical paradigm.

> successful theory in any science, theorists [of even all other branches of sciences] tried to imitate it; but that meant adopting what was thought to be Newton's philosophy of nature as well as his scientific method. If we think that the sciences have now rejected the Newtonian paradigm, we can nevertheless use this historical case to understand what it means to be dominated by a paradigm. (Brush 1976, p. 21)

In the above, we recognize this link between both Koyré and Kuhn and the basic choices of Newton's theory. In Kuhn, this link is even stronger than in Koyré. In fact, previous papers showed that the main notions of Kuhn's historiography – i.e. paradigm, normal science, accumulation of results, anomaly, crisis, incommensurability, revolution, etc. – correspond to the basic notions of that scientific theory, Newtonian mechanics, which has constituted the one paradigm of the entire period of time examined by him, respectively: inertial system, constant velocity, integral of a function *A(c,t)* of the scientific results produced by the scientific community *c* at the time *t*, force-cause, acceleration, discontinuity in the derivative of the same function *A(c,t)* with respect to one or other of the two variables, change of the reference system, etc. (Cerreta and Drago 1989, 1991).

In addition, a general remark by Ivor Grattan-Guinness stressed that a historian might translate the Newtonian notions into historiographical categories:

> […] In other, more mathematical words, the historical space [of the historiography of science] is insufficiently defined. I use the word 'space' deliberately, for I regard the historian as working in a space in a modern mathematician's sense: a multi-dimensional region, whose dimensions are determined by the historical and historiographic factors which he brings to his studies. Historical figures are like mass-points; influences between them are like forces of attraction and repulsion, and more general influences resemble fields. A community is thus a collection of mass-points, usually in some sort of equilibrium, but vulnerable to substantial disturbance. I find the analogy useful, although I do not take it further and play Cauchy, for example, and try to set up the differential equations to represent the phenomenon. Cliometrics has not yet advanced so far. (Grattan–Guinness 1990, vol. 1, p. 6)

Cerreta and Drago's interpretation (Cerreta and Drago 1989) explains some surprising facts regarding Kuhn's historiography:

1. It had many forerunners (e.g. Planck, Fleck, etc.).
2. Kuhn's book was extraordinarily successful.
3. He excluded an application of his categories to the history of mathematics, since they translated the mathematical notions which govern Newton's dynamics into historical notions, which cannot represent abstract notions.
4. He introduced the Gestalt phenomenon as representing a unexplained historical phenomenon occurring within the scientific community, since a change of paradigm – in the above interpretation, an abrupt change of velocity – cannot be represented by means of the continuous mathematics of Newton's mechanics; hence, the discrete changes – in the parameters of the bodies – occurring during an impact have to be represented by means of an ad hoc idea.
5. His book almost ignored one important case study of classical physics, i.e. thermodynamics; it misinterpreted the birth of chemistry ("supramechanical aspects"?; Kuhn 1969, Chapter 9); actually, the mathematics of both is quite different from Newton's.

6. For the same reason – a basic difference in mathematics – he was unable to apply his categories to the case study of the birth of quantum mechanics (Kuhn 1978; Klein et al. 1979; Drago 2001, pp. 49–51).
7. In fact, Kuhn's account (Kuhn 1969) agrees with the common view shared by contemporary physicists, i.e. the basic problems in the development of classical physics have been all subsumed and solved by the theoretical resolution of the crisis that occurred in the first years of the twentieth century. Actually, they all share the belief that Newtonian mechanics is the sole paradigm of classical theoretical physics.

I conclude that:

> Kuhn's categories are a translation of the basic notions of Newtonian mechanics into some intuitively appealing notions of a historical nature, combined according to paradigmatic Newtonian dynamics.

In short, Kuhn translated into a historiographical framework the scientific framework of the theory, Newtonian mechanics, which dominated the development of science before the twentieth century. In addition, in his celebrated book, he very cleverly gave, through reference to a multitude of rapidly recalled historical case studies, a vivid picture of this mechanical-historical account; it was for these reasons that his book reached a large audience.

7.7 Foundational Remarks on the Historical Categories of Both Koyré and Kuhn

Let us recall both Koyré's and Kuhn's translations of Newton's choices and rejections of the alternatives of the two dichotomies.

They were translated in Koyré's mind into two scientific notions – finite and space – and two historical processes-geometrization and dissolution; these notions were then combined into the formula representing the categories for interpreting historical processes.

Kuhn's implicit translation included a previous step. In Kuhn's mind, a physical theory in its entirety was dominant, i.e. the theory which was dominant in the historical period at issue. As a second step, some constitutive notions of Newtonian dynamics – i.e. reference system, force, etc. – were extracted and then suitably translated into historical notions, normal science, anomaly, etc., or sociological notions, i.e. the scientific community and its Gestalt switches. Both Koyré's and Kuhn's translations may be represented through a common schema (see a representative table in Drago 2001, p. 52).

Koyré's translation obtained two characteristic features which proved to be appropriate categories for interpreting the birth of modern science to the extent that they, on one hand, summarize the basic notions of the new science and, on the other, correspond to the basic choices of the final theory to which this historical process led, i.e. Newton's mechanics.

Kuhn's translation, on the other hand, obtained a detailed historical dynamics of a Newtonian kind, Through it he depicted a historical development in which of course one theory was dominant over all others, hence a historical development of a paradigmatic kind.

These translations make apparent the unprecedented novelty of the "new historiography". Kuhn's and Koyré's analyses introduced into history of science the recourse to an interpretative schema which was to become an unavoidable method from then on. Moreover, since both schemas implicitly involved some philosophical aspects of the foundations of science – above all the basic Newton's choices – they suggested deep and wide-ranging interpretations of the history of science. Furthermore, since Newton's mechanics had played a dominant role in the science of the previous centuries, both Koyré's and Kuhn's historiographies, which essentially are linked to it, obtained a unprecedentedly high degree of faithfulness to the historical development of science; as a fact, no other historiography on the same subjects has yet been produced.

Let us proceed to a further comparison of the two authors. Koyré's categories translated both Newton's positive choices and in opposition his rejections; this complete representation of all four choices suggested to him an account of the conflictual nature of the historical events of his case study. Kuhn's translation of Newton's only positive choices, on the other hand, cannot explain any revolution; nor is the only conflict considered by his book – between Priestley and Lavoisier – could be represented as an interaction.

Let us now consider a further point. Which of the two interpretative schemas has been the more productive one with respect to the long effort by historians to discover the foundations of science?

Remarkably, Koyré introduced into the historiography of science the account of a conflict, i.e. the conflict between ancient and modern science. In addition, his categories represented this conflict as a mutual opposition of the two parts of his interpretative formula.

Kuhn's book reduced the search for the foundations of science to suggesting an interesting sociological viewpoint which however was a partial novelty because it corresponds to the scientific-cultural domination of one theory – Newtonian mechanics – over all other physical theories. Last but not least, while at a first glance this book confirmed previous Koyré's idea of representing science as a conflict – even its title announces that the book presents scientific revolutions, i.e. the highest kind of a scientific conflict – nevertheless an accurate reading of the book does not reveal such revolutions; rather Kuhn presented a peaceful development of science over two centuries (so peaceful that it included chemistry, too!). In fact, Kuhn's translation ignored Newton's basic rejections – just as the assumption of a paradigm leads to ignoring any alternative! Hence, through his categories, he illustrated a one-dimensional, merely progressive history of classical physics. For instance, he could not give an adequate account of the period of the French Revolution, when mutually conflicting scientific theories were born (Gillispie 1959). Thus, Koyré's previous hint concerning the basic conflicts within the foundations of science was cancelled by Kuhn. However, since Kuhn's categories ignored any conflict, it was easier for him to explain the main historical events than it was for previous historiographies, in particular Koyré's, which took into account that science results from a debate, if not an open conflict, among scientists and their scientific views. This is one more reason for the success of Kuhn's book (1969).

In conclusion, Kuhn's historiography constituted under several aspects a regressive move with respect to the search for foundations of science.

7.8 Interpretation of the Further Historiographies by Means of the Two Dichotomies

Apart from Koyré's and Kuhn's historiographies, the others appear to belong to two classes, i.e. those whose categories in some ways correspond to the Newtonian choices and those whose categories correspond to the opposite choices. These two kinds of historiographies are illustrated and then their main representatives are listed in a Table (Drago 2001, pp. 53–58).

Here it is enough to recall the most salient historiographies. Among the historiographies of science of the former class, the structuralist one plays an important role. It is the only historiography, apart from Koyré's, for which mathematics is basic for interpreting the history of science. However, this mathematics is the most abstract one – set theory, whose choices are of course AO and AI –; its claim is to obtain an understanding of science foundations and its ambition is to obtain an a priori comprehensive view of the history of the whole of science. It states that physics – or rather, the physics that is taught at university – is reducible to its mathematical structure, which in its turn is interpreted in terms of that theory which had for a century played a dominant role in mathematics, i.e. set theory; in sum, this historiography is drawn from a single assumption that the mathematics of set theory is fundamental for axiomatizing all scientific theories. Hence, its basic notions lack both physical meanings and philosophical meanings – apart from the philosophical meanings conveyed by both set theory and some hardly definable "theoretical terms" (Holger 2013) – of the theory interpreted by means of set theoretical models (Stegmueller 1979; Balzer et al. 1987). Of course, the choice of set theory as basic implies that this historiography chooses AI and AO, i.e. the Newtonian choices. In fact, the structuralists built their categories by following a similar track to Kuhn's. They chose as their paradigm set theory; their method is to axiomatize the theory, in the sense of applying a predicate – summarizing the entire theory through models expressed by means of set theory – to the theory represented through set theoretical models of the experimental reality. Since their basis is essentially a mathematical one, no historical notions are included by their categories. In sum, the main characteristic feature of structuralist historiography is to represent the Newtonian paradigm in a transcendent way of mathematical nature; this high degree of generality allows it to suggest detailed interpretations of the structure of each classical physical theory and moreover to suggest some links among the several theories (Balzer et al. 1987).

However, since the basic role claimed by set theory for mathematics as a whole is disputable, this kind of historiography represents an a priori assumption about the foundations of science. Moreover since this theory is characterized by the basic choices *AI* and *AO*, the representation of the actual history of science results a partial and idealized view – see, e.g. the case of chemistry (Drago 1995; Scerri 1997). In addition, it is blind with respect to all scientific conflicts.

The historiographies of Mach (1883), Bogdanov (1920), Scott (1970), Thackray (1970), Williams (1971) and Prigogine and Stengers (1977) belong to the latter class; however, subsequently to Mach, the link between the foundational choices and the categories of these historiographies is ever less clear in the next authors; moreover, these categories do not take into account the viewpoint of Newton's dominant paradigm. An exception is Scott's account of the three-century-long debate about the conflict between atomism and conservation laws. Although he did not offer an interpretative history but only a historical description of the long series of events concerning this debate, he reiterated Koyré's representation of the history of science as a conflict among some basic notions. No doubt, this conflictual representation constitutes a superior historical view, since it is the only one which in principle can explain why classical physics underwent the crisis of the early twentieth century, during which Einstein looked for a "reconciliation" of the theories at issue (Einstein 1905, p. 891).

7.9 Koyréian Historical Categories for the History of the Exact Sciences in Nineteenth Century

Now, I will confirm the great relevance of Koyré's historiography. I will extend the import of my previous interpretation of Koyré's characteristic formula to suggesting a new version of it which is suitable for interpreting the history of science after Newton.

Let us consider the first scientific theories whose fundamental choices are manifestly different from Newton's (i.e. chemistry, Lazare Carnot's mechanics, Sadi Carnot's thermodynamics). At present, the physical theories of the two Carnots (Gillispie and Pisano 2014) are valid only partially, because L. Carnot's mechanics makes use of equations and concepts which have to be in part interpreted (Drago 2004) and S. Carnot's thermodynamics relies on the obsolete notion of caloric (Ibidem). The third theory, on the other hand, is valid even at present (it has been "subsumed" by both physical chemistry and quantum chemistry, but these facts do not concern us here). By putting the history of chemistry in relation to its two fundamental choices, one obtains some suggestions at the level of Koyré's categories.

Let us start from the rejections. Chemists saw AO as the typical organization of Newtonian mechanics; this theory explains all phenomena by means of a crucial principle, the force-cause (or even the cause – God), whose mathematical function gives rise, through differential equations, to all the laws composing the theory. Thus, the force-cause summarizes and is a good representative of the AO of Newton's mechanics. In addition, from the viewpoint of their essentially operative activity, the chemists saw the calculus relying on AI as so abstract as to judge it to be a fictitious representation of reality. One may characterize the historical process of loss of relevance of the key notions of Newton's theory concerning the organization, with the word "Evanescence […]".

About the positive choices, we notice that chemistry is a theory aimed at solving a basic problem, how to explain the organization of matter; hence, it is a PO theory, and its basic notion is "matter" rather than "space" – as it was during the birth of modern science and also in Koyré's historiography; hence, the choice PO may be translated by the word "matter", understanding the problem to explain the organization of matter.

In addition, chemistry supported the view of matter as constituted by atoms, which suggested a discrete mathematics, which corresponds to the choice PI. Thus, we naturally compose the phrase "discretization of the matter". We obtain the following table (Table 7.1).

I avoid illustrating confirmations coming from the historical accounts of classical chemistry, Lazare Carnot's mechanics, Sadi Carnot's thermodynamics, Faraday's electromagnetism and kinetic theory of gases, since these confirmations are shown in Drago (1994 pp. 676–678). These confirmations lead to say that the new Koyréian formula applies well to the nineteenth-century theories, which are alternative to the Newtonian theory.

It is very remarkable that the same formula is substantially valid beyond this period since it grasps the essential features of the two great theories of the first half of the twentieth century. Quantum mechanics introduced an extreme process of discretization, concerning not only matter but also radiation; moreover, it certainly eliminated the notion of force-cause (e.g. recall quantum jumps). Special relativity abandoned the notion of force since force is not invariant under Lorenz transformations; moreover, this theory represents the "discretization of matter" in an abstract, algebraic way; for the first time, it founded a physical theory on a transformation group rather than on a differential equation. (General relativity instead led to the extreme consequences of Koyré's formula for Newton's theories; it "geometrized" even mass, through the use of an infinitesimal metric representing the choice AI,

Table 7.1 The relationship between the two dichotomies and the historical categories

	Newtonian mechanics	Koyré	Chemistry	Koyréian formula
AO	+	space	–	force-cause
PO	–	dissolution	+	matter
AI	+	geometrization	–	evanescence
PI	–	finite cosmos	+	discretization

which is opposed to the choice PI of the discrete world presented by quantum mechanics.) These last notes show that the Koyréian formulas give an account of the history of physics from Newton's time until the theories of the first half of the twentieth century, including the theoretical conflict between quantum mechanics and special relativity (and also between the latter and general relativity). Hence, they represent the greatest possible advancement of Koyré's studies one can hope to obtain through an improvement of the original formula.

7.10 Conclusion

In the above, a re-visiting of the entire history of the "new historiography" was at stake.

Through his analyses of all the original texts relating to the birth of modern science, Koyré introduced a new historiography of science which for the first time was of an interpretative kind according to philosophic-scientific categories.

Unfortunately, Koyré's innovations have been considered by subsequent historians as constituting no more than an attempt at a well-founded historiography. Kuhn's historiography received the most attention; it was however less profound because its interpretative categories correspond to the basic notions of the Newtonian paradigm of the history of theoretical physics suitably translated into historico-sociological notions. Indeed Kuhn's study of the revolution in the Newtonian paradigm, the birth of black-body theory, represented the failure of Kuhn's previous categories. Hence, Kuhn added no innovation to the search of the foundations of science. A group of historians called "structuralists" suggested a historiography which seemed a great novelty, although it assumed again an unexplained paradigm, i.e. set theory. Some other historiographers tried to apply Lakatos' methodology to a rational reconstruction of the history of statistical mechanics, a theory considered (also by Kuhn) to be a first important change in the Newtonian paradigm; but they were unsuccessful, because an accurate inspection of the foundations of this theory would be necessary (de Regt 1996, pp. 31–32).

Despite the fact that these historians have all produced numerous innovations, at present, the historiography of science has entered a blind alley. One first reason is that few subsequent historians of science have taken up the methodological suggestions of the previous innovators. A second reason is that these innovators, apart from the structuralists, have been accused of being irrationalist (one may suspect that actually this charge implicitly assumes the Newtonian paradigm as an indispensable framework for historians). As a matter of fact, readers' interest in the history of science was lessened considerably, as a consequence of:

1. The exhaustive analyses of all the most interesting study cases by means of the new historiography.
2. The wealth of suggestions by the new historiographers (3) the appeal for the advent of a new philosophical insight into the foundations of science (see, e.g. de Regt).

Hence, may reasonably suggest that without a decisive philosophical answer to the question what the foundations are (possibly by reflecting upon the philosophical advances suggested by historians' works), the historiography of science is no longer able to produce interesting accounts on major case studies.

3. Therefore, the second part of Lakatos' motto, recalled above ("[...] History of science without philosophy of science is blind") posed the crucial problem of what relationship there should be between the first and the second, a problem that requires a conclusive solution.

In other words, the time has come to no more perform historical analyses of an important case study in the hope to discover the foundations of science by mere ingenuity and hence to consider as definitely closed the period of the "new historiography", whose most important, though implicit, aim was the recognition of the foundations of science; rather, the time has come to produce historiographies whose categories, by avoiding any paradigmatic theory, refer to a declared conception of the foundations of science. Already 30 years ago, the structuralists declared at the outset of their analyses their commitment to a particular foundation of science, yet they unsatisfactorily considered it as constituted by almost only mathematical and logical notions.

In the above, I have presented my conception of the foundations of science, the two basic dichotomies, as obtained from the categories of the major historians. They have already been applied to obtain accounts of the main historical case studies, the black-body case included (Drago 2013a, b). Moreover, their translations into both Koyré's and Koyréian categories make it possible to interpret all the main case studies of the history of physics also through these categories of subjective kind.

According to these dichotomies, a retrospective view suggests that Koyré's work casts a powerful light on the entire historiography of science. He offered a master's account of the case study of the birth of modern science. After almost 60 years, of all subsequent historiographies, this historiography was the one that came closest to an understanding of the foundations of science. Moreover, the previous Sect. (7.8) showed that an interpretation of Koyré's categories suggests how to improve them in new categories for a detailed interpretation of the main case studies of classical physics – i.e. the same aim as that of Kuhn (1969) – and, in addition, the theories born after the crisis of the early twentieth century, the case study to which Kuhn (1978) unsuccessfully tried to apply his categories. Moreover, it was a great merit of Koyré that he, through his studies of historical nature, was more effective than the previous philosophers of science and subsequent historians of science in undertaking the search for the foundations of the science – a task that would have been more appropriately carried out by the former.

For these reasons Koyré's novelty may be called a true revolution in the history of science.

Acknowledgement I am grateful to Prof. David Braithwaite for having revised my poor English and to an anonymous referee for some important suggestions.

References

Balzer W, Moulines CU, Sneed JD (1987) An Architectonic for Science: The structuralist Program. Reidel, Dordrecht.

Bishop E (1967) Foundations of Constructive Analysis. Mc-Graw Hill, New York.

Bogdanov A (1920) Nauka i rabochyi klass. Moscow [*Id.* (1970) Italian Translation: La Scienza e la classe operaia. Bompiani, Milano, 1970).

Burtt EA (1924) The Metaphysical Foundations of Modern Physical Science. Routledge and Kegan, London.

Brush SG (1976) The Kind of Motion we call Heat. North–Holland, Amsterdam.

Cerreta P, Drago A (1989) The Conceptual Structure of "The Structure of Scientific Revolutions" by T.S. Kuhn. In Krafft F, Scriba CJ (eds). XVIII International Congress of History of Science: Science and Political Order. Abstracts' section. Steiner, Stuttgart (1st–9th August Hamburg–Munich).

Cerreta P, Drago A (1991) Matematica e conoscenza storica. La interpretazione di Kuhn della storia della scienza. In Magnani L (ed). Conoscenza e Matematica. Marcos y Marcos, Milano, pp. 353–364.

Cohen IB (1966) Alexandre Koyré (1982–1964). Commemoratio. Isis 57:157–165.

Coumet E (1987) Alexander Koyré: La révolution scientifique introuvable? History and Technology 4:497–529.

Crombie A (1953) Robert Grosseteste and the Origins of Experimental Science, 1100–1700. The Clarendon Press, Oxford.

de Regt H (1996) Philosophy of the Kinetic Theory of gases. British Journal of Philosophy of Science 47:31–62.

Drago A (1991) Le due opzioni. Una storia popolare della scienza. La Meridiana, Molfetta–Bari.

Drago A (1994) Interpretazione delle frasi caratteristiche di Koyré e loro estensione alla storia della fisica dell'ottocento. In Vinti C (ed). Alexandre Koyré. L'avventura intellettuale. ESI, Napoli, pp. 657–691.

Drago A (1995) Il caso della teoria chimica come rivelatore dei limiti della interpretazione strutturalista della scienza. In Amat di Sanfilippo P (ed). Proceedings VI Convegno Fondamenti e Storia della Chimica. Rendiconti della Accademia Casse di Scienze XL 113/29:269–285.

Drago A (1996) Mathematics and alternative theoretical physics: The method for linking them together. Epistemologia 19:33–50.

Drago A (2001) The several categories suggested for the "new historiography of science": An interpretative analysis from a foundational viewpoint. Epistemologia 24:48–82.

Drago A (2004) A new appraisal of old formulations of mechanics. American Journal of Physics 72:407–409.

Drago A (2013a) The Relationship Between Physics and Mathematics in the XIXth Century: The Disregarded Birth of a Foundational Pluralism. In Barbin E, Pisano R (eds). The Dialectic Relations Between Physics and Mathematics in the XIXth Century. Springer, Berlin, pp. 159–179.

Drago A (2013b) The emergence of two options from Einstein's first paper on quanta. In Pisano R, Capecchi D, Lukesova A (eds). Physics, Astronomy and Engineering. Critical Problems in the History of Science and Society. Scientia Socialis Press, Siauliai University, pp. 227–234.

Dummett M (1977) Elements of Intuitionism. The Clarendon Press, Oxford.

Einstein A (1905) Zur Elektrodynamik bewegter Körper, Annalen der Physik, 17:891–921.

Enriques F (1919) Il significato della critica dei principi nello sviluppo delle matematiche. Scientia 12:176–177.

Feferman S (1998) In the Light of Logic. The Oxford University Press, Oxford.

Feyerabend P (1965) Problems of Empiricism. In Colodny RG (ed). Beyond the Edge of Certainty: Essays in Contemporary Science and Philosophy. Prentice–Hall, NJ, pp. 145–260.

Gillispie CC, Pisano R (2014) Lazare and Sadi Carnot. A Scientific and Filial Relationship. 2nd edition. Springer, Dordrecht.

Gillispie CC (1959) The Encyclopédie Française and the Jacobin Philosophy of Science. In Clagett M (ed). Critical Problems in the History of Science. The University Wisconsin Press, Madison, pp. 255–289.
Grattan–Guinness I (1990) Convolutions in French Science. 1800–1840. Birkhäuser, Berlin.
Holger A (2013) Theoretical Terms. In Zalta NE (ed). Stanford Encyclopaedia of Philosophy. Via: http://plato.stanford.edu/entries/theoretical-terms-science/
Husserl E (1962 [posthumous]) The Crisis of European Sciences and Transcendental Phenomenology. Basic Books, New York.
Klein MJ, Shimony A, Pinch TJ (1979) Paradigm Lost? Review. ISIS 70:429–440.
Koyré A (1939) Études Galiléens. Hermann, Paris.
Koyré A (1957) From the Closed World to the Infinite Universe. The University of Maryland Press, Baltimore.
Koyré A (1965) Études Newtoniens. Gallimard, Paris.
Kuhn TS (1957) The Copernican Revolution. Planetary Astronomy in the Development of Western Thought. The Harvard University Press, Cambridge–MA.
Kuhn TS (1969 [1962]). The Structure of Scientific Revolutions. Chicago University Press, Chicago.
Kuhn TS (1970) Reflections on my Critics in Critcism and Growth of Knowledge. In Lakatos I, Musgrave A (eds). London, Cambridge University Press, 231–278.
Kuhn TS (1978) Blackbody Theory and the Quantum Discontinuity, 1894–1912. The Clarendon, Press, Oxford.
Jammer D (1954) Concepts of Space. Harper, New York.
Lakatos I (1976) Proofs and Refutations: The Logic of Mathematical Discovery. The Cambridge University Press, Cambridge.
Lakatos I (1978) The Methodology of the Scientific Research Programs. The Cambridge University Press, Cambridge, pp. 102–137.
Mach E (1883) Die Mechanik in ihrer Entwickelung. Brockhaus, Leipzig [English Translation: *Id.*, (1960) The Science of Mechanics. Open Court, Lasalle].
Markov A (1962) On Constructive Mathematics. Trudy Mathematicieskie Instistut. Steklov, 67:8–14 [English Translation: American Mathematical Society Translations 98/2:1–9].
Mastermann M (1970) The Nature of a Paradigm. In Lakatos I, Musgrave A (eds) Criticism and the Growth of Knowledge. The Cambridge University Press, Cambridge, pp. 59–89.
Naughton J (1982) Revolution in science: 20 years on. New Scientist 95:270–275. Via: http://books.google.it/books?id=MdyCKvgw4ksC&pg=PA374&dq=margaret+mastermann+kuhn+paradigm&hl=it&sa=X&ei=RW9wVPqNIYWfygOuhIGIDw&ved=0CCIQ6AEwAA#v=onepage&q=margaret%20mastermann%20kuhn%20paradigm&f=false
Naughton J (2012) Thomas Kuhn: the man who changed the way the world looked at science, The Guardian, 9 August. Via: http://www.theguardian.com/science/2012/aug/19/thomas-kuhn-structure-scientific-revolutions
Panza M (2001) La révolution scientifique les révolutions et l'histoire des sciences. Comment Ernest Coumet nous a libérés de l'héritage d'Alexandre Koyré. Revue de Synthèse, 2/3/4:411–424.
Prawitz D (1977) Meaning and Proof. Theoria 43:2–40.
Prigogine I, Stengers I (1977) The new alliance. Scientia, 112:319–332.
Scerri E (1997) Has the periodic table successfully axiomatized? Erkenntnis 47:229–247.
Scott WL (1970) The Conflict between Atomism and Conservation Theory 1644 to 1860. Elsevier, New York.
Shapiro A (1984) Experiment and mathematics in Newton's Theory of Color. Physics Today, 37:34–43.
Stegmueller W (1979) The Structuralist View of Theories. Springer, Berlin.
Thackray A (1970) Atoms and Powers: An Essay on Newtonian Matter–Theory and the Development of Chemistry. The Harvard University Press, Cambridge, MA.
Williams P (1971) Entry: Michael Faraday. In Gillispie CC (1970–1980) (ed). Dictionary of Scientific Biography. IV. Charles Scribner's Sons, New York, p. 531.

Chapter 8
Alexandre Koyré's Essential Features of the Scientific Revolution

Daria Drozdova

Abstract In this chapter I am going to (a) examine the logical connections between various descriptions of the Scientific Revolution proposed by Alexandre Koyré and (b) propose an attentive and detailed reading of texts written by Koyré in different periods of his life in order to identify various aspects of his interpretation of the revolution in thought that occurred in early modern Europe. His most famous description of the Scientific Revolution (the dual characterization) indicates two aspects of the process that led to the emergence of classical physics: "destruction of the Cosmos" and "geometrization of space". However, Koyré frequently used other expressions for characterization of the period, such as "mathematization of nature" or transition "from the world of more or less to the universe of precision" and "from the closed world to the open universe". One could expect that Koyré would try to reduce his initial dual characterization to one single formula. I argue here that, on the contrary, the duality of description had a special meaning which permits us to keep in focus the complexity of the intellectual change that occurred during the seventeenth century, when a new science was rising from a new conception of reality and a new world-view was emerging from the new science.

Keywords Scientific Revolution • Intellectual revolution • Dual characterization • Destruction of the Cosmos • Geometrization of space • World-views • Mathematization

8.1 Introduction

The larger part of Alexandre Koyré's historical research was dedicated to the Scientific Revolution of the seventeenth century. From the early 1930s, he tried to establish and express the "deepest meaning and aim" of the radical intellectual transformation that gave birth to modern science. He intentionally refused to search

D. Drozdova (✉)
National Research University Higher School of Economics,
Staraya Basmannaya 21/4, 105066 Moscow, Russian Federation
e-mail: ddrozdova@hse.ru

R. Pisano et al. (eds.), *Hypotheses and Perspectives in the History and Philosophy of Science*, DOI 10.1007/978-3-319-61712-1_8

for causes and origin of the Scientific Revolution and believed that the phenomenon first had to be described and only then explained. Koyré's most famous description of the Scientific Revolution was expressed through a dual characteristic, i.e. the "destruction of the Cosmos" and the "geometrization of space". The first appearance of this description took place in 1935, when Koyré published the first article of *Études galiléennes – Au l'aurore de la science moderne. La jeunesse de Galilée.* He used this characteristic continuously until his very last publications.

This characteristic is rightly regarded as the most accurate expression of his position on the Scientific Revolution (Coumet 1987; Drago 1994). Nevertheless, there are some changes and variations in the description that emerge in the development of Koyré's thought. The aim of this chapter is to distinguish and elucidate Koyré's different descriptions of the Scientific Revolution, such as the "mathematization of reality", the "infinitization of the Universe" and the "transition from the world of more or less to the world of precision". In this context, two main questions arise:

First, whether variations in description tend to present some new aspects of the process or they are simply correcting and defining the features already announced.

Second, whether all the characteristics can be reduced to a single one, for example, to the mathematization of nature (Cohen 1994).

Answering the latter question, I argue that the duality of characterization has a special and unreducible meaning. It allows us to focus on two levels of changes, which Koyré used to distinguish. The first one is a transformation in scientific theories and methods – a process that affects a very limited social group of natural philosophers, mathematicians and engineers. The most evident expression of this strictly *scientific* revolution is a formulation of the principle of inertia and elaboration of Newtonian mechanics. The second level is a transformation of a world-view, acknowledging a new image of the Universe, unlimited and even infinite, which affects the reasoning and imagination of the wider population (even if it is still a limited group of the educated class).

The main thesis of Koyré is that the transformation in scientific theory has a metaphysical basis and implies a change in the representation of physical reality. The very subject of Koyré's study is the intellectual revolution, that is, spiritual and mental transformation, "which changed the very framework and patterns of our thinking and of which modern science and modern philosophy are, at the same time, the root and the fruit" (Koyré 1957, p. vii). In what follows, I examine the interplay of various descriptions of the Scientific Revolution proposed by Koyré in order to demonstrate that he recognizes and keeps bonded, but separated, two different levels of change that occurred in that period.

I claim that through "geometrization of space," he described an aspect that has more significance for scientific thought. "Geometrization of space" allowed natural philosophers to develop new ideas of motion, recognize the principle of inertia and unify the world in homogeneous space. All of his was very important for the development of a new physical theory. With the "destruction of the Cosmos", by contrast, Koyré described the process that affected the general representation of the human world. This transformation of the image of the world had an enormous influence on the imagination of the common man and created an anxiety before the boundless, empty spaces of the new Universe. The dual descriptive characteristic that Koyré proposed reflects the complexity and ambiguity of this complex revolution.

8.2 Destruction of the Cosmos and Geometrization of Space

Alexandre Koyré considered the Scientific Revolution of the seventeenth century to be a radical transformation of scientific theory, which was accompanied by a radical transformation of the basic philosophical concepts. During this process, a new ontology of the material world arose, which changed the image of the Universe. From the very beginning, Koyré characterized the process of developing a new physics by appealing to two phenomena: the *destruction of the Cosmos* and the *geometrization of sp*ace.

These two characteristics first appeared in the article *Au l´aurore de la science moderne. La jeunesse de Galilée* published in *Annales de l´Université de Paris* in 1935. This article later became the first part of *Études Galiléennes*. In the beginning, Koyré discussed various descriptions and explanations of the deep intellectual transformation in early modern Europe that led to the emergence of a new scientific culture. Sometimes the rise of modern science was explained by a "deep spiritual upheaval" that involved the most fundamental human attitudes. In this interpretation, modern man aimed at the domination of nature *(vita activa)*, whereas ancient man sought contemplation *(vita contemplativa)*. Koyré was opposed to this kind of interpretation because he believed that such an approach expressed a certain blindness to the practical and technical ideas of previous centuries and to the alchemical and magical desire to transform the material world. He also criticized a position, which claimed that scientific progress expressed the triumph of empirical research over empty theoretical speculations. Koyré argued that the new science was based not on *experience* but on *experiment,* which is inconceivable without a theoretical basis. Essentially, it was the decision to interpret nature in mathematical, geometrical terms that gave the foundation to modern experimental science. Koyré also saw as insufficient the description of a new science that took into account only its theoretical innovations. He argued that we could easily identify classical physics with its basic concepts such as momentaneous velocity, the principle of inertia, the three laws of Newtonian mechanics and so on. But the main question, as Koyré argued, was not the postulates of the new theory, but the conditions that made the new conception of motion natural and "self-evident".

In this context, Koyré offered, for the first time, his own description of the processes that underlay the rise of modern science:

> We believe that the intellectual attitude of the classical science can be characterized by two connected features: the geometrization of space and the destruction of the Cosmos, that is, the disappearance from science of all considerations based on this concept and substitution of the abstract space of Euclidean geometry for the concrete space of pre-Galilean physics. This substitution one permits to formulate the principle of inertia. (Koyré 1935–1936; see also 1966a [1939], p. 15)

While the thesis on the geometrization of space seems to be quite clear, the meaning of the "destruction of the Cosmos" needs to be clarified. This idea came to Koyré in the early 1930s, when he commented the works of Copernicus, Galileo and Kepler during his seminars in *École Pratique des Hautes Études*. In the presentation of the course of 1933–1934, Koyré wrote:

> While Copernicus, the heir of the Pythagorean and Neoplatonic tradition of metaphysics of light, elaborates the astronomical system based on his vision of the Cosmos as harmony, series of regular bodies, etc. (as we have shown in our edition of *De Revolutionibus*), this kind of consideration totally disappears from Galileo's thought. Moreover, while the cosmological ideas are the integral part of Kepler's argumentation, these ideas have no place in Galileo's reasoning. (Koyré 1986, pp. 42–43)

And 1 year later, he added:

> The study of the correspondence between Kepler and Galileo totally confirms the previously found difference in their styles of thought: Kepler reasons in terms of the Cosmos whereas Galileo completely ignores this concept. (Koyré 1986, p. 44)

Later, Koyré concluded that the idea of the Cosmos had not disappeared completely from Galileo's thought. However, he maintained the conviction that the rejection of the Cosmos (i.e. disappearance of arguments based on the notion of harmony, hierarchy or structural order) was the most important characteristic of the new scientific thinking.

Once stated, Koyré used these characteristics continuously until his very last publications (Koyré 1943, pp. 403–404; 1965 [1950], p. 6–7; 1955, p. 107; 1957, p. viii). Nevertheless, there were some changes in the description. Although the basic characteristics remain unchanged – it was always the destruction of the Cosmos and the geometrization of space – the descriptions slightly changed. For example, in the article *Galileo and Plato* (1943), Koyré literally repeated the passage from *Études Galiléennes* and then explained:

> [...] The dissolution of the Cosmos means the destruction of the idea of a hierarchically-ordered finite world-structure [..] a qualitatively and ontologically differentiated world, and its replacement by [...] an open, indefinite and even infinite universe, governed by the homogenous universal laws; a universe in which, contrary to the traditional opposition of the two worlds of Heaven and of Earth, all things are on the same level of Being. (Koyré 1943, p. 403)

Then, in his most famous book, *From the Closed World to the Infinite Universe*, Koyré presented an even more elaborate account:

> […] This scientific and philosophical revolution can be described roughly as bringing forth the destruction of the Cosmos, that is the disappearance, from philosophically and scientifically valid concepts, of the conception of the world as a finite, closed, and hierarchically ordered whole (a whole in which the hierarchy of value determined the hierarchy and structure of being, rising from the dark, heavy and imperfect earth to the higher and higher perfection of the stars and heavenly spheres), and its replacement by an indefinite and even infinite universe which is bound together by the identity of its fundamental components and laws, and in which all these components are placed on the same level of being. (Koyré 1957, p. 2)

These metaphysical transformations contributed to the formation of the new physics and new astronomy. The destruction of the Cosmos and the unification of the Universe made it possible to transfer physical reasoning to celestial bodies, while the geometrization of space contributed to a new understanding of movement.

However, we want to draw attention to some changes in how Koyré introduced and described these characteristics. Initially, in *Études Galiléennes*, they seemed to represent not processes but the initial and final stages of transformation. Koyré described "the destruction of the Cosmos" as a sort of abandonment of a previous intellectual practice: "the disappearance in science of all considerations based on that notion" (Koyré 1943, p. 404). In place of the conception of the structured Cosmos came an idea of geometric space, which became the organizing principle for the unity and coherence of the physical world. Therefore, "the geometrization of space" was considered as an outcome of the process.

However, in his following works (starting with *Galileo and Plato*, 1943), Koyré preferred to describe this transition as a two-level, parallel process: a hierarchical Cosmos was replaced by a homogeneous Universe, and Aristotelian physical space was replaced by geometric abstraction. This is clearly seen in the parallel structure of the sentences by which Koyré explains it:

> the destruction of the cosmos, that is, the *substitution* for the hierarchically structured finite world of the Aristotelian tradition of the infinite universe bound together by the uniformity of its fundamental components and laws; and (ii) the geometrization of space, that is, *substitution* for the concrete physically structured place–space of Aristotle of the abstract, isomorphous, and infinite dimension-space of Euclidean geometry now considered as real. (Koyré 1955, p. 107. Author's *Italic*)

Koyré's parallelism also implies that these two processes are strongly connected. The geometrization of space necessarily leads to the destruction of the ancient Cosmos: in a homogeneous space of Euclidean geometry, the radical separation of the Earth from the other celestial spheres is unthinkable, since the Earth cannot be allocated to a unique place. The concept of the ordered Cosmos does not allow, on the other hand, for an abstract geometric space. Thus, these two articulations can be seen as two aspects of one process.

We suppose that Koyré's double articulation of a single process has a special meaning. It enables us to focus on two levels of change that occurred during the Scientific Revolution, which we describe as *scientific* revolution and *intellectual* revolution. The geometrization of space is more important for scientific thought and the development of a new physical theory of classical science. It allows for the development of new ideas of motion by which physical principles such as inertia become natural and self-evident. The destruction of the Cosmos, and the recognition of the Universe as infinite and homogeneous, affects the general view of the world in which man lives. This transformation of the image of the world is not only scientific, but it also covers much wider social strata. That permeates the imagination of the common man and creates a feeling of anxiety in front of the endless expanses of an empty new Universe, the evidence of which fills the literature of the seventeenth and eighteenth centuries (Nicolson 1950; Seidengart 2006).

8.3 Mathematization of Nature and Mathematization of Science

Koyré, however, was not fully satisfied with this dual characterization. He was constantly looking for other synthetic formulas. For example, in other texts, he specified that the emergence of a new science involved not only the destruction of the Cosmos and the geometrization of space but also the infinitization of the Universe (Koyré 1957, 1971 [1949]). He also pointed out that along with these processes, the concept of motion process was supplanted by the concept of motion state. But first and foremost, Koyré believed that the general sense of the dual characteristics could be reduced to a more concise formula: "the mathematization (geometrization) of nature and, therefore, the mathematization (geometrization) of science" (Koyré 1943, 1965 [1950]).

In his research on Cartesian and Galilean physics, Koyré paid a lot of attention to the theme of mathematization or, more precisely, geometrization of nature. It was clear for him that the basic content of this new natural science consisted precisely in creating a "mathematical physics", i.e. the study of nature, based on exact mathematical reasoning and mathematical methods. This led him to conclude that the deep meaning of the Scientific Revolution of the seventeenth century consists in the mathematization of physical reality (Koyré 1966b [1957]).

However, this characteristic by itself was not anything new. Even before Koyré, there was no doubt that modern science was based on mathematics. In his own time, there were many philosophers who described the Scientific Revolution in terms of "mathematization". The special role of mathematics in the rise of modern science was emphasized, for example, by Cassirer. In the 1920s, the theory of the mathematization of the world was developed by Edwin Burtt (1925); the geometrization of space by Galileo was a special theme of The *Crisis of European Sciences* by Edmund Husserl (1936); Martin Heidegger described the essence of modern science as a mathematical project (*Entwurf*) that enframed nature for measure and calculation (*Die Frage nach dem Ding*, 1935–1936). Dutch mathematician and historian Eduard Jan Dijksterhuis (1924, 1950) stressed in his writings that true scientific knowledge was the knowledge of the quantitative relationships between physical phenomena, the essences of which were not available to us. Science could not know the essence of things, but it could describe them using the language of mathematics. Therefore, in the history of science, special attention should be paid to the birth of mathematical methods (Cohen 1994).

For Koyré, mathematization was something more substantial than a mere adoption of language that helped us to express and to organize facts. Koyré first considered the mathematization of nature as a transformation of the relation between mathematics and physical reality. The birth of a new physical science was accompanied by a new vision of the world according to which mathematical forms were not a mere external framing but the true inner essence of material reality. The mathematization of science (i.e. the application of mathematical methods of research to nature and the elaboration of experiments based on measurement) was a result of this mathematization of nature.

Even if Koyré proposed the mathematization to be an expression and synthesis of his previous dual characterization, the description of the Scientific Revolution in terms of mathematization states both more and less than dual characterization alone. It states less, because the emphasis on the mathematical side of science put out of focus the cosmological and anthropological transformation that occurred during the Scientific Revolution. It states more, because the mathematization of nature had wider meaning than the geometrization of space. Of course, the geometrization of space was essential to the principle of inertial motion, but it was not enough for the full elaboration of a new mechanics. Koyré demonstrated that the most challenging problem for the fathers of mathematical physics was the mathematization of time and of time-agent causality. The extreme geometrization of physical processes did not adequately describe the actions of causes or the factor of time. Restricting the Scientific Revolution to only "the geometrization of space" would not adequately appreciate the effort it took to create a new, mathematical physics. Therefore, the "mathematization of nature" expresses more than "the geometrization of space" alone.

Nevertheless, Koyré's description of the Scientific Revolution in terms of the mathematization of nature does not seem to be complete and satisfying. The mathematization had not been a single and uniform process, equally understood by all participants. For example, Galileo and Descartes, who both made undeniable contributions to the development of mathematical physics, were inspired by the ideal of accuracy and reliability that was inherent to mathematical knowledge. Both of them wanted to achieve such precision in other spheres of knowledge, especially in the physical sciences. At the same time, they were moved by very different ideals. Galileo's application of precise mathematical and geometric reasoning was based on the idea of the perfect embodiment of geometric shapes in matter. Descartes' mathematization of physics was subordinated to the desire to achieve absolute certainty in scientific thinking, which was organized according to a deductive model of mathematical proof. Galileo's mathematization was, firstly, a geometrization of nature, and it affected the ontological representation of the internal structure of the material world. Descartes offered a model of mathematization that we can call rational and formal, and it first required an epistemological adjustment.

The mathematization of nature and the mathematization of science seem to be very important processes for the seventeenth-century investigation of nature. Koyré often emphasized that the unity of physics and mathematics that came into being during the seventeenth century subsisted till the most recent times. All the theoretical transformations in the physics of the seventeenth century cannot be compared to the Scientific Revolution, because they don't change the most essential feature of the modern natural science, i.e. its mathematical character. Nevertheless, the "mathematization of nature" seems to cover only scientific level of transformation, and the revolution in the world-view is moved out of attention.

8.4 From the World of the "More or Less" to the Universe of Precision

The mathematization of nature and of science, discussed previously, had one important consequence. Koyré characterized it as a substitution of the "empirical world of the more or less" (*monde de l'à-peu-près, du plus ou moins*) by the "rational universe of precision" (*univers de la précision*) (Koyré 1956). This characteristic allowed an emphasis on some general consequences of the mathematization of nature, its impact on the perception of the world of everyday life and its importance for the emergence of modern technology as well as experimentation based on measurement.

This description appeared in Koyré in the late 1940s when he began studying the history of technology (Koyré 1971a [1948]; 1971b [1948]), even if in Koyré's earliest works on Galileo we can find an initial sketch of this idea. Describing Galileo's Platonism and his efforts to create a mathematical physics, Koyré indicated that Platonic realism led to a reconsideration of empirical reality on the basis of an ideal, mathematical reality (Koyré 1966a [1939], p. 207). In this context, Koyré emphasized that the qualitative world of Aristotelian science was deprived of certainty because it did not conform to the rigidity and precision of mathematical concepts. With the advent of post-Galilean physics, the nature of the world radically changed. The uncertainty of sensorial data and of essential qualities was replaced by the certainty of figures and numbers.

The mathematization of nature, the replacement of physical places by abstract geometrical space and the formation of a new ontology of the physical world in which there is no place for sensible qualities, values and meaning, all these signified that "classical science [...] has substituted a world of quantity for that of quality, [...] a world of being for the world of becoming and change" (Koyré 1965 [1950], p. 8). In fact, Koyré argued, modern physics applied rigid and stable mathematical and geometrical concepts to physical reality. However, to make this possible, the conception of the physical world itself had to be changed because the Aristotelian world as well as the world of everyday life is not the world of mathematical precision. Therefore, Koyré concludes that:

> [...] the deepest meaning and aim of [...] the whole Scientific Revolution of the seventeenth century [...] is just to abolish the world of the "more or less", the world of qualities and sense perception, the world of appreciation of our daily life, and to replace it by the (Archimedean) universe of precision, of exact measures, of strict determination. (Koyré 1965 [1950], pp. 4–5)

The world of everyday life as well as the Aristotelian world is a domain of variable, imprecise approximation. Is it possible, for example, to indicate the precise dimension of a living being? Koyré believed that we cannot even pose such a question: "a horse obviously is bigger than a dog and smaller than an elephant, but neither horse nor dog or elephant have a precisely determined dimension. There is always some level of approximation and of "more or less" (Koyré 1971a [1948], p. 342). In the Aristotelian world, all physical beings were similar to animals. They did not have an exact shape or size. They escaped from dominion of exact numbers and measure.

Therefore, mathematics could not be applied to living and mobile beings. Mathematics belonged to the realm of abstraction, to an ideal world that was separated from the empirical reality of senses. Only elimination of this gap made mathematical physics possible.

However, Koyré recognized that not the entire ancient world was subject to imprecision. Precision could be applied only to the sublunary world. The heavens, instead, obeyed mathematical laws since ancient times. Astronomy sought hidden geometric regularity in the motions of the planets and tried to explain the apparent motion of celestial bodies through a system of perfect circular orbs. This approach was considered possible and justified. Thus, the transition to the "universe of precision" was not an unprecedented synthesis of mathematics and reality, but in the descent of mathematical accuracy from the heavens to Earth. This required the unification of nature and the abolition of the division of the Cosmos into two ontologically distinct areas. It was also a result of the geometrization of space.

The transition from the world of more or less to the universe of precision had two very important consequences: the emergence of modern experiments and of modern technology. As we indicated earlier, Koyré did not accept the idea that the birth of modern science was due to modern man's new orientation towards practical activity. He repeatedly stressed his opposition to an empiricist epistemology which gave special attention to the practical, empirical foundations of modern European science. Koyré recognized that the use of empirical data is an important aspect of modern natural science, but he argued that Galilean mechanics, although pretending to be an experimental science, remained largely *a priori*, since it was based on the idea of the mathematical structure of the physical world, which could not be discovered by experience. Koyré constantly insisted that the modern science was born as a result of the new ontological unification and theoretical reconsideration of the world and not out of pure observation and generalization of facts:

> [...] one must not forget that observation and experience, in the sense of brute, common-sense experience, did not play a major role – or if it did, it was a negative one, the role of obstacle – in the foundation of modern science. The physics of Aristotle, and even more that of the Parisian Nominalists, of Buridan and Nicole Oresme, was, as stated by Tannery and Duhem, much nearer to common sense experience than those of Galileo and Descartes. (Koyré 1943, p. 402)

The process of ontological unification of the world was itself bidirectional. On the one side, mathematics came down from heaven to Earth, and on the other side, physical reasoning started to be applied to celestial objects. The physical world became perceived as being built on numbers and figures, and physical objects acquired geometrical shape and accuracy. This made possible the application of measurements to physical processes, so that functions and numbers revealed substantial information about the internal structure of the world. The emergence of experimental science, according to Koyré, was a consequence of this process, not its cause. Modern experimentation was based on accurate measurement and became possible only because modern science assumed precision as a central principle: "it asserts that the real is, in its essence, geometrical and, consequently, subject of rigorous determination and measurement" (Koyré 1953, p. 225).

Even if this position has its own credibility, it could be reasonably criticized. For example, Italian historian Luca Bianchi proposed the opposite reading of this transformation (Bianchi 1990). He argued that the desire of precision could be a real obstacle for the development of experimental science, since any deviation of measurement from the exact (but unknown) value can be considered false. Therefore, it was not the ontological acquisition of precision that made the experimental knowledge possible but rather the recognition of the legitimacy of approximate and probable knowledge. This point seems to be very reasonable. However, it did not change the significance of the core of Koyré's claim that the physics of the seventeenth century changed the very nature of approximate knowledge, making this itself the subject of measurement and numerical evaluation.

Furthermore, Koyré insisted that the new universe of precision was also responsible for the rise of the modern technology, which for the first time appeared in the seventeenth century, when scientific reasoning and calculation started to be applied to technical problems. He did not accept the theory that modern science arose due to a "practical turn", influenced by development of civil and military constructions, or in a response to the necessity of resolving practical needs of transport or ballistics (Hessen and Grossmann 2009). Koyré insisted that civil and military engineering had been evolving for many years by their own laws, which differed significantly from the principles of scientific thought (Koyré 1971b [1948]). Technical thought was focused more on preserving tradition, whereas the development and improvement of technology were accomplished through a long series of trials and error.

Koyré declared that it would be unfair to underestimate the technical genius of previous centuries and explain the absence of modern technology in the Middle Ages or in antiquity by a lack of "practical attitude" (Koyré 1971b [1948]). It was sufficient to recall the achievements of the architects and engineers of ancient Rome, who erected huge basilicas and amphitheatres, laid roads and built bridges and aqueducts. Their war machines – ballistae and catapults – were not inferior in the destructive power to the guns of the sixteenth century. The Middle Ages were also accompanied by technical and material development: a yoke was invented which expanded the use of horses in agriculture, a rudder was created which made transatlantic travel possible, and new architectural forms arose which required new construction skills. Therefore, it was inaccurate and unfair to attribute to previous epochs disdain for technical art.

According to Koyré, the Scientific Revolution did not arise from technical development. On the contrary, the emergency of modern technology, i.e. application of scientific thought to technical problems, became possible only as a result of a general transformation of the conception of physical reality. The physical world became the universe of precision, and, therefore, technical calculations and projects could be made in it. Geometrical forms could not only be *found* in bodies, but they could be *made* there. Consequently, a technical device, a machine, could be designed and engineered, represented in drawings and then embodied in matter. Technology began to rely not on the empirical tradition but on theory and calculation. Scientific theory was becoming capable of directing technical ideas. It was at this moment that empirical technique, based on the experience of generations, was replaced by technology (i.e. technical science and scientific techniques). The imposition of thought into matter also led to the emergence of the scientific instrument that Koyré calls "incarnation of reason and thought" (Koyré 1971a [1948]), as well as the calculated improvement of technical devices.

To sum up, the transition from the world of more or less to the universe of precision represented the new aspect of the Scientific Revolution and clarified the meaning of "mathematization of nature". The shift from empirical experience to mathematical experimentation showed that scientific and extra-scientific consequences emerged from the mathematization of nature, the geometrization of space and unification of the Universe. The Scientific Revolution produced not only a more accurate and confident knowledge of the material world but also the possibility of effective and calculated action on it. Koyré argued that mathematically organized physical reality, to which precision and certainty are inherent, became available to measurement, calculation and design. Thus, experiments became possible, in which the scientist received answers in the form of numbers or numerical patterns.

8.5 From the Closed World to the Infinite Universe

The characterization of the Scientific Revolution as a transition from the world of the "more or less" to the universe of precision ignored the cosmological aspect of changes that took place in the seventeenth century. Cosmology came into the focus later, when Koyré described the Scientific Revolution of the seventeenth century as "the story of the destruction of the Cosmos and the infinitization of the Universe". This emphasis on the transformation of cosmological ideas and images grew from the fact that Koyré estimated the spiritual transformation of the seventeenth century to be a fundamental process:

> [...] as the result of which man – as it is sometimes said – lost his place in the world, or, more correctly perhaps, lost the very world in which he was living and about which he was thinking [...]. (Koyré 1957, p. 2)

We should note that the characteristic of the Scientific Revolution as a transition *From the Closed World to the Infinite Universe* was far from typical in Koyré's writings when compared with previous characteristics that were regularly repeated in different works. However, it appeared in the title of one of the most famous and widely read of his works and thereby became a hallmark of his historiography. In essence, it corresponded to a previously mentioned thesis about the infinitization of the Universe, which was accompanied by the abolition of the idea of Cosmos as an ordered structure of solid heavens to which the stars and planets were attached.

Koyré recognized the importance of the infinitization for scientific thought quite early. Already in the first part of *Études Galiléennes* (*À l´aube de la science classique*, 1935–1936), he stated that a gradual destruction of the concept of a finite world occurred in the mind of Galileo. His Cosmos blurred its boundaries and extended indefinitely:

> The center of the Universe is still there. But the sphere of the Cosmos increases, becomes indefinite, loses its circumference. It would be enough that it becomes infinite so the slighted traces of the ancient Cosmos and of privileged direction would disappear in the homogeneous space. (Koyré 1966a [1939], pp. 72)

The infinitization of the Universe is essentially linked to the geometrization of space. The infinite Universe has no centre and no periphery. There is no qualitative or ontological distinction between places and directions; they are all ontologically equal, so that natural motion becomes inconceivable. In the infinite Universe, all places are perfectly equivalent and perfectly natural for all bodies which imply the abolishment of a hierarchical organization of the Cosmos and lead to a reconsideration of the traditional conception of local motion as process with a goal. Consequentially, the infinite Universe permits us to reconsider traditional ideas of motion and formulate the principle of inertia (i.e. to recognize the theoretical possibility of an unlimited rectilinear motion of a body that once set in motion can conserve its direction and speed eternally).

If Koyré's other characteristics (the mathematization of nature and the transition to the world of precision) are directly related to "the geometrization of space", this later characteristic particularly develops the theme of "destruction of the Cosmos". It exposes the influence of scientific conceptions on the general world-view, on the everyday images of the world and of man's place in it. If geometry and its application are internal issues for the scientific community, the destruction of the boundaries of the Universe concerns not only natural philosophers but also the wider society.

Koyré's history of the infinitization of the Universe in *From the Closed World to the Infinite Universe* (1957) was an "internal" and "external" history at the same time. It referred to both the conditions and consequences of the Scientific Revolution. The Scientific Revolution required a new ontology of the physical world, but it also gave rise to global changes in world-view. The destruction of the image of the "closed world" (the finite Cosmos) occurred, in fact, in two stages. The first was associated with the development of a new physical theory. Construction of a new mechanics and astronomy required new concepts of space and world. This new cosmic view was born in the minds of certain intellectuals such as Bruno, Descartes and Newton and allowed them to make decisive steps towards a new scientific theory. The second phase of infinitization was associated with the public recognition of the new image of the world. The effectiveness of the new mechanics compelled the wider community to accept the hypotheses about the world on which it was built. For the principle of inertial motion, it was necessary to see the world as a system of mathematical bodies moving in infinite space. Accepting this principle inevitably made the new world-view the only one possible.

The mutual dependence of the new physical theories and the new conception of the infinite Universe brings an element of tragic inevitability to the story of the

Scientific Revolution. It is impossible to accept the new physics without also accepting its ontological and cosmological implications. It is impossible to construct aircraft or spaceships while believing in the solid heavens. It is impossible to use modern machines and mechanisms while believing in natural places and natural motion. Our technological and our mechanical progress make the image of the infinite and homogeneous Universe more credible. However, the consequences of this new cosmology are even more dramatic, since the recognition of the infinite Universe gave rise to serious philosophical and theological problems, as Koyré showed us in his classic works. In particular, early modern European thinkers had to reconsider not only their key ontological concepts but also their theological ideas. Infinity was traditionally a divine attribute, an indication of self-sustaining being which does not require any external causation. When the visible world acquired this quality, previously assigned only to God, the whole discourse of the theology of creation was irreversibly changed.

Thus, Koyré used to describe the scientific and spiritual revolutions of the seventeenth century as a transition from the conception of the world as a finite, ordered and anthropocentric Cosmos to the idea of an infinite and homogeneous Universe. This description added some nuances to the ideas of mathematization and revealed a particular aspect of the ongoing transformation, which not only changed the self-image of man but also lead to a radical revision of non-scientific theories as well as to the disappearance of traditional system of philosophical and theological thought.

8.6 Conclusion

We have considered several types of description of the Scientific Revolution, which had been proposed by Alexandre Koyré from the 1930s till the 1950s. The first and the most important of them was the dual characterization, which described the Scientific Revolution in terms of the destruction of the Cosmos and the geometrization of space. The "destruction of the Cosmos" expressed a change in ideas about how the world was organized. This description was logically related to the infinitization of the Universe, which had a significant impact on the world-view of modern men. The "geometrization of space" represented instead a mathematical line, a particular aspect of mathematization of nature, which led to the creation of modern technology and experimental science.

These two aspects of the Scientific Revolution – philosophical and scientific – being closely related should not be reduced to each other. Their distinction helps to preserve and to express the complexity of the changes that have occurred in the intellectual world of the seventeenth century, and the ability to describe these changes is part of Alexandre Koyré's invaluable contribution to historiography of early modern science.

References

Bianchi L (1990) L'esattezza impossibile: scienza e "calculations" nel XIV secolo. In Bianchi L, Randi E (ed). Le verità dissonanti. Laterza, Roma–Bari, pp. 119–50.

Burtt EA (1925) The Metaphysical Foundations of Modern Physical Science. A historical and critical essay. Keagan Paul, Trench. Trübner, London.

Cohen HF (1994) The Scientific Revolution: A Historiographical Inquiry. The University of Chicago Press, Chicago–London.

Coumet E (1987) Alexandre Koyré: La révolution scientifique introuvable ? History and Technology 4:497–529.

Dijksterhuis EJ (1924) Val en worp. Een bijdrage tot de geschiedenis der mechanica van Aristoteles tot Newton. Groningen, Noordhoff.

Dijksterhuis EJ (1950) De mechanisering van het wereldbeeld Meulenhoff, Amsterdam [English Translation: *Id.* (1969) The Mechanization of the World Picture. The Oxford University Press, Oxford]

Drago A (1994) Interpretazione delle frasi caratteristiche di Koyré e loro estensione alla storia della fisica dell'Ottocento. In Vinti C (ed). Alexandre Koyré: l'avventura intellettuale. ESI, Napoli.

Hessen B, Grossmann H (2009) The Social and Economics Roots of the Scientific Revolution, Freudenthal G, McLaughlin P (eds). Springer, Dordrecht.

Husserl E (1936) Die Krisis der Europäischen Wissenschaften und die Transzendentale Phänomenologie. Ein Einleitung in die Phänomenologische Philosophie. Philosophia 1:77–176 [English Translation: *Id.*, (1970) The Crisis of European Sciences and Transcendental Phenomenology. An Introduction to phenomenological philosophy. Evanston, IL.

Koyré A (1935–1936) À l'aurore de la science moderne : la jeunesse de Galilée. Annales de l'Université de Paris, 10:540–551; 11:1933–1956.

Koyré A (1943) Galileo and Plato. Journal of the History of Ideas 4:400–428.

Koyré A (1953) An Experiment in Measurement. Proceedings of the American Philosophical Society, 97:222–237.

Koyré A (1955) Influence of Philosophic Trends on the Formulation of Scientific Theories. The Scientific Monthly 80:107–111.

Koyré A (1956) The Origins of Modern Science: A New Interpretation. Diogenes 4:1–22.

Koyré A (1957) From the Closed World to the Infinite Universe. The Johns Hopkins University Press, Baltimore.

Koyré A ([1950] 1965) The Significance of Newtonian Synthesis. In Koyré A (1965) Newtonian Studies. The Harvard University Press, Cambridge, MA, pp. 3–24.

Koyré A ([1939] 1966a) Études Galiléennes. Hermann, Paris.

Koyré A ([1957] 1966b) Gassendi et la science de son temps. In Koyré A (1966) Études d'histoire de la pensée scientifique. Presses universitaires de France, Paris, pp. 284–296.

Koyré A ([1948] 1971a) Du monde de l' « à-peu-près » à l'univers de la précision. In Koyré A (1971) Études d'histoire de la pensée philosophique. Éditions Gallimard, Paris, pp. 341–362.

Koyré A ([1948] 1971b) Les philosophes et la machine. In Koyré A (1971) Études d'histoire de la pensée philosophique. Éditions Gallimard, Paris, pp. 305–339.

Koyré A ([1949] 1971) Le vide et l'espace infini au XIVe siècle. In Koyré A (1971) Études d'histoire de la pensée philosophique. Éditions Gallimard, Paris, pp. 37–92.

Koyré A (1986) De la mystique à la science. Cours, conférences et documents (1922–1962). Redondi P (ed). E.H.E.S.S., Paris.

Nicolson MH (1950) The Breaking of the Circle: Studies in the Effect of the "New Science" Upon Seventeenth Century Poetry. The Northwestern University Press, Evanston.

Seidengart J (2006) Dieu, l'univers et la sphère infinie. Penser l'infinité cosmique à l'aube de la science classique. Albin Michel, Paris.

Chapter 9
Koyré, Cassirer and the History of Science

Massimo Ferrari

Abstract The history of science is a highly promising field of inquiry for contemporary scholarship. Alexandre Koyré and Ernst Cassirer are among the major protagonists of the twentieth century and offer an interesting perspective in understanding the development of the history of science in relation to the history of philosophical ideas. The aim of this paper is to show several crucial aspects of both Koyré's and Cassirer's work, particularly concerning the role of Platonism in the rise of modern science and the concept of "scientific revolution" underlying their historical analysis. According to the author, it is nonetheless impossible to distinguish Koyré's and Cassirer's own reconstructions of modern science from their respective philosophical backgrounds, which belong to very different philosophical traditions such as Husserlian phenomenology and Marburg neo-Kantianism.

Keywords Koyré • Cassirer • Scientific revolution • Platonism • Galileo

9.1 Introduction

Alexandre Koyré is acknowledged as one of the greatest pioneers of the history of science in the twentieth century. His main contribution to this field of research before its professionalization in the post-war period (Kuhn 1997, p. 105) can be divided into two different areas of inquiry: on the one hand, we have the understanding of the scientific revolution as a crucial breakthrough in the history of modern thought; on the other hand, we have the more general question concerning "the relevance of philosophy for the history of science" (Cohen 1994, p. 84). Whereas the former has played an essential role in the framing of the received view of Koyré's work as a historian, the latter has been long underestimated. In contrast, the deep connection between these two features constitutes the veritable uniqueness of Koyré's intellectual enterprise and, at the same time, is the reason why to this day

M. Ferrari (✉)
Department of Philosophy and Science Education, Torino University,
Via S. Ottavio 20, 101024 Torino, Italy
e-mail: massimo.ferrari@unito.it

R. Pisano et al. (eds.), *Hypotheses and Perspectives in the History and Philosophy of Science*, DOI 10.1007/978-3-319-61712-1_9

his work remains, as Thomas Kuhn once said, an extraordinary "historiographical revolution" (Kuhn 1970b, p. 69).

This characterization is also applicable, to some extent, to the eminent figure of Ernst Cassirer as well. His outstanding analyses of the historical origins of modern thought he developed in the first decades of the twentieth century continue to represent a highly original contribution to the rise of history of science as a historical discipline (Ferrari 2015). Nonetheless, Cassirer's great achievements in this field have not received the recognition they deserve. There are two principal reasons for which Cassirer's peculiar way of understanding history of science has been neglected. Firstly, Cassirer's view of the scientific revolution is not so clearly formulated as Koyré's; that is, it rests on a wide philosophical framework articulated as a sophisticated theory of knowledge in a Kantian sense. Secondly, Cassirer's philosophical perspective is bound to the Marburg school of neo-Kantianism, a philosophical trend which was highly influential in Germany until the First World War but whose eclipse seems more and more overwhelming in the following age of decline of neo-Kantian philosophy in general. Consequently, it is clear why a reconsideration of Cassirer as a historian of science would require a deeper look into his philosophical insights which have unfortunately been ignored, or even totally forgotten, for a long time. Thus, in the case of Cassirer, we are witness to an "adventure of ideas" (as Alfred North Whitehead would say) worthy of being closely revisited – better still if we look at Koyré's work at the same time.

9.2 Alexandre Koyré's History of Science and Its Philosophical Background

In 1951 Koyré drafted a research project whose focus was the crucial turn represented by the scientific revolution, from Galileo to Newton. The inquiry into the rise of modern scientific thought, Koyré pointed out, must begin with a more general assumption regarding both the essential "unity of human thought" and the "structure" underlying the work of the founders of the modern image of universe (Koyré 1973, p. 11). Hence, according to Koyré, the history of science requires a careful examination of the philosophical and religious thought from which arises the conceptual mutation worked on by the greatest protagonists of the mathematization of nature (*Ivi,* p. 13). Interestingly enough, Koyré was referring especially to Jacob Boehme, to whom he had already devoted a penetrating book in 1929.[1] The "shoemaker of Görlitz", who was praised by Hegel was for Koyré an illuminating example of the close relationship existing between both the mystical and alchemic traditions and the origins of modern cosmology (Koyré 1973, p. 11). Similarly, Koyré argued, the

[1] Koyré 1979. The book is dedicated to Léon Brunschvicg and Étienne Gilson. Before his "conversion" to history of science, Koyré had already composed, during his early stay in Paris, several essays on the German mystical currents in the sixteenth century (Koyré 1971b). See also Koyré (2016, pp. 61–69).

evolution of scientific thought would not have been possible without the background of philosophical, metaphysical and religious "trans-scientific ideas", as he called them (*Ivi,* p. 12).

Koyré thus aimed at conceiving of the history of science as a discipline connected to a more general framework. In 1961, on the occasion of the Oxford Colloquium about *Scientific Change*, with his early intellectual apprenticeship with Edmund Husserl in Göttingen surely in mind, Koyré attempted to emphasize that "the science of our time as well as the science of the Greeks is *theoria*, i.e., the search of truth". (Koyré 1973, p. 399). Once again, Koyré stressed how this point precisely constitutes the central issue of the history of science and in that sense – as he remarked in his exemplary book on the astronomical revolution – it represents a kind of "phenomenology of human thought" (Koyré 1961, p. 18). Hence, Koyré was convinced that science needs a "philosophical substructure", a "philosophical horizon" representing nothing but the "absolute necessity" without which science could not achieve its epoch-making results (Koyré 1971a, p. 253, p. 255). Koyré then expressed a very provocative opinion in a conference delivered in Boston in 1954 at the yearly meeting of the *American Association for the Advancement of Science*. While Koyré did have some doubts towards Edwin Burtt's book on *The Metaphysical Foundations of Modern Physical Science*,[2] he did express criticism of the positivistic view of science endorsed by Bacon, Comte, Mach and eventually the Vienna Circle (*Ivi*, p. 254). Of course, it was not the first time that Koyré so strongly criticized the "virus of the empiricist and positivist epistemology" which had also "infected" the history of science. Koyré's conclusion was, at any rate, very clear. It is quite impossible to separate science from philosophy, and, in turn, scientific theories always have a considerable influence on the philosophical ideas. The development of scientific thought needs to be explained within a framework of fundamental principles, of axiomatic evidences that are normally considered entirely foreign to philosophy, so that we no longer have a purely historical process in vacuo (*Ivi*, p. 256).

In one of his most famous books, *From the Closed World to the Infinite Universe*, Koyré expressed with particular emphasis both his methodological and, in a broader sense, philosophical assumptions. There are many ways, Koyré argued, to provide an interpretation of the reasons why something like the scientific revolution occurred in the history of human culture. For instance, we can either deal with the radical change in the usual relationship between theoria and praxis, or with the replacement of the medieval ideal of *vita contemplativa* with the secularized view of *vita activa*. For Koyré, "these characterizations are by no means false"; nevertheless, the deep and fundamental process represented by modern scientific change is rooted in an all-encompassing new way of thinking rather than merely in some historical circumstances external to it. In a passage worthy of being quoted at length, Koyré added:

[2] *Ivi,* p. 255. Koyré disagrees with Burtt for having considered the metaphysical assumptions of scientific theories as mere "scaffoldings" that allow for the theories' constructions. According to Koyré, by contrast, modern science rests not provisionally but intrinsically on metaphysical, permanent assumptions.

> This scientific and philosophical revolution – it is indeed impossible to separate the philosophical from the purely scientific aspects of this process: they are interdependent and closely linked together – can be described roughly as bringing forth the destruction of the Cosmos, that is, the disappearance, from philosophically and scientifically valid concepts, of the conception of the world as a finite, closed, and hierarchically ordered world [...], and its replacement by an indefinite and even infinite universe which is bound together by the identity of its fundamental components and laws, and in which all these components are placed on the same level of being. This, in turn, implies the discarding by scientific thought of all considerations based upon value-concepts, such as perfection, harmony, meaning and aim, and finally the utter devalorization of being, the divorce of the world of value and the world of facts. (Koyré 1957, p. 2; see also Koyré 1965, pp. 5–6)

Koyré has repeatedly insisted on this essential feature of the scientific revolution. The "destruction" of the cosmos and, accordingly, the "geometrization" of space are, in his mind, the main aspects that allow for the "radical intellectual 'mutation'" that Koyré himself describes as "scientific revolution", finding its first pioneer in Galileo.[3] Koyré had already used the term "scientific revolution" in the first pages of his *Études galiléennes* (1939), where he referred to eminent authors such as Pierre Duhem, Émile Meyerson, Léon Brunschvicg, and Ernst Cassirer, drawing attention to the "philosophical fruitfulness" of the inquiry into the "developments (and the revolutions) of scientific ideas" (Koyré 1939, p. 5). As a matter of fact, Koyré also used another expression that he had found in Gaston Bachelard's work, i.e. the "'mutation' of the human intellect", thanks to which the crucial ideas developed by the greatest geniuses of scientific thought had become available to the scientists of the following era. In particular, Koyré insisted that the "scientific revolution" during the seventeenth century was due to a process of "mutation" involving a "deep intellectual transformation, of which moderns physics – or more precisely classic physics – is at once the expression and the fruit" (*Id.*, p. 6). A few years later, the same claim will be repeated even more sharply by Koyré in his essay "Galileo and Plato", published in 1943 in the *Journal of the History of Ideas*. It was thanks to this significant study that the term "scientific revolution" made its entrance into the scientific community, especially in the United States, and became henceforth a canonical expression for historians of science (Koyré 1943, p. 400). Moreover, Koyré's paper summed up the main results of his previous *Études galiléennes* and can still be considered as a kind of manifesto of a new conception of the history of science, making Koyré its universally acknowledged founder. And it was also during this glorious period, i.e. between the late 1930s and the early 1940s, that Koyré definitively outlined his ambitious research project aimed at reconstructing the scientific revolution inaugurated by Galileo and completed by Newton's impressive physical system. As Koyré wrote in 1951, at stake was the long process of the "infinitization" of the Cosmos and, at the same time, the "deep intellectual transformation" lying at the origins of an irreversible change in the "framing of our thought" (Koyré 1973, p. 13).

---

[3] Koyré 1943, p. 400. "The dissolution of the Cosmos [...]: this seems to me to be the most profound revolution achieved or suffered by the human mind since the invention of the Cosmos by the Greeks" (*Ivi,* p. 404).

Koyre's research was therefore developing a history of the sciences centred on the "influential metaphysics" – as the subsequent philosophy of science would have said – constituting its framework, although in his mind the fieldwork of the scholars should remain focused primarily on the *explication de textes*, as emphasized with some reservations by Kuhn in 1970 (Kuhn 1970b, p. 68). According to this perspective, in the *Études galiléennes*, Koyré had undertaken a careful examination of Galileo's efforts in order to discover both the law of falling bodies as well as the principle of inertia (although Koyré estimated that the latter had not been formulated by Galileo but only later by Descartes). The focus of Koyré's original step-by-step reconstruction of the development of scientific ideas and their interconnection was surely based on the great lessons of Meyerson, Brunschvicg, Duhem and, to some extent, Cassirer. Nonetheless, what constitutes the most important aspect of Koyré's historical method is his recourse to some epistemological assumptions that enable him to take into account both the medieval theories of the impetus and the rise of modern dynamics. Firstly, Koyré believed that imaginary experience plays a decisive role in allowing a scientist to explain "the reality through the impossible"[4]; secondly, he was convinced that physics, whenever it is well done, is a priori physics; and, finally, he did not hesitate to stress how Galileo was committed to Platonism, since it seems quite impossible to conceive of mathematical physics outside a Platonic context. As suggested by Koyré, from a purely epistemological standpoint, the mathematical theory precedes the experience. Accordingly, Koyré repeats time and time again that Galileo was a "good" Platonic who had even provided Platonism with "an experimental proof" constituting, at its essence, "the revenge of Platonism". Koyré would go on to add that "Metaphysics with regard to science is Platonism" (Koyré 1939, p. 266, p. 269, pp. 273–281).

Yet Koyre's main interest was the historical description of the scientific revolution during the seventeenth century that culminated in Newton's work. The triumph of this revolution cannot be simply explained through some external factors such as the rise of the bourgeoisie or a new attitude towards human life. On the contrary, Koyré suggested that so great a change in modern culture was due to, and was a consequence of, the "new *theoretical*, that is, the new *metaphysical* approach to nature that forms the content of the scientific revolution of the seventeenth century, a content which we have to understand before we can attempt an explanation (whatever this may be) of its historical occurrence" (Koyré 1965, p. 6). But what is even more striking here is the fact that Koyré also speaks of a "new ontology" (*Ivi*, p. 7), an expression recalling evidently Husserl's concept of "regional ontology" and, especially, the (ontological) "region of nature". In this sense, Newton's "new ontology" replaces, once and for all, the quality with the quantity, the world of change and becoming with the world of being, where motion and rest are no longer processes but must be properly understood simply as a *status* belonging to being, "as permanent and indestructible at rest" (*Ivi*, p. 9, p. 23). By using Husserl's late terminology, with which he was also well acquainted, Koyré maintained here that the new science implies the loss of the "lived world", of the *Lebenswelt*. Nonetheless,

[4] On this topic see the recent contribution of Ferrarin (2014, pp. 90–116).

it is not really a loss but a peculiar conquest, which, instead of leading science to a pure realm devoid of any metaphysics, underpins Newton's insight into absolute space and absolute time. This implies the omnipresence of God – God who through his will has created our world with its rigorously mathematical laws:

> The belief in creation as the background of empiricomathematical science – that seems strange. Yet the ways of thought, human thought, in its search for truth are, indeed, very strange. *Intinerarium mentis in veritatem* is not a *right* line. That is why the history of this search is so interesting, so passionate.[5]

Last but not least, it is also for these reasons that in 1952 Koyré defends the concept of the scientific revolution by opposing the thesis of the essential *continuity* in the development of scientific thought endorsed by Pierre Duhem in the early history of history of science and reformulated, in later years, by Alistair C. Crombie. Koyré did not agree with Crombie's evaluation of Grossatesta as a veritable pioneer of the scientific revolution, which would establish a substantial continuity between the Medieval Age and modern science.[6] Using the terminology he had elaborated throughout his career as a historian of science, Koyré affirmed that, in contrast to Crombie (and Duhem), scientific revolution signifies a theoretical revolution based on mathematical Platonism and, therefore, requires the rebirth of said Platonism. Mathematics, Koyré stated, "is more than a formal way of being able to put facts in order; it is the key to the understanding of nature" (Koyré 1973, p. 82). This core idea is of great significance, since in 1956 Koyré already seemed aware of two possible consequences that historians of science could find in Crombie's contributions. In one respect, Koyré refused the possibility of giving an account of the rise of modern science based on the discovery of new tools: the firearms, Koyré poignantly said, have demolished medieval castles, but the dynamics of that age was not modified in the modern sense (*Ivi*., p. 75). Historical and social changes by no means represent the necessary presuppositions of scientific revolutions. To put it in other words: an *external* history of science is not only insufficient but is fundamentally misleading.[7] On the other hand, Koyré efficaciously brought into question the wide-

---

[5] Koyré 1965, p. 114. See also the remarks about Henry More in Koyré (1957, p. 126), where Koyré underlines that More's great merit consisted in having provided "some of the most important elements of the metaphysical framework" of modern science. More "succeeded – Koyré states – in grasping the fundamental principle of the new ontology, the infinitization of space, which he asserted with an unflinching and fearless energy". An interpretation of More that is very similar to that of Koyré was already offered by Burtt (1954, pp. 141–142). But the first historian interested in both More's theory of space and metaphysics was Ernst Cassirer (see below footnote 14). Note that Koyré 1935 nourished indeed another opinion about More's relevance in opening the path to Newtonian science (see below footnote 24).

[6] Koyré refers in particular to Crombie (1953). Crombie's interpretation of modern science as, to certain extent, a linear development of Medieval science was already illustrated in his well-known and influential book *Augustine to Galileo* (Crombie 1952), which Koyré quotes in his essay (Koyré 1973, p. 62, footnote 2). Koyré's criticism towards both Duhem's and Crombie's thesis concerning the "continuity" in history of science has been recently questioned by Biard (2016).

[7] An attempt to interpret the scientific revolution as the outcome of a more general change, both social and historical in early modern time, was provided at the end of the 1940s by Butterflied (1949).

spread empirical and experimental understanding of modern science. In his opinion, this was precisely the case with Galileo, since his scientific method very clearly shows "the primacy of the theory over the facts" (Koyré 1973, p. 83). Here Koyré speaks not only as a historian of science but as a scholar who in his extensive work has offered, from the start, a philosophical view of the conceptual structure of science.

Thus, from a philosophical point of view, Koyré's characterization of the modern scientific revolution calls into question the meaning of that "intellectual mutation" which implies, first and foremost, the change of epistemological paradigms (as Kuhn would say) inherited from the past, in particular from the Aristotelian tradition which greatly influenced medieval science. Though Cassirer, Brunschvicg, Meyerson and Bachelard are indeed all important for Koyré's understanding of the scientific enterprise, we must bear in mind that his philosophical apprenticeship was nourished by Edmund Husserl's early phenomenology. A young man, Koyré, was in Göttingen, in the Mecca of mathematics, as Husserl launched the so-called transcendental turn of phenomenology around 1910–1911, and since he did not share this view – quite similarly to Adolf Reinach and Max Scheler – he abandoned the German milieu and moved to Paris (Zambelli 1998, 1999, 2016, pp. 57–74). This disappointment had great consequences for Koyré, since from these early days he disagreed with the heritage of transcendental philosophy, including the Kantian sense as well.

Not accidentally, it was Meyerson (to whom Koyré will dedicate his *Études galiléennes*) who represented for Koyré a very important epistemological perspective (Meyerson 2009, pp. 226–253; Bensaude Vincent 2016). On the occasion of the German translation in 1931 of Meyerson's main work – *Identité et réalité*, published in 1908 – Koyré devoted an essay to him which thoroughly documents his agreement with some of the most original conceptions endorsed by Meyerson. According to Meyerson, human reason rests undoubtedly on a priori structures as proposed by Kant but is at the same time indissolubly connected with the historical, as well as the a posteriori, empirical side of scientific knowledge, which is in turn a "collective thought" worthy of being considered from an historical point of view (Meyerson 1951, pp. xv–xvi, 463–464). For Koyré, Meyerson's great merit consisted indeed in his profound account of the history of science, a research field that Meyerson had practiced in a way quite similar to Dilthey's and Weber's studies regarding human sciences (Koyré 1931, p. 202). Nonetheless, Koyré was somewhat sceptical about Meyerson's view of reason, which seemed "baroque" to him because of its characterization as an independent structure outside of history aimed at finding the "identity", as opposed to the "irrational" which has to be rationalized.[8] While Meyerson emphasized the unavoidable role played by the given and, at least, the "irrational" reality which compels the mind to find unchangeable, a priori structures, Koyré was rather interested – as he will state in 1961 – in grasping the differences of human

[8] This objection is formulated, albeit with some precaution, by Koyré (1946).

thought in historical development of science (Koyré 2016, p. 201). Beyond this, it should be noted that Koyré considered Meyerson as a well-grounded alternative to both idealism and the Kantian heritage (Cassirer is not explicitly mentioned, but it seems highly probable that Koyré also had this very influential neo-Kantian in mind).[9] Meyerson, Koyré added, was in no way a philosopher aiming at constructing, as well as forming, reality through both the categories and the activity of mind. On the contrary, reality, according to Meyerson, is something already given and, broadly speaking, "irrational", which science has to rationalize along with the endless process of knowing. Finally, and even more importantly, Koyré suggested that Meyerson was quite close to Husserl, though he had no contact with phenomenological philosophy. However, in Koyré's opinion, Meyerson shared the same spirit of phenomenology, i.e. with regard to "a phenomenological explication both of the typical and essential approach and way of thinking characterising science as such" (Koyré 1931, p. 203).

The extensive spectrum between Husserl and Meyerson guides Koyré from mathematics to philosophy and from phenomenology to the history of science. However, there are still many other features of Koyré's work as a historian of philosophy and, later, as a historian of science that must be taken into account. In order to comprehend Koyré's own understanding of science – and of modern science in particular – in its indissoluble connection with mathematics, one cannot disregard the role played by his teacher at the *École normale supérieure de Paris*, Léon Brunschvicg. At issue, here is the very intricate relationship with the Kantian author of the famous book on history of mathematics or, more precisely, on the "mathematical philosophy", which was first published in 1912 (and was held in high esteem by Koyré). In the later period of his intellectual life, in 1962, Koyré stressed, and not accidentally, that his great admiration for Brunschvicg's *Les étapes de la philosophie mathématique* did not exclude a fundamental objection to a Kantian account of mathematics. Koyré did not believe, contrary to both Brunschvicg's and Cassirer's neo-Kantian point of view, that numbers are the result of an activity of mind, but rather that they belong to another realm or, in Husserlian terms, to the pure sphere of "essences". Only God has created the numbers, Koyré affirmed, although for the rest mathematics depends on the "invention" of the human mind (Koyré 1963, p. 11). But, such an opinion would have never been accepted by the neo-Kantian Cassirer. And this would likely be of importance if we were to attempt to compare Koyré's and Cassirer's way of considering science and its history.

[9] Indeed, Meyerson was in utter disagreement with the Marburg School and, specifically (as we shall see below), with Cassirer's rejection of the concept of substance (Meyerson 1951, pp. 443–445, p. 491). Note that Koyré, for his part, was totally in contrast with Marburg neo-Kantianism in what concerns the interpretation of Kant's thing in itself, which he considered not an infinite task *à la* Cohen (*unendliche Aufgabe*) but rather as the realistic, "metaphysical" ground in perceiving appearances (Koyré 2016, p. 228).

9.3 Ernst Cassirer: A Neo-Kantian Historian of Science

Koyré published his *Études galiléennes* in 1939. A year later Ernst Cassirer emphasized, in his article "Mathematical Mystique and Mathematical Science of Nature", that not only is the history of science a crucial issue for both the philosopher and the historian of philosophy but that the question of the *origins* of the exact sciences represents a philosophical focus which cannot be ignored (Cassirer 1940, p. 284). At this time Cassirer was already acquainted with the *Études galiléennes*, to which he refers explicitly. No wonder, therefore, that Cassirer, echoing Koyré's use of the term, speaks on this occasion of *revolution* in science:

> The history of human knowledge – Cassirer states – repeatedly shows us new, particular ages (the more important ones, to be sure), in the course of which knowledge doesn't simply increase its extent as much as change both its overall conceptual tools and its sense. Instead of a mere quantitative growth, there suddenly appears a qualitative "change" (*Umschlag*). Rather than dealing with an evolution, we are dealing with an unexpected *revolution*. The very ideal of exact knowledge of nature arises from just such a revolution. (Ivi., p. 285)

Nevertheless, it is worth noting that for Cassirer, a conceptual "revolution" in no way signifies a sudden break from the previous scientific age. On the contrary, Cassirer argues that it would be "misleading" to consider the rise of modern natural science as being totally independent from its medieval roots. "We are never truly dealing with an interruption in the continuity", he states (*ibidem*). Both continuity and discontinuity are, therefore, the two faces of scientific progress, although Cassirer does underline that the "jump" accomplished by scientific thought in the modern age would not have been possible in *vacuo*. Yet, like Koyré, Cassirer questions the usefulness of the concept of continuity as featured in Duhem's historical account of the development of science from the Middle Ages to Galileo. Whereas, in his influential works on the history of mechanics before Galilean science, Duhem meritoriously points out the undeniable importance of the theory of *impetus*, in Cassirer's judgement it is "audacious and doubtful" to place the prehistory Duhem which is dealing with on the same level as the rise of the new science which represents, in Cassirer's mind, both an enormous change from mathematical and empirical standpoints as well as the birth of a very different image of the universe (Ivi., p. 286).

Cassirer's insight into the scientific revolution, provided in his essay from 1940, only represents the late achievement of his previous admirable work on the development of philosophical and scientific thought in the modern era. At the very beginning of his career, Cassirer was already committed to the historical and systematic reconstruction of the rise of modern science. This is well documented by his first book, published in 1902, which is devoted to the interpretation and critique of the "scientific foundations" of Leibniz's system, breaking away from the traditional image Leibniz as the author of a "metaphysical novel", i.e. the *Monadology* (Cassirer 1902). This investigation of Leibniz as a seminal mathematician and physician of his time constitutes Cassirer's first attempt to grasp the scientific roots of modern science and philosophy. The next step in his project devoted to the "prehistory of pure reason" is even more impressive – the masterful book *The Problem of Knowledge in the Philosophy and Science of the Modern Time,* which first appeared in 1906–1907. These two volumes serve as a testament to Cassirer's highly fruitful insight into the development of epistemology in its connection with, and in its reliance on, the modern mathematical science of nature (Cassirer 1910–1911).

This central feature of Cassirer's work is closely related to his commitment to Marburg neo-Kantianism which is usually either scarcely considered by scholarship or solely remembered as more of biographical, rather than properly philosophical, backdrop to Cassirer's thought. While there are certainly many important differences that gradually emerge along the way, from his early neo-Kantianism to the ultimate outcome of his *Philosophy of Symbolic Forms*, Cassirer remains faithful to at least the essential methodological premise of the neo-Kantianism formulated by Hermann Cohen and Paul Natorp. According to Cohen, whose book *Kant's Theory of Experience* from 1871 is doubtless the "Bible" of the Marburg School, transcendental philosophy rests on the *Faktum* of the mathematical science of nature. This "fact", as Cohen suggests, is both historically determined and steadily changing and demands an analysis which uncovers the conditions of its possibility, thereby discovering the synthetic principles and epistemological foundations of mathematical science itself (Cohen, 1885). This is the reason for which Cohen maintains that the commonly used term "theory of knowledge" is misleading, whereas the proper description of Kant's reformulated project would be the "critique of knowledge" (*Erkenntniskritik*) (Cohen 1883, pp. 4–6). Thus, transcendental philosophy deals neither with the constitution of the human subject nor with his ability to know but rather with the "metalevel" of philosophical and epistemological reflection on the a priori conditions of scientific knowledge. In short, the "critique of knowledge" aims to uncover the a priori presuppositions and foundations of scientific thought beginning with the historically determined "fact" of natural science. This is precisely what Cohen, and the Marburg School in general, call the "transcendental method".[10] In particular, Marburg neo-Kantianism was devoted to an investigation of the history of mathematics, especially mathematical science, which aimed to show how infinitesimal analysis, non-Euclidean geometries, modern logic and profound transformations in physics at the turn of the twentieth century had deeply changed the *Faktum* to which transcendental philosophy refers. As demonstrated by Natorp's studies of Hobbes, Copernicus, Galileo and Leibniz, as well as of Descartes' theory of knowledge, this was also related to the ambitious project of revisiting the history of philosophy with regard to its relationship with the development of science or to the changing "fact" of science.[11]

It was Cassirer who first understood the true relevance of the historical and alterable dimensions of this "developing fact" for a Kantian epistemological project. As a result, he bound, more deeply than his Marburg predecessors had, the fate of

---

[10] Cohen 1885, pp. 93–110. An illuminating overview is offered by (Natorp 1912).

[11] I am referring especially to Natorp (1882a, b, c, 1985). The importance of Natorp's early historical work is extensively illustrated in Sieg (1994).

critical philosophy to its relationship with the development of the exact sciences. Cassirer thus identified the sole enduring task of a critical inquiry based on the transcendental method as the:

> […] continually renewed examination of the fundamental concepts of science, […] which simultaneously involves a thorough subjective self-examination of the critique itself. (Cassirer 1907, p. 37)

But if the "fact" of science is "in its nature a historically developing fact" (Cassirer 1910–1911, I, p. 14), then philosophical reflection on the forms of knowledge that underlie this "fact" and make it possible must be characterized by a fundamental *dynamism* – a dynamism which is intrinsic to the transcendental method that can be extended to all areas of both cultural and scientific objective forms:

> The 'fact' of science - Cassirer says in his *Introduction* to the first volume of *The Problem of Knowledge* – is, and will of course remain, in its nature a historically developing 'fact'. If this insight does not yet appear explicitly in Kant, if his categories can still appear as *finished* 'core concepts of reason' in both number and content, the modern development of critical and idealistic logic [here he is referring to Cohen's *Logik der reinen Erkenntnis*] has made this point perfectly clear. By the *forms of judgment* are meant the unified and active *motivations* (*Motive*) of thought which course through the manifold particular formations and are continually put to use in the generation and formulation of new categories. (Ivi., pp. 14–15)

Thus, Cassirer's impressive reconstruction of the problem of knowledge in modern times from an historical and systematic standpoint is the result of his neo-Kantian apprenticeship and, at once, the greatest testament to his highly original approach to the epistemological reflection on the scientific *Faktum* with which the transcendental method deals. On the one hand, Cassirer's main idea is that science and philosophy must be mutually connected; modern philosophy and modern science constitute a unique whole, and, more precisely, the understanding of the problem of knowledge must consider both philosophers such as Descartes or Leibniz as well as scientists such as Galileo, Kepler or Newton. According to Cassirer, the traditional history of philosophy has, for the most part, neglected the essential ways in which the rise of modern science contributed to the deep changes that have occurred in philosophical thought. In the early modern age, scientists and philosophers worked together on a new image of both nature and the universe, which also entailed a radical break from the previous conception of man. For Cassirer, the final outcome – and the final goal – of this history is Kant's critical philosophy. One can say that Cassirer envisages here a kind of "history of pure reason" in the Kantian sense, which is based on – as Cassirer emphasizes – the strict collaboration between the epistemological standpoint and historical enquiry (Ivi., p. VII). On the other hand, Cassirer intended to continue the ambitious project laid out by a young Natorp in his early book on Descartes' theory of knowledge, namely, to outline the prehistory (*Vorgeschichte*) of Kant's critical philosophy through a philosophical and historical examination of its sources in the philosophy and scientific thought of Descartes, Galileo, Kepler and Leibniz, the founders of the idealistic tradition (in the sense of the "logical idealism" of the Marburg school), whose origins Cohen, and later

Natorp himself, saw in Plato's theory of ideas.[12] Surely, it cannot be denied that Cassirer goes far beyond the original conception of the history of philosophy and the history of science endorsed by Cohen and Natorp. Nevertheless, it would be quite impossible to outline Cassirer's own achievement in this field without taking into account his former apprenticeship in Marburg and the enduring influence of neo-Kantianism on his work. Insofar as it is plausible to speak of a neo-Kantian tradition in the history of science in the first decades of the twentieth century, Cassirer is surely both its most representative interpreter and its first promoter.

One of the most influential heroes of the philosophical story told by Cassirer is undoubtedly Galileo, the great father of modern science and the modern concept of scientific law. The fundamental question here involves how Galileo conceived of scientific experience in its relationship to mathematical procedures and, more generally, to reason as such. In short, Cassirer's explanation of this crucial issue is centred, broadly speaking, on the Kantian account of knowledge as the result of the spontaneous activity of reason, both in organizing empirical phenomena and in submitting them to purely functional mathematical laws. Given the fact that, according to Galileo, experience is grounded in mathematics, it is possible, as Cassirer suggests, to evaluate the fundamental role played by mathematical laws within the tradition of Platonism, which, in turn, was interpreted within the framework of "logical idealism", formulated by Cohen in the early days of the Marburg School. As we shall go on to see, this kind of Platonism views mathematics as being applied to experience, thereby providing a complete "functionalization" to the conceptual equipment of scientific thought (Ivi., pp. 324–325, pp. 356–358). Thus, in Cassirer's eyes, Galileo is the first scientist able to grasp the fundamental concept of function and, consequently, to overcome the metaphysical concept of substance.

From its very outset, Cassirer's work on the history of science has not only deeply connected with the history of epistemology but has always rested on the firm belief that – as stated in the introduction to *The Problem of Knowledge* – modern science is also a cultural form, a way in which the spirit of early modern culture shows one of its most typical characteristics. According to Cassirer, science is bound to the various "intellectual energies" which have contributed to the rise of the early modern age, from humanism to the scientific revolution. In this sense, science represents the core – as Cassirer calls it – of the "theoretical self-awareness" of a new era in human culture (Ivi., p. xi). This understanding constitutes a kind of leitmotiv in Cassirer's work: even in the later period of his intellectual life he lays emphasis on the break, represented most notably by Galileo, whose main achievement was the essential transformation of scientific thought thanks to a new concept of truth which resulted in an "ethics of science" (Cassirer 1937; see also Cassirer 1942a, p. 53, p. 64).

---

[12] See in particular Natorp (1994). The decisive role played by the epistemological interpretation of Plato's theory of ideas within the Marburg neo-Kantianism is undoubtedly crucial to understanding its interpretation of the modern mathematical science of nature. A fine survey is available in the noteworthy book by Lembeck (1994). See also Servois (2004).

In his momentous book on Renaissance thought published 20 years after the *Erkenntnisproblem*, Cassirer offers a more detailed account of the intrinsic relationship between the rise of modern science and the heritage of Renaissance culture. *Individuum und Kosmos in der Philosophie der Renaissance* is a work composed in connection with the milieu of the Warburg Library and influenced by the image of the Renaissance which Aby Warburg himself had elaborated in his fascinating analysis of the rebirth of paganism and ancient astrological beliefs in the early fifteenth century (Cassirer 1927). Here Cassirer underlines the wide context of symbolic forms (religion, art, mythical thought), which constitute the cultural background enabling the rise of the modern scientific image of the universe, from Nicholas Cusano to Giordano Bruno. For Cassirer, both a new sentiment of life as well as the increasing emancipation of natural science from the dark power of magic and astrology made it possible to conceive of nature in a new light, namely, as the object of mathematical measurement rather than something which could only be approached purely qualitatively. Therefore, in broad terms, the scientific worldview is the outcome of a new image of man, now placed at the centre of the world as a Prometheus unbound.

Yet, one has to question whether Cassirer's highly sophisticated reconstruction of the origins of modern science during the Renaissance can be interpreted in the sense of a revolutionary break or as a kind of linear progress towards the final goal of pure reason. To be sure, Cassirer was convinced that reason essentially means – quite similarly to Cohen's own insight – *continuity*, namely, that reason is the continuous process thanks to which the development of scientific thought does not involve a relativism which is opposed to the universality of the "logical functions of knowledge" (Cassirer 1910–1911, I, p. 13). Nevertheless, Cassirer is fully aware that scientific progress is not simply cumulative. In the first pages of *The Problem of Knowledge,* Cassirer suggests very clearly that in the "critical periods" during which scientific knowledge changes its fundamental standpoints, we are not witness to a mere "quantitative growth" but, on the contrary, to a "strong dialectical contradiction" between the different insights at issue. "A concept earlier considered as untenable – Cassirer adds – can later become both a means and a necessary condition of knowledge". In other words, what has been a basic principle of empirical knowledge is not universally and unconditionally valid; on the contrary, it seems quite possible that its previous fundamental function in explaining phenomena will be overthrown by a new conceptual framework, which may even make the former one appear "absurd" (Ivi., p. 4; my emphasis). Thus, in this context, it would not be an exaggeration to suggest that Cassirer's conception of the conflict between opposed principles of knowledge anticipates a sort of "transition from one paradigm to another" as Thomas Kuhn put it. What is still more interesting, however, is the fact that Cassirer clearly provides a concept of scientific revolution that is akin to the Koyré's one. According to Cassirer, a scientific revolution implies, quite similarly to Koyré, a radical "intellectual mutation", thereby constituting a "new way of thinking", in Kantian terms. But, the main difference between Cassirer and Koyré consists both in the origins and in the structure that they believed to be characteristic of this new attitude towards human reason.

9.4 Koyré, Cassirer and Scientific Revolutions

Since the publication of *The Problem of Knowledge,* Cassirer has assumed an eminent place within the history of science.[13] His subtle, documented analyses concerning the origins of both scientific thought and philosophical development in connection to mathematical sciences can still be considered a fundamental contribution to contemporary historical scholarship. At the beginning of the twentieth century, Cassirer was among the first to offer an extensive account of scientific and philosophical insights by (to quote only the most representative figures) Leonardo, Galileo, Kepler, Newton, Descartes, Leibniz, Euler and Kant, viewing the history of science as an essential aspect of the history of philosophy and, above all, the history of epistemology.[14] Nonetheless, the neo-Kantian philosophical background of Cassirer's work has led scholars to underestimate his quite original contribution and is responsible, for the most part, for the fact that, to this day, Cassirer is seen as more of a pure philosopher with a marked interest in the history of philosophy, rather than as a true historian of science. It is a little wonder, then, that in the introduction to his book *The Metaphysical Foundations of Modern Physical Science* (1924), Edwin Arthur Burtt wrote:

> Professor Cassirer [...] has done work on modern epistemology which will long remain a monumental achievement in its field. But a much more radical historical analysis needs to be made. (Burtt 1954, p. 29)

One can share Burtt's desire for a "more radical historical analysis", but it would be difficult to maintain that Cassirer was unfamiliar with proper historical inquiries. Interestingly enough, we may add, an outstanding scholar such as Arthur O. Lovejoy was also acquainted with Cassirer, at least with Cassirer as a Leibniz editor.[15] More generally, the American community of historians of science during the 1940s manifested a high regard for Cassirer's studies on Renaissance thought, as well as on Leibniz, Newton and Galileo, among others, which he published in the time of the harrowing years of his exile in the United States.[16] This is significant inasmuch as Cassirer not only contributed to the very influential *Journal for the History of Ideas*, but the

---

[13] Surprisingly enough, Cassirer is only incidentally quoted by Cohen 1994, p. 11.

[14] See, in particular, the section of the *Problem of Knowledge* devoted to the long scientific and philosophical road "from Newton to Kant" (Cassirer 1910–1911, II, pp. 372–397). Here Cassirer deals with thinkers who, at that time, were little known such as Henry More. In addition, he bases his inquiry on the handbooks on mechanics published in the age of Newton's triumph. Obviously, the history of science has since made enormous progress in this field of research, but Cassirer's outstanding work remains a pioneering example that even today demands closer consideration.

[15] On Cassirer and Lovejoy, see Meyer 2006, pp. 234–235. In his opus magnum, Lovejoy never refers to Cassirer's *Problem of Knowledge* or any other works by him, though Lovejoy does quote the two volumes of Leibniz's writings edited by Cassirer and Artur Buchenau (Leibniz 1904–1906) and (Lovejoy 1936, p. 349, footnote 1). Moreover, Lovejoy had already published a review of this Leibniz edition (Lovejoy 1906). The question of whether Lovejoy was indeed influenced in some way by Cassirer's interpretation of Leibniz remains still open for discussion.

[16] See in particular Cassirer 1942a, b, 1943.

volume presented to George Sarton includes one of his final essays, on *Galileo's Platonism*, which – as we shall see – represents the ultimate achievement of his long research activity in this field (Cassirer 1946).

Moreover, it has not been stressed enough how Cassirer was well known among the leading figures of philosophy and the history of science in France. Brunschvicg, Meyerson and Koyré are particularly worthy of being considered here because of their acquaintance with Cassirer's works concerning the historical development of modern science. It is no wonder, then, that in his masterful history of mathematical philosophy, Brunschvicg repeatedly refers to Cassirer's *Problem of Knowledge*,[17] while also endorsing his plea for an historical analysis of mathematical and scientific thought based only on what Brunschvicg calls "the historical method" (*la méthode historique*; Brunschvicg 1993, p. 3). According to Brunschvicg, as well as Cassirer, at issue is a view of scientific knowledge as a kind of historical dynamics. It is no accident, therefore, that in 1936 Brunschvicg will contribute to the volume devoted to Cassirer on the occasion of his 60th birthday with a paper focusing that at the turn of the twentieth century the exact sciences (probability calculus, thermodynamics, theory of relativity) exhibited a new form – the "form of history". History, therefore, was no longer a mere appendix to knowledge but rather an intrinsic, immanent feature of scientific thought. For Brunschvicg, this perspective, very similar to Cassirer's own, was essentially tied to a philosophy of mind (*philosophie de l'esprit*) that opens the domain of reason to the dimension of historical development.[18]

In contrast, Meyerson does not agree with Cassirer's neo-Kantian standpoint. In his review of *The Problem of Knowledge* published in 1911 in the *Revue de Métaphysique et de Morale*, Meyerson insists on the independence of "objective reality" from any epistemological framework; the permanent connection between epistemology and ontology is, according to him, the missing aspect of Cassirer's conception of science and, more generally, the essential assumption on which science is based.[19] Despite this, Meyerson praises the enormous achievement that Cassirer has offered to scholarship with his masterful work and its careful historical reconstruction of modern science. Moreover, Meyerson remarks in his review that Cassirer's excellent book is the outcome of an "immense knowledge" (*immense savoir*), embracing not only the history of philosophy but also the history of science in its many different aspects and topics. Therefore, in Meyerson's opinion, Cassirer's work represents both a great novelty and a veritable model for scholarship devoted

---

[17] Brunschvicg 1993, especially p. 262; see also p. 205, where Brunschvicg quotes Cassirer's book on Leibniz and laments the fact that both Bertrand Russell and Louis Couturat failed entirely to understand "l'exactitude fondamentale et la profondeur" of this illuminating work.

[18] Brunschvicg 1936. Koyré himself had been invited to collaborate to the volume in honour of Cassirer. Unfortunately, he did not accept because he had no contribution ready for publication (Zambelli 2016, p. 183).

[19] I refer to Meyerson (1911). See also Meyerson (1951, p. 439), where he states that epistemology cannot be separated from ontology, as both constitute a fundamental unity. Meyerson's work is analysed at length by Fruteau de Laclos (2009; see also Fruteau de Laclos 2014 and Jorland 1981, pp. 23–70).

to analysing scientific thought from an historical standpoint, a standpoint – we might add – which at the time was not as familiar to philosophers or historians of philosophy as it is today.

Thus, both Meyerson and Brunschvicg are the French philosophers who, at the beginning of the twentieth century, contributed to the acknowledgement of Cassirer as an eminent historian of science.[20] For his part, Koyré often refers to Cassirer, especially in his *Études galiléennes*, albeit he disagrees with him when he comes, in particular, to the question of Galileo's Platonism.[21] Although Koyré (along with Duhem, Brunschvicg and Meyerson) recognizes Cassirer's great accomplishment in having acknowledged in his writings both the extraordinary importance of the modern scientific revolution and its relevance for philosophy (Koyré 1939, p. 11), it is nonetheless remarkable to witness the *differences* that emerge from a comparison between Koyré's and Cassirer's interpretation of the history of science. On the one hand, unlike Cassirer, Koyré tries to capture the inner process of the growth of scientific thought "by comprehending its development – as he said in 1951 – in the course of its own creative activity" (Koyré 1973, p. 14). In other words, Koyré attempts to explore the specific procedure by which a scientist, e.g. Galileo, Descartes or Newton, builds a scientific theory by composing, so to speak, the individual, abstract pieces upon which said theory depends. Koyré, on the other hand, is not committed to a Kantian way of thinking, and it is no accident that he was deeply influenced by Meyerson, as has been suggested. According to Koyré, scientific thought must be illuminated by considering both its intrinsic conceptual instruments and reality as it is, without referring to a kind of transcendental subject that constitutes it. This explains why Koyré refuses Cassirer's interpretation of Galilean Platonism in so far as Cassirer considers Plato as he would be Kant ("Cassirer a kantisé, si l'on peut dire, Platon").[22] By no means does this suggest that Koyré's account of the scientific revolution does not share a disciplinary background with that of Cassirer or that his view is indebted either to old positivism or to Ernst Mach's conception of the history of science. On the contrary, Koyré and Cassirer are both firmly convinced that mathematical Platonism, and not experimental outcomes as such, have been at the core of the exact science since the days of the rise of modern scientific thought.[23] The question is, rather, how Platonism should be understood and in what way mathematics plays a pivotal role in grounding physical science. According to Koyré, mathematics is geometry above all else. He goes on to identify the very origins of the modern scientific image of the world with both the "geometrization" of space and the "destruction of the Cosmos" inherited by the Greeks. For Cassirer, by contrast, mathematics essentially amounts to the quantification of

[20] An overview of this issue is offered by Seidengart (1995).

[21] A minor, but not marginal, disagreement regards specifically the status of the law of inertia as Galileo and Descartes conceived of it (Koyré 1939, pp. 8–9, pp. 90–91).

[22] Koyré 1939, p. 215. Koyré adds: "Galileo's Platonism means also, in Cassirer's eyes, that Galileo has given the predominance both of function and law over being and substance (*ibidem*).

[23] "It is the pure thought without mixture and not the sense experience that lies at the core of Galileo's 'new science'" (Koyré 1973, p. 210).

natural phenomena through infinitesimal calculus, a topic largely overlooked by Koyré. Thus, in Cassirer's mind, the main feature of mathematics is that of applied mathematics in the broader, transcendental Kantian sense, rather than belonging solely to a realm of pure *mathemata*, as Koyré seems to believe, in agreement with Husserl.

Appropriately, the specific role played by Platonism and, to certain extent Neo-Platonism, in the origins of modern science, from Galileo's revolution onwards, is a question that came up time and again.[24] As Cassirer writes in his essay, "Mathematical Mystique and Mathematical Science of Nature", in which he refers to Koyré, Burtt, Edward Strong and Dietrich Mahnke, among others, it is crucial to simply illustrate what Platonism means and why it can be considered the essential framework within which the scientific revolution is rooted. Interestingly enough, Cassirer distanced himself from both Burtt's and Strong's interpretations. On the one hand, Cassirer argues that Burtt overestimates the "Neo-Platonic speculations as well as the Pythagorean mystique of numbers" which he believes significantly influenced Galileo's mathematical physics. On the other hand, Cassirer maintains that, for his part, Strong did not fully understand how insufficient the empirical angle of Galileo's inquiries was to find the new science. Cassirer, however, clearly points out the distinction between the mere "symbolic" aspects constituting the wide cultural background of scientific thought and their properly rational components that lie at the core of both the mathematical and empirical investigations of nature. Accordingly, the main feature of Galileo's theory of science consists of the "correlation" between experience and thought, which in no way means that mathematics can be reduced to a simple "technical application", as Strong suggests.[25]

Hence, it is not by accident that Cassirer – both as a great historian of scientific thought and as a neo-Kantian philosopher – devotes his last intellectual energy to discussing once more the question of the significance of Galileo's Platonism. In his contribution to the volume presented to George Sarton, Cassirer takes into account Koyre's "excellent article" on Galileo and Plato (Koyré 1943), stressing his essential agreement with his French colleague. Koyré had criticized Burtt's interpretation by distinguishing between two kinds of Platonism: the first being a purely mathematical Platonism, while the second is strictly connected to the mystical-speculative tradition that flourished within the Florence Academy, to which Galileo was entirely foreign.[26] According to Koyré, Galileo was, therefore, involved in the founding of exact science through a straightforwardly oriented mathematical Platonism, a cir-

[24] See, for instance, Koyre's critical review of Cassirer's book on the Cambridge Platonist (Cassirer 1932), where Koyré contends that Henry More as well as the other exponents of the School of Cambridge was rather connected to Florentine Neo-Platonism representing, Koyré maintains, a "reactionary" orientation (Koyré 1935, p. 146).

[25] Cassirer 1940, p. 290, p. 297. According to Eugenio Garin, Cassirer is wrong in proposing this distinction, since "mathematical Platonism and mystique of numbers" are quite similar (Garin 2007, p. 312).

[26] Koyré 1973, p. 212 and Koyré 1943, p. 425, footnote 64. In this footnote Koyré refers to the distinction between two different traditions of Platonism that Brunschvicg had rightly proposed (Brunschvicg 1993, pp. 67–70). See, by contrast, Burtt 1954, p. 68.

cumstance which Koyré provocatively summarized by stressing – as we have remarked above – that "the new science is for him [Galileo] an experimental proof of Platonism".[27] In answering Koyré, Cassirer emphasizes a modified account of this story or, at the very least, a more sophisticated point of view. In his mind, Galileo's Platonism represents a *third* kind of Platonism, one that is neither metaphysical nor mystical or even simply a mathematical Platonism. Quite differently from Koyré, Cassirer begins with a stimulating historical reconstruction, through which he attempts to describe the *physical* Platonism underlying Galileo's scientific revolution. What is at stake here is the use of the method of geometrical analysis as a basic conceptual tool of natural science which enables Galileo to conceptualize empirical phenomena while avoiding any commitment to either the Aristotelian tradition or traditional Platonism:

> Galileo [...] acted and spoke – Cassirer suggests – as a faithful disciple of Plato's. He followed the same method that Plato used in his "Meno" [...] Galileo simply transferred the method of "problematical analysis" that had stood its ground in the history of geometrical and astronomical thought [...] He had to deviate both from the principles of Platonism and Aristotelianism. He accepted Plato's hypothetical method but he gave this method a new *ontological* status; a status which it had never possessed before.[28]

It has often been remarked that, in his later work, Cassirer distances himself irremediably from his former neo-Kantianism. Yet, after reading this essay on Galileo, one cannot help but feel impelled to modify this view. Cassirer seems to still be in agreement with the interpretation offered by Paul Natorp more than 60 years earlier in his seminal article *Galileo as Philosopher* (*Galilei als Philosoph*), which essentially builds on Hermann Cohen's way of rethinking Plato through the lens of critical idealism.[29] What Cassirer has never overlooked, despite his extraordinary historical awareness might disguise it, is the focus on the crucial theoretical matter regarding the conditions of mathematical science, i.e. the very idea of mathematics as the structure of pure thought (as Cohen would say) which constitutes the reign of physical laws. This is another reason for which Cassirer is remembered as an eminent but, unfortunately, outdated historian of science.

[27] Koyré 1943, p. 428. Koyré's interpretation of Galileo is discussed with shrewdness by Galluzzi (1994).

[28] Cassirer 1946, p. 351. One must remember that Cassirer clearly disagrees with John Randall's opinion, according to which the Aristotelianism of Padua would have exercised a decisive influence on Galileo.

[29] Natorp 1882b. In particular, Natorp underlined that Galileo's references to Plato could be considered as proof of his greater commitment to the mathematical laws of "mechanics" than to a purely geometrical Platonism (Ivi., p. 206).

9.5 Conclusion

Thomas Kuhn once noted that Cassirer undoubtedly had a significant influence on the subsequent history of science in spite of his "profound [...] limitations" (Kuhn 1977, p. 108, p. 149). One should not underestimate such an acknowledgement, but it seems even more significant that Kuhn's teachers were rather Meyerson, Koyré, Anneliese Maier and Hélène Metzger, historians who, as Kuhn indicates, were not influenced by Kantianism or neo-Kantianism (demonstrated, in particular, by the case of Meyerson).[30] Yet, the picture is more complicated than the accepted reconstruction might suggest. Studies in the history of science were, without doubt, more philosophically oriented during their golden age of pioneering work than they have become since the field's professionalization. In this sense, both Cassirer and Koyré testify to the fact that every account of the history of science requires, to some extent, a philosophy of science or, at the very least, philosophical assumptions regarding revolutions, paradigms, continuity and discontinuity in the development of scientific thought. This is the fundamental consideration articulated by both Cassirer and Koyré throughout their careers, despite their many differences regarding their own way to understand history of science as a "philosophical history of science" (Zambelli 2016, p. 262). It, therefore, proves to be a compelling reason to reassess not only their respective contributions to contemporary debates but also the remarkably original nature of their historical account of scientific reason. The space for further investigations is still open and seems to promise a stimulating view on our recent intellectual history.

Acknowledgements I would like to warmly thank Shimon Shemtov and Natalia Iacobelli for having helped me with the English corrections of this paper.

References

With the exception of *The Problem of Knowledge*, the works of Ernst Cassirer are quoted from (1998–2009) *Cassirer Gesammelte Werke. Hamburger Ausgabe*. Recki B (ed). 26 Vols. Meiner Hamburg (henceforth EWC).

Bensaude Vincent B (2016) Koyré, disciple de Meyerson. In Seidengart 2016, pp. 185–204.

Biard J (2016) Koyré et le problème du vide au Moyen Âge: remarques sur le continuisme et le discontinuisme. In Seidengart 2016, pp. 125–145.

Brunschvicg L ([1912] 1993) Les étapes de la philosophie mathématique. Blanchard, Paris.

Brunschvicg L (1936) History and Philosophy. In Klibansky and Paton 1936, pp. 27–34.

Burtt E A (1954 [1924]) The Metaphysical Foundations of Modern Physical Science. Anchor Books, New York.

Butterfield H (1949) The Origins of Modern Science. Bell and Sons, London.

[30] Kuhn 1970a, p. VI. Regarding Kuhn and Meyerson, see Friedman (2002), especially pp. 31–34 as well as Friedman (2005, p. 80). Finally, Friedman (2010) focuses on Cassirer and Kuhn. The topic Kuhn and Koyré is discussed by Ferrari (2016).

Cassirer E (1902) Leibniz's System in seinen wissenschaftlichen Grundlagen. Elwert, Marburg (ECW, vol. 1).
Cassirer E (1907) Kant und die moderne Mathematik (Mit Bezug auf Bertrand Russells und Louis Couturats Werke über die Prinzipien der Mathematik). Kant–Studien 12:1–49 (ECW, vol. 9, pp. 37–82).
Cassirer E (1910–1911) Das Erkenntnisproblem in der Philosophie und Wissenschaft der neueren Zeit. 2nd edition. 2 Vols. Bruno Cassirer, Berlin.
Cassirer E (1927) Individuum und Kosmos in der Philosophie der Renaissance. Teubner, Leipzig und Berlin (ECW, Vol. 14).
Cassirer E (1932) Die Platonische Renaissance in England und die Schule von Cambridge. Teubner, Leipzig (ECW, Vol. 18).
Cassirer E (1937) Wahrheitsbegriff und Wahrheitsproblem bei Galilei. Scientia 62:121–130, 185–193 (ECW, Vol. 22, pp. 51–72).
Cassirer E (1940) Mathematische Mystik und mathematische Naturwissenschaft. Lychnos 4: 248–265 (ECW, Vol. 22, pp. 248–305).
Cassirer E (1942a) Galileo Galilei: A New Science and a New Spirit. American Scholar 12:5–19 (ECW, Vol. 24, pp. 53–66).
Cassirer E (1942b) G. Pico della Mirandola. A Study in the History of Renaissance. Journal of the History of Ideas 3:123–144, 319–346 (ECW, Vol. 24, pp. 67–113).
Cassirer E (1943) Leibniz and Newton. Philosophical Review: 366–391 (ECW, Vol. 24, pp. 135–159).
Cassirer E (1946) Galileo's Platonism. In Montague 1946, pp. 277–297 (ECW, Vol. 24, pp. 355–354).
Cohen FH (1994) The Scientific Revolution: A Historiographical Inquiry. The University of Chicago Press, Chicago.
Cohen H (1883) Das Prinzip der Infinitesimal-Methode und seine Geschichte. Dümmler, Berlin.
Cohen H (1885) Kants Theorie der Erfahrung, 2nd enlarged edition. Dümmler, Berlin.
Crombie AC (1953) Robert Grossetesta and the Origins of Experimental Science 1100–1700. The Clarendon Press, Oxford.
Crombie AC (1952) Augustine to Galileo. Falcon Press, London.
Ferrari M (2015) Ernst Cassirer and the History of Science. In Tyler and Luft 2015, pp. 11–29.
Ferrari M (2016) Alexandre Koyré et Thomas Kuhn : l'histoire des sciences et les révolutions scientifiques. In Seidengart 2016, pp. 221–244.
Ferrarin A (2014) Galilei e la matematica della natura. ETS, Pisa.
Friedman M (2002) Kuhn and Logical Empiricism. In Nickles 2002, pp. 19–44.
Friedman M (2005) Ernst Cassirer and Philosophy of Science. In Gutting 2005, pp. 71–83.
Friedman M (2010) Ernst Cassirer and Thomas Kuhn: The Neo-Kantian Tradition in the History and Philosophy of Science. In Makkreel and Luft 2010, pp. 177–191.
Fruteau de Laclos F (2009) L'épistémologie d'Émile Meyerson. Une anthropologie de la connaissance. Vrin, Paris.
Fruteau de Laclos, F (2014) Émile Meyerson. Les Belles Lettres, Paris.
Galluzzi P (1994) Gli studi galileiani. In Vinti 1994, pp. 241–261.
Garin, E (2007) Rinascite e rivoluzioni. Movimenti culturali dal XIV al XVIII secolo. Laterza, Roma–Bari.
Gutting G (2005) (ed.) Continental Philosophy of Science. Blackwell, Oxford.
Jorland G (1981) La science dans la philosophie. Les recherches épistémologiques d'Alexandre Koyré. Gallimard, Paris.
Klibansky R, Paton HJ (1936) (eds) Philosophy and History. Essays Presented to Ernst Cassirer. The Clarendon Press, Oxford.
Koyré A. ([1929] 1979) La philosophie de Jacob Boehme. Vrin, Paris.
Koyré A (1931) Die Philosophie Émile Meyerson. Deutsch-französische Rundschau 4:197–217.
Koyré A (1935) Review of Cassirer (1932). Review de l'histoire des religions 111:145–148.
Koyré A (1939) Études galiléennes. Herman, Paris.

Koyré A (1943) Galileo and Plato. Journal of the History of Ideas 4:400–428.
Koyré A (1946) Les « Essais » d'Émile Meyerson. Journal de psychologie normale et pathologique 39:124–128.
Koyré A (1957) From the Closed World to the Infinite Universe. The John Hopkins Press, Baltimore.
Koyré A (1963) Commémoration du cinquantenaire de la publication des *Étapes de la philosophie mathématique* de Léon Brunschvicg. Bulletin de la Société française de philosophie 57:10–14.
Koyré A (1965) Newtonian Studies. Chapman & Hall, London.
Koyré A (1971a) Études d'histoire de la pensée philosophique. Gallimard, Paris.
Koyré A ([1955] 1971b) Mystiques, spirituels, alchimistes du XVI siècle allemande. Gallimard, Paris.
Koyré A (1973) Études d'histoire de la pensée scientifique. Gallimard, Paris.
Koyré A (2016) De la mystique à la science. Cours, conférences et documents 1922–1962. Nouvelle édition revue et augmentée par P. Redondi. E.H.E.S.S., Paris.
Kuhn T (1970a) The Structure of Scientific Revolutions. 2nd edition. The University of Chicago Press, Chicago.
Kuhn T (1970b) Alexandre Koyré & History of Science. On an Intellectual Revolution. Encounters 34:67–69.
Kuhn T (1977) The Essential Tension: Selected Studies in Scientific Tradition and Change. The University of Chicago Press, Chicago.
Leibniz G W (1904–1906) Hauptschriften zur Grundlegung der Philosophie. Cassirer E, Buchenau A (eds) 2 vols. Dürr'sche Buchhandlung, Leipzig.
Lembeck KH (1994) Platon in Marburg. Platonrezeption und Philosophiegeschichtsphilosophie bei Cohen und Natorp. Königshausen & Neumann, Würzburg.
Lovejoy AO (1906) [Book Review] Hauptschriften zur Grundlegung der Philosophie. The Philosophical Review 15:437–438.
Lovejoy AO (1936) The Great Chain of Being. A Study of the History of an Idea. Harvard University Press, Cambridge (Mass.).
Makkreel AK, Luft S (2010) (eds) Neo-Kantianism and Contemporary Philosophy. Indiana University Press, Bloomington and Indianapolis.
Meyer Th (2006) Ernst Cassirer. Eller & Richter Verlag, Hamburg.
Meyerson É ([1908] 1951) Identité et realité. Vrin, Paris.
Meyerson É (1911) L'histoire du problème de la connaissance de M. Cassirer. Revue de Métaphysique et de Morale 19:100–129.
Meyerson É (2009) Lettres françaises. Bensaude Vincent B, Telkes–Klein E (eds). CNRS Éditions, Paris.
Montagu M F Ashley (1946) (ed) Studies and Essays of Science and Learning. Offered in Homage to George Sarton on the Occasion of his Sixtieth Birthday 31 August 1944. Schuman, New York.
Natorp P ([1921] 1994) Platos Ideenlehre. Eine Einführung in den Idealismus. 2nd edition, Verlag, Meiner.
Natorp P (1882a) Descartes' Erkenntnistheorie. Eine Studie zur Vorgeschichte des Kritizismus. Elwert, Marburg.
Natorp P (1882b) Galilei als Philosoph. Philosophische Monatshefte 18:193–229.
Natorp P (1882c) Die kosmologische Reform des Kopernikus in ihrer Bedeutung für die Philosophie. Preussische Jahrbücher 41:355–375.
Natorp P (1912) Kant und die Marburger Schule. Kant-Studien 17:193–221.
Natorp P (1985) Leibniz und der Materialismus. Studia Leibnitiana 17:3–14.
Nickles T (2002) Thomas Kuhn. The Cambridge University Press, Cambridge.
Seidengart J (1995) Cassirer et la philosophie des sciences en France. Rivista di storia della filosofia 50:753–783.
Seidengart J (2016) (ed) Vérité scientifique et vérité philosophique dans l'œuvre d'Alexandre Koyré. Suivi d'un inédit sur Galilée. Les Belles Lettres, Paris.

Servois J (2004) Paul Natorp et la théorie platonicienne des idées. Presses Universitaires du Septentrion, Lille.
Sieg U (1994) Aufstieg und Niedergang des Marburger Neukantianismus. Die Geschichte einer philosophischen Schulgemeinschaft. Königshausen & Neumann, Würzburg.
Tyler J T, Luft S (2015) (eds) The Philosophy of Ernst Cassirer. A Novel Assessment. De Gruyter, Berlin–Boston.
Vinti C (1994) (ed) Alexandre Koyré. L'avventura intellettuale. ESI, Napoli.
Zambelli P (1998) Alexandre Koyré. Alla scuola di Husserl a Gottinga. Giornale critico della filosofia italiana 78:303–354.
Zambelli P (1999) Alexandre Koyré im „Mekka der Mathematik". Naturwissenschaften Technik Medizin 23:208–230.
Zambelli P (2016) Alexandre Koyré in incognito. Olschki, Firenze.

Chapter 10
Alexandre Koyré and the History of Science as a Species of the History of Philosophy: The Cases of Galileo and Descartes

Stephen Gaukroger

Abstract It was Koyré who, more than anyone else, was responsible for the transformation of the history of science from an antiquarian discipline, whose role was to chronicle gradual advances towards modern science, into one in which deep conceptual issues could be raised about the nature of space, time and matter, the role of experiment and theory, the role of scientific instruments, the place of mathematics in physical theory and so on. But Koyré's transformation of the discipline into something with real conceptual content was, I believe, achieved at a cost. He quite rightly drew attention to and explored—indeed in some respects pioneered—the metaphysical and epistemological dimensions of natural-philosophical or 'scientific' questions. But on a number of occasions, he went beyond this, effectively translating the latter into metaphysical and epistemological questions. This is counterproductive, for it blinds us to the nature of the questions under investigation, and I want to look at two cases—Galileo's discussion of free fall and Descartes' rejection of a vacuum—where this feature of his approach is more evident and most deleterious. In both cases, the underlying assumption seems to be that the most fundamental concepts in physical theory are ultimately philosophical or philosophically driven ones: Koyré effectively reduces natural philosophy to epistemology and metaphysics.

Keywords Cosmology • Descartes • Galileo • Historiography of science • Hydrostatics • Idealisations • Kinematics

S. Gaukroger (✉)
Sydney University, Sydney, NSW, Australia
e-mail: stephen.gaukroger@sydney.edu.au

R. Pisano et al. (eds.), *Hypotheses and Perspectives in the History and Philosophy of Science*, DOI 10.1007/978-3-319-61712-1_10

10.1 Galileo on Falling Bodies

Koyré construed the seventeenth-century scientific revolution in physics and astronomy in terms of a conflict between a mathematically orientated Platonism and an experimentally orientated Aristotelianism, where the former completely triumphed. On this account, Galileo, who more than anyone else stands at the head of the scientific revolution for Koyré, followed the platonistically inspired procedure of subordinating empirical reality to a realm of mathematical idealisations. Galileo's account of falling bodies, on this account, is an idealisation which abstracts from real bodies so as to exhibit a simple mathematical relation between time and distance traversed. Experiment has no role to play in this process, as he construes it:

> [Galileo] was obliged to drop sense perception as the source of knowledge and to proclaim that intellectual, and even *a priori* knowledge, is our sole and only means of apprehending the essence of the real. (Koyré 1968, p. 38)

Galilean physics is the physics of a world of mathematical idealisations, as indeed is all mathematical physics, on Koyré's view (Koyré 1968, 1978). For Koyré, the mathematisation of physics is effectively a translation of physics into the platonic realm of numbers and geometrical figures. And the rationale for this is ultimately an epistemological one, for physical theory is, for Koyré, ultimately a form of applied epistemology.

But in the *Two New Sciences*, Galileo explicitly defends his account of falling bodies from being a mathematical idealisation, and his account integrates theoretical and experimental considerations in a novel and extremely fruitful way, which goes well beyond anything that can be captured in terms of Platonist/Aristotelian or rationalist/empiricist distinctions (for details see Gaukroger 1978, chapter 6). Galileo provides in *Two New Sciences* the first modern full-scale kinematical treatment of motion: in particular, he presents and justifies laws concerning free fall and projectile motion. These he presents in the form of mathematical descriptions of what happens in a void. Now the motion of bodies in a void is something we never experience and something to which we have no direct access. The motions of bodies in resisting media *is* something we regularly experience, yet these motions differ from the motions those same bodies would undergo in a void. Galileo's law of free fall tells us that all bodies undergo a uniform acceleration in a void, but this is clearly not the case in a resisting medium. At first sight, therefore, the law appears to suffer from two drawbacks: it appears to tell us something about a situation which may never occur, and it appears not to tell us about situations which do normally occur. Hence there seem to be problems both about the relevance of the law and about whether it could receive any evidential support. Since we do not experience bodies falling in a void, for example, we cannot arrive at Galileo's law inductively. On the other hand, since, if the law holds, its truth is contingent—bodies may just as easily have fallen at different rates in a void or may have fallen with an unaccelerated or non-uniformly accelerated motion—it is impossible that a priori arguments will lead us to the law. To maintain that the law is a hypothesis open to empirical tests is of no real help either. First, the problem simply reappears at a different level.

The situation described in the law does not naturally occur and cannot be experimentally induced, so in what way is it open to empirical test? And, even if the situation *could* somehow be experimentally induced, that would still leave the problem of how the law could be at all relevant to the case of bodies falling in resisting media. Second, the presumption behind the hypothetical construal is that the theory itself is somehow developed at a purely conceptual level and then tested empirically to determine whether it is true or not. But, experiment is an integral part of scientific discovery, not an added extra.

This is, at least, the view presented and defended in Galileo's own account of how the law is to be established (Galilei 1974, pp. 65–108). Galileo's project is to discover the relations that hold between a body falling in a void and that body falling in a resisting medium. He takes the fall of bodies in a resisting medium as his starting point and then describes a series of experiments, including thought experiments, designed to decide what factors are operative in determining the rate of fall of a body and how these factors operate.

He deals first with the traditional Aristotelian view that rate of fall is directly proportional to absolute weight. He has two arguments against this. The first is empirical: if two bodies made of the same material but of different absolute weights are dropped simultaneously from the top of a tower to the ground, they arrive at the ground simultaneously. The second is a thought experiment. If we drop, say, two lead spheres of different weights, then on the Aristotelian account, the heavier will fall faster. But, suppose we tie the spheres together. The slower one would then surely slow down the faster one, and the faster one would speed up the slower, so that the resultant speed would be somewhere between the two original speeds. But the aggregate weight is greater than the weight of the heavier body. Hence, rate of fall cannot be directly proportional to absolute weight. Now Aristotle had also held that the rate of fall is inversely proportional to the density of the medium. Against this another thought experiment is proposed. If we let the density ratio of water to air be n:1, where $n > 1$ (since the specific weight of water is greater than that of air), and take a body which falls in air but floats in water (e.g. a wooden sphere), and say that this has a rate of fall of one unit in air, then it would follow that it has a rate of fall of $1/n$ units in water. But we have already said that it floats—i.e. would rise and not fall, in water. So rate of fall cannot be inversely proportional to the density of the medium.

Next, Galileo makes an important generalisation. Instead of thinking simply in terms of rate of fall determined with respect to one body in two media or with respect to two bodies in the one medium, he considers the case of any body in any medium. First, he describes an experiment which shows that the ratio between the rates of fall of bodies is not the same as the ratio of their specific weights. Gold and lead, which fall at approximately the same rate in air, behave quite differently in mercury, the former sinking, the latter rising to the surface. This experiment indicates that differences in rate of fall of bodies diminish as the density of the medium decreases. This prompts him to ask what would happen in the limiting case of a void: he raises the possibility that in such a case the rate of fall of all bodies would be equal. But, until we know the precise connection between speed, specific weight and resistance, we will not be able to establish this. Galileo therefore proposes an experiment in which the buoyancy effect of the medium can be distinguished. The buoyancy effect is the ratio between the specific weight of the body and that of the medium. The problem is to determine precisely what effect this ratio has on rate of fall. He compares the buoyancy effect of two media (air and water) on two bodies (ebony and lead). Given the specific weights of these substances, the buoyancy effect can easily be calculated. It turns out that the buoyancy effect varies much more radically than the specific weight of the body: if we let the specific weights of air, water, ebony and lead be 1, 800, 1000 and 10,000, respectively, then it turns out that whereas the buoyancy effect of air has a negligible effect on rates of fall of ebony and lead, the buoyancy effect of water on ebony is huge (it loses four-fifths of its effective weight), whereas its effect on lead is very small (less than one-tenth). It is the specific weight *ratios* that determine rate of fall, not the specific weights themselves. Since a void has no specific weight, it cannot bear a ratio relation to the specific weight of the falling body, i.e. this ratio which determines differences in the rate of fall cannot be operative in the case of a void, and so we must conclude that all bodies—whatever their specific weight—fall in a void with the same 'degree of speed' (i.e. as it turns out, degree of uniformly accelerated motion). This conclusion is particularly important since on the basis of the equality of rates of fall of all bodies in a void, we can proceed, at least in principle, to calculate the differences in speeds between any two bodies in any media by determining the amount by which the theoretical speed in a void will be diminished by comparing the specific weight of bodies with that of the media. To this end, Galileo takes us through (largely unsuccessful) experiments to measure the specific weight of air.

There remains one problem. Bodies falling in media do not in fact accelerate uniformly. Neither specific weight nor the buoyancy effect can account for this since they are both constant (the latter being a constant for any particular body and any particular medium). This leads Galileo to invoke a form of resistance to fall which is distinct from the buoyancy effect: the friction effect. The friction effect increases with the acceleration of the body since larger and larger amounts of resisting medium have to be traversed, and equilibrium is reached when the body ceases to accelerate because of friction. This state of equilibrium occurs much earlier in rarer bodies, not because the friction effect bears a direct relation to specific weight but because the buoyancy effect is much greater in bodies of lower density, and hence their motion is already greatly retarded. For this conclusion to go through, however, two things have to be shown: first, that the friction effect is greater for rarer bodies, and, second, that it increases with speed. In order to show the first, we need to isolate the friction effect experimentally from the buoyancy effect. Free fall does not allow us to do this. Galileo suggests rolling two bodies, e.g. one of cork and one of lead, down a plane which is gently inclined so as to make the motions as slow as possible and thereby to reduce the buoyancy effect. The trouble here is that the more gentle the incline, the greater the surface friction, which would interfere with our isolating the friction effect (which is totally different from surface friction since it is an effect of the medium). He resolves these difficulties by proposing an ingenious experiment in which a cork and a lead sphere are suspended on threads of equal length and set in oscillation. The periods of oscillation remain identical for both spheres, but in the case of the cork sphere, the amplitude of swing is considerably reduced very quickly. This cannot be due to the greater specific weight of the lead causing it to move faster: we can begin the experiment by swinging the cork through a greater arc so that it initially moves faster, but the same thing will happen. Moreover, since the buoyancy effect is simply the specific weight ratio, it cannot be due to this either. It must therefore be due to the friction effect. Finally, all that remains to be shown is that the friction effect increases with speed. Again no direct experiment on freely falling bodies is possible because of the great distances that would be involved and the difficulties in measurement that would ensue. Hence the consequence that bodies projected at artificially high speeds will be retarded until they reach their natural maximum speed for that medium is of crucial importance, since an experimental situation in which this can be tested can be realised in a relatively easy and straightforward way. We simply fire a gun vertically downwards from a great height and measure the penetration of the bullet into the ground. We then fire the gun close to the ground and measure that penetration. The first is less than the second, which means that the bullet has been retarded.

In sum, Galileo shows, by means of a series of actual experiments and thought experiments, that rate of fall bears a complex relation to specific weight, buoyancy effect and friction effect. By determining exactly how these factors are related to one another, he is able to determine what happens when the medium is removed entirely, and this forms the content of his law of fall. This is a clear example of a case in which experiment plays a necessary role in the formulation of a theory. Any attempt at epistemological reduction is bound to fail: the issues on which the successes of Galileo's formulation of the law hinge go far beyond epistemological considerations.

10.2 Descartes on the Void

Galileo formulates his law of falling bodies in terms of what happens in a void. Descartes, by contrast, rejects any analysis of the motion of bodies based on how they behave in a void. Like many other commentators, Koyré assumes that what motivates Descartes' rejection of the idea that there are empty spaces in the universe are a priori epistemological and metaphysical arguments about the void deriving from Aristotle and the Stoics, arguments about the very possibility of a void. Here is what he says in *From the Closed World to the Infinite Universe*:

> The void according to Descartes, is not only *physically impossible*, it is essentially impossible. Void space—if there were anything of the kind—would be a *contradictio in adjecto*, an existing nothing. Those who assert its existence, Democritus, Lucretius and their followers, are victims of false imagination and confused thinking. They do not realise that *nothing* can have no properties and therefore no dimensions. To speak of ten feet of void space separating two bodies is meaningless: if there were a void, there would be no separation, and bodies separated *by nothing* would be in contact. (Koyré 1957, p. 101)

The question here is why, in the context of physical theory, Descartes rejects the existence of a void. Is it because, having reflected on the nature of space, he has decided that there is a logical contradiction in supposing there to be such a thing as empty space? Descartes does indeed think there is a contradiction in the notion of empty space, but the question is whether this epistemological/metaphysical consideration is a rationalisation of a conception of the conditions under physical processes that occur which he has arrived at through pursuing physical theory or whether it is what drives his attempts to think through the motion of bodies in terms of the displacement of fluid media. That the latter cannot be the answer is evident from the fact that Descartes' persistent and consistent attempts to think through physical problems in terms of the motion of bodies in dense media begin in 1619, nearly 20 years before he offers the argument for the impossibility of a void. What drives Descartes' rejection of a void is not a philosophical view about the impossibility of a void but a particular way of thinking through physical problems, namely, in terms of hydrostatics. It was in hydrostatics that Descartes first learned his physical vocabulary, and throughout his life he thought of physical action in terms of bodies being acted upon or acting upon the surrounding medium.

Descartes' route to dynamics was via statics/hydrostatics. Since statics does not deal with moving bodies, but does deal with forces, the aim is to develop the treatment of forces used to describe stationary bodies into the realm of moving bodies, e.g. by asking how these forces are modified or supplemented when the body begins to move in a particular way. The advantage of this approach was that statics had been pursued since Archimedes in a precise, quantitative, geometrical fashion, which was exactly how the seventeenth-century natural philosophers wished to pursue dynamics. Descartes' pursuit of the hydrostatic route is evident in the conceptual vocabulary, particularly in the talk of tendencies to motion rather than motions proper, and in the kinds of models and analogies—in which pressure, beam balances and states of equilibrium predominate—that he uses to think through problems in dynamics and kinematics (Gaukroger 2000). Two examples bring out the flavour of this approach: his cosmology and his account of the laws of collision.

His hydrostatic model in cosmology was initially developed in *Le Monde*. This was the dominant physical model of the solar system in the mid-seventeenth century, offering a far more attractive picture, to mechanists, of how a heliostatic system might function than anything in the versions of Galileo or Kepler, both of which require action at a distance. In this account, physical effects are produced by means of vortical motions in an all-encompassing fluid. One of the aims of the model was to account for the stability of planetary orbits, and this is achieved in a way that invokes no 'occult' forces, such as gravity, because it acts by contact between the bodies affected. We must imagine the universe to be a spatially extended region which is wholly occupied by matter—what we can call material extension. The constituent matter has an initial motion (provided by God), and as a result of this motion, it breaks up into large pieces of matter (planets), middle-sized pieces (liquid and gaseous matter including the atmosphere and interplanetary ether) and small effectively formless matter filling up the interstices between the other parts of matter and making up light and heat. If one allows rotation of various parts of this material extension, which form individual solar systems, then Descartes believes that all he needs to establish the rotation of the planets around the sun are his theory of matter, centrifugal force and rectilinear inertia. In such a system, he argues, the large parts of matter will be flung outwards, and the small matter will be pressed into the centre. The large clumps of matter will be arranged according to their size, the larger being the further out, because larger bodies will be able to realise their tendency to follow a rectilinear path more effectively, and so will describe a larger circle, which more closely approximates to a straight line, than smaller parts of matter. This will result in an ordering of large bodies in the solar system which matches the ordering of the known planets and what he takes as their estimated sizes: Mercury, closest to the Sun, then Venus, Earth, Mars, Jupiter and finally Saturn. The planets are swept along circular paths in and by the dense fluid that fills the interplanetary spaces, and their orbits are stable because if a planet were to move closer to the Sun, it would encounter more rapidly moving matter and would be squeezed back, whereas if it moved away from the Sun, it would meet larger pieces of matter moving more slowly which would retard its motion and cause it to fall back into its original orbit. Moreover, the smallest matter is squeezed into the centre where it rotates rapidly. This central region, which we call the Sun, radiates fine matter in all directions because of the centrifugal forces at its surface. This is how light is propagated throughout the solar system (Descartes 1998, pp. 21–75).

Descartes also seems to have a statistical notion of force underlying what might ordinarily be regarded as his kinematics. If anything can be identified as kinematics in Descartes, it is surely his rules of collision, but the content of these rules does not seem to be guided kinematically. This can be illustrated by considering Rule 4 (*Principia*, II, Art. 49), which is, by kinematic standards, a very peculiar rule. It tells us that if a moving body collides with a 'larger' (more massive) stationary body, then the smaller body, no matter how slight the difference in size, and no matter how quickly it is moving, will never dislodge the larger one but will rebound off it, its 'quantity of motion' being conserved.

If one looks to Descartes for a rationale of the law, what one finds is a physical claim—that a smaller moving body colliding with a larger stationary one cannot affect the state of the larger body—filled out in quasi-scholastic natural-philosophical terms. There are two important premises in Descartes' treatment. The first is that rest has as much reality as does motion: rest is not simply a 'privation' of motion as the Scholastics had argued. The second is that rest and motion are opposed to one another: they are modal contraries. We must therefore think of the interaction of the bodies in terms of the smaller having a particular quantity of motion and the larger having a particular quantity of rest. These are opposing states, so the bodies will be in dynamic opposition, and Rule 4 therefore describes a contest, as it were, between a larger body at rest and a smaller body in motion. The bodies exercise a force to resist changes of their states, and the magnitude of this force Descartes considers to be a function of their size. A body in motion cannot, for that reason alone, have more force than one at rest, nor can greater speed confer greater force upon it. Either of these would undermine the ontological equivalence of rest and motion that Descartes wants to defend. Now, bearing this in mind, we can ask what happens when the smaller moving body collides with a larger stationary one. Clearly they cannot both remain in the same state in collision, so there will have to be a change state. And since the smaller or 'weaker' body can hardly change the state of the larger or 'stronger' one, it is the smaller one that has its state changed (the direction of its motion is reversed), the larger body remaining unaffected in the process.

This account explains why it has to be an all-or-nothing matter. We might be tempted to ask why the smaller body should not move the larger one if the smaller body had sufficient speed or if the difference in size were very marginal. The answer to the first question is that the speed of the smaller body is irrelevant to the outcome of the collision. The answer to the second is that, because of the irrelevance of speed, the only remaining factor is size. Still, it does seem somewhat peculiar that the outcome would be the same irrespective of whether the difference in size were very significant or whether they were almost exactly the same size. The peculiarity is removed immediately once we think of the situation in terms of statics, however. Think of the bodies as occupying the two pans of a beam balance. The arm will always be tipped down on the side of the heavier, no matter how slight the difference in weight. That this is indeed the reasoning behind Descartes' account is made clear in a letter to Hobbes in which Descartes responds to Hobbes' claim that the extent to which a body is moved is proportional to the force exerted on it, so that even the smallest force will move a body to some extent. Descartes replies:

> His assumption that *what does not yield to the smallest force cannot be moved by any force at all* has no semblance of truth. Does anyone think that a weight of 100 pounds in a balance would yield to a weight of one pound placed in the other pan of the scale simply because it yields to a weight of 200 pounds. (Descartes 1998, p. 73)

What Rule 4 seems to do is to reduce the question to one of statics, by removing considerations of speed. And the means by which it does this is through the principle of the ontological equivalence of motion and rest. Descartes' statement of this equivalence has often been seen as an important move in the direction of a proper understanding of the principle of inertia, as a step on the road from seeing rest simply as a privation of motion, to treating rest and uniform rectilinear motion as being dynamically on the same footing, as being states that require no force for their maintenance. The principle of the ontological equivalence of motion and rest, which in a physical context such as the rules of collision amounts to a dynamical equivalence, is in fact a step in a completely different direction for Descartes. The ontological/dynamical equivalence of motion and rest means that what holds for rest holds for motion. Statics tells us about the behaviour of bodies at rest: perhaps it can be built upon to deal with bodies in motion, if motion can somehow be seen to be a variation on rest. Descartes does not abandon the conceptual apparatus of statics even in his treatment of collision: on the contrary, he seems to be trying to build on it.

10.3 Conclusion

Once we consider Descartes' cosmology and impact rules in terms of a hydrostatics/static model, we can begin to see that his way of doing physics is completely different from that of Galileo and Newton. Koyré misses this because he takes a philosophical rationalisation of a view that Descartes had held on physical grounds for the previous 20 years as if it drove Descartes' project. And what he misses is, I believe, one of the most significant conceptual parting of ways in the seventeenth-century physical theory: the rift between hydrostatic/statical models for dynamics and kinematic ones. Just as, in the case Galileo's problem of free fall, he completely misses what is at stake in the first successful attempt to translate a physical problem into mathematical terms, for Galileo, by thinking through the role of experiment very carefully, goes to great lengths to make sure that the problem of free fall remains a genuine physical problem and not a mere mathematical idealisation. To have pointed out so clearly and emphatically the central importance of epistemological and metaphysical considerations in the development of early modern natural philosophy is Koyré's greatest achievement: to attempt reduce natural philosophy to metaphysics and epistemology is to detract from that achievement.

References

Descartes R (1998) The World and Other Writings, ed. and translated by S. Gaukroger. The Cambridge University Press, Cambridge.

Galilei G (1974) Two New Sciences. Translated by Stillman Drake. The University of Wisconsin Press, Madison

Gaukroger S (1978) Explanatory Structures. Harvester Press, Brighton.

Gaukroger S (2000) 'The Foundational Role of Hydrostatics and Statics in Descartes' Natural Philosophy'. In Gaukroger S, Schuster J, Sutton S (eds). Descartes' Natural Philosophy. Routledge, London.

Koyré A (1957) From the Closed World to the Infinite Universe. The Johns Hopkins University Press, Baltimore.

Koyré A (1968) Metaphysics and Measurement. Harvard University Press, Cambridge–MA.

Koyré A (1978) Galileo Studies. Harvester Press, Brighton.

Chapter 11
Is Descartes' Theological Voluntarism Compatible with His Philosophy?

Glenn A. Hartz and Patrick K. Lewtas

Abstract In 1922, Alexandre Koyré writes that Descartes' doctrine of theological voluntarism – according to which God creates and controls logic and normativity – is incompatible with every bit of his philosophy. We agree and explain how voluntarism undermines the major arguments (including the *cogito*) in the *Meditations* by making them *logically* or *normatively circular*. Thus clear and distinct ideas are not useful as premises in an argument for God's existence except on the assumption that God exists and has already made them "true." However, Koyré also claims that Descartes protected his system from disaster by abandoning voluntarism toward the end of his life. We argue that this is not correct, as an exhaustive look at the texts shows Descartes affirming voluntarism unambiguously from 1630 right to the end of his life. We point out the flaws in other attempts to save Descartes from himself and some theological and philosophical reasons which may have led Descartes to hold the doctrine. We claim that, while voluntarism will have to be ignored in an overall understanding of Descartes' thought, it has genuine philosophical significance because a "voluntarist demon" represents the most formidable skeptical threat possible.

Keywords *Cogito* • God • Koyré • Hyperaspistes • Irrationalism • Logical circle • *Meditations* • Normative circle • Regis • Reason • Skepticism • Voluntarism

11.1 Introduction

From 1630 to 1649, Descartes endorses theological voluntarism, the claim that God's will determines aspects of reality typically thought independent of it. In particular, God does not act based on his knowledge of the good or the true, but his acts

G.A. Hartz (✉)
Ohio State University, 1760 University Drive, Mansfield 44906, The United States
e-mail: hartz.1@osu.edu

P.K. Lewtas
Department of Philosophy, American University of Beirut, Beirut, Lebanon
e-mail: plewtas@gmail.com

R. Pisano et al. (eds.), *Hypotheses and Perspectives in the History and Philosophy of Science*, DOI 10.1007/978-3-319-61712-1_11

determine what, if anything, is good and true. Descartes describes his view as "the creation of the eternal truths," where the good and the true are created ex nihilo and with complete indifference by God. Their existence is as contingent on his will as is that of Eve. Indeed, the eternal truths are explicitly said to be created in the same way that everything else is (CSM 3, p. 25; AT 1, pp. 151–152).

A "Cartesian" study, in our sense, is an interpretation which expounds the philosophical system as proposed by Descartes himself. The interpretation need not endorse all of it, of course. But a Cartesian interpretation retains the integrity of the original system and explicitly notes any deviations therefrom.

By contrast, a "cartesian" is willing to pick and choose among the canonical doctrines according to his or her philosophical needs and interests. The goal is to build the best argument or theory even if it means adding some doctrines or redacting some texts.

Here we are concerned only with the Cartesian. Often a creative solution to a problem for Descartes can be found by pruning or enriching the corpus, thus arriving at a broadly cartesian position. But such a project is of no interest to us here. We aim to press the question of whether voluntarism is compatible with Descartes' thought in its unmodified form.

First, we examine the texts which show that Descartes was throughout his life committed to theological voluntarism. Second, we ask whether this doctrine is compatible with the main arguments of the *Meditations* and Descartes' natural philosophy. We argue that it is not and that all attempts to establish their compatibility fail. Finally, we evaluate various interpretive strategies for mitigating the negative effects of voluntarism within the Cartesian system, and some replies which have been offered on Descartes' behalf.

11.2 Voluntarism

Many medieval philosophers (e.g., Ockham) and some moderns like Gassendi advocated "normative voluntarism," the view that divine will sets normative facts like the nature of perfection. Yet these thinkers often explicitly held that God cannot violate the law of noncontradiction (Ockham 1957, p. 25 *Quodlibeta* [1324–1325] VI, p. 6; Gassendi 1658–1964 I, p. 308). This preserved the core of reason.

Descartes' voluntarism knows no such limits. It embraces both normative voluntarism and "logical voluntarism," according to which logic and mathematics also rest on divine decree. Descartes' voluntarism therefore leaves no proposition necessary in any ultimate sense. Through his incomprehensible power, Descartes claims, God can make contradictories true together (CSM 3, p. 235; AT 4, p. 118).

Yet, as we will see, it is sometimes claimed that the mature Descartes did not hold voluntarism and that his system should not be held hostage to it. Thus we must examine the textual evidence in some detail.

On 6 May 1630, Descartes wrote to Mersenne:

> Passage 1: As for the eternal truths, [...] they are true or possible only because God knows them as true or possible. They are not known as true by God in any way which would imply that they are true independently of him. [...]. So we must not say that if God did not exist nevertheless these truths would be true; for the existence of God is the first and most eternal of all possible truths and the one from which alone all others proceed. [...]. [S]ince God is a cause whose power surpasses the bounds of human understanding, and since the necessity of these truths does not exceed our knowledge, these truths are therefore something less than, and subject to, the incomprehensible power of God. (CSM 3, pp. 24–25; AT 1, pp. 149–150)

Eighteen days later, he added:

> Passage 2: God established the eternal truths ... by the same kind of causality as he created all things. [...] [God] was free to make it not true that all the radii of the circle are equal—just as free as he was not to create the world. (CSM 3, p. 25; AT 1, pp. 151–152)

In 1638 he again mentions the doctrine to Mersenne:

> Passage 3: You ask whether there would be real space, as there is now, if God had created nothing. [...] I believe that our intellect can reach the truth of the matter, which is, in my opinion, that not only would there not be any space, but even those truths which are called eternal – as that 'the whole is greater than its part' – would not be truths if God had not so established [...]. (CSM 3:102-03; AT 2:138)

In the Meditations' *Fourth Objections* (CSM 2, pp. 152–153; AT 7, pp. 217–218), Arnauld had challenged the orthodoxy of the *cogito*. Descartes supposes, Arnauld says, that "powers" such as thinking cannot exist apart from a substance that thinks, even though the doctrine of the Eucharist holds that accidents or "modes" like extension can exist apart from a substance. Descartes responds in the *Fourth Replies*:

> Passage 4: [M]y saying that modes are not intelligible apart from some substance for them to inhere in should not be taken to imply any denial that they can be separated from a substance by the power of God; for I firmly insist and believe that many things can be brought about by God which we are incapable of understanding. (CSM 2, p. 173; AT 7, p. 249)

Thus he allows the power of God to trump even the *cogito*. Hence its necessity is only "necessity*" (i.e., a diminished form of necessity) since God is under no obligation to legitimate it.

Later in the *Replies* (*Sixth Replies* § 8), Descartes furnishes one of his most detailed statements of voluntarism:

> Passage 5: If some reason for something's being good had existed prior to his preordination, this would have determined God to prefer those things which it was best to do. But on the contrary, just because he resolved to prefer those things which are now to be done, for this very reason, in the words of Genesis, "they are very good"; [. . .] [T]here is no need to ask how God could have brought it about from eternity that it was not true that twice four make eight, and so on; for I admit this is unintelligible to us. Yet [. . .] I do understand, quite correctly, that there cannot be any class of entity that does not depend on God. [...]. Hence we should not suppose that eternal truths "depend on the human intellect or on other existing things"; they depend on God alone who, as the supreme legislator, has ordained them from eternity. (CSM 2, p. 294; AT 7, pp. 435–436)

Another text is found in a letter of August 1641. Hyperaspistes had argued against voluntarism with a clever thought experiment:

> [L]et us suppose *per impossibile*, that [God] never thought of a triangle; yet suppose you are in the world as you now are: would you not agree that it was true that the three angles of a triangle equal two right angles? (CSM 3, p. 194)

Descartes responds:

> Passage 6: There is no force in what you say about the triangle. [W]hen God or the infinite is in question, we must consider not what we can comprehend – for we know that they are quite beyond our comprehension – but only what conclusions we can reach by an argument that is certain. (CSM 3, p. 194; AT 3, p. 430)

Later, to Mesland on 2 May 1644, Descartes writes:

> Passage 7: [T]he power of God cannot have any limits [. . .] God cannot have been determined to make it true that contradictories cannot be true together, and therefore [...] he could have done the opposite. [But] even if this be true, we should not try to comprehend it, since our nature is incapable of doing so. I agree that there are contradictions which are so evident that we cannot put them before our minds without judging them entirely impossible, like the one you suggest: 'that God might have brought it about that his creatures were independent of him'. But if we would know the immensity of his power we should not put these thoughts before our minds, nor should we conceive any precedence or priority between his intellect and his will; for the idea which we have of God teaches us that there is in him only a single activity, entirely simple and entirely pure. (CSM 3, p. 235; AT 4, pp. 118–119)

Finally, to Arnauld on 29 June 1648, he sends this:

> Passage 8: But I do not think that we should ever say of anything that it cannot be brought about by God. For since every basis of truth and goodness depends on his omnipotence, I would not dare to say that God cannot make a mountain without a valley, or bring it about that 1 and 2 are not 3. (CSM 3, pp. 358–359; AT 5, pp. 223–224)

Given the dates and provenance of these texts, the evidence is clearly in favor of Descartes' lifelong adherence to normative and logical voluntarism. In particular, his allegiance is unwavering during the *Meditations* period (1641–1642) even though the doctrine is not explicitly mentioned in the *Meditations* themselves.

We will now examine the impact of both types of voluntarism on the *Meditations'* arguments, claiming that they engender two corresponding circles.

11.3 The Logical and Normative Circles

The *Logical Circle* arises because of Descartes' logical voluntarism. When all the metaphysical assumptions of a logically circular argument are made explicit, at least one premise assumes "God exists" (without which there would be no such thing as argumentation) even though the argument concludes "God exists."[1] Descartes' voluntarism entails that logic exists only because God exists. That makes all of Descartes' arguments for God's existence (relying as they do on logic) logically circular.

[1] Note that the logical circle undercuts the traditional one. For on voluntarism God must "first" make circularity a flaw in argumentation. The fact that the traditional circle is an *objection* depends on the assumption that God exists.

Descartes' normative voluntarism engenders a second circle – the *normative circle*. When all the metaphysical assumptions of a normatively circular argument are made explicit, at least one premise assumes "God exists" (without which the argument could not rely on, say, norms of perfection) even though the argument concludes "God exists." Suppose for the nonce that Descartes abandons his more radical logical voluntarism. Even with the resulting diminished doctrine in play, the normative circle ensnares some of the arguments for the existence of God – namely, those which appeal to normative facts.

11.4 The Meditations' Principal Arguments

How do these circles affect the *Meditations*' main arguments?

In *3* (using italicized numbers to abbreviate *Meditations* 1–6*)* and elsewhere, the *clear and distinct ideas argument* says that, as all clear and distinct ideas are true and as I have such an idea of a perfect, non-deceiving God, a perfect, non-deceiving God must exist. That argument is logically and normatively circular. God must exist to make logic and argumentation and meaning in the first place. And God must exist to create perfection and make non-deception an instance of it.

3's *trademark argument* claims that, as I have a clear and distinct idea of God's having all perfections and as every such idea must be caused by a God who has those perfections, a perfect God must exist. *5*'s *ontological argument* maintains that, as all clear and distinct ideas are true, and as I have such an idea of a perfect God, and as existence is a perfection, a perfect God must exist. Both arguments are logically circular just because they're arguments. And, since each appeals to perfection, they're normatively circular as well.

In *6*, Descartes advances the *physical world argument*. According to it, given that a truthful God's existence has been established by the three arguments above and that my propensity to believe in a physical world would be a deception if it were unfounded, the physical world must exist. But, as every one of those previous arguments is logically and normatively circular, this argument inherits those circularities.[2]

11.5 The *Cogito*

Things are not so bad, one wants to say, as long as the centerpiece of the work – the *cogito* – emerges unscathed. Indeed, it seems well placed to avoid these circularities. It is not an argument for God's existence and does not explicitly appeal to normative facts to establish its conclusion.

[2] Of course, the problems are compounded when one notes that some of the same arguments appear in the *Principles*, so parts of that work are also beset by the circularities.

Formulate it as follows:

Cogito Argument

(a) If a demon is deceiving me about my existence at a given time *t*, I would have to be a being who is thinking at *t* but does not exist at *t*.
(b) There is no being who is thinking at *t* but does not exist at *t*.
(c) A demon is not deceiving me about my existence at *t* – that is, I exist at *t*.

The demon who is being addressed here (in *2*) has stepped into God's shoes in *1* (CSM 2, p. 15; AT 7, p. 22). The *cogito* is directed *to him.* Yet within the larger framework of Descartes' thought, that demon has been granted voluntarist powers. When a "malicious demon of the utmost power and cunning" takes over for God, it replaces a *voluntarist* God. So it is a *voluntarist demon.* Descartes writes that the demon is using "all his energies in order to deceive me" (CSM 2, p. 15; AT 7, p. 22) and is "deliberately trying to trick me in every way he can" (CSM 2, p. 18; AT 7, p. 26).

A voluntarist agent can make it the case that *any proposition* – no matter how necessary it seems – is not true. In terms of the argument above, he can make (b) ("There is no being who is thinking at t but does not exist at t") not true. So, when addressed to that sort of creature, the *cogito* argument fails. Descartes writes

> [...] let him deceive me as much as he can, he will never bring it about that I am nothing so long as I think that I am something. [...]. I must finally conclude that this proposition, *I am, I exist*, is necessarily true whenever it is put forward by me or conceived in my mind. (CSM 2, p. 17; AT 7, p. 25)

Descartes' deceiver is limited in no way whatever – hence can easily make it the case that Descartes thinks that he is something yet does not exist. No one can get leverage on *this* imp. Thus Descartes' "necessarily" is actually "necessarily*."

(Note that we say "not true" rather than "false." In *1*, when the demon appears, Descartes says that, e.g., his belief that he has hands will no longer be credited. Instead he will consider himself as "falsely believing" that he has hands (CSM 2, p. 15; AT 7, p. 23). With a voluntarist demon installed, however, Descartes needs to eliminate the "falsely believing" bit as well. For the demon might not have *made* falsehood or beliefs. The nightmare that looms before him is that *none* of the rational apparatus might exist. "Falsely believing" might be as much a conceptual illusion as the phantasm of his hands is a perceptual one.)

This general sort of argument, claiming that voluntarism undercuts the certainty of the *cogito*, first appeared (as far as we can tell) in Pierre-Daniel Huet's *Censura Philosophiae Cartesianae* I–6 (Huet 2003 [1689, 1694], pp. 75–77):

> What if we were to say that, even if it is given as true that he who thinks exists, it might yet also be true that he who thinks does not exist. For Descartes's view is that God can bring it about that contrary and contradictory propositions can be simultaneously true, from which it follows that he who thinks can both be and not be. (Huet 2003 [1689, 1694], pp. 75–76)

(Note that if Huet had noticed Passage 4 [to Arnauld, *Fourth Replies*], where Descartes gives up the certainty of the *cogito* to preserve voluntarism, the objection would be absolutely unanswerable. Huet could then simply add: "as Descartes himself says. QED.")

The reaction to Huet's argument on the part of early Cartesians will provide a nice transition to our discussion of some of the responses to Descartes' voluntarism in the secondary literature.

11.6 Some Responses to the Charge of Logical Voluntarism

The task of answering Huet point by point fell to Pierre-Sylvain Regis (1632–1707). In 1691, he published his *Réponse*, quoting Huet and responding to each charge. He advances (i) and (ii) below in the course of answering Huet's argument against the *cogito*. After discussing them, we present additional defenses offered by other Cartesians:

(i) *Denial*. Advocates of denial simply claim Descartes never endorsed logical voluntarism. In Regis' final paragraph directed at Huet's argument, he gives voice to it:

> Huet […] attributes a view to Descartes that he does not hold, which is that God can make things even that contradict [*repugne*], which he never taught, or if he did teach it someplace, it is only by suppositions that he himself called 'exaggerated' [*extravagantes*]. (Lennon 2008, p. 155; Regis 1691, p. 20)

Huet lost no time in pointing out in a later edition of his *Censura* (1694) the many passages in which Descartes stated and endorsed the doctrine. He adds:

> Since they who dare to deny that this was advanced by Descartes bite their tongues, they finally take refuge by saying that if these things were asserted by him, it was by a kind of fiction that goes beyond the bounds of reason. But when these things are repeatedly and seriously asserted by him and are reinforced by arguments […] we judge them to be, not fictions, but considered positions drawn from the very core of his philosophy. (Huet 2003 [1689, 1694] pp. 76–77)

That should have been the end of the matter for future Descartes scholarship. Nevertheless, even today one finds occasional instances of denial, as when Stephen Gaukroger writes:

> […] the doctrine of eternal truths does not appear again after the early 1630s in Descartes' writings, not because he has abandoned it, but because it is a metaphysical doctrine which has now [in the *Meditations*] been reformulated epistemologically, in terms of hyperbolic doubt. […]. In its metaphysical version it is insoluble, for all one can do is point to the chasm dividing divine and human understanding: the doctrine of the divine creation of eternal truths is a dead end. The epistemological reformulation of the problem does yield a solution, however, and far from being a dead end it turns out to be the key to the defence of Cartesian natural philosophy. (Gaukroger 1995, p. 317)

Gaukroger's rescue does not work historically. (Indeed, in correspondence Gaukroger has abandoned the claim.) But it also backfires as a defense of Descartes' philosophy. For once one admits that the metaphysical doctrine is a "dead end," it is clear that if such a doctrine were present during the later period, the *Meditations* would also be a dead end rather than a defensible natural philosophy.

(ii) *Logical voluntarism is hyperbolic*. This defense of Descartes' system likens the doubt emanating from voluntarism to that engendered by the demon. It is based partly on Descartes' claim at the end of *6* (CSM 2, p. 61; AT 7, p. 89) that the doubts he had been entertaining are hyperbolic, exaggerated, and laughable, thus implying that they can be ignored. Regis – again answering Huet's argument – attempts to classify the doubt arising from logical voluntarism as similarly "exaggerated" or hyperbolic – hence clearing the *cogito* of any threat from that quarter.

Yet in the passage from *6*, Descartes is not referring to doubts engendered by *God*: they seem rather to do with the supposition that Descartes is being deceived by a demon.

Nevertheless, one can understand Regis' desire to liken the vexing doctrine to one which Descartes denigrates. Regis gets some help in this from Picot's embellished French translation [1647] of *Principles* I, 7 (the original text of this passage from 1644 is found at CSM 1, pp. 194–195; AT 8A, p. 7). Regis quotes the translation as if it were all penned by Descartes himself:

> For it is a contradiction to suppose that what thinks does not, at the very time when it is thinking, exist. [*Picot's addition*: and, notwithstanding the most exaggerated suppositions (*extravagantes suppositions*), we cannot help but believe that 'I am thinking, therefore I am', is true.] (Lennon 2008, p. 153; Regis 1691, p. 19)

Regis then feels free to represent Descartes as classifying voluntarist doubts as exaggerated, thus freeing the *cogito* of their strictures.

It is an ingenious maneuver, especially as Descartes was happy with the translation and apparently added some of the extra material himself.

Nevertheless, nothing of the sort appears in any of the documents of the *Meditations* (1641) or the original text of the *Principles* (1644). In particular, if this were Descartes' view in 1641, Passage 4 would be quite different. For why would Descartes give up the *cogito* for a metaphysically dispensable, hyperbolic supposition? He should be there telling Arnauld that the *cogito* stands after all because the supposition of a God who can do anything is laughable anyway.

(Of course, by 1647 – when the embellished translation appeared – Descartes had reason to distance himself from the demon. He had been accused of blasphemy because the demon's power seemed to rival God's (Janowski 2000, Chapter 2)).

(iii) *Abandonment*. It is sometimes said that, while Descartes once held logical voluntarism, he eventually abandoned it. Thus Alexandre Koyré writes:

> Absolute indifference, excessive omnipotence, which Descartes seems to profess, actually is compatible neither with his physics, nor his psychology, nor his metaphysics, nor his theory of knowledge.
>
> In fact, Descartes was aware of the innumerable difficulties which this doctrine implies. […] The resolute affirmation […] is replaced by a much more hesitant and uncertain conception, and, towards the end of his life, we see him abandoning this excessive doctrine definitively and returning to the classical theory. (Koyré 1922, pp. 20–22)

While the claim of incompatibility is correct, the solution offered is not. Koyré offers as evidence of the withdrawal Passage 7 (to Mesland, 1644) and two late texts to More and Clersellier (5 February and 23 April 1649, respectively).

Passage 7 is one of the strongest statements of voluntarism. It ends with Descartes' saying that we really should not talk about a priority of attributes in a simple being. But, that is obfuscation, not retraction. Descartes has just said things compatible only with the claim that God's will is prior to his intellect. (Descartes here seems to catch a glimpse of the fact that his doctrines of divine simplicity and voluntarism are incompatible. That is, if God is simple, then voluntarism can't be stated: voluntarism requires a priority of attributes, but simplicity proscribes it.)

The remark to More says that it is no defect in the power of God if he cannot do something "we perceive" to be "altogether impossible." But, he ends by saying "I know that God can do more things than I can encompass within my thought" (CSM 3, pp. 363–364; AT 5, pp. 273–274). That is not a denial of voluntarism. Finally, we are unable to find anything in the post to Clersellier which indicates Descartes is changing his position. For example, he characteristically says of God "it is not for me to set any limits to his works" (CSM 3, p. 377; AT 5, p. 355).

(iv) *Descartes is not a voluntarist.* Margaret Osler maintains that Descartes' mature theory is not a form of voluntarism at all, but a "kind of intellectualism" (Osler 1994, p. 120). She writes:

> Descartes' theory of the creation of the eternal truths served as the metaphysical foundation for his philosophy. On this foundation he erected his theory of knowledge – [...] the *cogito*, his proof for the existence of God, and the criterion of clear and distinct ideas. (Osler 1994, p. 135)

She also claims this "indubitable kernel of certainty" supported his natural philosophy since, e.g., Descartes appeals to a perfection of God (immutability) when establishing laws of nature in the *Principles* (Osler 1994, pp. 136–138).

How did Osler arrive at this unusual position? Early on she says she will define "voluntarism" and "rationalism/intellectualism" in a nontraditional way. Typically they are explicated "in terms of the relationships between two of God's attributes, his intellect and his will" (Osler 1994, p. 17). But, Osler writes:

> The difference between rationalist and voluntarist theology can be understood in terms of the distinction between *potentia absoluta* and *potentia ordinata*. [T]he difference can be expressed by asking how binding the created order is on God's present and future acts. Rationalists or intellectualists were prepared to accept the existence of some necessity in the world, while voluntarists regarded the present order as utterly contingent. (Osler 1994, p. 18)

Intellectualists include Peter Abelard and Thomas Aquinas and voluntarists Occam and Gassendi. With voluntarism, she writes:

> [...] all human knowledge of nature must be ultimately empirical. Moreover, nothing about the creation can ever be known with certainty; God's absolute power renders our knowledge fallible and, in this sense, only probable. (Osler 1994, p. 20)

Some interpreters, Osler claims, have taken Descartes to be a voluntarist, but they have overlooked the fact that

> Descartes proceeded to argue that God has created us in such a way that we can have a priori knowledge of the eternal truths and that his own nature prevents him from changing what he once created freely. [...]. God's ability to intervene in the created world is limited by his initial act of creation. [...]. It is with regard to his ordained power that Descartes was an intellectualist: He accepted the existence of some necessity in the world, something that the voluntarists could never accept because of their emphasis on the utter contingency of the world. (Osler 1994, p. 130)

For support she quotes Descartes' claim to Mersenne (1630) that the eternal truths are "inborn in our minds" (CSM 3, p. 23; AT 1, p. 145). And, using the same letter, she adds that "divine immutability provides Descartes justification for the necessity of the eternal truths that God created" (Osler 1994, p. 131).

But this is the same Descartes who would, 11 years thence, say that we don't know whether the eternal truths are *true*. One cannot rescue texts from the 1640s from the morass of voluntarism by quoting a passage written well before them and in seeming unawareness of the issues they raise.

Osler's most crucial oversight, in relation to Descartes' mature position, is that there is no guarantee that *our* "eternal truths" are the eternal truths which God believes – or that God even *has* beliefs. God's immutability can be of no solace to fallible human knowers so long as they don't know whether God really is immutable and whether, say, the principle of noncontradiction really is true. And similarly for every vestige of reason – mathematics, geometry, the natural light, the *cogito*, and the laws of nature. Certainty about all of it is gone. Rather than Descartes' doctrine of the eternal truths providing Descartes with "justification for the necessity of the eternal truths," it undermines *all* necessity.

11.7 Why Did Descartes Endorse Voluntarism?

Why did Descartes hold so unhesitatingly to voluntarism when it reduced his foundations to quicksand? We offer two rationales – one sociological and one philosophical.

A look at the context shows that throughout his life, Descartes was aware of the need to keep his works off the "Index" of forbidden books. Thus he abandoned *The World* in 1633 because it would have presented a Copernican view of the universe. He noted well the fates of Bruno and Galileo and so had every incentive to construct a view that would not draw the charge of heresy. In the theologico-political atmosphere he was working in, voluntarism was an instant and fail-safe escape. For if Descartes said that at the end of the day reason – no matter how spectacular – is subservient to faith, he could draw the sting from any ecclesiastical accusers (Janowski 2000, Chapter 6). In particular, the miracle of the Eucharist seemed central in the debate – as in Passage 4, where Descartes invokes voluntarism to dodge Arnauld's challenge (Gasparri 2007).

Moreover, any theist in the broadly Christian tradition would in the end have to uphold "mysteries" like the Incarnation and Trinity despite the fact that these claims break logical rules. Leibniz, for instance, will hold forth a stringent concept of necessity for the Identity of Indiscernibles – only to give it up to accommodate the mysteries of the Christian religion (Leibniz 1985 [1710], p. 104, *Preliminary Discourse*, §§ 55–56). So what's the difference, one might wonder, between Descartes, who is candid from the outset about living with an attenuated "necessity*," and Leibniz – whose "necessity" turns out to be "necessity*" anyway?

In the *Discussion with Burman*, there is a passage which shows thoughts like this were not far from Descartes' mind. Comparing theological reasoning with the geometrical sort, he says:

> But these [theological claims] are truths which depend on revelation, and so we cannot follow or understand their mutual connection in the same way. And certainly theology must not be subjected to our human reasoning, which we use for mathematics and for other truths, since it is something we cannot fully grasp; and the simpler we keep it, the better theology we shall have. [...]. However, we can and should prove that the truths of theology are not inconsistent with those of philosophy, but we must not in any way subject them to critical examination. (CSM 3, pp. 350–351; AT 5, p. 176)

Our philosophical rationale is this. Conceptually, one can see voluntarism as arising when foundationalism is pushed to a pathological extreme. Descartes wants to find beliefs which are certain and on which the rest of knowledge can be built. But the search for foundations must end somewhere. Descartes holds that, while logical and mathematical claims are usually taken (by rationalists like Plato and Aquinas, and later Leibniz) to be reliable, certain, and properly basic, they are not. Like the rest of our beliefs, they can be doubted. So Descartes digs down to find foundations for *them*. At this point the only player in the field capable of underpinning them is God. But, the account will be circular if God is *already* rational. So you need a nonrational God to underpin rationality. And while this avoids circularity, it is deeply problematic and paradoxical. For it casts doubt on the reliability of the very concepts needed to make knowledge, foundations, and certainty possible.

11.8 Voluntarism in Contemporary Commentary

We now consider the status of voluntarism in some contemporary interpretations of Descartes.

First, Thomas Lennon has translated and provided detailed comments on Huet's *Censura*. He expounds Huet's "Trump Argument" (see above Sect. 11.5) thus:

> If Descartes (or any of his followers defending him) produces an argument that appears to demonstrate some proposition (e.g. 'I exist'), Huet cites the doctrine [affirming the possibility of true contradictions] and asserts that the denial of the proposition might also be true and therefore that the Cartesian can have no certainty about the proposition. It is a trump argument in the sense that it putatively bests every Cartesian position and every card played in its defense. (Lennon 2008, pp. 148–149)

That is beautifully put. It seems to show that Descartes' quest for certainty is doomed. But Lennon believes that Huet is not ultimately successful in this argument:

> The response that Regis should have made, but did not, is that while the doubting one's existence is possible, it is nonetheless unreasonable even on that supposition that God is a deceiver. Or, more precisely, if the cogito is taken to be a pragmatic tautology, then doubting it is impossible, even if its truth, like the truth of everything else, depends on God's indifferent will. (Lennon 2008, p. 154)
> […]. When Descartes asserts that God exists, or that twice two is four, or even that he exists when he thinks, he does not at the same time think that God could have brought about the opposite. The argument of the *Meditations* is that taking such a possibility as a basis for doubt would be, under the circumstances developed there, *unreasonable*. […]. It is also Descartes's conviction that *anyone* who meditates as he suggests there will arrive at the same conclusion. […]. While God could have made it true both that while I think I exist and while I think I do not exist, there is no reason to believe that He has done so and every reason to believe that He has not done so. In short, Huet's appeal to the doctrine is utterly gratuitous and unreasonable. (*Ivi*, p. 155)

First, Lennon's suggested reply is not available to Regis. For the scope of Descartes' voluntarism is not just the principle of noncontradiction, but *all* of reason. A reply appealing to what is "unreasonable" in response to a "pragmatic tautology" begs the question against Huet.

Second, in 1641 Descartes cannot exempt from voluntarist-inspired doubt propositions such as that twice two is four or that he exists when he thinks. In Passage 5 (1641) he admits that "twice four is eight" might not be true if God declared otherwise, and in Passage 4 (also 1641) he explicitly places the claim "I exist when I think" among those which God can trump. Far from being gratuitous, Huet's objection is a deadly indictment of a system of thought committed at once to the reliability of reason and its utter collapse.

John Cottingham – arguably the Dean of Anglophone Descartes scholars – discusses the issues of reason and God in several places. For example, in an article on "God" (Cottingham 1993, pp. 69–72), Cottingham writes:

> A fundamental paradox, which lies at the heart of Descartes' theocentric metaphysics, is that the very being who is invoked as the guarantor of truth and of the reliability of the perceptions of the intellect is also frequently declared by Descartes to be beyond our human grasp. (Cottingham 1993, p. 71)

After citing some of the texts, Cottingham adds:

> Descartes' position in these and similar passages is, however, tolerably consistent, and relies on a crucial distinction between knowing something and fully grasping it. […]. The infinite perfections of God […] cannot be fully encompassed or grasped by the human mind […]. But nevertheless we can attain to sufficient knowledge of the divine attributes to enable us to be sure at least of those aspects of the divine nature which need to be established to validate knowledge: we can prove that God exists, and that he is no deceiver. (CSM 2, p. 48; AT 7, p. 70; Cottingham 1993, p. 71)

Descartes' position "in these passages" is consistent enough. But that was never in question. We want to know whether his position *within his larger system of thought* is consistent. It is not. For within that larger system – where the epistemic apparatus itself has been doubted – how will anyone attain "sufficient knowledge of the divine attributes" to prove God's existence? With voluntarism in place, Descartes is not justified in "proving" anything.

Later, in an article with the sweeping title, "The Role of God in Descartes's Philosophy," Cottingham says nothing about the paradoxes of voluntarism. Cottingham instead tries to distance Descartes from them. He writes that Descartes' philosophy "should not be confused with the very different, faith-based religious approach" found in Pascal and Kierkegaard. According to Cottingham, such an approach is "light years away from the Cartesian worldview" (Cottingham 2008, 299–300). In Descartes' philosophical system, he says:

> God is central, but it is a God who is established by reason, and who underpins the rationality of a system of science and morality. […]. Given the assurance of a rationally ordered universe, and a supremely benevolent creator, we can be sure we have the means at our disposal to achieve knowledge of the true and the good. […]. (Cottingham 2008, pp. 299–300)

Cottingham's claim that Descartes' God is "established by reason" is precisely backwards. According to every relevant text, *reason is established by God.*

Moreover, while there are differences between fideism and voluntarism, they have in common the subjugation of what is objective (truth, logic, reason, intellect) to what is subjective (the will, human or divine). There are two forms of irrationalism: fideism because, as in Kierkegaard, if one believes a contradiction with enough "subjective inwardness," it becomes true; and voluntarism because God can make a contradiction true just by willing it so.

By underreporting the facts, Cottingham joins a host of Descartes scholars who would preserve the myth of Descartes-the-consistent-rationalist.

That is a hopeless enterprise. The nice, neat idealization they invent to save Descartes' system from collapse is destined to be smashed to bits on the stones of Passages 1–8.

One may wonder: So what should interpreters do in light of this? Answer: admit that the doctrine will have to be largely ignored in a fruitful exposition of Descartes' thought. The resulting system would not need to be otherwise alien to the one Descartes proposed. It could be genuinely Cartesian as long as the loss of voluntarism – and any other problematic doctrines – is explicitly acknowledged.

We note, however, that even with voluntarism included, Descartes' thought contains some great philosophy. As we have argued (Hartz and Lewtas 2014), when one leaves voluntarism in Descartes' system, his demon scenario can be used to state the logically limiting case *for* skepticism. When the demon becomes a voluntarist agent, he is a skeptical force than which none greater is possible. No one can doubt more than everything.

11.9 Conclusion

Contemporary Anglophone philosophers became interested in Descartes' voluntarism largely because of Frankfurt's seminal article (1977). In its wake, Curley (1984), Bennett (1994), and others offered ways to rescue Descartes' system. Unfortunately, Curley and Bennett offer cartesian defenses. Curley imports into the system a distinction which Descartes would reject – between "higher and lower order eternal truths" – while Bennett redacts the passages so that they seem to be offering modal epistemology rather than modal metaphysics.

At the outset of his article, Frankfurt presented a battery of questions:

> It is problematic just what the doctrine is, what Descartes thought it implies, what motivated him to adopt it, how he would have met the rather plausible charge that it is incoherent, what his main arguments for it are, where those arguments and the doctrine itself fit into the general scheme of his reasoning, why he did not discuss the doctrine in his systematic accounts of his philosophy and his science, whether it actually does make a veiled appearance in some of those accounts, how it bears upon his attempt to validate reason, what its relation is to his physics, and so on. (Frankfurt 1977, p. 37)

We have explained what Descartes' voluntarism is, what it implies, and some motivations.

Descartes could easily rebut Frankfurt's charge of "plausible incoherence" by noting that it begs the question: the concepts of plausibility and coherence are themselves at issue.

The arguments for voluntarism that Descartes presents are largely theological and concern the proper understanding of omnipotence.

We would say that voluntarism does *not* fit into Descartes' general scheme of reasoning, but would also note that he defiantly holds to it even at this cost.

Descartes *did* discuss voluntarism in the "systematic accounts of his philosophy." Historiographical principles would require one to take the *Meditations* along with the *Replies* as a unit, and voluntarism is endorsed quite explicitly in the *Replies*.

Voluntarism "validates reason" by making it subservient to God.

As to the physics, it replaces the absolute necessity of the laws of nature with an attenuated necessity* held hostage to God's power.

Ever since Descartes wrote his first letters to Mersenne, Descartes' voluntarism has scandalized philosophers. Judging from what interpreters have said, Descartes' voluntarism is his most hated doctrine. Maybe because it deprives philosophers of the autonomy, authority, and ultimacy of their discipline. Maybe because it marks a betrayal of the very spirit of reason. Watching Descartes throw it all away is more than we can bear.

Acknowledgments We warmly thank readers of earlier drafts of this paper, including Rick Groshong, Mogens Laerke, Cody Baith, Maddie Collins, and audiences at two American Philosophical Association meetings and at Syracuse University.

References

Descartes R (1964–1976) = abbreviated 'AT'

Descartes R ([1985] 1991) = abbreviated 'CSM'

Bennett J (1994) Descartes's theory of modality. The Philosophical Review 103:639–67.

Broughton J, Carriero J (2008) A Companion to Descartes. Blackwell, Oxford.

Cottingham J (1993) A Descartes dictionary. Blackwell, Oxford.

Cottingham J (2008) The role of God in Descartes's philosophy. In Broughton 2008, pp. 288–301.

Curley EM (1984) Descartes on the creation of the eternal truths. The Philosophical Review 93:569–97.

Descartes R (1964–1976). Œuvres de Descartes. 12 Vols. Adam C, Tannery P (eds). Vrin–C.N.R.S, Paris.

Descartes R ([1985] 1991) The philosophical writings of Descartes. 3 Vols. Translated by Cottingham J, Stoothoff R, Murdoch D [for Vol. 3, also Kenny A). The Cambridge University Press, Cambridge.

Frankfurt H (1977) Descartes on the creation of the eternal truths. The Philosophical Review 86:36–57.

Gasparri G (2007) La création des vérités éternelles dan la postérité de Descartes. Revue philosophique de la France et de l'étranger 132:323–36.

Gassendi P (1658–1964) Opera Omnia. 6 Vols. Anisson L & Devenet JB, Lyon [reprint: Friedrich Fromman Verlag, Stuttgart–Bad Cannstatt].

Gaukroger S (1995) Descartes: an intellectual biography. Clarendon Press. Oxford.

Hartz GA, Lewtas, PK (2014) Descartes' metaphysical scepticism. Revue Roumaine de Philosophie 58:79–89.

Huet PD (2003 [1689, 1694]). Against Cartesian Philosophy. Censura Philosophiae Cartesianae. Translated by Lennon TM. Humanity Books, Amherst.

Janowski Z (2000) Cartesian theodicy: Descartes' quest for certitude. Kluwer, Dordrecht.

Koyré A (1922) Essai sur l'idée de dieu et les preuves de son existence chez Descartes. Éditions Ernest Leroux, Paris.

Leibniz GW (1985 [1710]) Theodicy. Open Court, LaSalle.

Lennon TM (2008) The plain truth: Descartes, Huet, and skepticism. Brill, Leiden.

Ockham W (1957) Philosophical writings. Boehner P (ed). Nelson, Edinburgh.

Osler M (1994) Divine will and the mechanical philosophy. The Cambridge University Press, Cambridge.

Regis PS (1691) Réponse au livre qui a pour titre P. Danielis Huetii [...] Censura philosophiae cartesianae, servant d'éclaircissement à toutes les parties de la philosophie, surtout à la métaphysique. Par Pierre–Sylvain Regis. Jean Cusson, Paris.

Chapter 12
The Posterity of Alexandre Koyré's *Galileo Studies*

Gérard Jorland

Abstract Alexandre Koyré's *Galileo Studies* have been epoch-making in the history of science. Every Galilean scholar, since their publication just before the Second World War, has introduced his or her own work by taking a position in relation to that of Koyré. Even the return to the pre-Koyrean scholars, from Favaro to Wohlwill, has taken Alexandre Koyré as its starting point. This chapter aims to show how Alexandre Koyré's main theses concerning Galileo's experiment and Platonism were challenged through the experimental and then the sociological turns in the historiography of science of the last three decades of the previous century. Nonetheless, his theses could still be maintained, albeit with due qualifications.

Keywords Galileo • Drake • Naylor • Platonism • Aristotelism • Sociological turn • Experimental turn

12.1 Koyré's *Galileo Studies*

Throughout his career as a historian of science, Alexandre Koyré made Galileo a point of reference. These references span three decades, from his 1933–1934 seminar at the 5th section of the *École Pratique des Hautes Études* in Paris (Koyré 1935) to one of his last publications, his 1960 paper on "Newton, Galilée et Platon," which appeared in the journal of the 6th section of the *École Pratique des Hautes Études* in Paris (Koyré 1960b). He was to die 4 years later. Although he republished his first two papers on Galileo together with an as yet unpublished essay in a three-volume set under the title of *Études galiléennes* (Koyré 1935–1936, 1937c, revised and reprinted as Koyré 1939)[1] just before the Second World War, this was not the work

[1] A third essay on Galileo and the law of inertia was first published as the third part of Koyré (Koyré 1939). The *Études galiléennes* has been republished in 1966 by Hermann in one volume as number XV of the collection "Histoire de la Pensée" created by Alexandre Koyré himself.

G. Jorland (✉)
Paris École des Hautes Études en Sciences Sociales (EHESS), Paris, France
e-mail: jorland@ehess.fr

R. Pisano et al. (eds.), *Hypotheses and Perspectives in the History and Philosophy of Science*, DOI 10.1007/978-3-319-61712-1_12

that made his reputation in the Anglophone world. In fact, it was translated into English only in 1978, 14 years after his death, at the very moment when his claims about Galileo were being called into question in every quarter. Rather, Alexandre Koyré's work on Galileo became known to English-speaking scholars through his collected English papers, published in 1968 under the title *Metaphysics and Measurement* (Koyré 1968a). Out of six papers, four were devoted to Galileo. The book was reviewed by Thomas Kuhn (1970). According to Kuhn, there had been a twofold historiographical revolution in history of science: first, a context-free philosophical history of science, followed by a contextualized social history of science. Whereas the latter was still "in embryo" at the time of the review, he credited Koyré with the former: "More than any other single scholar, Koyré was responsible for the first stage of the historiographical revolution" (Kuhn 1970, p. 67).

Thomas Kuhn rebuked Koyré for his failure to take into account the role of *observation* in Galileo's science, although he praised him for showing "conclusively" that contemporary apparatus was inadequate for *measurement*. In fact, Kuhn contended that one could not speak of quantitative or numerical experiment before the nineteenth century; in the sixteenth and seventeenth centuries, experiments were qualitative—not only Galileo's experiments but those of Descartes and even Newton as well. If Galileo had made the experiments he described, *pace* Koyré, they were not quantitative but qualitative or non-numerical. In addition, they were qualitative not because natural philosophers were not looking for numerical results, but because they could not obtain them for lack of precision apparatus. What these qualitative experiments provided were observations (Kuhn 1961). If this qualification was accepted, Kuhn was willing to concede that Koyré was right:

> The transformation of the classical sciences during the Scientific Revolution is more accurately ascribed to new ways of looking at old phenomena than to a series of unanticipated experimental discoveries. (Kuhn 1976, p. 46)

From his first to his last publications on Galileo, Koyré had insisted that Galileo was a Platonist. All of his arguments for a Platonist Galileo were articulated at the outset: the perfectly smooth and hard plane and the perfectly spherical and hard sphere postulated by Galileo were to be found not in the real world of everyday experience, but "beyond," in the Platonic ideal world of mathematical forms (Koyré 1966b, p. 77; 1978, p. 37).[2] These mathematical forms embodied the "essence" of the phenomena that explained how they happened to be as they were (Koyré 1966b, pp. 155–156; 1978, p. 109).[3] Since experiments dealt with real phenomena, they could not fathom that essence; at most they might prevent being led astray (Koyré 1966b, p. 155; 1978, p. 107). The kind of experiment most suitable to these mathematical or ideal objects was the thought experiment. Koyré did his best to show that Galileo's experiments were not real experiments actually performed, but rather thought experiments that he had simply imagined:

[2] Koyré quoted Galileo from *De Motu*'s manuscript (fl. 16th, pp. 298–299) *à propos* the proposition that a body resting on a plane can be moved by the smallest force ever.

[3] For this interpretation, Koyré had nothing to quote.

> Galileo [wrote Koyré] did not refrain from telling and presenting us as actually realized experiments that he remained content to imagine (Koyré 1966a, p. 198).

Koyré goes so far as to make of Galileo an innatist:

> It has not been noticed enough that for Galileo the fundamental ideas of science (the mathematical ideas) are innate ideas. (Koyré 1937a, p. 44)

Although he acknowledged that the word could not be found in Galileo's writings, he was able to give two references to Plato's theory of reminiscence in the *Dialogo* (Koyré 1966b, pp. 286–287; 1978, p. 207; Galileo 1897, p. 183, p. 217). If Galileo had adopted the form of the dialogue, it was, asserted Koyré, with the same pedagogical purpose as Plato, to stage the theory of reminiscence. One knows everything, but does not know that one knows. To learn means to remember and to teach or to explain is to pave the way to remembrance. But if one knows everything from the start, if what one has to discover can only be found within oneself, the experiment becomes pointless: it cannot teach anything not known beforehand. It can at best vindicate prior knowledge. This is the background to Koyré's provocative statement:

> Good physics is made a priori. Theory precedes fact. Experience is useless. (Koyré 1943b, p. 13)

In his first post-*Études galiléennes* paper, published in English while a refugee in the United States during the Second World War, Koyré expatiated on Galileo's mathematical Platonism (Koyré 1943a) and later on Galileo's cosmological Platonism.[4] Moreover, he argued for the non-experimental, a priori character of Galileo's science in two other papers, one also published in English (Koyré 1953) and the other at the end of his life (Koyré 1960a).

Both of Koyré's main claims have been called into question on the occasion of yet another twofold historiographic turn in the history of science: the experimental turn of the seventies and the sociological turn of the nineties. These turns have been nurtured by the bringing in the forefront of the historical research of what Gerald Geison has called "the private science," those leftover papers of scientists scrutinized for the hidden truth of their published work. We will see how Koyré's work on Galileo has weathered these historiographical movements. But first, we must cast a glance at Galileo's *Nachlass*.

12.2 Galileo's Private Science

For Stillman Drake, Thomas Settle's 1961 repetition of Galileo's experiment on inclined planes, in response to Koyré's claim that Galileo could not have obtained his results with the apparatus he described, proved only that Galileo *could* have

[4] Koyré (1960b) could show that Galileo was even more Platonist than Plato since he attributed to Plato a story of the genesis of the planetary system that was of his own invention.

done so, not that he actually *did* (Drake 1973b, p. 147–148). To prove that he actually did so, Stillman Drake examined Galileo's unpublished manuscript notes on motion which "contained only calculations or diagrams" and "could not satisfactorily be accounted for except as representing a series of experiments" (Drake 1973b, p. 148). The year before, Stillman Drake had set himself the task of ordering the notes on motion from the 1600s to the 1630s that Galileo had filed to be used in his 1638 *Discourses* (Drake 1971). This task was overtly aimed at refuting Koyré's analysis of Galileo's achievements.[5] Eight years later, Stillman Drake published a facsimile of these notes in his new ordering, as compared to the original Galilean Ms. 72 of the *Biblioteca Nazionale Centrale di Firenze*.

Drake followed two criteria for his chronological ordering of the notes: the watermarking of the paper and the handwriting. The differences in both, assessed either separately or altogether, could be matched to Galileo's letters and be dated accordingly. This technique suffers from several shortcomings of which Stillman Drake was very well aware (Drake 1979, pp. XI–XII). This technique offers only a gross overall chronology; it does not specify the proper date of each note. The minute ordering of the notes follows a story, the story of Galileo's discovery of his laws of motion as told by Stillman Drake. He acknowledged that, having changed his story, he altered his dating accordingly (Drake 1979, p. XXI). The ultimate story he told ran as follows: first, the experimental determination of the law of free fall proportional to the square of the time, then the law of inertia, and last of all, the parabolic projectile trajectory (Drake 1979).[6] Koyré had claimed Galileo could not have formulated the law of inertia since it presupposes rectilinear motion, whereas for Galileo the only possible inertial motion was circular motion since, relative to the Earth's center, it is neither an ascent, for it would be therefore retarded, nor a descent, for it would be therefore accelerated (Koyré 1966b, pp. 159–341, 1978, pp. 127–275). Drake objected that:

> If Galileo never stated the law in its general form, it was implicit in his derivation of the parabolic trajectory of a projectile, and it was clearly stated in a restricted form for motion in the horizontal plane many times in his works. (Drake 1964, p. 123)

---

[5] On the overall contrast between Alexandre Koyré and Stillman Drake, see Prudovsky 1997.

[6] "Galileo, in the last part of his final work *Discourses on Two New Sciences* (1638), proved that the trajectory of a projectile traveling through a nonresisting medium is a parabola. The proof is simple and straightforward, once it is known that the vertical distance the object falls from rest is proportional to the square of the time elapsed and that the projectile's horizontal velocity will remain uniform" (Drake 1975a, p. 160). [...] "As we have mentioned, two laws must be known in order to derive the parabolic trajectory. One is the law of free fall, which states that the distance an object falls from rest is proportional to the square of the time elapsed, and which Galileo had discovered in 1604 by a combination of luck and mathematical reasoning. The other law needed was a restricted principle of inertia that gives a relation between the motion imparted to a body and the behavior of the body after it is free from the initial impulse. Such a principle was precisely what Galileo was testing when he discovered the parabolic trajectory. Luck played no part this time. Galileo's discovery of the parabolic trajectory was a case of serendipity: the discovering of something other than what the seeker had set out to find" (*Ivi*, p. 161).

Although he himself quoted Galileo on the impossibility of experimentally demonstrating that a body could be set in motion by a "minimum force" for lack of a plane "actually parallel to the horizon, since the surface of the earth is spherical, and a plane cannot be parallel to such a surface" (Galilei 1960, p. 68), when we move away from that point where the plane touches the sphere, we move up relative to the center of the Earth rather than straight.

Bringing together folios 117r, 116, 175, and 114, which satisfied his two criteria, Stillman Drake thought that he could reconstitute the experiment that Galileo would have devised to test the law of inertia. In the middle of folio 117r, a horizontal line is drawn in landscape orientation, divided into four unequal segments whose ratios two by two are about the same as the figures underneath each of these segments and whose overall sum is computed above the line on the upper right of the folio. Now, Drake interpreted this as an actual experiment on the deceleration of a rolling ball along a grooved plane in equal times. He repeated it himself and "obtained similar relations over a distance of 6 feet" (Drake 1973b).[7] In order to equalize the times, he used a metronome, an instrument unavailable to Galileo. This was the main argument used by Koyré to cast doubt on Galileo's ability to perform this kind of experiment. Any repetition of such an experiment with the help of a contemporary device for measuring time misses the point made by Koyré. However, there is much more evidence to urge against Drake's interpretation. First, as he acknowledged, since "the figures vary greatly for balls of different materials and weights and for greatly different initial speeds" (Drake 1973b), this experiment seems to be pointless. Moreover, since the ball would not be at rest at the end of the last time interval, the addition on the upper right corner would be meaningless. Nevertheless, Drake did not comment on this addition, although his interpretation should account for *all* the features of the document.

According to Drake, this folio meant to give the general setting of the experiment on the law of inertia. Folio 116v was pursuing the experiment further by allowing different speeds according to the law of free fall. On this folio, a vertical line was graduated with markings of 300, 600, 800, and 1000. One centimeter below the 300 mark, a horizontal line was drawn from either side of the vertical line. This vertical line was extended by a dotted line down 828 points[8] to another perpendicular horizontal line. Galileo mentions that this figure measures the height of the table represented by the former horizontal line. From the intersection of the plain vertical line and the table, five parabolas point to the lower horizontal line, their intersections being marked 800, 1172, 1328, 1340, and 1500 from left to right. Galileo commented to the last four figures that they should be, respectively, 1131, 1306, 1330,

[7] In other words, the ball would have rolled over 2 feet in the first unit of time, over 1,60 in the second, over 1,25 in the third, and over 1 foot in the last, corresponding to Galileo's figures in whatever units.

[8] On the evidence of another folio, Stillman Drake deduced that the point was the unit distance of Galileo's proportional compass, viz., 0,938 mm which makes the height of the table at nearly 80 cm.

and 1460 and gave the respective differences between the actual and the expected figures. It is thus tempting to see this folio as a report of an experiment.

For Drake, it was plainly such a report that folios 175 and 114 helped to interpret. Since it was a case of a fall followed by a horizontal movement, the fact that this horizontal movement happened when the fall, and therefore its cause, had ceased to operate, it was perfectly suited for a test of the law of inertia. He maintained this interpretation despite the fact that all other Galilean scholars were at odds with him on this score and interpreted it rather as an experiment on the law of free fall. All of them affirmed the same reading of Galileo's theory of movement for which the circular motion is the sole natural movement and thus the sole inertial movement—or, in Koyré's words, that Galileo regards *inertial mass* and *gravitational mass* as being essentially the same (Koyré 1960a, p. 62). On that point, that Galileo could not formulate a law of inertia since horizontal motion, being neither natural nor violent, is unthinkable, much less testable, Koyré seems to be vindicated. Before we turn to these experiments, a last word of warning from Pierre Costabel might well be relevant:

> The survival and transmission of rough notes, the jottings of a mind at work, is largely a matter of luck. By their very nature, these notes were meant to be ephemeral and when they reach us they can only speak inarticulately. As partial records of a discussion that an author carried on with himself, they are by themselves broken chains of thought. Rigor is necessary to acknowledge their fragmentary nature, caution to resist the desire to fill the gaps with logical connections, and objectivity to limit our pronouncements to what the documents actually disclose. (Costabel 1975, p. 177)

12.3 The Experimental Turn

Koyré questioned Galileo's experiments on two very different grounds. He raised doubts as to whether Galileo had ever performed the experiments that he described and, if he had, whether he had obtained the experimental results that he claimed. If Koyré clung to his latter doubts, he acknowledged from time to time that Galileo could have performed his experiments but persisted in his skepticism that he could have obtained precise measurements from them.

There can be no doubt nowadays that Galileo did perform his experiments, at least some of them, and there should not have been any doubt even for Koyré himself. For instance, the *passe-vins* experiment of the first day of the *Discorsi* turns out exactly as Galileo described it despite Koyré's denial (Galilei 1638, p. 116; Koyré 1960a, pp. 82–84). It was already common practice among distillers at that time (Galilei 1995, p. 252, n. 34), and it is quite easy to replicate (MacLachlan 1973; Thuillier 1983, p. 453). Even without replicating it himself, Koyré should still have given credit to Galileo, since Galileo did not mention it in support of a preconceived idea; quite the contrary, he described it as a puzzle that he did not know how to solve. Moreover, Koyré did not entertain any suspicion as to the reality of Mersenne's experiments. A contemporary of Galileo, Mersenne did not possess better experimental apparatus, and as a mathematician, he too could have been labeled a "Platonist." Since Mersenne performed the same experiments as Galileo at about the same time and with the same inclines, there is no reason to question Galileo's experiment on free fall, and nor did Koyré do so.

The question is thus whether Galileo experiments on free fall were precise enough to count as measurements or not.

Drake and his school have repeated these experiments as apparently described in Galileo's notes and claimed that the results were genuine measurements and furthermore that they were not only regulative but even constitutive of Galileo's science. Other Galilean scholars refrained from going so far and vindicated Koyré's doubts. However, even Drake's school endorsed Koyré's question: they did not claim that Galileo performed all of the experiments that he described, but only those that gave results conforming to established physical laws. The experiments on free fall were thus deemed genuine and exact to the point of calling Galileo "the grandfather of experimental science" (Drake 1973b, p. 305), whereas his experiments with the pendulum were judged imaginary.

Koyré wondered about Galileo's ability to obtain exact measurements of free fall because of two features of Galileo's experimental apparatus: first, its design and, second, his timing device (Koyré 1953, p. 224). We shall see that on both counts Koyré's doubts were legitimate.

In his 1961 *Science* paper repeating Galileo's free fall experiment (Settle 1961) as described in the third day of the *Discorsi* (Galilei 1638, pp. 212–213), Thomas Settle reproduced Galileo's apparatus, except on two points: he used a billiard ball and a steel ball instead of Galileo's bronze one, and he allowed them to roll over the edge of the groove instead of in the trough, which he did not line. Therefore, the results obtained with the steel ball were not comparable with those obtained with the billiard ball. Settle explained the discrepancy by their different rolling circumferences: they were not of the same diameter but were rolled over the same groove width (Settle 1961, p. 23). But the most important departure from Galileo's experimental setup was in the timing device used. Whereas Galileo had a large pail with a slender tube affixed to its bottom through which a narrow thread of water ran into a little beaker that was weighed after each descent down the incline (Galilei 1974, p. 170), Settle fixed a stopper on the slender tube and used a graduated beaker. Moreover, he checked the rate of flow against his hand watch. With these improvements, he eliminated two main sources of measurement errors: the rate of flow and its reading. Koyré's doubts about the ability of Galileo to obtain exact measurements were left unanswered.

Ronald Naylor has pointed out this flaw in Settle's reconstruction if its purpose was to dispel Koyré's doubts. Naylor reproduced as faithfully as possible Galileo's apparatus, both its design, including the lining of the groove, and its timing device (Naylor 1974, p. 127–134). Allowing the steel, instead of bronze, balls to run inside the groove as well as along its edges, he found out that there was a significant difference between the two kinds of run, the ones along the edges giving "the least satisfactory results in terms of consistency of running times" (Naylor 1974, p. 130). Moreover, his results were at odds with Settle's. Whereas Settle concluded that "all this turned out quite well" (Settle 1961, p. 22), with the experimental results fitting Galileo's times-square law, Naylor asserted that "it does not appear plausible that Galileo confirmed his theory with the accuracy claimed, using this experiment" (Naylor 1974, p. 133). This experiment would have been an *idealized* experiment, according to Naylor (1974, p. 134). However, as Drake remarked, Settle's and Naylor's experiments, whatever one may think of their outcomes, proved only that Galileo *could* have made the experiment that he described, not that he *did* make it. To ascertain that he did, one had to turn to his notes on motion.

Drake set the pace. Although most Galilean scholars differed in their interpretation of the experiment sketched on folio 116v, they endorsed Drake's reconstruction of the underlying apparatus from folios 175 and 114v. He imagined that the upper vertical line, bearing the marks from 300 to 1000, was the vertical height of "a plane tilted about 60° resting near the edge of a table on which the ball rolled briefly before falling to the floor" (Drake 1973b, p. 150). This apparatus was supposedly devised to test the inertial horizontal motion that Galileo needed, together with the times-square law of free fall, to deduce the parabolic trajectory of projection. As for the law of free fall, Drake had first proposed that Galileo could have deduced it from his odd number rule arrived at either empirically, by rolling a ball down an incline and observing that the length of the spaces traversed in each pulse beat were the successive odd numbers times the first one, or arithmetically, by looking for the arithmetic progression in which the ratio of the first two terms would be as the ratio of their sum to the sum of the following two terms, and so on, since the spaces traversed in equal time increase uniformly during natural acceleration (Drake 1969, pp. 195–197).

After discovering the notes on motion, Drake proposed a third reconstruction of Galileo's discovery of the law of free fall, relying this time on folio 152r (Drake 1973a). On the basis of ad hoc square numbers, his usual triangular graph of time, distance and speed, and the mean proportion, Galileo would have arrived at the times-square law and the parabolic trajectory. However, Drake's reading of a scribble at the top of the folio was challenged by Winifred Wisan (Wisan 1974, p. 201–215). Her reading led her to interpret the folio as "an unsuccessful attempt to use the correct law of fall in a derivation of the fundamental relation between accelerated and uniform motion" (Wisan 1974, p. 210), that is to say Galileo's double-distance rule of uniform motion compared his double-distance rule of accelerated motion. Although her interpretation has the advantage on Stillman Drake's to take into account all of the features of folio 152r, it has been challenged, on its turn, by Naylor (1977a).

For Naylor, this folio represents a turning point in Galileo's work on motion, since it documents the passage from the notion of speed proportional to distance to the notion of speed proportional to the square root of distance, reciprocally of distance proportional to the square of velocity, without any reference to time. Moreover, not only had Galileo established the new relationship between velocity and distance but also that its diagrammatic representation is a parabola. Naylor explains this passage by a change of "conceptual scheme" concerning speed marked by a change in wording, from *grado di velocità* to *gradus velocitatis*, which have different and incompatible physical meanings: "degrees of speed are consumed or they may be speeds at which bodies travel, they cannot be both." Underneath the technicalities, Naylor appears close to Koyré in their emphases on conceptualization, as opposed to the mere serendipity or empiricism.

Drake proposed a fourth explanation of Galileo's discovery of the law of free fall, not as a response to Wisan's critique of his former third explanation but as a consequence of the experimental turn that he gave to his study of Galileo's scientific revolution. The law of free fall was no longer discovered through pure reasoning but through an experiment (Drake 1975b). Drake read folio 107v as preserving the data of this experiment, an experiment that he did not reconstruct but rather invent from beginning to end, since there is no indication that Galileo even thought of such an experiment. The only evidence that Drake could extract from folio 107v was a column of eight numbers together with their tentative mode of computation: respectively, 4, 8, 13, 19, 27, and 35 times the length of a ruler of 60 units plus a certain number of units. He interprets these eight numbers as the measured distances traveled by a ball rolling down an incline in equal units of time from one to eight, given in a second column to the left of the former one. Drake imagines that Galileo determined his units of time by the even beat of a song. The rationale for measuring time this way was that Galileo belonged to a family of professional musicians and was himself a proficient lute player; on the other hand, this way of measuring time made the experiment unworthy of publication. But Mersenne, about the same time, has just done that and published it. A third column on the extreme left would record the square of the successive units of time, from 1 to 64. When multiplied by the unit distance, viz., 33, these figures give approximately the distance traveled per unit of time recorded in the right column. The distance was thus proportional to the times squared.

From Galileo's data and the law of free fall down inclines, Drake determined the length and the height of the incline that Galileo could have used. He performed the experiment singing at a tempo of two notes per second "Onward, Christian Soldiers" and releasing the ball at the first note while marking the position of the ball on the incline at the following notes, the second, the fourth, the sixth note, and so on, pictures being taken at the same time. The distance between two adjacent marks was calculated:

> The ratios of the successive intervals were found to agree closely with the figures recorded by Galileo on folio 107v. (Drake 1975b, p. 98)

Drake did not give up folio 152r altogether; on the contrary, he proposed the following link between folio 152r and folio 107v, both chronological and thematic: the experiment of folio 107v provided Galileo with the times-square law of free fall, which helped him overcome the contradiction that would have stopped his reasoning on folio 152r. This alleged contradiction derived from Drake's controversial reading of Galileo's scribble. This entire construction was meant to explain Galileo's blunder in his 1604 letter to Sarpi, in which he tries to deduce his times-square law from a principle of proportionality of speed to distance: in his 152r note, Galileo had shown that the graph of this relation was a parabola, whereas in his 107v note, it was implied by his triangular graph (Drake 1975b, p. 104; see also Drake 1974).

Naylor criticized Drake's interpretation of folio 107v as recording a free fall experiment (Naylor 1977a, pp. 373–377). First, he objected that three other experiments differing from Drake's could approximate Galileo's data as well. Second, he doubted that Galileo could have used an incline as small as Drake's, both in height and in length, the two being related: the steeper the incline, the longer it must be to allow for the measurement of the time of descent. But, the less steep is the incline, the more friction the ball encounters, and therefore, the measurements deviate most from the times-square law. Third, he denied that such an experiment could be as accurate as Drake's. To vindicate his criticism, Naylor conducted a series of experiments with Settle, each working independently of the other in order to control for the personal factor (Naylor 1980, pp. 373–377). They let the length and the height of the incline vary. They also reformatted Drake's results for the sake of comparison with their own. As for the timing device, they did not sing but instead beat time with the foot or counted loudly after training with a metronome. The results were at odds with Drake's:

1. First, comparison of the results obtained by the experimenters "shows that personal factors do not affect the outcome of these experiments in any significant ways."
2. Second, the data of folio 107v "cannot relate to measurements made in an inclined plane experiment."
3. Third, a rule linking time and distance cannot be discovered by this means.
4. Fourth, "the correlations between times and distances in all six experiments are poor compared to those claimed by Galileo in the *Discorsi*" (Naylor 1980, p. 377).

A controversy ensued between Drake and James MacLachlan on the one side and Naylor on the other. The issue was whether the discrepancies between the observed experimental data and the computed theoretical data were due to the competence of the experimenter (Drake, MacLachlan) or to the conditions of the experiment (Naylor).[9] This returns us to the Galileo–Mersenne situation as described by Koyré (Koyré 1953, pp. 94–94).

Although Pierre Costabel doubted that folio 116v referred to an experiment, most Galilean scholars were convinced that it did. Costabel's doubts stemmed from the fact that the observed data values were greater than the expected ones, even though they should have been smaller, given all the causes that slowed the rolling ball, such as friction, air resistance, and the like (Costabel 1975). He instead interpreted it as an experiment in mathematics devised by Galileo to test his principle of proportionality between speed and distance traveled that underpinned his times-square law in the 1604 letter to Sarpi. As Koyré (of whom he was a close associate) might have written, Costabel commented that this experiment in mathematics was not "mere speculation, for struggling with mathematical reality is not the same thing as inventing abstract theories" (Costabel 1975, p. 186). "Mathematical reality": such a formulation would have come to the support of the Koyrean thesis of a

[9] MacLachlan (1982), Naylor (1982, 1983a, b), Drake (1981, 1982, 1983).

Platonist Galileo. Unfortunately, Costabel could not reconstitute more than scraps of his supposed experiment in mathematics.

The other Galilean scholars agreed that folio 116v reported a free fall experiment but diverged as to its purpose. For Naylor, it was "the most important of the recently published manuscripts" (Naylor 1974, p. 107). He reconstructed the experiment just as Drake had imagined it (Naylor 1974, pp. 109–113). However, he varied the angle of the incline, from 15° to 40°; the data obtained were in very close agreement with one another and with the observed data of folio 116v.

By the Merton Rule, Galileo knew that the ratio of speeds was proportional to the ratio of times, as well as to their squares; by his law of chords, he knew that the ratio of speeds squared was proportional to the ratio of distances traveled. Since the distances traveled were the heights of fall, the ratio of speeds squared was proportional to the ratio of heights of fall. Since the horizontal distance was proportional to the product of speed and time as well as their squares, the time being the time of fall from the table and so constant in this experiment, the ratio of the horizontal squared was proportional to the speed squared and therefore to the ratio of the heights of fall. Thus, the comparison of the data of the heights of fall and of the corresponding data of the horizontal distances was a test of the law of free fall (Naylor 1974, p. 112). Galileo would have computed the expected values from this relationship, starting from his double-distance rule applied to the first value taken as a basis. He would then have compared the observed to the expected values and would have been appalled by their discrepancies. Naylor concluded:

> […] even the refinement of this experiment suggested in the *Discorsi* would not provide as good a confirmation of the times–squared law as the folio 116v experiment. Having spent considerable time reconstructing both experiments, I found it difficult to avoid the conclusion that the folio 116v experiment was simpler and provided better confirmation of the law of free fall. (Naylor 1974, p. 117)

Furthermore, both Winifred Wisan and David Hill agreed with Ronald Naylor on what the experiment recorded on folio 116v was about, viz., the law of fall and the double-distance rule. However, they departed from him and from each other on the purpose of the experiment. They based their interpretation on the rest of the folio, the computations, and the graph on its lower half. Deciphering Galileo's computations, Winifred Wisan showed that they made use first of Galileo's principle of speed proportional to distance of fall and then, when he saw that it failed experimentally, of speed proportional to the square root of distance. Moreover, she concluded that "it is probable then that the main result of the experiment on folio 116v was simply to confirm doubts about the validity of experimental results" (Naylor 1974, p. 126). Deciphering Galileo's computations in more detail and showing that he did not change his principle of computation but only his mode of computation, David Hill claimed that folio 116v recorded a successful experiment of the times-squared law that Galileo arrived at immediately by his law of chords (Hill 1986). By changing slightly the basic figure of all his computations, Galileo could reduce drastically the discrepancies between expected and observed data. There would therefore be no reason for him to doubt the validity of experimental results. But Hill did not explain why, if Galileo's experiment was that good, he never published it.

Thus, if it is undoubtedly true that Galileo made experiments on the law of free fall, contrary to what Koyré had claimed, the analysis of Galileo's private science did not obviate the objection that the experiment described in the *Discorsi* did not give results precise enough to sustain the law of free fall because of the experimental conditions of the time, as Koyré had claimed. The good experiment would rather have been the one reported in folio 116v, but which could not satisfy Galileo because it disproved his double-distance rule. In any case, his times-squared law of free fall would not have been arrived at experimentally but theoretically from his law of chords.

However, Galileo did not perform all the experiments that he described, a fact acknowledged even by Drake's school. MacLachlan had distinguished between *real* experiments that were actually performed, *imaginary* experiments that could have been performed but were not, and *thought* experiments that cannot be performed (MacLachlan 1973, p. 374). He reconstructed Galileo's two pendulum experiments on the fourth day of the *Discorsi* and observed that if one could have been a real experiment, the other could only have been imaginary (MacLachlan 1976, p. 178–181). Both were supposed to prove that the period of a pendulum was independent of its amplitude. The first experiment used bobs of different material and weights differing a hundredfold from each other, both of which were drawn aside at the same vertical angle; the second used bobs of the same material and the same weight but drawn aside at very different vertical angles, respectively, 10° and 80°. MacLachlan observed that the experiment with the same bobs could have been real since he obtained the same results as Galileo had: one swing ahead in one hundred for the larger amplitude. Although Galileo claimed that the oscillations of both bobs would diminish at the same rate, which is untrue, MacLachlan observed that the oscillations at greater amplitude decreased much more rapidly at the beginning (Naylor 1977b). In any case, the experiment with the bobs of hugely different weights could not have been performed and was thus imaginary, not only because the heavier bob made significantly fewer oscillations, but because it could not swing a thousand times, as Galileo had claimed. Therefore, if it is not true that Galileo did not perform *any* experiment, it is at least true that he did not perform *some* experiments and that *none* were exact experiments, but rather qualitative or "adorned" experiments.

If Koyré's second thesis concerning Galileo's Platonism had been challenged from the outset (see, for instance, Geymonat 1957), a new turn in the historiography of science, this time a sociological turn, brought about under the influence of the new sociology of science, substituted an Aristotelian Galileo.

12.4 The Sociological Turn

The opening paper of the Max Planck Institute für Wissenschaftsgeschichte in Berlin (MPIWG)'s *Galileo in Context* by Wolfgang Lefèvre challenged Koyré's claim that the birth of modern science involved replacing the everyday world of Aristotelian natural philosophy with a Platonist mathematical science. Lefèvre maintained that the context of Galileo's science was the "world of craftsmen and engineers" or, in more epistemological wording, the transformation of mechanics "from mere knowledge about art into the paradigm for physics" (Lefèvre 2001, p. 12, p. 25). To substantiate his claim, Lefèvre had to show how the theoretical problems that Galileo tried to solve stemmed from the technical practical life, more specifically how the law of free fall and the trajectory of projectiles were problems posed by the available technology. However, these problems had already been raised by Aristotle and by the medieval science of motion. It is there that Galileo found them, not in the workshop. When Lefèvre looked inside the workshop for the tool, the machine, or the process that could have shaped Galileo's way of investigating the problem, he found nothing worth noticing. The sawing-machine pendulum of Jacques Besson that he cited and reproduced is merely a drawing, not an actual machine (Lefèvre 2001, pp. 20–21). It is doubtful that Galileo had it in mind when he investigated his law of the pendulum. As for the pulse-clock pendulum of the Venetian physician Santorio, it was described—not built, just described, thus not part of practical experience—in 1602. Scholars have assumed that it was an application of Galileo's law. Nor does it seem plausible to assume, as Lefèvre did, that, on the contrary, Sanorio's device was the source of Galileo's law, since "Galileo discovered the pendulum law at the beginning of his career," already in 1588 according to Drake (Lefèvre 2001, p. 20; Drake 1978, p. 21). As for Galileo's law of chords, Lefèvre wondered whether it could not have been the invention of a practical engineer (Lefèvre 2001, p. 23). But, he is left with the Marxist-like generalities about economic prosperity, commerce and trade, new means of communications, and new educational institutions (Lefèvre 2001, p. 24). Much simpler would have been Archimedes' method of exhaustion.

If Koyré's thesis that Galileo's Platonism went beyond mere use of mathematics is debatable, his most profound insight is that Archimedes was the genuine predecessor of Galileo. If this is granted, then there has been a discontinuity in the history of physics, a kind of fold that legitimates linking a scientific revolution to Galileo's name: the successful mathematization of dynamics paralleling what Archimedes had done in statics. On both points, Lefèvre agreed with Alexandre Koyré:

> Modern dynamics, which goes back to Galileo, is a genuine novelty of modern times, whereas modern statics goes back to Archimedes. The way in which Galileo treated problems of dynamics is — in principle, regarding the type of treatment — comparable with the way in which Archimedes treated problems of statics. (Lefèvre 2001, p. 15)

Another member of the MPIWG group, Matteo Valleriani, has described Galileo as an artist–engineer, then as a military engineer–scientist, and even as an Aristotelian military engineer–scientist (Valleriani 2010), although of the three main educational paths for artist–engineer Valleriani listed—the workshop, the Abaco School, and the Accademia del Disegno—Galileo followed none. Rather, he studied medicine at the university and then switched to a private training in mathematics under Ostilio Ricci. Because Galileo gave a class on fortification—to which he contributed nothing (Valleriani 2010, p. 86)—among many other topics, and because he sold sectors of his own making to military officers that he lodged and trained at his home as paying guests, he counts as a military engineer–scientist. Yet when he turned his telescope into binoculars to determine the longitude at sea and tried to sell it to officials as a private entrepreneur, he was not commissioned to develop it, and it proved a failure. Valleriani sees Galileo as an Aristotelian engineer–scientist although as an engineer, "he did not reflect much on the materiality of the machines" and "he was too abstract and ignorant of real and relevant aspects" (Valleriani 2010, p. 113, p. 140). Moreover, he was allegedly an Aristotelian although he met with the opposition of the engineers and artisans of the Arsenal of Venice because he "superseded the Aristotelian vision" (Valleriani 2010, p. 153) and despite the fact that his thermoscope was at odds with Aristotle theory of heat (Valleriani 2010, pp. 188–190).

In his introduction to the *Science in Context* volume, Jürgen Renn described Galileo as "Engineer–Scientist, Artist, and Courtier." This was a judgment on Galileo, not a contextualization. To contextualize Galileo, one must look at what it meant to be a mathematician at the end of the sixteenth century and the beginning of the seventeenth century, since he made a living from teaching mathematics. If one does that, one will find all the facets of Galileo's public life: university professor, astronomer, instrument maker, engineer, and even astrologer (Biagioli 1989). Moreover, he strove to be recognized as a philosopher and he achieved his aim. To complete his contextualization, one would have to look at what it meant at that time to be a philosopher and what it meant to him. For Galileo, the answer is straightforward: it legitimated him in his fight against the Aristotelians such as his friend Cremonini. As a philosopher, he was entitled to write his *Dialogo sopra i due massimi sistemi del mondo* (hereafter *Dialogo*) against his Aristotelian adversaries in the manner of Plato.

Despite the evidence, the MPIWG group has maintained that Galileo remained an Aristotelian. They have shown minutely how the Aristotelian concept of motion that requires a cause, either efficient or final, and its correlated distinction between natural and violent motion, which Galileo retained throughout his career, was an obstacle on his road toward the foundation of classical mechanics (Damerow et al. 2004, pp. 135–278). That Galileo could not rid himself of the entire Aristotelian scheme of thought and that these conceptual residues posed difficulties does not mean that he remained an Aristotelian. It is not a case of all or none. The ontology of Galileo diverges greatly from that of Aristotle: besides the conception of a mathematized nature, it suffices to mention the only one natural movement toward the center of the earth, the only one quality of weight, the existence of the void, and the atomistic constitution of matter instead of the Aristotelian four elements.

Whereas Drake had reconstructed Galileo path of discovery starting from the law of fall, then proceeding to horizontal motion, and finally arriving at the parabolic trajectory, the MPIWG group argued that Galileo had followed the reverse path, going from the parabolic trajectory to the horizontal motion and last to the law of fall (Renn et al. 2001). They wanted to show that Galileo's science of motion originated in practical life as Aristotle's allegedly had, remote from any Platonist pure innate ideas. Galileo's point of departure was allegedly the observation of catenary with its quasi-parabolic form, which could be a model for the artillery (Renn and Damerow 2003; Büttner et al. 2003). Here again we see the Aristotelian military engineer–scientist. Decomposing the parabolic trajectory into a neutral horizontal uniform motion and a vertical accelerated motion, since by the first lemma of the fourth day of the *Discorsi*, the vertical distances from the vertex are proportional to the square of the horizontal distances, and the vertical distances of the fall are proportional to the square of the times elapsed measured along the horizontal distances, whence the times-square law of fall (Renn et al. 2001, p. 53, pp. 127–128). However in the text from the *De Motu* that they quoted in the very same page (*Ivi*, p. 53), the neutral motion, that is neither natural nor violent, is not a horizontal motion, but a circular motion around the center of the Earth.

The thesis of an Aristotelian Galileo had previously been advanced by William Wallace (Wallace 1986). Building upon the notes taken by Galileo during his youth on the Jesuit commentaries on the Aristotelian texts, mainly the *Posterior Analytics* and its emphasis on causal explanation, and what he called the "progressive Aristotelians" who were not opposed to the use of mathematics in the study of nature, he argued, as the MPIWG group would do, that Galileo retained this principle of causal explanation throughout his career. Therefore, Galileo could espouse the Copernican cosmology if and only if he could explain the tides causally by the rotation and the revolution of the Earth, which he did in the fourth day of the *Dialogo*. But since this explanation remained inconclusive because at odds with an experiment on a barge and then only speculative (Wallace 1984, p. 345), Galileo allegedly realized that the motion of the Earth had not been established in terms of causal explanation and therefore he "would not have perjured himself" when he abjured the Copernican world system. He could accept "on faith" and "in clear conscience" that the Earth does not move since he "had failed to prove the opposite" (Wallace 1986, p. 28).

However, even if its source was Aristotle, adherence to the principle of causality, or even to its weaker version, the principle of sufficient reason, does not ipso facto make one an Aristotelian. It is a principle of rationality that governs all scientific knowledge. Scientists have from time to time suspended their adherence to this principle when they were convinced of their description of a phenomenon without being able to identify its cause, gravitation being a case in point. As for Galileo's notes on Aristotelian logic, the expert on Galileo's Latin *juvenilia* Raymond Fredette's analysis is illuminating:

> These notes are those of a young man who has abandoned medicine to study mathematics, a field which feeds his imagination to concoct a very ambitious project: establish with theoretical rigour and experimental precision that Aristotle's basic principles in natural philoso-

> phy cannot be held true anymore. [...] Before embarking on his ambitious project of criticizing Aristotle, Galileo makes sure he knows and understands his target as well as he possibly can. (Fredette 2001, pp. 177–178)

Galileo might well have thought that he had not proven the motion of the Earth and yet nevertheless has been convinced of the truth of the Copernican system. Following Emil Wohlwill, among others, Koyré had pointed out that Galileo's theory of motion originated within Galileo's Copernican cosmology. To quote only one Galilean scholar, Winifred Wisan, the link between the two is the brachistochrone problem:

> If Galileo could prove that bodies naturally descend faster along a circular path than along a straight one, this proof would support the view that the earth itself might have a natural circular motion. [But to solve the brachistochrone problem he] had to create a new science of motion based on new propositions [such as] the squared law and his postulate that descending bodies acquire equal speeds in descent through equal vertical distances [...] the law of chords which established that times of descent are equal along chords from the highest or lowest points of a vertical circle [...] the theorem on accelerated motion according to which the times of descent along planes of equal heights are proportional to the lengths of those planes [and the] principle of horizontal inertia as an approximation of circular inertia. (Wisan 1984a, pp. 270–271; see also Wisan 1984b, pp. 41–49)[10]

12.5 Conclusion

To conclude, one can say that Galileo did perform experiments, but their results were not precise enough to be reliable, as shown by the divergent results arrived at by modern scholars who repeated them, either as he described them or as they conjectured that he had performed them. Galileo's experiments were qualitative and not quantitative; therefore, Galileo could not have arrived at his mathematical law of falling bodies experimentally. At the end of his life, when everything was said, Galileo could write to his friend Baliani:

> But, returning to my treatise on motion, I argue *ex suppositione* about motion defined in that manner, and hence even though the consequences might not correspond to the properties of the natural motion of falling heavy bodies, it would little matter to me, just as the inability to find in nature any body that moves along a spiral line would take nothing away from Archimedes' demonstration. But in this, I may say, I have been lucky; for the motion of heavy bodies, and the properties thereof, correspond point by point to the properties demonstrated by me of the motion as I defined it. (Galilei 1639, pp. 12–13)

Together with the conviction that mathematical forms such as triangles or circles or conic sections were embedded in nature like the regular polyhedrons as the shapes of the four elements of Plato's *Timaeus*, the reliability of geometrical reasoning, occasionally despite the experiments, constituted the basis of Galileo's Platonism. But Galileo's atomism was at variance with Platonism, as Koyré well knew. Therefore, when he called Galileo a Platonist, he could not mean that Galileo was a

[10] See also, for different accounts of the Copernican basis of Galileo's theory of motion, Naylor (2003) and Clavelin (2004, pp. 555–565).

self-conscious follower of Plato, but rather that, willy-nilly, what he achieved was the accomplishment of Platonism.

References

Biagioli M (1989) The Social Status of Italian Mathematicians, 1450–1600. History of Science. 27/1:41–95.

Büttner J, Damerow P, Renn J, Schemmel M (2003) The Challenging Images of Artillery. Practical Knowledge at the Roots of the Scientific Revolution. In Lefèvre W, Renn J, Schoepflin U (eds). The Power of Images in Early Modern Science. Birkhäuser, Basel, pp. 3–27.

Clavelin M (2004) Galilée copernicien. Albin Michel, Paris.

Costabel P (1975) Mathematics and Galileo's Inclined Plane Experiments. In Righini Bonelli ML, Shea WR (eds). Reason, Experiment and Mysticism in the Scientific Revolution. Macmillan Press, London, pp. 177–187.

Damerow P, Freudenthal G, McLaughlin P, Renn J (2004) Exploring the Limits of Preclassical Mechanics. A Study of Conceptual Development in Early Modem Science: Free Fall and Compounded Motion in the Work of Descartes, Galileo, and Beeckman, 2nd ed. Springer, Dordrecht.

Drake S (1964) Galileo and the Law of Inertia. In Drake 1999b, pp. 121–133.

Drake S (1969) Galileo's 1604 Fragment on Falling Bodies (Galileo Gleanings XVIII). In Drake 1999b, pp. 187–207.

Drake S (1971) Galileo Gleanings XXI: On the Probable Order of Galileo's Notes on Motion. In Drake 1999b, pp. 171–184.

Drake S (1973a) Galileo's Discovery of the Law of Free Fall. In Drake 1999b, pp. 248–264.

Drake S (1973b) Galileo's Experimental Confirmation of Horizontal Inertia. Unpublished Manuscripts (Galileo Gleanings XXII). In Drake 1999b, pp. 147–159.

Drake S (1974) Galileo's Work on Free Fall in 1604. In Drake 1999b, pp. 281–291.

Drake S (1975a) Galileo's Discovery of the Parabolic Trajectory. In Drake 1999b, pp. 160–170.

Drake S (1975b) The Role of Music in Galileo's Experiments. Scientific American 232/6:98–104.

Drake S (1978) Galileo at Work: His Scientific Biography. The University of Chicago Press, Chicago.

Drake S (1979) Galileo's Notes on Motion. Supplemento agli Annali dell Istituto e Museo di Storia della Scienza. Vol 2. Giunti, Firenze.

Drake S (1981) Alleged departures from Galileo's law of descent. Annals of Science 38:339–342.

Drake S (1982) Analysis of Galileo's Experimental Data. Annals of Science 39:389–397.

Drake S (1983) Comment on the above Note by R. H. Naylor. Annals of Science 40:395.

Drake S (1999a) Essays on Galileo and the History and Philosophy of Science. Vol 1. The University of Toronto Press, Toronto.

Drake S (1999b) Essays on Galileo and the History and Philosophy of Science. Vol 2. The University of Toronto Press, Toronto.

Fredette R (2001) Galileo's De Motu Antiquiora. Notes for a reappraisal. In Montesinos J, Solis C (eds). Largo Campo di Filosofare. Eurosymposium Galileo 2001. Fundación Canaria Orotava de Historia de la Ciencia, La Orotava, pp. 165–181.

Galilei G (1890–1909) Le Opere di Galileo Galilei Edizione Nazionale sotto gli auspice di Sua Maestà il Re d'Italia. XX Vols. Favaro A (ed). Tipografia di G. Barbèra, Firenze.

Galilei G (1632) Dialogo sopra i due massimi sistemi del mondo. In Galilei 1890–1909, VII, 21–520.

Galilei G (1638) Discorsi e dimostrazioni matematiche, intorno à due nuove scienze. In Galilei 1890–1909, VIII, 4–458.

Galilei G (1639) [Letter] Galileo a Gio. Battista Baliani in Genova. Firenze, 7 gennaio 1639. In Galilei 1890–1909, XVIII, pp. 11–13. [English Translation: Wallace 1981, p. 144].
Galilei G (fl. 16th) De Motu. In Galilei 1890–1909, I, pp. 243–420.
Galilei G (1960) On Motion and On Mechanics. Translated by Drabkin IE, Drake S. The University of Wisconsin Press, Madison.
Galilei G (1974) Two new sciences: including centers of gravity & force of percussion. Translated, Introduction and Notes by Stillman Drake. The University of Wisconsin Press, Madison.
Galilei G (1995) Discours concernant deux sciences nouvelles. Trad. Clavelin M. PUF, Paris.
Geymonat L (1957) Galileo Galilei. Einaudi, Torino.
Hill DK (1986) Galileo's Work on 116v: A New Analysis. Isis 77/2:283–291.
Koyré A (1935) Études sur Galilée. Annuaire 1934–1935. École Pratique des Hautes Études. Section des sciences religieuses, pp. 53–54.
Koyré A (1935–1936) À l'Aurore de la science moderne : La jeunesse de Galilée. Annales de l'Université de Paris 10/6:540–551; 11/1:32–56.
Koyré A (1937a) Galilée et Descartes. Travaux du IXe Congrès international de philosophie-Congrès Descartes. Études cartésiennes, 2ème partie. Hermann, Paris, pp. 41–46.
Koyré A (1937b) Galilée et l'expérience de Pise : à propos d'une légende. Annales de l'Université de Paris 12/5:441–453. [Reprinted: Koyré 1966, pp. 192–201].
Koyré A (1937c) La loi de la chute des corps : Galilée et Descartes. Revue Philosophique de la France et de l'Étranger 123(5/8):149–204.
Koyré A (1939) Études galiléennes. Vol 1: À l'aube de la science classique. Vol 2: La Loi de la chute des corps. Descartes et Galilée. Vol 3: Galilée et la loi d'inertie. Hermann, Paris.
Koyré A (1943a) Galileo and Plato. In Koyré 1968a, pp. 16–43.
Koyré A (1943b) Galileo and the Scientific Revolution of the Seventeenth Century. In Koyré 1968a, pp. 1–15.
Koyré A (1953) An experiment in measurement. In Koyré 1968a, pp. 89–117.
Koyré A (1960a) Galileo's Treatise De Motu Gravium: The Use and Abuse of Imaginary Experiment. In Koyré 1968a, pp. 44–88.
Koyré A (1960b) Newton, Galilée et Platon. Annales. Histoire, Sciences Sociales 15/6:1041–1059. [Reprinted: Koyré 1968b, pp. 243–265].
Koyré A (1966a) Études d'histoire de la pensée scientifique. PUF, Paris.
Koyré A (1966b) Études galiléennes. Hermann, Paris.
Koyré A (1968a) Metaphysics and Measurement. Essays in the Scientific Revolution. Chapman & Hall, London.
Koyré A (1968b) Études Newtoniennes. Gallimard, Paris.
Koyré A (1978) Galileo Studies. Harvester Press, Hassocks.
Kuhn T (1961) The Function of Measurement in Modern Physical Science. In Kuhn 1977, pp. 178–224.
Kuhn T (1970) Alexandre Koyré & the History of Science. On an Intellectual Revolution. Encounter 34/1:67–69.
Kuhn T (1976) Mathematical versus Experimental Traditions in the Development of Physical Science. In Kuhn 1977, pp. 31–65.
Kuhn T (1977) The Essential Tension. The Chicago University Press, Chicago–London.
Lefèvre W (2001) Galileo Engineer: Art and Modern Science. Science in Context 14/1:11–27.
MacLachlan J (1973) A Test of an 'Imaginary' Experiment of Galileo's. Isis 64/3:374–379.
MacLachlan J (1976) Galileo's Experiments with Pendulums: Real and Imaginary. Annals of Science 33:173–185.
MacLachlan J (1982) Note on R. H. Naylor's error in analysing experimental data. Annals of Science 39:381–384.
Naylor RH, Drake S (1983) Discussion on Galileo's Early Experiments on Projectile Trajectories. Annals of Science 40/391–396.
Naylor RH (1974) Galileo and the Problem of Free Fall. The British Journal for the History of Science, 7/2:105–134.

Naylor RH (1977a) Galileo's Theory of Motion: Processes of Conceptual Change in the Period 1604–1610. Annals of Science 34:365–302.
Naylor RH (1977b) Galileo's Need for Precision: The 'Point' of the Fourth Day Pendulum Experiment. Isis 68/1:97–103.
Naylor RH (1980) The Role of Experiment in Galileo's Early Work on the Law of Fall. Annals of Science 37:363–378.
Naylor RH (1982) Galileo's Law of Fall: Absolute Truth or Approximation. Annals of Science 39:384–389.
Naylor RH (1983a) Galileo's Early Experiments on Projectile Trajectories. Annals of Science 40:391–394.
Naylor RH (1983b) Letter to the editor. Annals of Science 40:396.
Naylor RH (2003) Galileo, Copernicanism and the Origins of the New Science of Motion. The British Journal for the History of Science 36/2:151–181.
Prudovsky G (1997) History of science and the historian's self-understanding. The Journal of Value Inquiry 31:73–76.
Renn J, Damerow P, Rieger S, Guilini D (2001) Hunting the White Elephant: When and How did Galileo Discover the Law of Fall? Science in Context 14/1:29–149.
Renn J, Damerow P (2003) The Hanging Chain: A Forgotten 'Discovery' Buried in Galileo's Notes on Motion. In Holmes FL, Renn J, Rheinberger HJ (eds). Reworking the Bench. Research Notebooks in the History of Science. Kluwer, Dordrecht, pp. 1–24.
Settle TB (1961) An Experiment in the History of Science. Science 133:19–23.
Thuillier P (1983) Galilée et l'expérimentation. La Recherche 14/143:442–454.
Valleriani M (2010) Galileo Engineer. Springer, Dordrecht.
Wallace W (1981) Prelude to Galileo. Reidel, Dordrecht.
Wallace W (1984) Galileo and his sources. The Heritage of the Collegio Romano in Galileo's science. The Princeton University Press, Princeton, NJ.
Wallace W (1986) Reinterpreting Galileo on the Basis of his Latin Manuscripts. In: Wallace W (ed) Reinterpreting Galileo. The Catholic University of America Press, Washington, pp. 3–28.
Wisan WL (1974) The New Science of Motion: A Study of Galileo's De motu locali. Archive for History of Exact Sciences 13(2/3):103–306.
Wisan WL (1984a) Galileo and the Process of Scientific Creation. Isis 75/2:269–286.
Wisan WL (1984b) Galileo's De Systemate Mundi and the New Mechanics. In Galluzzi P (ed). Novità celesti e crisi del sapere: atti del Convegno internazionale di studi galileiani. Barbèra, Firenze, pp. 41–49.

Chapter 13
Alexandre Koyré, Kepler's Reader Without Prejudices. Harmony of the World, Music of the Heavens

Anna Maria Lombardi

Abstract In his *La révolution astronomique*, published in 1961, Alexandre Koyré for the first time foresaw the important role played by musical interests in Kepler's cosmological model. Overcoming some widespread prejudices, he built his analysis of the mental paths practiced by the German astronomer on an accurate reading of original sources. In this way he was able to recover important steps in the Scientific Revolution and to enlighten the history of scientific thought. Following in Koyré's footsteps, I will focus on the finding of Kepler's third law. This astronomical law for the German astronomer represents in part the successful end of a 23-year-long path and in part the empirical measure of the Harmony of the Spheres. It is a long journey, which committed Kepler for a great part of his life, accompanying him through personal and scientific events. Going forward, I will offer a general picture of the presence and the role of harmonic studies in Kepler's work, focusing particularly on Kepler's own translation and commentary to Ptolemy's *Harmonica*, set as an appendix to *Harmonice mundi*, and on the differences between Kepler's edition and the other ones.

Keywords Kepler • Harmony of the Spheres • Kepler's third law • Astronomical Revolution

13.1 Introduction

Alexandre Koyré, as a pioneer, in his studies gave importance to the discipline of music as a key to understanding some crucial passages in the Scientific Revolution of the seventeenth century. Up until his work, the history of science had minimized the diffused and strong involvement in music by most of the Scientific Revolution protagonists (such as Galileo, Descartes, Newton), describing it almost as a hobby,

A.M. Lombardi (✉)
Dipartimento di Fisica, Università degli Studi di Milano, Via Celoria 16, 20133 Milano, Italy
e-mail: annak.lombardi@gmail.com

R. Pisano et al. (eds.), *Hypotheses and Perspectives in the History and Philosophy of Science*, DOI 10.1007/978-3-319-61712-1_13

as something not affecting the mainstream of their research. So, the contributions that were recognized as scientifically significant were simply extracted from the books that contained them, without any concern of a connection.

Koyré, in a very innovative way, decided to avoid prejudices and to start from a deep reading of their whole works, in order to understand the cognitive mechanisms underlying the discoveries and the changes of those revolutionary years. This intuition paved the way to a range of great historical studies, enhancing the role of music in the first years of modern science. As an example, we can quote a paper by Stillman Drake:

> [...] to look for roots of modern science in natural philosophy before the Copernican period is semantically an idle enterprise, the meaning of the word "science" being different before and after the seventeenth century. Yet roots of Galileo's physics did exist; those are to be found not in past philosophy but in the practices of musicians of his own time, just as roots of Kepler's cosmology are to be found in music theory. (Drake 1992, p. 4)

In particular, in regard to Kepler's astronomical third law, Koyré suggested that the studies about harmonies had to be considered as responsible for the discovery of the unusual, fractional exponent of the law, even if such historians as Robert Small or Jean Baptiste Delambre had previously considered its finding accidental.

This intuition encouraged me to investigate Kepler's work and the way he arrived to his astronomical laws, with a special regard to his third law, and, more generally, to search for connections between music and science in Kepler's writings.

13.2 The Harmony of the Spheres

Today the word harmony has a general meaning, to suggest a proportion, an aesthetically pleasant entity. The original meaning was quite technical, referring to different kinds of mathematical algorithms that originated proportions. The science of harmonics dates back to Pythagoras of Samos, sixth-century BC. He discovered that the ratios obtained from some simple mathematical reasoning (nowadays, such as an algorithm) could be heard by ear, through an instrument called a monochord, a single string stretched on a soundboard. In fact, using the numbers obtained by particular proportions, called harmonies, he could produce sounds that were heard as pleasant when played together, producing the phenomenon called consonance. So, mathematical harmonies were identified with musical harmonies. The monochord was the instrument that allowed one to hear and to perceive with human senses the beauty of divine mathematics.

The same harmonies were searched also in the skies. It must be remembered that, up to Copernicus, the Earth was motionless in the center of the cosmos and it was impossible to measure the real planetary radii. One needed some metaphysical a priori extra conditions. For Aristotelians, orbital sizes were determined by the horror vacui, so that each sphere was exactly resting on the other.

Another solution was to impose that the same ratios, determined by the mathematical and musical harmonies, had to be found between the planetary radii. This is what is called "Harmony of the Spheres." Both philosophers and mathematicians gave their contributions to this theory, which crossed centuries up until Kepler's age.

We can remember, as examples, Plato with his *Timaeus*, Aristotele with his *De Caelo* or *Metaphysics*, Cicero's *Somnium Scipionis*, Boethius' *De Institutione Musica*, and Kircher's *Musurgia Universalis* or Gaffurio's *Practica Musicae*. Different authors searched for universal harmony in different parameters; some imagined cosmos as a lyre, with a correspondence between planetary radii and strings' lengths; others focused on planets' speeds, each corresponding to the vibrations of a string; others used aspects – the respective positions of more planets in the sky – to create their own celestial harmony.

Kepler was initiated to the Harmony of the Spheres during his theological studies in Tubingen and it pervaded his whole work. We must note that Kepler used the same language, geometry, both in his musical and astronomical studies. This common language made it possible for him to design a unique cosmological picture, in which both music and astronomy could find their place. The idea of a Universe expressing harmony in every single part is a hinge of Kepler's thought, and he describes it in all his works. The faith in the possibility to find these regularities accounts for Kepler's obstinacy in his search for natural laws.

13.3 Johannes Kepler and the Harmonice Mundi

Johannes Kepler (1571–1630) was an astronomer and mathematician who gave important contributions to the Scientific Revolution of the sixteenth and seventeenth centuries.

Even if he is well known for the three astronomical laws taking his name, his major merit is related to his efforts to connect physics to astronomy, to build a celestial physics which paved the way to Newtonian dynamics. Though he is considered one of the first protagonists of modern science, Kepler is still steeped in classical culture, as results in an evident manner from the models he builds to represent the architecture of his scientific system.

To achieve his results, he follows roads full of philosophical, theological, metaphysical suggestions, and this is the reason why his works are hardly understood by modern readers. This paper presents one of these roads, the one connected with the search for a Harmony of the World.

Since his childhood in Wurttemberg, Kepler was involved in practice and theory of music, in church, and in school. He continued to deepen his skills during his student life from primary school to the theological faculty of Tubingen.[1] Here he built

[1] Kepler's education and experiences as theorist and practicing musician have been recently analyzed by Peter Pesic (Pesic 2014, p. 75).

up his concept of God, as a geometer, an architect, and a musician, according to contemporary culture. To search for God meant to look for his footprints as mathematical proportions, id est, harmonies, either in planetary parameters, in practical musical, or in the physiology of human body, both in microcosmos and in macrocosmos. Therefore, music becomes an investigative tool to explore the world, in order to find regularities and laws, in a time when a scientific law was still a concept to come.

In this attempt to recover the universal harmony of creation, Kepler reached a first success in his youth, when he was a teacher in Graz and faced two similar sets of issues. The first set is related to astronomy: why is there a certain number of planets, and why do they have just those distances from the center of the cosmos? The second one regards music: why is there a certain number of consonances, and why do those consonances correspond just with those distances from the bridge of a monochord?

Kepler applied a similar model to solve both problems. Taking as a basis the philosophy of Nicholas of Kues, who assimilated human knowledge of nature to a series of polygons inscribed into a circle which represents the divine knowledge, Kepler was able to build a model founded on the regular polyhedra (the five Platonic solids) inscribed in a sphere to answer the astronomical questions and a model founded on the constructible regular polygons inscribed in a circle to solve the musical issues (Lombardi 2008, pp. 15–40).[2]

Kepler wrote in his *Mysterium Cosmographycum* how, while teaching the cycle of the major conjunction between Saturn and Jupiter, a flash of inspiration hit him.[3] On the blackboard, he was drawing the succession of points, each representing a distinguished conjunction in the sky and forming almost 60° with the former, thus outlining a regular triangle slowly moving into a circle (Figs. 13.1 and 13.2).

Kepler's idea was that maybe this was the geometrical way to define astronomical spheres, which he believed to be no more concrete, but just mathematical entities. Each polygon would be circumscribed in an internal circle and inscribed in an external one. The difficulty of choosing among infinite regular polygons was soon dissolved by Kepler, as he decided to turn to a solid space, with polyhedra inscribed and circumscribed in three-dimensional spheres. In fact, there are only five regular polyhedra, well known since antiquity. They were protagonists in Euclid's *Elements* and have been identified by Plato in his *Timaeus* as the four constitutive elements of the Universe combined with the fifth, the dodecahedron, representing the Universe as a whole (Fig. 13.3).

The success of this model in Kepler's eyes is not to be confined to metaphysical reasons. First of all, according to Kepler the number of Platonic solids allowed a choice between Ptolemaic and Copernican astronomical systems: if the Earth were in the center of the cosmos, in Kepler's days you would have had seven planets, as you would have counted even the Moon as a planet, while with the Sun in the center, you couldn't count the Moon, so reducing the planets to six. But with just five solids, you can account only for six planets.

---

[2] On the web you find an animation explaining the astronomical model via: http://www.mogi-vice.com/Scaricamento/Keplero-MC.zip.

[3] A complete account is given in a letter from Kepler to Maestlin, August 2, 1595.

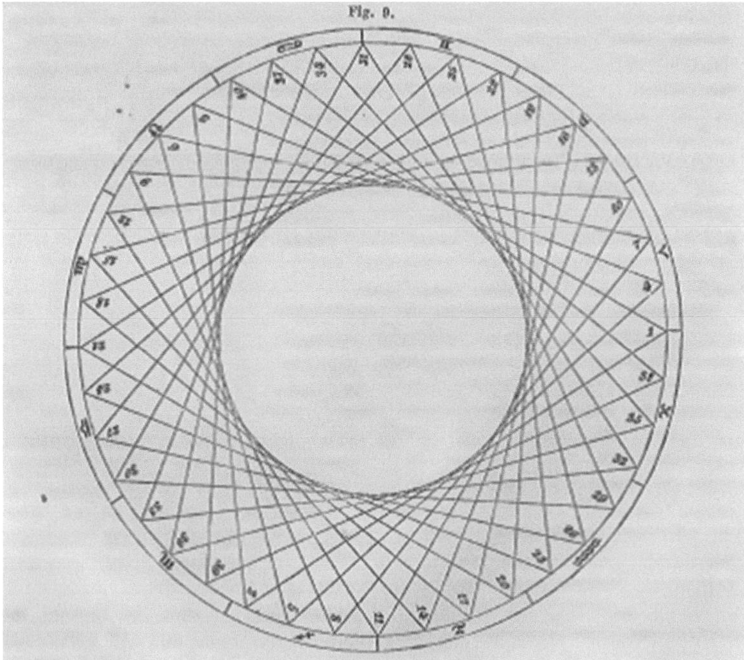

Fig. 13.1 Saturn and Jupiter meet in our sky approximately every 20 years, in a vertex of an equilateral triangle that seems to move around. This succession delineated two circumferences: an incircle and a circumcircle (Kepler 1858–1871, I, p. 108). Source: Public Domain. Retrieved via: https://archive.org/details/operaomniaedidit01kepluoft

Moreover, the reciprocal ratios between the sizes of the sphere radii, resulting from this embedding of polyhedra, showed an unexpected and rewarding correspondence with the ratios between the orbital radii, measured from the Sun. Little differences accounted for the shapes of the orbits, so that a sphere had a definite thickness, depending on eccentricity of that planet. Here again the model supported the Copernican cosmology (Figs. 13.4 and 13.5).

As we can read in a letter to his correspondent Herwart von Hohenburg, Kepler succeeded in applying a similar model to deal with the consonance problem. It consisted in explaining why only some pairs of notes, when produced at the same time, originate consonant intervals, id est, pleasant harmonies. Simple solutions had been found before the Renaissance, when musical theory provided just few consonances: the Fourth, the Fifth, and the Octave, respectively, represented by the ratios $\frac{3}{4}, \frac{2}{3}, \frac{1}{2}$. It is enough to say that all ratios you can obtain with the first four numbers originate consonant intervals in music.

Circumstances evolved after the musical revolution of *Ars Nova*, when also the Third and the Sixth, both major and minor, entered in the ensemble of consonances. There was a need for a different musical theory, which explained the diffused interest among mathematicians about music. Kepler, victorious in his astronomical

Periodi.	Anni ante Christum.	A rerum origine.	Personae insignes.	Res coincidentes: tu lector cave a trigonis effectas dixeris.
1	4000	000	Adam.	Creatio mundi.
2	3200	800	Enoch.	Latrocinia, urbes, artes, tyrannis.
3	2400	1600	Noah.	Diluvium.
4	1600	2400	Moses.	Exitus ex Aegypto. Lex.
5	800	3200	Esaias.	Aera Graecorum, Babyloniorum, Romanorum.
6		4000	Christus Dominus.	Monarchia Romana. Reformatio orbis.
	Post Christum			
7	800	4800	Carolus Magnus.	Imperium Occidentis et Saracenorum.
8	1600	5600	Rodolphus II.	Vita, fata et vota nostra, qui haec disserimus.
9	2400	6400		Ubi tunc nos et modo florentissima nostra Germania? Et quinam successores nostri? an et memores nostri erunt? Siquidem mundus duraverit.

Fig. 13.2 In his *De Stella Nova*, Kepler proposed a table recording successive entries of the major conjunction in Igneous Trigon. A full turn needs about 800 years to complete, welcoming the dawn of the reign of a great man (Jesus around 0 AD, Charlemagne around 800 AD, Rudolph II, Holy Roman Emperor, around 1600) and the birth of a new star. According to Kepler, this conjunction and this star, and not a comet, were the reasons for the Magi's quest (Kepler 1858–1871, II, p. 636). Source: Public Domain. Retrieved via: https://archive.org/details/joanniskeplerias02kepl

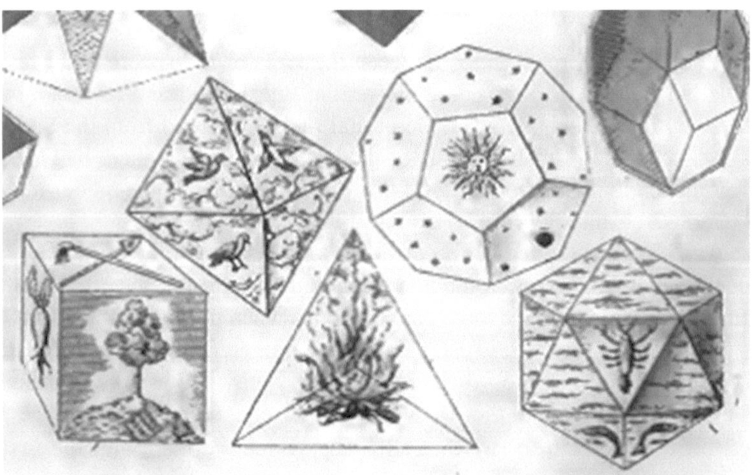

Fig. 13.3 The five Platonic solids, as presented in Kepler's *Harmonice Mundi*. According to Plato, air was associated with the octahedron, fire with the tetrahedron, the whole Universe with the dodecahedron, Earth with the cube, and water with the icosahedron (Kepler 1858–1871, V, p. 119). Source: Public Domain. Retrieved via: https://archive.org/details/operaomniaedidit05kepluoft

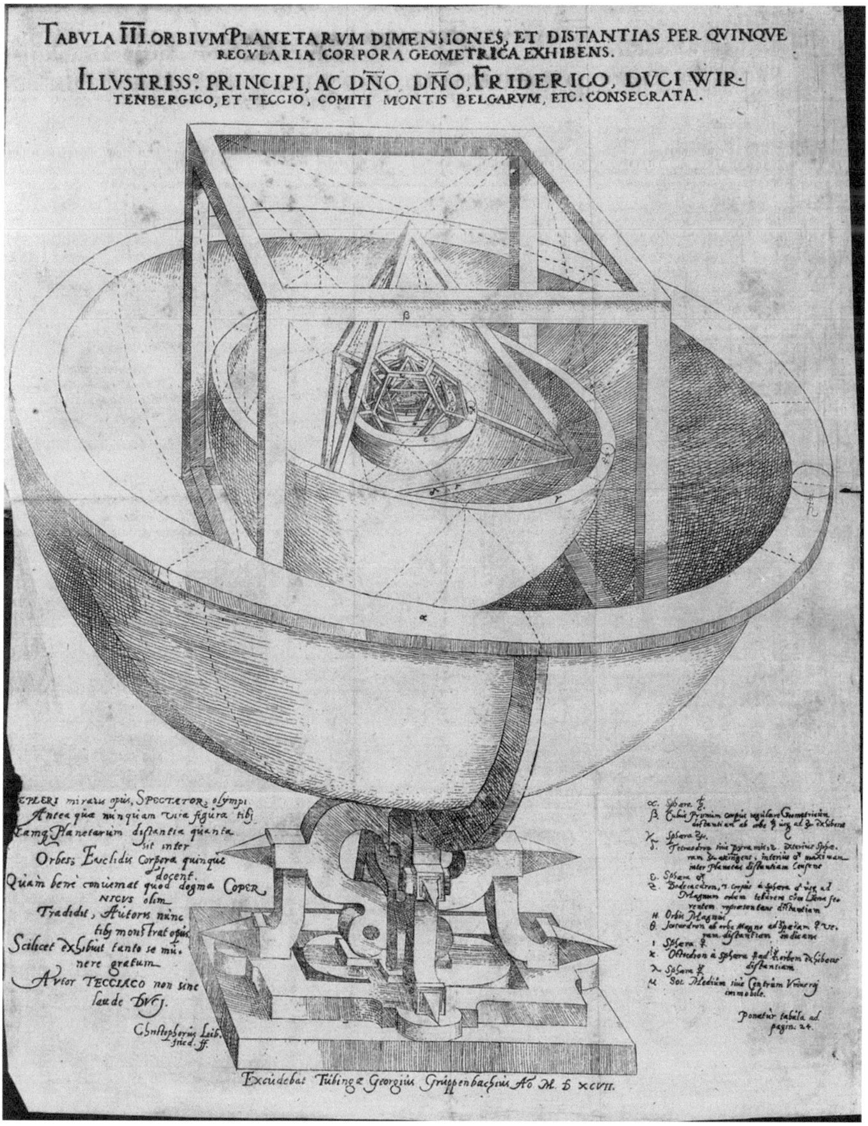

Fig. 13.4 The cosmological model in *Mysterium Cosmographycum.* The five Platonic polyhedra are nested one within the other, so that each one is circumscribed to the sphere in which the former polyhedron is inscribed. Five polyhedra account for six planetary spheres. The sphere of Mercury is inscribed in a octahedron, which is inscribed in the sphere of Venus, inscribed in an icosahedron, inscribed in the sphere of Earth, inscribed in a tetrahedron, inscribed in the sphere of Mars, inscribed in a dodecahedron, inscribed in the sphere of Jupiter, inscribed in a cube, and inscribed in the sphere of Saturn. If one measures the distances between the geometrical spheres from the center of the model, he finds that they are in good approximation proportional to the effective distances of the planets from the Sun (Kepler 1596, Tabula III). Source: Public Domain, CC BY-SA 4. Deutsche Fotothek. Retrieved via: http://www.deutschefotothek.de/documents/obj/80780811

Fig. 13.5 A zoom on the center of the system, proposed by Kepler himself in *Mysterium Cosmographycum* (Kepler 1858–1871, I, p. 214). Source: Public Domain. Retrieved via: https://commons.wikimedia.org/wiki/File:Kepler-solar-system-2.png

model, tried to apply it to consonance, without any results. However, as we read in the letter, Kepler grasped a great difference: while planets move in a solid space, music can be described in a plane space. It is the same geometry which rules the two worlds, but, if his planetary model needs polyhedra inscribed in a sphere, his musical model needs polygons inscribed in a circle.

Here Kepler faced a problem: we only have five regular polyhedra, while we have infinite regular polygons. To find the ratios, you divide each polygon in a "part" (a number of sides smaller than half the total number of sides), a "remaining," and the "whole." For example, in a pentagon, you have 1, 4, 5 or 2, 3, 5, respectively. The ratios between the whole and the remaining are the ratios corresponding to consonant musical interval. As you obtain an infinite number of ratios, Kepler needed an additional constriction to select the right polygons. He considered as allowable only constructible polygons, which can be constructed with a straightedge and a compass; then he ordered the polygons according to its sides and discarded all the ratios obtained by a preceding polygon which could not have been constructed.

Although this model could be considered faulty and groundless to us, it had the merit of being the only one able to explain the consonance of Thirds and Sixths, both major and minor. It worked even better than the widespread Zarlino's *Senario* which, on the basis of an alleged superiority of the number six, could explain all consonances but the minor Sixth, whose ratio is $5/8$.

Kepler was highly satisfied with both solutions, but he was even more pleased in that the two series of issues had been described by almost the same model, to highlight the deep connection between musical and astronomical harmonies.

However, while this model could explain the numerousness and the orbital size of the planets, it gave no reason for the motion of the planets. To get over this picture of a static cosmos, Kepler began an exhausting hunting, which will see its end more than 20 years later, with the finding of Kepler's third law.

13.4 The Third Law

While nowadays they are presented as a unique, compact corpus, the three astronomical Kepler's laws were originally introduced in two dramatically different ways.

The first and second laws appeared in a classical treatise on astronomy, the *Astronomia Nova* of 1609. However, the third law had to wait 10 more years, and it was published only in 1619 in a more complex text, *Harmonice Mundi*,[4] devoted to planets, music, and astrology.

In a book sounding prescientific to its modern readers, the presence of this astronomical law, which in the light of Newton's work will become fundamental, claims for a kind of justification, and different hypotheses were proposed.

Eduard J. Dijksterhuis, though admitting that Kepler had been looking for such a law for about 20 years, believes that the third law was written inside the *Harmonice* even if it had nothing to do with this book.

Max Caspar, in his wonderful biography of the German astronomer first published in German in 1948, writes that Kepler "found his law simply by trial, by comparing the powers of the known values of the distances with the periodic times" (Caspar 1993, p. 287).

Owen Gingerich (Gingerich 1975) gives us a more deep reconstruction. Since 1596, he writes Kepler had highlighted a fact already present in Copernicus' *De Revolutionibus*: the farther the planet was, the longer its orbital period was. However, once observed that $T_1/T_2 = r_1/r_2$ could not exactly describe the proportion between radii and periods, Kepler tried to adjust it and obtained a formula that could be approximated with $T_1/T_2 = (r_1/r_2)^2$. But now the law was wrong in the opposite sense. According to Gingerich, at this point Kepler began writing the squares and the cubes of radii and periods in a series of tables and, accidentally, he noticed that the values of two columns were in correspondence.

It seems reasonable to ask ourselves if some exponents were more meaningful than others in Kepler's eyes. I. B. Cohen, in his *Newtonian Revolution* wrote:

> [...] it should be noted that Kepler's discovery apparently resulted from a purely numerical exercise and insofar differed from his discovery of the area law and of the law of the elliptic orbits, both of which were presented originally (and may have been discovered) in association with a definite causal concept of solar force and a principle of force and motion. (Cohen 1983, p. 22)

---

[4]The complete title of this work *is Harmonices Mundi Libri V*, i.e., *The Five Books of the Harmony of the World*, usually referred to as *Harmonice Mundi* or *The Harmony of the World*. "Harmonices" is the genitive of the Greek term for "theory of harmony," which Kepler has taken over in his Latin (Caspar 1993, p. 288).

Cohen does not elaborate the issue, pretending that "Kepler is astonishingly silent as to how he came upon this law," but he refers to the works of Delambre, Small, and Koyré (Cohen 1983, p. 297).

Robert Small mentions what he calls "various trials," all the attempts to find out a relationship between radii and periods of planets orbits, first comparing the values and then the squares and the cubes of both the series of data (Small 1804, pp. 298–299). At the end, according to Small, Kepler discovered "with the highest delight and even astonishment" the right law. The reconstruction proposed by Delambre (Delambre 1821, p. 356) is quite similar. We can read that Kepler wrote something like

$$T_1 / T_2 = (r_1 / r_2)^x$$

trying first for x = 1, 2, 3, …

After that, according to Delambre, Kepler attempted with simple fractional exponents, finding 3/2 to be the right one.

The third historian mentioned by I. B. Cohen, as we said, was A. Koyré.

13.5 La Révolution Astronomique

In 1961, A. Koyré published *La révolution astronomique*, a series of three studies – devoted to Copernicus, Kepler, and Borelli – to illustrate the transformation of the key concepts at the foundation of astronomy in sixteenth and seventeenth centuries. According to the author, it's a history of important steps in scientific thought, the history of men looking for truth, mostly founded on original sources.

Three sections are dedicated to Kepler and his astronomy; the third is titled "From celestial physics to cosmic harmony" and is divided in three chapters. In the second one, "The Harmonice Mundi," a good space is granted to Keplerian musical reasonings.

In regard to Kepler's third law, Koyré reports the two accounts by Small and Delambre that we have seen before. Then he asks himself if it is plausible that Kepler could arrive at his law by "trial and error." It would have not been, he writes, a very difficult job. Kepler, from theoretical deductions, could attempt with time proportional to distance or time proportional to the square of the distance. Seeing both were unsatisfactory, Kepler could have tried the sesquialteral ratio, which is in the middle. Koyré writes that these are "quite natural steps in the development of thought." Yet he is not convinced.

> We are not accustomed to such simple developments from Kepler… Furthermore, when he says that he first had the idea on 8 March 1618, and abandoned it because it was not confirmed by his calculations – which could mean that the calculations he had made (and in which he had doubtless made a mistake) were not confirmed by the sesquialteral [sesquialter, ed.] ratio – but that he came back to it on 15 May of the same year, when he found it to agree perfectly not only with empirical facts but also *with his present studies (of harmony)*, he seems to imply that these later considerations were responsible for setting him on the way towards discovery of his law. (Koyré 1992, p. 339)

So, let us embrace the hypothesis that Kepler could have been influenced by his harmonical interests in his search for the right exponent, and, as recommended by Koyré, let us explore the original sources.

13.6 The Harmonical Road

Maybe the harmonical road does not convince modern readers as the soundness of its demonstrative passages, but it unfolds to us an amazing intellectual enterprise.

As A. Koyré highlights, in his previous works and especially in *Mysterium Cosmographycum* and *Astronomia Nova*, Kepler had successfully designed a Universe regulated by mathematical laws, which had a physical meaning.

However, Kepler's celestial physics gave only a limited answer. "It was too static," Koyré writes, because it gave reason of sizes, but not of velocities. Therefore, Koyré goes further, we have not to consider God "merely a geometer," but as "a musician" (Koyré 1992, p. 327).

To support the thesis that the presence of the third law in this book is not accidental, I want to point out how the *Harmonice Mundi* is an accurately designed work, whose architecture is delineated all along its path.

It's even possible to refer to December 14, 1599. In a letter written to his patron Herwart von Hohenburg, Kepler announces that he is preparing a new work, *Harmonice Mundi*, organized in five books. Moreover, the fifth book is titled: "De causis motuum periodicorum – astronomicus" (Kepler 1858–1871, V, p. 30).

The actual title of book V, printed 17 years later, assures, if not the discovery of a law, a firm determination to inquire for it. In the index we can read: "the fifth is astronomical and metaphysical, on the most perfect harmonies of the celestial motions, and on the origin of the eccentricities in the harmonic proportions" (Kepler 1997 [1619], p. 1), and, with a slight change in the effective title of the chapter: "On the most perfect Harmony of the heavenly Motions, and on the origin from the same of the Eccentricities, Semidiameters and Periodic Times" (Kepler 1997 [1619], p. 387). We can imagine he became more accurate in this second version because in the meanwhile between the index and the fifth book, Kepler did find his law.

We can note another relevant change, which attests the path followed by Kepler. In the index we can read about an appendix containing "a comparison of the work with book III of the *Harmony* of Claudius Ptolemy..." and even more, at the beginning of the fifth book, we read that this appendix will contain a translation of Ptolemy's book III, a completion of its gaps, and a comparison with Kepler's theories. This purpose will be disregarded, as we will see later, as Kepler will provide only a brief summary of Ptolemy's *Harmonica*. Evidently, Kepler had changed his judgment about a possible analogy between Ptolemy's and his own works, but even this failure attests a strong faith in the existence of a relation between musical harmonies and planetary parameters.

Again, in the introduction to book V, Kepler exposes his project: "that the whole nature of harmony, to its full extent, with all its parts, as expounded in Book III, is to be discovered among the celestial motions" (Kepler 1997 [1619], p. 389).

Focusing on the third law, we note how the necessity to find a law connecting distances of the planets from the Sun (semidiameters) and periodic times had been present in Kepler's mind for years.

Copernicus, in his *De Revolutionibus*, had already pointed out that an important difference between the Ptolemaic and his own system was that the planetary periods calculated with the Sun in the center of the cosmos show to be ordered, as the more distant the planets are from the Sun, the longer the periods are:

> In this arrangement, therefore, we discover a marvellous symmetry of the universe, and an established harmonious linkage between the motion of the spheres and their size, such as can be found in no other way. (Copernicus 1978 [1543], p. 22)

Kepler's contemporary astronomers, among them Robert Fludd, had predicted a proportion between orbital radii, but, as we read in *Harmonice*, Kepler had already proved the inconsistency of this claim on the basis of the accurate data collected by Tycho Brahe. It is interesting to observe that Kepler made his calculations either with a Ptolemaic or Copernican approach, as if the discovery of some harmonical relations could have been essential evidence to make one of the two systems prevail upon the other.

Kepler, instead, discovered a harmonical proportion between the velocities of the planets. As regards orbital radii, in Kepler's mind they were already determined by the model of *Mysterium Cosmographycum*, in which they were set by the proportion between Platonic solids. So, it seemed natural to him to look for a connection between orbital sizes and velocities and hence periodic times.

13.7 Harmony and Velocities

As we have seen, Kepler tested the existence of a harmonic proportion between planetary radii, but he proved it false in front of experimental data. Kepler's first attempt was addressed to ascertain a proportional relation between planetary shapes, id est, eccentricities, and the average radii of the individual planets. Even this hypothesis did not agree with data. Nevertheless, Kepler's persuasion on a musical cosmos was too strong to give it up, and he contemplated the idea to renounce spatial variables and to consider some parameters concerning time and precisely the velocities: "in ipsis motibus, non in intervallis." He introduced a proportion between planetary velocities (the angular paths traveled in a day) and musical notes (today we would say frequencies). Here Kepler goes back to Plato, which in his *Myth of Er* had assigned a musical note to each planet. However, Platonic orbits were dramatically different from Keplerian ones: those were perfectly circular and moved with a uniform velocity. Sitting on each sphere, a mermaid was singing perpetually the same note, so giving her contribution to celestial music, the Harmony of the Spheres.

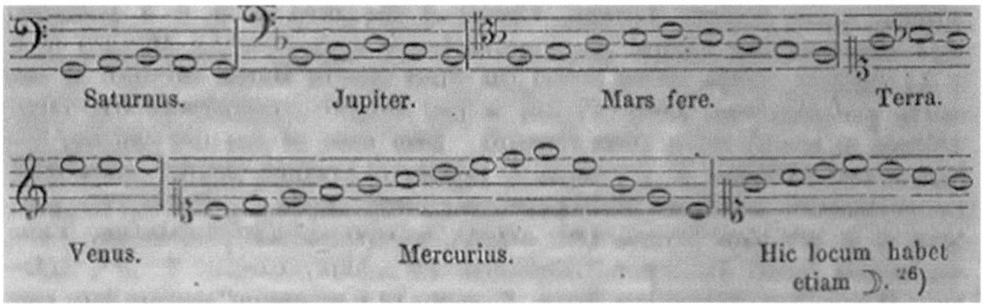

Fig. 13.6 In *Harmonice Mundi* Kepler assigns a range of notes to each planet. To Saturn, which has the slowest speed if one considers the angular daily velocity, he ascribes at the aphelion the deepest note, a grave G which in Kepler's time was the common pedal. When Saturn is at perihelion, it moves faster, so the notes change. And, for the other planets, the faster they travel in the skies, the more acute are the notes. This score is proposed by Kepler himself in the fifth book of his *Harmonice* (Kepler 1858–1871, V, p. 294). Source: Public Domain. Retrieved via: https://archive.org/details/operaomniaedidit05kepluoft

Keplerian orbits are elliptical and velocities are variable, as each planet moves faster as it is closer to the Sun and slower when it is farther. Kepler argues that it no longer makes sense to associate a single note to each planet but that he needs to consider a range of notes (Fig. 13.6).

The discovery that each couple of two near planets "sounds good" together gave Kepler an unexpected delight and a confirmation of the Harmony of the Spheres. Precisely, he observed that the sound associated to each minor velocity of a planet is consonant with the sound associated with the major velocity of adjoining planets. This fact, which in our view is a simple coincidence, to Kepler is a demonstration that the velocities are those parameters to be used in highlighting celestial harmonies. Next step is to find the mathematical relation between velocities and planetary radii, which according to tradition had to be related to harmonies.

A key role in this inquiry is played by the exponent involved in the law, 3/2 or *proportio sesquialtera*, which is at the foundations of modern occidental musical system realized by Pythagoreans, whose importance Kepler himself claims in his *Harmonice*.

The octave, the fundamental unit of music delimited by two frequencies, one being the double of the other, according to Pythagoreans, is divided in 12 small steps, known as semitones. These are defined, in modern terms, starting from a fundamental frequency f_0, multiplying at each step the lower frequency by 3/2, and, if the result is greater than 2 f_0, dividing it by two. So, sesquialter proportion can be considered the seed which originates our musical scale.

In the third book of *Harmonice*, devoted to musical intervals, Kepler anticipates that in the fifth book, he will use musical intervals to describe the rate of planetary motions:

> [...] and finally to construct from them the splendid edifice of the harmonic system, or musical scale. Its construction is not arbitrary, as some may suppose, not a human invention which may also be changed, but entirely rational, and entirely natural, so much so that God Himself the Creator has given expression to it in adjusting the heavenly motions to each other. (Kepler 1997 [1619], p. 158)

Moreover, later in the third book, he introduces each musical interval, its connection with a mathematical proportion, and its metaphysical meaning. But it is precisely for the sesquialter interval, and for none other, that he returns on the astronomical function of music, writing that this interval expresses a kind of defect, "[…] and God the Creator Himself has also expressed this defect in the planetary motions, as we will hear in the fifth book" (Kepler 1997 [1619], p. 184).

To discover that the relation between radii and periods was just a sesquialter proportion should have been like an enlightening revelation to Kepler.

> So strong was the support from the combination of my labor of seventeen years on the observation of Brahe and the present study, which conspired together, that at first I believed I was dreaming, and assuming my conclusion among my basic premises. But it is absolutely certain and exact that the proportion between the periodic times of any two planets is precisely the sesquialterate [sesquialter, ed.] proportion of their mean distances, that is, of the actual spheres. (Kepler 1997 [1619], p. 411)

The sense of "certain and exact" and of "precisely" is truly modern, as it relates to the comparison between data experimentally measured by Brahe and those deduced from the law. Kepler proposes to consider the ratios between maximum and minimum velocity of each planet, from which he obtains eccentricities, medium motions, and, by his third law, medium radii.

Here again we can touch Kepler's modernity, as he offers the reader the possibility to compare different models. In three distinct tables (Kepler 1997 [1619], p. 420, p. 486, p. 487), he juxtaposes observational data collected by Tycho Brahe about distances (at aphelion and perihelion) between planets and the Sun, from those deduced from the model of *Mysterium Cosmographycum* and those obtained by the harmonical model. The last ones are nearer to experimental data, so that the harmonical model results to be empirically authenticated.

13.8 Kepler, Editor of Ptolemy's *Harmonica*

To complete this analysis, it is interesting to consider Kepler's fascination with a work by Claudius Ptolemy, the *Harmonica*, devoted to musical intervals and mathematical ratios.

In 1599, Herwart von Hohenburg wrote to Kepler about a book by Ptolemy discussing the number of consonant intervals. In those years, as we said, Kepler was studying the problem of consonance, so he repeatedly asked his correspondent to lend him the *Harmonica*. He received a copy from Herwart, a Latin translation published in 1562 by Antoninus Gogavinus, in 1600, few months before going to work for Tycho Brahe. The new work absorbed him almost completely, by engaging him in that long war with Mars, which gave birth to the first and second astronomical laws. Only in 1607, Kepler wrote again to Herwart, complaining of the poorness of the Latin translation and asking his patron for a Greek transcription, hopefully more faithful to the original. Herwart sent to Kepler a manuscript in Greek with a commentary by Porphyry and another by the monk Balaam. The latter suggested for the first time that the last three chapters of the third book of the *Harmonica*, which sounded particularly interesting to Kepler for their focusing on connections between music and astronomy, were not original.

Kepler decided to prepare his own edition, with a Latin translation and a commentary, and in the index of the *Harmonice Mundi*, he promised, as an appendix, a translation of the Ptolemy's work, from the third chapter of the third book to the end. Yet, he never published it. Even though there is an appendix at the end of the *Harmonice*, it consists of only six pages, in which a brief comparison of Ptolemy's with his own work is integrated with a critical analysis of the ideas of a contemporary of Kepler, Robert Fludd, the author of the model known as the celestial monochord (Fig. 13.7).

Kepler's edition, preserved in the so-called Pulkovienses manuscripts, will be published only in 1864, in the fifth volume of the *Opera Omnia* edited by Christian Frisch (Kepler 1858–1871). The interest of Kepler is particularly focused on the chapters offering some relations between planetary parameters and musical harmonies, id est, from Chap. 8 to the end of the third book. Ptolemy, that in previous chapters had discussed some metaphysical issues about music and soul, first connects the Greek musical systems with zodiacal signs and aspects, even if Kepler comments that he does not approve the way he does it. It is in Chaps. 10, 11, and 12 that Ptolemy begins dealing with planetary motions, relating different movements in the sky with different musical concepts (melody, genres, modes). After the 13th chapter, on planetary aspects with respect to the Sun, Kepler finds what he considers a fundamental precedent to his work, with a substantial, dramatic difference: "whereas my celestial harmonies are formed by rays [radii, ed.], not at the Earth but at the Sun" (Kepler 1997 [1619], p. 503).

The last three chapters of the third book propose to give a key to interpret the harmony of celestial spheres. Going into the specifics, Chap. 14 gives a method to associate aspects and notes, Chap. 15 is on the way to calculate mathematical ratios from planetary movements, and Chap. 16 compares consonances or ratios between planets with musical ones.

It is easy to understand Kepler's desire to confront these chapters with his own work. Unfortunately, according to the tradition, only the titles were authentic. Kepler tried to reconstruct the chapters on the basis of the titles and of Ptolemy's astronomical and musical conceptions. Then, in the notes, which are longer than the text, Kepler comments on the mistakes that, according to him, result from Ptolemy's premises.

It is significant that, after curating this edition, he never published it. One can argue that, after this accurate analysis, he realized how he could not demonstrate a celestial harmony on the basis of Ptolemy's theories. In addition, once he found the third law, he didn't need to refer to Ptolemy anymore. However, now that he had deeply studied it, Kepler is intrigued by Ptolemy's *Harmonica* also for the concept of *vis harmonica*, which is here a need of natural phenomena to obey regular laws. In Kepler's age, the same concept of scientific law was a novelty. Who can assure that Nature will follow any law? Kepler, who at the time of this edition had already found his three astronomical laws, is obviously interested by this concept. He reports the title of third and fourth chapters – in italics – with his own remarks:

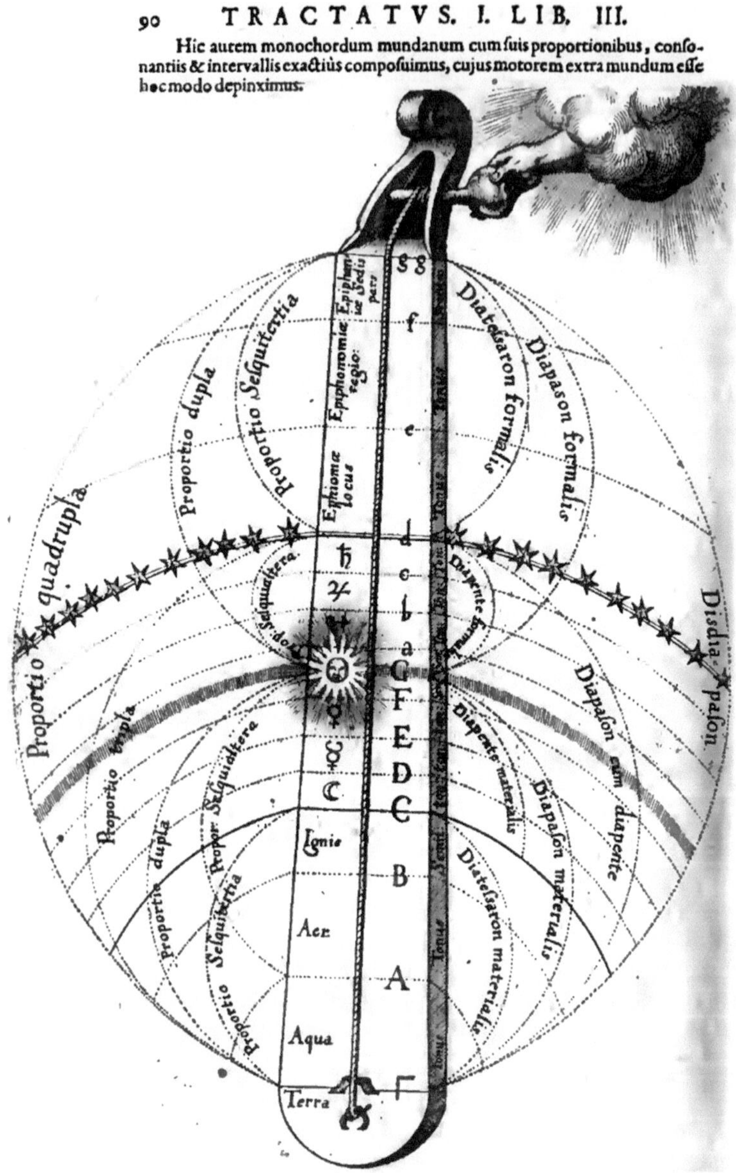

Fig. 13.7 The celestial monochord from Robert Fludd (1574–1637). The hand of God, coming out from the clouds, tunes the mundane monochord. This is a geocentric, Ptolemaic monochord: at the end we find the Earth, with its four elements, while on the fingerboard, the Moon, the Sun, and the planets follow one another in correspondence with frets. Their position on this board is determined by the condition that their distances on the string are proportional to their distances from the Earth in the skies. This instrument was supposed to produce a celestial harmony, but, as Kepler calculated, the notes assigned to the planets are not consonant at all (Fludd, 1617–1621, Vol. I, p. 90) (Source: Public Domain, CC BY-SA 4. Deutsche Fotothek. Retrieved via: http://www.deutschefotothek.de/documents/obj/88966469)

> Under what class of things the nature or force of harmony is to be placed, and the knowledge of it, showing that there is some principle, causal, formal, mental, or even divine, which links harmonies with things. Chapter IV: That the force of harmonic combination belongs to all things whose natures stand at a higher degree of perfection; and that that is chiefly apparent in human minds and in the heavenly revolutions. (Kepler 1997 [1619], pp. 500–501)

Lastly, we note how, also from the appendix to the *Harmonice*, we can deduce the modernity of Kepler as a scientist. He can adopt a classical, metaphysical model to explore the regularities of the cosmos, but he needs to discover quantitative laws and to compare theoretical predictions with experimental data. Kepler writes: "the Ptolemaic discoveries are compared with my own, and the difference is shown between the symbolism of Ptolemy and my own legitimate demonstrations" (Kepler 1997 [1619], p. 500). In addition: "What Fludd endeavors to teach as harmonies are mere symbolism" (Kepler 1997 [1619], p. 505).

13.9 Conclusions

At the dawn of the Scientific Revolution, Johannes Kepler discovered the law that states the relation between planetary radii and periods. The way the astronomer succeeded in discovering the law has often been declared as accidental. Alexander Koyré, by recognizing a major role to primary sources and to the reconstruction of Kepler's mental paths, was the first to foresee the key role of the scientist's contemporaneous musical interests.

A deeper analysis of *Harmonice Mundi*, of related Kepler's correspondence, of Kepler's edition of Ptolemy's *Harmonica*, and of the contribution of music to Kepler's astronomy emerges as effective and indispensable.

> *Itinerarium mentis in veritatem* is not a straight line. The road must be traversed, no matter how circuitous or mazelike; blind alleys must be negotiated; wrong paths must be retraced in order to discover the facts of the quest and hence the truth. Then, with Kepler, we can acknowledge that the ways by which the mind attains the truth are even more wonderful than the achievement itself. (Koyré 1992, p. 11)

Acknowledgments I want to express my gratitude to Fabrizio Castelli, who supported my work.

References

Caspar M (1993) Kepler. Translated and edited by Hellman CD. Dover, New York.

Coelho V (1992) (ed) Music and Science in the Age of Galileo. The Western Ontario Series in Philosophy of Science. 51. Springer, Dordrecht.

Cohen I B (1983) Newtonian Revolution. Cambridge University Press, Cambridge.

Copernicus N (1978 [1543]) Nicholas Copernicus on the Revolutions. Dobrzycki F (ed). Translation and commentary by Rosen E. Copernicus N – Complete Works. Vol. 2. Polish Scientific Publishers, Warsaw.

Delambre JBJ (1821) Histoire de l'astronomie moderne. Courcier, Paris.
Dijksterhuis EJ (1961) The Mechanization of the World Picture. Translated by Dikshoorn C. Oxford Clarendon Press, London.
Drake S (1992) Music and philosophy in early modern science. In Coelho 1992, pp. 3–16.
Fludd R (1617–1621) Utriusque cosmic maioris scilicet et minoris metaphysica. 2 Vols. Ære Johan–Theodori de Bry, typis Hieronymi Galleri, Oppenheim.
Gingerich O (1975) The origins of Kepler's Third Law. Vistas in Astronomy 18:595–601.
Kepler J (1596) Prodromus dissertationum cosmographicarum, continens Mysterium cosmographicum, De admirabili proportione orbium coelestium, deque Causis coelorum numeri, magnitudinis, motuumque periodicorum genuinis & proprijs, demonstratum, per quinque regularia corpora geometrica. Georgius Gruppenbachius, Tubingae.
Kepler J (1858–1871) Joannis Kepleri astronomi opera omnia. Frisch C (ed). 8 Vols. Heyder & Zimmer, Frankofurti et Erlangae.
Kepler J (1997 [1619]) The Harmony of the World by Johannes Kepler. Translated into English with an Introduction and Notes by Aiton EJ, Duncan AM, Field JV. Vol. 209. The American Philosophical Society, Philadelphia.
Koyré A (1992) The Astronomical Revolution: Copernicus, Kepler, Borelli. Dover, New York.
Lombardi AM (1997) Analisi armonica della terza legge di Keplero. Le Scienze 345:80–83.
Lombardi AM (2008) Keplero. Una biografia scientifica. Codice, Torino.
Pesic P (2014) Music and the Making of Modern Science. MIT Press, Cambridge–MA.
Raffa M (2002) La scienza armonica di Claudio Tolomeo. Saggio critico, traduzione e commento. EDAS, Messina.
Small R (1804) An Account of the Astronomical Discoveries of Kepler. Mawman, London.
Stephenson B (1994) The Music of the Heavens: Kepler's Harmonic Astronomy. Princeton University Press, Princeton.

Chapter 14
The History Between Koyré and Husserl

Rodney K.B. Parker

Abstract Alexandre Koyré (1892–1964) was a prominent member of the *Göttinger Philosophische Gesellschaft*, otherwise known as the Göttingen Circle. This group, who came together to study the work of Edmund Husserl, was responsible for the establishment and spread of the phenomenological movement. However, Koyré's place within this group and how his early training in phenomenology impacted his later works has not been fully explored. He left no autobiography. The accounts we do have tend to emphasize the impact of Adolf Reinach and realist phenomenology on Koyré's intellectual development and downplay the influence of Husserl and his transcendental phenomenology. After working with Husserl for roughly 3 years, Koyré submitted a draft–dissertation on the paradoxes of set theory. Husserl rejected the dissertation, and Koyré subsequently moved to Paris. Despite this change in location, Koyré kept in contact with his former Göttingen colleagues and, as I will show, never abandoned his phenomenological roots. Moreover, there is reason to believe that the historical–epistemological works that Koyré has become known for are a continuation of and a response to Husserlian phenomenology.

Keywords Phenomenology • Epistemology • Edmund Husserl • Conceptual frameworks • Mathematization of nature

14.1 Introduction

> It would be of inestimable value for the history and phenomenology of human thought were it possible to reconstruct step–by–step the development of the Copernican mind. Unfortunately, it is impossible to do so. Copernicus has left no autobiography describing his mental development […]. However, if we abandon any hope of writing a history of Copernican thought, we ought, nevertheless, to try and grasp its historical significance and nature; we ought to expose the hidden and acknowledged motives and incentives; and yet avoid any modernization. (Koyré 1973, p. 18)

R.K.B. Parker (✉)
Department of Philosophy, The University of Western Ontario, London, ON, Canada
e-mail: rparke4@uwo.ca

R. Pisano et al. (eds.), *Hypotheses and Perspectives in the History and Philosophy of Science*, DOI 10.1007/978-3-319-61712-1_14

Alexandre Koyré made his career as an historian and philosopher of science, writing extensively on the scientific revolution of the sixteenth and seventeenth centuries. The phrase "scientific revolution" itself was brought into the popular vernacular by Koyré in his *Études galiléennes* (Koyré 1939). Koyré writes that the study of the evolution and revolutions of scientific ideas provides us with a glimpse into the human mind at grips with reality. This struggle between human reason and the world "has sometimes led to a veritable 'mutation' in human thought", and of these *mutations* Koyré suggests that the "scientific revolution of the seventeenth century" was perhaps the most important since the invention of the concept of the *cosmos* by the Greeks (Koyré 1978, p. 1). Years later, in his overview of *The Astronomical Revolution*, Koyré states that the purpose of his historical studies is to trace the evolution and mutation of the fundamental concepts by means of which the human mind has endeavoured to bring to order and to save the phenomena and by means of which we have attempted to establish the "facts" that underlie and explain the chaos of perceptual appearances (Koyré 1973, p. 9). This is a project which aspires to be much more than a rigorous history of ideas and perhaps even more than a study of the ways that philosophical–conceptual frameworks influence scientific theories and vice versa.[1]

In his historical works, Koyré stressed the epistemological concern regarding the conceptual mutations between intellectual paradigms which make it difficult, if not impossible, to fully understand the work of past thinkers. This places limitations on what we can reasonably demand of history and anthropology as sciences. However, descriptions of these mutations can still reveal to us something about both the actual, concrete life of the mind and possibly the structures of consciousness that underlie this life. Philosophers such as Thomas Kuhn and Michel Foucault have expressed their indebtedness to the style of historical research conducted by Koyré in his ground-breaking work on figures such as Copernicus, Kepler, Galileo, Descartes, and Newton.[2] For his part, Koyré acknowledged his personal indebtedness to his mentors Émile Meyerson, Léon Brunschvicg, and Edmund Husserl, for shaping his approach to the study of history and for showing him the philosophical significance of such studies.

---

[1] We should note here that Koyré's historical interests went beyond natural science. He was interested in mutations in thought more generally, particularly during the French Enlightenment, as we see in his essays on the Marquis de Condorcet and Louis de Bonald.

[2] In *The Structure of Scientific Revolutions*, Kuhn credits Koyré with showing him for the first time "what it was like to think scientifically in a period when the canons of scientific thought were very different from those current today" (Kuhn 1996, p. viii). See also Kuhn (1970, pp. 67–69). Foucault identified a version of phenomenology found in the work of Koyré, along with Gaston Bachelard, Jean Cavaillès, and Georges Canguilhem, which laid the groundwork for his own studies on *épistème* and the archaeology of knowledge (Foucault 1998, pp. 465–467; Webb 2013, p. 16). Foucault singles out the importance of Koyré in the following passage: "Knowledge, reason, rationality, the possibility of elaborating a history of rationality [...] I would say here again we come across phenomenology, in someone like Koyré, a historian of science [...] [who] developed a historical analysis of the forms of rationality and knowledge in a phenomenological perspective" (Foucault 1988, p 23).

The importance of Husserl in this constellation of influences should not be overlooked, nor should it be marginalized. In a letter to Herbert Spiegelberg dated 10 December 1953, Koyré writes that he finds aspects of Husserl's work on the *life–world* interesting, yet somewhat troubling:

> The "world" of the primitive is *his* "Lebenswelt", it is perhaps *the* "Lebenswelt". It is real *for him*; but not for us. And not only not for us – it is not a *real* world. Husserl speaks about the relative right of *historicismus* – and certainly the "Greek" world, the "medieval" world are concepts that have good meaning. But what are the connections between these "worlds" and the world of the "Selbstverständlichkeiten"? I do not see that the transcendental phenomenology gives any answer. Or, it answers by postulating the transcendental unity of an inaccessible *Ego* – substitute for God or the *Weltgeist*. (see below)

He continues:

> [In response to your question regarding] how far I am still a phenomenologist – I do not know myself. I have been deeply influenced by Husserl, probably learnt from him – who did not know much about history – the positive approach to it; the interest for the objectivism of Greek and medieval thought, for the intuitive contents of seemingly purely conceptual dialectics; for the historical – and ideal – constitution of systems of ontology. I inherited from him the Platonic realism that he discarded; the antipsychologism and the antirelativism. But, probably, he would say that all that is very far from the meaning of phenomenology as philosophy. And that I have never understood it. Now I assume that he knew better than anyone else what "phenomenology" really meant. (Jorland 1981, p. 28 n. 1; Author's translation)[3]

This letter to Spiegelberg highlights the tension in Koyré's reception of Husserl. On the one hand, Koyré is reluctant to accept Husserl's transcendental *idealism*, which he believes amounts to a solipsistic idealism. This is reminiscent of the criticism of Husserl famously put forward by his Freiburg student Theodor Celms. Instead, Koyré adopts the Platonic realism that Husserl himself rejected as either naïve or ungrounded. On the other hand, Koyré credits Husserl with inspiring his approach to the study of history.

We find similar sentiments concerning Koyré's Husserl-inspired approach to history reported by Aron Gurwitsch. Gurwitsch recalls that Koyré once remarked:

> [...] even though Husserl was not a historian by training, by temperament, or by direction of interest, his analysis provides the key for a profound and radical understanding of Galileo's work. He submits [Galilean] physics to a *critique*, not (once again be it said) a criticism". (Gurwitsch 1974, p. 39)

[3] Koyré's remark here that Husserl did not know much about the history of philosophy should be read with a bit of caution. The common opinion of Husserl is that he did not know much about the history of philosophy, particularly early on in his career, owing to the fact that his formal training in philosophy was rather limited. This is not, however, to say that he had *no* knowledge of some of the major figures in the history of philosophy. If we consult Husserl's early writings, we can get a sense of the philosophical traditions and figures with which he was acquainted from roughly 1876 to 1891. The list includes Aristotle, Euclid, Leibniz, Spinoza, Berkeley, Hobbes, Locke, Hume, Newton, Kant, Schopenhauer, Bolzano, and Mill. Later in his career, Husserl became much more interested in the history of philosophy and the position of his own philosophical views within it but was never a historian of philosophy per se.

I argue that Koyré's work in the history and philosophy of science explores the evolution of intellectual frameworks from a phenomenological and *critical* perspective. Koyré's work can (and perhaps ought to) be read as an extension of Husserl's project, specifically the historical–teleological way into phenomenology that Husserl sketched in his *Crisis* writings from the 1930s.[4] While Koyré's work is historically oriented and takes on the admittedly impossible task of entering into the theoretical frameworks of past thinkers, the end goal here is not to paint a faithful representation of the past per se, but to study the life of the mind and the structures of consciousness. In doing so, perhaps we can unite the "worlds" of individual conscious subjects without having to uncritically posit the existence of the external world. Understood in this way, it is perhaps clear why Koyré "ended up being too much of a philosopher for French historians and too much of a historian for French philosophers" (Chimisso 2008, p. 136).

Unfortunately, the relationship – both personal and intellectual – between Koyré and Husserl has been misconstrued and underappreciated. As a result, the role Koyré played in the spread of Husserlian phenomenology is not fully understood nor is the influence of phenomenology on Koyré's historical–epistemological writings. In "Koyré phénoménologue", Gerard Jorland stresses that, even if we cannot consider Koyré to be a thoroughgoing phenomenologist, it is impossible to understand his work without taking into consideration the deep impact that phenomenology and the members of the Göttingen Circle had on his thinking (Jorland 1994, pp. 105–126). But rather than focusing on the relationship between Husserl and Koyré, Jorland focuses instead on Koyré's other instructor in phenomenology, Adolf Reinach. We find similar Reinach-oriented reconstructions of Koyré's phenomenological upbringing in the work of Paola Zambelli. Both commentators appear to have been led by the analyses of the eminent historian of the phenomenological movement, Karl Schuhmann, who tended to place more weight on Koyré's relationship with Reinach than on his relationship with Husserl.

There are numerous reasons as to why Schuhmann would have chosen to do this. First, as we see from the 1953 letter to Spiegelberg, Koyré did not follow Husserl's transcendental turn, remaining more closely aligned with Reinach and the realist phenomenologists. In addition, Koyré's noted interest in Plato seems to have been sparked by Reinach's lectures from SS 1910. However, the importance of Husserl often gets unduly overshadowed on such accounts. To claim that Reinach took priority over Husserl in influencing Koyré appears inaccurate based on the historical facts and surviving personal reflections on the events from Koyré's time in Hilbert and Husserl's Göttingen. It cannot be denied that a number of members of the Göttingen Circle felt a close personal and philosophical connection to Reinach. These students were even referred to by Theodor Conrad as the *Reinach-Phanomenologen* (Feldes 2013, p. 206), and their positive sentiments towards

[4] Husserl calls the *Crisis* a "teleological-historical reflection upon the origins of our critical scientific and philosophical situation" (Husserl 1970, p. 3), and in the "Foreword to the Continuation of the *Crisis*", Husserl states that he is attempting to give the reader a "teleological-historical way to the conception and the idea of transcendental phenomenology" (Husserl 1976, p. 435).

Reinach were further amplified following his untimely death in the First World War. Yet, even if Reinach made a strong impression on Koyré so too did Husserl. Long after his departure from Göttingen, Husserl continued to play a formative role in Koyré's thinking.

The second reason why Schuhmann places more weight on Koyré's relationship with Reinach than with Husserl may be the circumstances surrounding Koyré's departure from Göttingen. Or perhaps it is more correct to say that Schuhmann's own speculations as to *why* he left Göttingen lend themselves to a reading of Koyré where Husserl is marginalized, namely, that infighting in the Göttingen Circle caused him to leave for Paris. This hypothesis is also taken up in "Alexandre Koyré im 'Mekka der Mathematik': Koyré's Göttinger Dissertationsentwurf", where Zambelli speculates that "internal tensions" at Göttingen may have been the motivation behind the rejection of Koyré's dissertation (Zambelli 1999, p. 220). On this account, Koyré's decision to align himself within the Göttingen Circle as one of the *Reinach–Schuler* played a determining role in Koyré's dismissal. While this sort of "political" reasoning is prima facie plausible, there are good reasons for thinking that it is not quite true.

In this paper, I will begin to map out the history between Koyré and Husserl, from the reasons behind Koyré's arrival at and departure from Göttingen through to his final correspondence and visits with the Master. This involves considerations of both the personal relationship between the student and his teacher and the philosophical impact that Husserl had on Koyré. While Husserl claimed no expertise in the history of ideas, his phenomenological method deeply influenced Koyré's approach to history and his analysis of past metaphysical systems and their conceptual frameworks. In his own works, Koyré was not simply concerned with looking at texts within their historical context but with uncovering the story of the mind that they tell. I argue that Koyré's work follows Husserl in the study of consciousness and the critique of knowledge and should therefore be understood as a contribution to phenomenology.

More specifically, Koyré's project can be understood as a contribution to phenomenology in two ways. First, by looking at the transformation of human thought that took place during the scientific revolution, and the new way of conceptualizing or constituting the world, we can perhaps uncover the essential structures of consciousness and conditions of the possibility of knowledge. Understood in this way, the particular moments in the history of science are often Koyré's subject matter but not his primary concern. His concern is the evolution of scientific thought, the ontological frameworks which we utilize in understanding and interpreting phenomena, and what this tells us about consciousness and its relationship to the world. Second, we can understand Koyré's historical–epistemological writings as an investigation into the "historical a priori", which Husserl describes as "the universal source of all conceivable problems of understanding. The problem of genuine historical explanation comes together, in the case of the sciences, with 'epistemological' grounding or clarification" (Husserl 1970, pp. 372–373). An inquiry into the historical a priori would help to solve the phenomenological problem of the connections between, say, the Greek world and the modern world; the connection between such worlds as historical, cultural, and scientific entities; their connection to the world of the *Selbstverständlichkeiten*; etc. We might then compare his writings with Husserl's essay on "Der Ursprung der Geometrie als intentional–historisches Problem" (Husserl 1939), which deals with similar problems.

At bottom, I suggest that Koyré's is a phenomenological project of sorts. In doing so, I believe my reading of Koyré is, at least in large part, consistent with that of Jorland and that such a reading is only made more convincing when we clear up his relationship with Husserl and the phenomenological movement. Husserl is more important for the contextualization of Koyré than has traditionally been appreciated. For Koyré, the study of history can both act as a way into phenomenology and can contribute to phenomenology understood as a critique of knowledge, though perhaps only to phenomenological psychology rather than transcendental phenomenology understood as transcendental *idealism*. Like any nondogmatic follower, Koyré departs from Husserl in some ways, and where he does we might well consider the influence of Reinach, Scheler, Lévy-Bruhl, Brunschvicg, and others. But in other ways, we can discern Husserl's influence as formative for at least *part* of the basis of Koyré's critical historical–epistemological project.[5]

14.2 The Young Koyré in Hilbert and Husserl's Göttingen

Alexander Vladimirovich Koyransky (Александр Владимирович Койранский) was born in 1892 in Taganrog, Russia, into a prosperous Jewish family. As a youth he aligned himself with the Socialist Revolutionaries and was sympathetic to the 1905 revolution, perhaps as a response to the anti-Jewish pogroms taking place around him. In 1907, at the age of 15, Koyré was arrested for distributing revolutionary literature. The following year, he was imprisoned for conspiring to assassinate a government official. Koyré's father pleaded for his release into exile in order to study abroad, and Koyré was discharged near the end of August 1908 (Zambelli 2007, pp. 116–121). While in prison, Koyré is said to have acquired and read a copy of Husserl's *Logical Investigations* and upon his release left Russia in order to study philosophy and mathematics at Husserl's side (Jorland 1994, p. 107; Zambelli 1999, p. 208, p. 210).[6]

[5] In this sense, Zambelli is correct when she writes, "On metaphysical problems, Koyré had reconciled the perspectives of Lévy-Bruhl and Gilson with those of Husserl, privileging the latter perhaps out of loyalty to his phenomenological training" (Zambelli 2009, p. 3).

[6] We should note that Koyré must have read the *Logical Investigations* in German if the dates mentioned above are correct, since Semyon Frank's introduction to the Russian translation is dated October 1909, more than a year after Koyré's release into exile. Husserl's works were in circulation in Russia prior to the 1909 translation of the *Logical Investigations*. Nikolai Lossky, David Viktorov, and Georgii Chelpanov had all published on Husserl's philosophy prior to Koyré's imprisonment, though there is no evidence to suggest that Koyré would have been familiar with these works.

The exact date on which Koyré began his studies at the University of Göttingen is not known, though he arrived either in WS 1908/1909 or in the spring of 1909.[7] In his unpublished notebooks, Spiegelberg records Koyré as stating that he arrived after Husserl had delivered his lectures on the *Idea of Phenomenology* in the spring of 1907 but before Adolf Reinach took up his position as *Privatdozent* in the summer of 1909.[8] Koyré also reported to Spiegelberg that he had spent some time in Paris in 1908 before going to Göttingen.[9] It could also be the case that Koyré did not read Husserl while he was still in Russia at all and that he only learned of Husserl while in Paris in 1908. Though Husserl's phenomenology was not yet a widespread topic of discussion in France in 1908 (Dupont 2014, pp. 104–107), French mathematicians and logicians were aware of Husserl's work as early as the mid-1890s. Louis Couturat mentions Husserl warmly in *De l'infini mathématique* (Couturat 1896), whereas Georges-Henri Luquet, a student of Henri Bergson, published a sceptical review of Husserl's *Logical Investigations* in 1901.[10] However, even if Koyré's memory of travelling to Paris before Göttingen is correct, there is no evidence to suggest that this is when he first learned of Husserl or that he was familiar with the publications mentioned above. Sojourns in Paris aside, if Koyré did manage to arrive in time for the winter semester of 1908/1909, he does not appear to have attended Husserl's lectures that first term, since his name is missing from the class register (Schuhmann 1977, p. 122). Koyré's first course with Husserl would have been *Einführung in die Phänomenologie der Erkenntnis* the following spring. In these lectures Husserl proclaimed:

> Phenomenology and the phenomenological critique of reason is a comprehensive and laborious science, as comprehensive and laborious as any genuine science. And any genuine science grows from the bottom up only gradually. No one person creates a completed

[7] "Koyré gehörte von WS 1908–1909 bis 1912 und noch einmal im SS 1913 der Schule Husserls an [...] Leider ist die von Schuhrnann aufgefundene und sorgfältig analysierte Dokumentation von Koyrés Göttinger Zeit sehr spaerlich" (Zambelli 1999, pp. 210–211).

[8] "On the Göttingen period. Koyré [was a] late member of the Husserl circle; arrives after *Die Idee der Phänomenologie*. The important 'Schüler' (Reinach, etc) [were] not there at the time, at most [Wilhelm] Schapp" (Spiegelberg 1953, p. 52). According to the memorial written by Jean Hering, Koyré arrived in Göttingen sometime in 1910 (Hering 1965, p. 453). Similarly, in Husserl's *Briefwechsel*, Schuhmann states that Koyré arrived in WS 1909/1910 (Husserl 1994a, p. 358). But these reports appear to be mistaken.

[9] Koyré indicates in a letter to Herbert Spiegelberg, dated 10 August 1956, "I went to France as far back as 1908, then to Göttingen, then back to Paris, once more to Göttingen, and joined the French army during the first world war (1914)" (Schuhmann 1987, p. 162). A copy of this letter is in the Koyré archives in Paris.

[10] Schuhmann mentions Couturat in this context as well. In an endnote he writes: "Louis Couturat louait la *Philosophie der Arithmetik* (1891), le premier ouvrage de Husserl, pour « la finesse de ses analyses, la justesse et la subtilité de ses raisonnements, la richesse de son érudition » (De l'infini mathématique. Paris, Alcan, 1896, p. 331, n. 1). Husserl et Couturat furent en correspondance entre 1899 et 1904 (voir la section R II des Archives Husserl à Louvain). Il se peut donc que Couturat ait été le premier à attirer l'attention de Koyré sur Husserl. Couturat fut d'ailleurs aussi un des grands admirateurs de Hilbert" (Schuhmann 1987, p. 163). See his reviews of both Husserl and Scheler (Luquet 1901, pp. 414–419). While Luquet's review of Husserl is rather disparaging, his review of Scheler is positive.

> theory. Every new researcher adds only a new part of the structure. Accordingly, a science must already be fairly advanced before its architectonic can even be considered as a whole, can be represented as a unified, coherent doctrine. When Galileo put forward the idea of a pure physical science of nature…he had only humble beginnings. And it took the collected work of centuries to bring to full development this new science in a systematic series of coherent theories. Thus today what we have of a phenomenology are only beginnings and fragments. (Husserl 2005, pp. 100–101)

It was this radically new science, Husserlian phenomenology, which Koyré became a young researcher in.

Koyré had arrived in Göttingen at an incredibly important time. At the turn of the twentieth century, the University of Göttingen was renowned as the epicentre of mathematical research in Europe. It was also the recent birthplace of the phenomenological movement. The faculty in the department of mathematics played an active role in Husserl's appointment in Göttingen in 1901, just after the publication of his *Logical Investigations* (Husserl 1900–1901). While Husserl took up the position of Extraordinarius Professor in the department of philosophy, it seems that the offer came about due to David Hilbert, the chair of mathematics, and Felix Klein. Husserl was an accomplished scholar in the philosophy of mathematics and logic, but despite the efforts of his colleague Georg Cantor, he had been unable to secure a permanent position at the University of Halle. It seems that Husserl's former supervisor Carl Stumpf asked his friend Klein if there might be a spot for Husserl in Göttingen instead. Klein then arranged for a special position to be created for Husserl by Fritz Althoff, the famed Prussian Minister of Culture and Higher Education who was instrumental in aiding Hilbert to recruit the best researchers in mathematics. With Hilbert's endorsement, Husserl was invited to Göttingen. Upon his arrival, Husserl quickly aligned himself with Hilbert and became a member of the *Mathematische Gesellschaft*, to which he presented his now famous Doppelvortrag, "Der Durchgang durch das Unmögliche und die Vollständigkeit eines Axiomensystems", late in 1901.[11]

Despite the resistance Husserl met from some faculty members at Göttingen, particularly Georg Elias Müller and Julius Baumann (Husserl 2008, pp. xxvii–xxix), he began attracting participants from the *Mathematische Gesellschaft* to his lectures in philosophy. Shortly thereafter, and owing to the success of his *Logical Investigations*, Husserl also began to attract a number of students from Munich, who had originally been working with the psychologist Theodor Lipps. After Johannes Daubert rode his bicycle from Braunschweig to Göttingen in 1902 (Schuhmann 1977, p. 72), other members of the *Psychologische Verein* in Munich began making the trek to study with Husserl. These students invaded Husserl's lectures, eager to learn not about logic but about Husserl's phenomenology. By 1907, the Munich phenomenologist Theodor Conrad had established the *Göttinger Philosophische Gesellschaft*, known as the Göttingen Circle, along with a handful of Husserl's early followers (Schuhmann 1977, p. 103).

---

[11] Majer 1997, pp. 37–56; Schuhmann and Schuhmann 2001, pp. 87–123.

In his memoire on Koyré, Charles Gillispie writes that Husserl was the "idol of Koyré's schooldays" in Göttingen and notes that Koyré was close with Husserl and his family, especially Husserl's wife, Malvine, who had "mothered him a bit" (Gillispie 2006, p. 286). Owing to his early adoration for Husserl, Koyré became an active member of the recently formed Göttingen Circle. He began attending the lectures of Adolf Reinach, who acted as a mentor to the young students interested in phenomenology, and Max Scheler, who briefly lectured there from 1910 to 1911 after losing his teaching position in Munich. Scheler introduced Koyré and the other members of the Göttingen Circle to the work of Bergson.[12] The young Russian student was more than just a casual participant in the Göttingen Circle – he seems to have been a highly regarded member of the group. In a letter to Conrad from January 1911, Reinach writes:

> The philosophical society is flourishing, with about 15 members who are becoming quite acceptable under Gogo's [Dietrich von Hildebrand's] obviously skilled leadership [...] Koyré is becoming increasingly wiser, and Gogo is now also working properly. He, Scheler, Koyré, Karras, Kananow and I form a closer philosophical society where everyone presents their stuff.[13]

Koyré developed close and enduring friendships with a number of the members of the Göttingen Circle, including Theodor Conrad, Hedwig Conrad-Martius, Jean Hering, Hans Lipps, Alfred von Sybel, and Edith Stein.[14] This group of friends would later form the Bergzabern Circle of phenomenologists which emerged during the First World War and lasted until the late 1920s. Thus, phenomenology played a formative role in Koyré's academic career.

Taking full advantage of the close relationship between the *Philosophische Gesellschaft* and the *Mathematische Gesellschaft*, Koyré worked with Hilbert, Klein, and Husserl's former student Ernst Zermelo during his years in Göttingen. He eventually chose Hilbert to co-supervise his dissertation along with Husserl. Koyré's research focused on logical paradoxes, and in September 1912, Koyré published his first paper in *Revue de métaphysique et de morale* entitled "Sur les nombres de M. Russell" (Koyré 1912). In this paper, Koyré attacks the "logical

---

[12] Scheler had been deeply influenced by Bergson and had played a role in getting German translations of Bergson's work published. By 1911, both *Materie und Gedächtnis* and *Zeit und Freiheit* were available. Husserl is reported to have announced to the members of the Göttingen Circle in 1911: "We are the true Bergsonians!" (Hering 1939, p. 368 n. 1). Herbert Spiegelberg suggests that it was Koyré who introduced the group to the work of Bergson, but this is false. In fact, Spiegelberg undermines this point himself (cf. Spiegelberg 1969, p. 225, p. 236).

[13] "Die philosophische Gesellschaft floriert ziemlich, bei etwa 15 Mitgliedern und unter Gogos [Dietrich von Hildebrand] ganz annehmbarer und sichtlich geschickter werdenden Leitung [...] Koyré wird immer klüger, Gogo arbeitet jetzt auch ordentlich. Er, Scheler, Koyré, Karras, und Kannanow und ich bilden eine engere philosophische Gesellschaft, in der jeder seine Sachen vorträgt" (Reinach 1902–1907, Letter 87).

[14] Stein transferred to Göttingen to study with Husserl, at the behest of her cousin Richard Courant, in 1913, and met Koyré for the first time in the summer semester of that year. Courant had arrived in Göttingen around 1907 and worked with Husserl and Hilbert and attended the meetings of the Göttingen Circle when Koyré was a member.

definition of number" given by Gottlob Frege and Bertrand Russell (Russell 1992, pp. 58–59, pp. 450–454). With all this in mind, it is somewhat surprising that when Koyré presented a draft of his dissertation to Husserl sometime at the beginning of 1912, it was rejected. By fall, Koyré had left Göttingen and resumed his studies in France. The reasons behind Koyré's departure from Göttingen and his subsequent move to Paris are largely a matter of historical speculation.

14.3 *Insolubilia*: Koyré's Draft–Dissertation and His Departure from Göttingen

In WS 1911/1912, Koyré submitted drafts of two papers to Husserl which would serve as the basis for his dissertation: "Insolubilia" and "The Antinomies of Set Theory".[15] In this draft–dissertation, Koyré argues that the problems of set theory which had caused a crisis in the foundations of mathematics are simply new iterations of long-standing logical problems, namely, the *insolubilia* of the Scholastics, whose roots can be traced back to Antisthenes, the student of Gorgias and Socrates (Koyré 1999, p. 324). Historically, these logical paradoxes had been treated as nothing more than jokes aimed at mocking absurd philosophical positions or puzzles for the amusement of bored logicians. However, modern mathematics, particularly set theory, had revealed that these paradoxes are no laughing matter. Koyré's intent appears to have been to trace the development of these paradoxes from Scholastic thought to Frege and Russell on one hand and Cantor and Zermelo on the other. The title of the dissertation, *Insolubilia*, may be a direct reference to the work of the Scholastic logician Thomas Bradwardine. In the draft, Koyré also discusses ancient paradoxes, such as the Liar paradox and the counter-dilemma of Euathlus. A portion of this work dealing with Cantor's set theory and Russell's paradoxes had been presented to the Göttingen Circle (Zambelli 1999, p. 208) and eventually evolved into Koyré's "Bemerkungen zu den Zenonischen Paradoxen" (Koyré 1922a).

[15] Koyré had at least intended to submit these papers to Husserl early in the semester, as he suggests in a letter written to Hedwig Martius: "Was mich anbetrifft so hätte ich eigentlich für die Feriencuste? Heidelberg vorgezogen – aber wenn es schon beschlossen ist so muss ich mich fügen. Höchst wahrscheinlich wird es mir möglich sein Anfang Oktober etwa weg zu reisen folglich am 10. oder so in Strassburg zu sein. Aber das muss ich gestehen – ich bereite mich absolut nicht vor; ich lebe überhaupt in einer absoluten Arbeitseinstellung - das tiefe stille blaue Meer hat mich vollständig gefangen genommen. Ich segle, rudere atme die frische salzige Luft und geniesse das Heimsein. Von den geplanten Arbeiten wird sicher nichts zustande kommen. Sie waren ja doch selbst an der See – Sie wissen was das bedeutet" (Koyré 1911–1922, Letter to Hedwig Martius, date unspecified [sometime before 19 September 1911]). He also notes in this letter that he had spent the summer in Crimea and had been prohibited from corresponding (likely because he was in Russia illegally as an exile) and jokes that he had done some important work on the phenomenology of laziness: "Vor allem war ich eine Zeit lang gar nicht [in Göttingen], sondern in der Krim, wo ich mich der Phänomenologie der Faulheit widmete und sehr wichtige Vorstudien gemacht habe" (Ibid).

The draft–dissertation was inspired, at least in part, by Husserl's *Logical Investigations* and his lectures on logic from WS 1910/1911, both of which Koyré cites in the extant text dated 12 January 1912. Koyré also makes a passing reference to Reinach's work *Zur Theorie des negativen Urteils* (Reinach 1911) and appears to draw upon Koyré's notes from Reinach's lecture course on Plato from SS 1910. In March of 1912, Husserl informed Koyré that he could not accept the draft of *Insolubilia* as the basis for a thesis – at least not in its present form. However, his reasons for rejecting the work are unclear.[16] Schuhmann writes that Husserl rejected Koyré's thesis for no good reason, using a letter that Reinach sent to Theodor Conrad in the spring of 1912 as evidence to this effect (Schuhmann 1987, pp. 154–155). Reinach writes:

> On the matter of Koyré [...] nothing can be done. It seems very likely that, at bottom, there were considerations of a personal nature that were decisive for Husserl [...] Husserl believes that Koyré is arrogant and a bit immature [...] He seems to have decided, as a consequence, 'for the good of Koyré,' to not yet award him his doctorate. For Koyré, I am sorry about this entire unpleasant situation. Why does he not leave Göttingen?[17]

Koyré did leave Göttingen but not immediately. He participated in Husserl's seminar on Hermann Lotze, which began on 27 April 1912, before leaving for Paris (Schuhmann 1977, p. 169). What is more, Koyré returned to study in Göttingen for the summer semester the following year.[18]

Perhaps it was Reinach or the other phenomenologists closely aligned with him that advised Koyré to leave Göttingen and complete his dissertation somewhere else. This is a reasonable speculation based on the letter Reinach sent to Conrad cited above. Schuhmann makes the stronger suggestion that Koyré's departure from

---

[16] Husserl's notes on Koyré's draft–dissertation are kept in the Husserl archives in Leuven, under the signature A I 35/5–17. In these notes, Husserl refers to a conversation that he had with Koyré about the work: "Nach dem Gespräch mit Koyré scheint es mir aber doch, dass auch ein weiterer brauchbarer Begriff von Menge möglich und für den Mathematiker zu Anfang nötig ist. Es scheint, dass Koyré's Begriff darauf hinauskommt, dass ein Klassenbegriff als Gesamtheit der *A* überall berechtigt ist, wo das Partikular-urteil, es gibt ein *A* (wo ein ein positiver Begriff ist), wahr ist. Etwas, das nicht *A* ist, das gibt keinen berechtigten Begriff, da mit der Negation 'paradoxe Mengen' hineinkommen, die auf sich selbst bezogen wären. Doch muss das noch überlegt werden" (Husserl, A I 35/16b). There are also annotations on the extant copy of Koyré's draft–dissertation kept in the Koyré archives in Paris which may belong to Husserl.

[17] "Über K's Angelegenheit habe ich mich nicht weiter geäussert, weil ich in der Hauptsache Ihrer Keinung bin, und weil ich auch nicht glaube, dass man da irgend etwas machen kann. Dass in der Hauptsache persönl. Erwägungen für H. massgebend waren, scheint mir sehr wahrscheinlich zu sein. Natürlich persönl. Erw. ganz objektiver Art. H. hält K. wirklich für hochmütig und etwas unreif - was man auf Grund seiner etwas primitiven Psychologie ganz wohl verstehen kann. Er wird daraufhin 'zum Besten K's' beschlossen haben, ihn noch nicht promov. zu lassen. Die ganze unangenehme Sache tut mir für K. sehr leid. Warum geht er nicht aus G. fort?" (Reinach 1902–1907, Letter 91).

[18] "Die Unterlagen, die ich bezüglich Koyrés Pariser Laufbahn einsehen konnte, berechtigen jedoch zu der Behauptung, dass er erst 1912 als Student nach Paris kam und sich noch einmal von dort entfernte, um für das Sommersemester 1913 nach Göttingen zurückzukehren. Dies schliesst allerdings nicht die Möglichkeit weiterer Paris-Besuche aus, da Koyré von seiner reichen Familie unterstützt wurde, und dort sein Vetter Georges Lebedinsky lebte" (Zambelli 1999, p. 211).

Göttingen was primarily due to conflicts between Husserl and the other members of the Göttingen Circle after Husserl's turn to transcendental idealism. This is the hypothesis reiterated by Zambelli – that "internal tensions" in the Göttingen Circle led Husserl to reject Koyré's thesis and caused him to move to Paris – and which has become the standard view of Koyré's early relationship with Husserl. Reinach and the other realist phenomenologists felt that Husserl's idealism undermined the version of phenomenology presented in the *Logical Investigations* which they had been drawn to and caused his philosophy to fall prey to psychologism and relativism. While it is true that members of the Göttingen Circle criticized Husserl's phenomenology on these and other grounds, I find the view that Husserl rejected Koyré's draft–dissertation based on such disputes to be both dubious and unfair to Husserl. What is worse, this view has stifled discussions about the intellectual relationship between Koyré and Husserl.

Zambelli presents the standard view, which she attributes to Schuhmann, roughly as follows. Along with the other early members of the phenomenological movement, Koyré was a loyal follower of Husserl's project of phenomenology as a descriptive psychology and a science of the essences of thinking and knowing as presented in the first edition of the *Logical Investigations*. However, Koyré refused to accept the version of phenomenology which appeared for the first time in printed form in *Ideas I* (Husserl 1913), and this was the decisive factor in Husserl's choice to reject Koyré's draft–dissertation. Husserl's decision was based on personal reasons, rather than objective considerations of academic merit. Reinach took a hard position against Husserl's "pure" or "transcendental" phenomenology presented in *Ideas I*, and Koyré returned to Göttingen in the summer of 1913 to attend Reinach's philosophical colloquium for advanced students in which he voiced these concerns, despite the "trauma" he had suffered in Göttingen at the hands of Husserl the previous year, namely, the rejection of his draft–dissertation (Zambelli 1999, pp. 219–220). However, Zambelli is cautious not to fully endorse Schuhmann's reconstruction. She writes "Was Koyré, at twenty-years of age, the victim of internal tensions within the phenomenological school? This is only a hypothesis, discussions on which I leave to those more competent than myself".[19] Not being an expert in early phenomenology or the history of the phenomenological movement, Zambelli defers to the authority of Schuhmann and leaves it to future scholars to determine if this was in fact the case. However, she notes that Koyré never ceased to follow the theoretical developments of the phenomenological school, including the work of Heidegger.

[19] "Sollte der zwanzigjährige Koyré etwa ein Opfer der internen Spannungen innerhalb der phänomenologischen Schule geworden sein? Es handelt sich lediglich um eine Hypothese, deren Erörterung ich der Kompetenz anderer überlasse; hervorheben möchte ich jedoch erstens dass er nie aufhörte, die theoretischen Entwicklungen in dieser Schule mit grosser Neugier weiterzuverfolgen (einschliesslich des Falls Heidegger); zweitens, dass Koyré nach einen Jahrzehnt interessanter, bahnbrechender, jedoch vollkommen mathematikferner Untersuchungen über Häretiker und Mystiker zu sich selbst fand, als er – nach der grossen Versöhnung mit Husserl aus Anlass der Descartes-Gedenkfeier – dazu überging, die physikalisch-mathematischen Werke von Kopernikus, Galileo, Descartes und Newton zu studieren und diese in Beziehung zu metaphysischen und mystischen Voraussetzungen zu setzen" (Zambelli 1999, p. 220).

Zambelli also claims that Koyré's "reconciliation" with Husserl on the occasion of the Paris Lectures (Husserl 1998) led to a shift in his thinking about the relationship between science and metaphysics.

Building on Zambelli's cautionary notes, we can further challenge the standard view about the relationship between Husserl and Koyré and the story of the latter's departure from Göttingen. We should start with Husserl's reaction to the draft–dissertation. Based on Husserl's own notes on *Insolubilia*, there is nothing to indicate that Husserl rejected the work for personal or political reasons, that he was unfairly harsh, or that he refused to continue working with Koyré. From the evidence at hand, it seems that Husserl was of the opinion that Koyré needed more time in order to be granted his doctorate and that this judgement was not motivated by any ill will. It seems far more likely that Husserl thought that Koyré, at only 20 years old, needed more time to develop academically and that his scholarship was not yet at the level required for completing a dissertation. Schuhmann's reconstruction relies heavily on the testimony of Reinach, who was much more critical and resistant of Husserl than Koyré (Mulligan 1987, pp. 250–253). Husserl was indeed frustrated and disappointed by the fact that many of his students and colleagues did not follow him in his transcendental turn and instead remained bogged down in ontologism and realism (Spiegelberg 1969, p. 175) and expressed some disapproval towards the student colloquiums conducted by Reinach. But there is no good reason to believe that Husserl would have attempted to hold back Koyré's career as a result of disagreements over the nature and aims of phenomenology or departmental politics.

If we look at the content of his draft–dissertation, it makes perfect sense that Koyré would leave Göttingen in order to complete it. While both Husserl and Hilbert were accomplished logicians, their expertise was not in the history of logic per se. A large share of Koyré's proposed project dealt with connecting Greek and medieval logic with contemporary problems in the foundations of mathematics. Perhaps Husserl felt that while he and Hilbert could easily comment on those aspects of the dissertation dealing with contemporary issues in the foundations of mathematics and the philosophy of logic, they were unable to fairly assess the historical components. Husserl had no expertise in the history of philosophy, particularly ancient and medieval philosophy. Reinach had no background in this area of the history of philosophy either, and even if he did, he was unable to direct theses in his capacity as *Privatdozent*. It could well be that out of support for the project Koyré was advised by Husserl or Scheler to study in France. In France, Koyré could study with experts in Scholastic philosophy who could help him to ground his hypotheses regarding the nature of logical paradoxes and perhaps to trace the Scholastic roots of phenomenology. This does appear to have been part of the story behind Koyré's decision to leave Göttingen.[20] Moreover, if Koyré was interested in discussing the work

[20] "In 1913 [Koyré] went to Paris to undertake studies in the old Scholastic philosophy, in which he, like Brentano and others, had discerned some foreshadowing of phenomenological themes" (Hering 1965, p. 453). In a letter to Roman Ingarden written on 28 February 1921, Husserl mentions that young scholars in France are required to learn Scholastic philosophy. Husserl also suggests that Koyré's "Remarks on Zeno's Paradoxes" which are to appear in the next edition of

of Bergson in his dissertation, then it would make more sense for Koyré to move to Paris where he could work with Bergson or his followers than to go anywhere else.[21]

Given Koyré's involvement in the Göttingen Circle and his relationship with numerous outstanding members of the mathematics department, it is odd that he would abandon Göttingen simply because his progress was temporarily put on hold by Husserl's rejection. And if internal tensions in the Göttingen Circle and his decision to align himself with the realist phenomenology of Reinach caused Koyré to leave, why would he not then go to Munich to study with Alexander Pfänder – who was close to Reinach and was both a logician and a realist phenomenologist – rather than Paris? Moreover, if Koyré did leave to pursue the same dissertation project elsewhere, unmoved by Husserl's criticisms, which Reinach felt were unfair, then he did not fare any better, since it was not until Husserl published his "Bemerkungen zu den Zenonischen Paradoxen" in the 1922 edition of the *Jahrbuch für Philosophie und phänomenologische Forschung* that any of those ideas appeared in print. Rather than speculating that internal tensions in the Göttingen Circle caused Husserl to reject the thesis and Koyré to leave Göttingen, we should more plausibly assume that Koyré was advised, for the sake of his ideas, to move to Paris and that having family in Paris facilitated this move.

We might next note that Husserl's transcendental turn had taken place well before the publication of *Ideas I*. This change in Husserl's philosophy dates back to at least 1905 and was first publically announced in his 1907 lectures on the *Idea of Phenomenology*, though it may have been known by those closest to Husserl before this. We noted earlier that Koyré arrived in Göttingen just after these famous 1907 lectures, and we know that Husserl's lecture courses and seminars thereafter reflected this shift in thinking. Thus if Husserl's influence on Koyré extended beyond what he wrote in the *Logical Investigations*, then it is fair to say that Koyré was influenced at least in part by Husserl's transcendental phenomenology, since this was the form of phenomenology he was trained in by Husserl. If we are to begin comparing Koyré's work to that of Husserl, we should start by comparing it to these 1907 lectures and to the content of Husserl's lecture courses from that period.[22] At this point in his career, Husserl begins describing phenomenology as a *critique of knowledge* and how it can aid us in understanding the results of the natural sciences (Husserl 1994b, p. 19). Husserl writes that if we

---

the *Jahrbuch für Philosophie und phänomenologische Forschung* should be dedicated to Reinach, since Reinach had suggested that he write on the topic: "[Koyré] lebt jetzt dauernd in Paris und ist dort in der Habilitation begriffen. Eine kleine Arbeit über die Zenanisehen Argumente (dem Andenken Reinachs gewidmet, von dem er die Anregung dazu empfing) erscheint auch im nächsten Jahrbuch. Er arbeitet in Paris mit Hering zusammen, hauptsächlich historisch (über Scholastik), weil dies zunächst in Paris von jungen Philosophen verlangt wird" (Husserl 1968, p. 18).

[21] We should note, however, that Bergson is only mentioned once and in passing in the drafts Koyré submitted to Husserl.

[22] A similar project should be undertaken with respect to Reinach's influence on Koyré, comparing Margarete Ortmann and Winthrop Bell's notes from Reinach's lectures, along with Reinach's essay "Concerning Phenomenology" (Reinach 1969) and Koyré's unpublished notes from Reinach's Plato course, with Koyré's own writings.

> [...] disregard the metaphysical purposes of the critique of knowledge and attend solely to its task of clarifying the essence of knowledge and known objectivity, then it is a phenomenology of knowledge and known objectivity, which forms the first and fundamental part of phenomenology in general. (Husserl 1994b, p. 19)

He defines phenomenology as a science of phenomena understood as *cogitations* and above all as the specifically philosophical method and attitude of thought. Most importantly, in these lectures Husserl introduces the phenomenological reduction as an "epistemological reduction" (Husserl 1994b, p. 30). While Koyré did not attend these lectures specifically, this was certainly the version of phenomenology that was presented to him in Husserl's seminars and the one which shaped his thought concerning historical–epistemological analyses and the relationship between metaphysical–conceptual frameworks and theory-laden interpretations of the "objective" world in the natural sciences (Koyré 1955).

The source of much of our understanding of the history between Koyré and Husserl is the work of Schuhmann and Spiegelberg. The further elaborations on this relationship conducted by Jorland and Zambelli have relied heavily on these. Unfortunately, I find that the supposed negative relationships Husserl had with his students, Koyré included, have been exaggerated in these works. They also tend to focus on the importance of the realist phenomenology which had its roots in Munich while downplaying Husserl's transcendental phenomenology. It is worth noting that Spiegelberg wrote his doctoral dissertation in Munich under the direction of Pfänder, and Schuhmann had a well-known bias against Husserl's transcendental phenomenology, reportedly stating that Husserl never wrote anything of philosophical value after the *Logical Investigations*. Schuhmann also stressed the critical comments, both philosophical and personal, made by Husserl's former students in their private correspondence, creating the picture of a hostile environment in Göttingen and of Husserl as a sort of villain. Aside from the extreme cases of Leonard Nelson and Theodor Lessing,[23] Husserl seems to have had good relationships with his students, even if he was not much of a supervisor, as Edith Stein held (Ingarden 1962, pp. 155–175). The moanings of students and colleagues in personal correspondence needs to be taken with a grain of salt. Schuhmann and Spiegelberg may not have adequately kept their biases out of their historical reconstructions of the phenomenological movement.

To suggest that Reinach had more of an influence over Koyré's intellectual development than Husserl would be an exaggeration. Each of Husserl, Reinach, and Scheler played a role in shaping Koyré's thought, and by Koyré's own testimony, Husserl had the greatest effect on him. We should trust Gillispie's assertion that Husserl was the idol of Koyré's schooldays. While Reinach and Conrad were responsible for many of the incredibly important social aspects of the phenomenological movement, these should not overshadow the importance of Husserl. We might say that without Husserl there would have been no phenomenology, and without Reinach, there would have been no movement. We might also be thankful that

[23] The case of Nelson led to somewhat of a falling out between Husserl and Hilbert, dubbed "The Nelson Affair". For information on the dispute between Husserl and Lessing, see Baron 1983.

Husserl, for the good of Koyré, rejected the draft of *Insolubilia*, for the trajectory of Koyré's career, and his contributions to the history and philosophy of science would have been far different – perhaps non-existent – had he not done so.

Finally, we should note that from the documents that are available, it seems that Koyré maintained a friendly relationship with Husserl after leaving Göttingen. This is not what we would expect from a former student who had been callously tossed aside by his teacher, as Schuhmann would have us believe. In a letter dating from 21 May 1922, Koyré laments his inability to attend Husserl's London Lectures, "Phänomenologische Methode und phänomenologische Philosophie". Koyré even goes as far as to state that he suspected that Husserl's lectures would be "infinitely more important" than the lectures given by Albert Einstein in Paris earlier that year (Husserl 1994a, p. 357).[24] We also know that in the summer of 1922, Koyré spent 3 weeks with Husserl in Freiburg (Husserl 1968, p. 24). There is also the interesting series of interactions between Koyré and Husserl starting with Husserl's journey to France for his famous Paris Lectures through to his authoring of *The Crisis of European Sciences and Transcendental Phenomenology*. Having cast doubt on the standard view that Koyré's decision to leave Germany was the result of conflicts between Husserl and an opposing faction within the Göttingen Circle, we will now move on to discussing the relationship that Koyré had with both Husserl and the other members of the early phenomenological movement after his departure from Göttingen.

14.4 From Paris to Bergzabern and Back Again

By WS 1912/1913, Koyré had settled in Paris and enrolled at the *École Pratique des Hautes Études* (EPHE), graduating with a degree in philosophy in 1913. Bergson had been teaching at the nearby Collège de France, but was on leave from 1912 to 1914, and retired in 1920 (Bergson 2002, p. x).[25] Though it is reported that Koyré met Bergson, it is unclear when. Nevertheless, 1912 was probably the peak of Bergsonian fever in Paris, and Koyré could have studied under the tutelage of one of the many followers of Bergson located there. After spending SS 1913 back in Göttingen, Koyré returned home to Odessa, as evidenced by letters he sent to Scheler and Hedwig Conrad-Martius on October 5th and December 3rd, respectively. Sometime in 1914 Koyré returned again to Paris, and after the outbreak of the Great War, he enlisted in the French Foreign Legion, fighting primarily for the

[24] This is the earliest remaining piece of correspondence that we have between Husserl and Koyré following the latter's move to Paris. However, it is clear that Koyré is responding to an invitation sent to him by Husserl to attend the London Lectures, so there are certainly letters that are missing.

[25] It is also unfortunate that Henri Poincaré, whom Koyré had an interest in at this time and who had been teaching at the Sorbonne, passed away in July of 1912, and thus the two were unable to work together.

French near the German border, and later in Russia, serving as a part of a volunteer regiment after the October Revolution (Gillispie 1973, p. 483).

Zambelli magnificently reconstructed Koyré's involvement in the First World War. In her paper "Segreti di Gioventù. Koyré da SR a S.R: Da Mikhailovsky a Rakovsky?" (Zambelli 2007), she argues that Koyré served from 1917 to 1919 as a clandestine undercover agent for the French, acting as an informant from within the Tsarist army in Odessa. He was eventually captured by French forces in 1919 and imprisoned in Istanbul for 6 months, returning to France in 1920 (Zambelli 2007, pp. 131–138).[26] During the war, Koyré was also able to make some contact with his former Göttingen classmates. In a letter to Conrad-Martius dated 4 April 1918, he writes:

> I hope that Hans [Theodor Conrad], and all of our friends were spared during the war. I cannot write much. Most heartfelt greetings to Hans, Reinach, Husserl, and all the others.[27]

It is clear from this note that Koyré was unaware that Reinach had already fallen on the battlefield, and it also suggests that he harboured no sore feelings towards Husserl.

After the war, Koyré became a naturalized French citizen and began studying at the Sorbonne under Victor Delbos,[28] André Lalande,[29] Léon Brunschvicg, and François Picavet. He began writing on Scholastic philosophy and the history of science and undertook research on Descartes and Anselm under Picavet. According to Gillispie, although he "did not become as familiar with any of his teachers in Paris as he had with Husserl and his family", Koyré nevertheless "felt at ease in the cooler climate of French civilization" (Gillispie 2006, p. 286). What historians have failed to note is that even after his return to Paris Koyré continued to travel to Germany to visit his former Göttingen classmates, participating in the newly formed

---

[26] Koyré appears to have suggested that this was the case to Roman Jakobson. See Zambelli 2007, p. 150.

[27] "Ich hoffe dass Hans gesund ist, und dass alle unsere Freunde vom Kriege verschont blieben. Ich kann Ihnen leider nicht viel schreiben. Herzlichsten Gruss an Hans, an Reinach an Husserl und alle anderen" (Koyré 1911–1922, Letter to Conrad-Martius, 4 April 1918). After the war, Koyré wrote to Scheler from Milan before settling again in Paris. In his response on 17 July 1920, Scheler indicates that Koyré had been in contact with Conrad-Martius sometime in 1914: "Seit mir Frau Conrad von einem Brief erzählte, den Sie 1914 schrieben, habe ich Nichts mehr von Ihnen oder über Sie gehört". It is likely that the letter Scheler has in mind is one which Koyré sent to Conrad-Martius on 3 December 1913. In this letter, Koyré apologizes for not having written to the Conrad in so long and that he hopes to be able to leave Russia and visit them very soon: "Schon mehr als zwei Monate lebe ich wie auf dem Bivanac. Jeden Tag hoffe ich endlich abreisen zu können, jeden Abend hoffe ich die letzten Hindernisse beseitigt zu haben und jeden Morgen entdecke ich neue und wieder neue. Russland ist das schönste Land der Welt, dies steht fest, aber es ist oft recht unangenehm ein Russe zu sein und in diesem schönen Lande zu wohnen. Aber das unter strengster Discretion. Nicht mal Hering dürfen Sie dies Geheimnis verraten!" (Koyré 1911–1922, Letter to Conrad-Martius, 3 December 1913).

[28] Delbos was familiar with Husserl's philosophy. See Delbos (1911).

[29] Husserl had known Lalande as early as 1906, when he was asked to contribute to Lalande's *Vocabulaire philosophique* (Schuhmann 1977, p. 97, p. 123).

Bergzabern Circle. The shift to Scholastic philosophy and the history of science was not a new beginning for Koyré but the continuation of the phenomenological project that he had started in the years leading up to the war.

The Bergzabern Circle was comprised of Theodor Conrad, Hedwig Conrad-Martius, Alfred von Sybel, Hans Lipps, Edith Stein, Jean Hering, and Koyré. This group of former members of the Göttingen Circle began meeting during the First World War in Conrad's house in Bad Bergzabern, near the Alsatian border of France. Koyré, Lipps, and von Sybel were unable to attend the earliest meetings due to their ongoing involvement in the war, but the group continued to gather until sometime in the late 1920s, primarily during the summer and autumn months, to discuss phenomenology and developments in the phenomenological movement. Joachim Feldes writes, "Financially supported by Winthrop Bell, the circle attempted to build a kind of institute, one that included a library and an archive, thus carrying out a plan conceived by Hering and Reinach before the war, but in order to react in an appropriate and effective way against Heidegger whose influence on students and behaviour toward Husserl they saw as a 'dark spot' in the phenomenological movement" (Feldes 2013, p. 205). Thanks to Zambelli, we know approximately when Koyré participated in the meetings of the Bergzabern Circle. Koyré's passport indicates that he visited Germany on 27 May 1920, 1 August 1923, 8 July 1924, 21 November 1924, 31 March 1925, 4 June 1925, and 21 September 1925 (Zambelli 2007, p. 110, n. 4).[30] While a number of their friends from the Göttingen Circle had died during the war along with Reinach (such as Rudolf Clemens, Fritz Frankfurter, and Heinrich Rickert Jr.), and while Husserl had moved to Freiburg and Scheler to Cologne, this group dedicated themselves to keeping the phenomenological movement, as a community of researchers, alive.

The members of the Bergzabern Circle seem to have been divided with respect to the position they took towards Husserl's transcendental turn. Hering denied that all of the members of the former Göttingen Circle refused to accept Husserl's transcendental reduction and thus the turn to transcendental phenomenology. He suggests that those who had not come from Munich, including Koyré, accepted the transcendental reduction. These students never called into question the epistemological value of transcendental considerations or the importance of the analyses of constitutional problems of consciousness. Their main complaint was Husserl's insistence on the metaphysical primacy of consciousness, his insistence that all other *being* is secondary to the *absolute* existence of consciousness and that all being is nothing more than *being for* consciousness. Like their realist phenomenologist contemporaries, they interpreted such statements as endorsements of metaphysical or subjective idealism (Hering 1959, p. 27).[31] Husserl, however, denied that his

[30] Aside from these visits, Koyré was collaborating with Conrad-Martius and Stein during this period on the German edition of his essay on Descartes, *Descartes und die Scholastik* (1923a). Though their names do not appear in the published text, Conrad-Martius and Stein had helped to translate and edit the book (Stein 2005).

[31] Similarly, in his notes from an interview with Hering, Spiegelberg writes that: "[Hering] denies that he and [the] older phenomenologists refused to accept the 'transcendental reduction.'

position entailed such idealism and insisted that his students had misunderstood his transcendental phenomenology. In his later works, Husserl claimed that only someone who "misunderstands either the deepest sense of intentional method, or that of transcendental reduction, or perhaps both, can attempt to separate phenomenology from transcendental idealism" (Husserl 1999, p. 86).

Koyré's exact position on this issue is not entirely clear, though we might guess that his position was close to that of Hering. In a meeting with Koyré and Hering sometime before 1933, Spiegelberg notes that he discussed the topic of transcendental phenomenology with them. Koyré interjected that transcendental phenomenology is "genuine phenomenology".[32] In addition to Hering's later remarks that not all of Husserl's students were hesitant to make the transcendental turn along with Husserl, a text by Hering written during the Great War has also surfaced which supports this claim. In "Phänomenologie als Grundlage der Metaphysik?" (Hering 2015), Hering, with reference to *Ideas I* §62 [*Epistemological Anticipations – The 'Dogmatic' and the Phenomenalogical Attitude*], suggests that he is *not* convinced that Husserl's transcendental phenomenology leads to subjectivism and argues that "a factual science of transcendentally reduced experiences" might serve as a foundation of metaphysics.[33] We find similar defences of Husserl's transcendental phenomenology given by Hering in his essay "Sub specie aeterni. Réponse à une critique de la philosophie de Husserl" (Hering 1927). It may be that Koyré, like Hering, remained closer to Husserl in his conception of phenomenology than has been previously thought.[34]

Hering's reports also give us reason to believe that the historical–epistemological works that Koyré has become known for are a continuation of and a response to Husserlian phenomenology. He writes that Koyré's intent was to demonstrate "how phenomenology sheds new light on historical problems" and that his *Essai sur l'idée de Dieu et les preuves de son existence chez Descartes* (Koyré 1922b) and *L'idée de Dieu dans la philosophie de saint Anselme* (Koyré 1923b), the works that established Koyré as a serious historian of philosophy, are the first manifestations of this project.[35] Hering expands on this in his article "Phenomenology in France".

Correlativity of *Sein-Bewusstsein* [was] always admitted, but not that *Sein* [is] secondary to *Bewusstsein*. Husserl's refusal to see the equivocation ('obession')" (Spiegelberg 1953, p. 45).

[32] "Last meeting with him and Hering before 1933; discusses about "transcendental phenomenology" (oder "echte Phenomenologie" [genuine phenomenology] – Koyré's interjection)" (Spiegelberg 1953, p. 53).

[33] "Die von Husserl (§ 62 der Ideen) nur als wenig bedeutsam erwähnte "Tatsachenwissenschaft von den transzendental reduzierten Erlebnissen" ist also die Grundlage der Metaphysik" (Hering 2015, p. 43.)

[34] Hering writes that for Reinach phenomenology was a method of doing philosophy, for Husserl it was a branch of philosophy proper. "Husserl était avant tout préoccupé de la mettre au service d'une discipline fondamentale qu'il appelait phénoménologie, et qui était destinée à surmonter radicalement le malaise épistémologique des temps modernes" (Hering 1939, p. 368).

[35] "After the war from 1919–1925 in Paris. Koyré's first struggles until position at École des Hautes Études; his enthusiasm. Demonstration of how phenomenology sheds new light on historical problems (Descartes, Anselm)" (Spiegelberg 1953, p. 46). Years later, Husserl would name Koyré,

According to Hering, what is both original and distinctly phenomenological about Koyré's work is that he abandons the historical prejudice that:

> A philosophical system could be understood in the same measure in which it can be explained by the historical influences the thinker had undergone, as well as by his intellectual temperament. On the contrary, the phenomenologist thinks that the desire of explaining everything (in that way) means the danger of understanding nothing at all. For important though the historical and psychological method may be for showing the route traversed by a philosopher (and many times in spite of the influences it underwent!), in order to reach the point where he came in contact with truth, it would be vain thus to attempt to explain truth itself as well as its recognition by the philosopher. In the end, it is only the errors (as well as the truths borrowed from others but not really thought) that could be explained (according to phenomenologists) in studying the influences of the surroundings. It is for that reason that the historical studies of Mr. Koyré, whether they deal with Plato, St. Anselm, Descartes, Jacob Boehme, or Galileo, put us in immediate relation not only with the era of those philosophers but also with a certain field of philosophical problems itself. *Philosophica philosophice interpretanda*: rarely has that motto been taken so seriously as in the phenomenological movement, the more so because the problems again discovered by it often proved identical to those treated by the great thinkers of the past. (Hering 1950, p. 71)

Hering's suggestion here is that Koyré's historical studies are phenomenological insofar as they deny that philosophical systems can be reduced to a set of historical facts. While contextualization is important for understanding changes in human thought as historical moments, what the phenomenologist is primarily concerned with is what these changes reveal to us about consciousness and its relationship to the world. Interestingly, the publication of these first works from Koyré's tenure at the Sorbonne coincides with his visit with Husserl in Freiburg in the summer of 1922 and the inclusion of his reworked "Bemerkungen zu den Zenonischen Paradoxen" (Koyré 1922a) in Husserl's *Jahrbuch*.[36]

At the same time, the influence of realist phenomenology from within the Bergzabern Circle on Koyré is undeniable. Koyré endorsed the view that the aim of a systematic philosophy is to express or render intelligible phenomena and our knowledge of them, on the basis that the object of philosophical reflection is some mind-independent reality. According to Gillispie, it follows from this that Koyré was a Platonist and that "the best introduction to the unity of view and value inspiring the whole body of his work is his beautiful essay *Discovering Plato*" (Gillispie 2006, p. 285). In his writings on Plato, which draw on his notes from Reinach's lectures on Plato from SS 1910, Koyré compares the Platonic doctrine of knowledge as recollection with the Cartesian notion of innate ideas. He points out that for Plato knowledge of the Forms is not *originally* innate in the human soul. The Forms exist apart from and prior to the soul and enter into the mind from the outside. To know is to grasp the Forms or the essence of states of affairs. Reinach's position appears,

---

along with Etienne Gilson, in the *Cartesian Meditations* for uncovering the deep influence of Scholastic philosophy on Descartes (Husserl 1999, pp. 23–24).

[36] We might note that a fellow member of the Bergzabern Circle, Hans Lipps, published a review of the essay on Zeno's paradoxes in the *Göttingisehe gelehrte Anzeigen* in 1924. While engaging with the Bergzabern Circle, Koyré wrote a review of Max Scheler's *Wesen und Formen der Sympathie* (Koyré 1925).

at least at first blush, to be quite close to this (save the notion of a realm of the Forms). He believes that the objects of philosophical investigation are necessary and intelligible a priori relations between states of affairs that are realized in the world and which exist independent of cognition.

We might hesitate in labelling Reinach as a Platonist (as we should also hesitate in labelling Husserl simply as an idealist rather than a transcendental idealist). Hedwig Conrad-Martius denied that Reinach was a Platonist in any strict sense (Conrad-Martius 1921, pp. xxvi–xxvii), while Hering and Koyré affirm that Reinach was a Platonist of some sort. Reinach's Platonism aside, the position sketched above appears to be a form of Platonic realism, and hence why Koyré claims to have inherited from Husserl, by way of the *Logical Investigations*, the realist phenomenology he discarded. Yet Koyré's phenomenology is not concerned primarily with the objects and phenomena which the natural sciences are obligated to take as *real*[37] but the relationship between consciousness and its objects. Koyré writes that "The ways of the Mind like the ways of God are mysterious and remarkable", and it is precisely the life of this mysterious mind which he seeks to expose (Koyré 1973, p. 66). This appears to be at least in the same spirit as Husserl's programme of phenomenology as a critique of reason rather than a metaphysical project necessarily grounded in Platonism per se.

On 12 September 1932, Koyré gave a talk before the *Société thomiste* in Juvisy in which he stated that: "Phenomenology is, in its basic inspiration, Cartesian and Platonic. What weds it to the philosophies of the Middle Ages is its objectivism, its method (heuristic) of making distinctions, the analysis of essences, and ontologism. As a result, it is closer to the Augustinians than the Aristotelians and closer to Scotism than Thomism" (Feuling 1932, p. 73). At this same conference, Koyré allegedly agreed with Edith Stein – who was also in attendance – that Husserl never managed to convince any of his old students of the necessity to arrive at transcendental idealism. However, based on what we have shown above, it appears to be the *idealism* that many of the Göttingen students worried over, not the transcendental aspect of Husserl's mature thought.

14.5 Koyré and the Reception of Phenomenology in France

Koyré's relationship with Husserl in the late 1920s through to Husserl's death is quite well documented. The gap in history beginning with Koyré's departure from Göttingen ends when Husserl came to Paris to lecture in February 1929. Having missed Husserl's London Lectures, Koyré made sure he was in attendance for the

[37] "Nevertheless, astronomy is much more than a purely mathematical doctrine: the Sun, Moon, and planets are *real* objects; and even though Ptolemaic astronomy gives us complete satisfaction from the purely geometric and practical points of view, it is not so when we come to cosmology. The astronomer is then obliged to treat his circles and orbits as real objects in real space. Certain difficulties then arise" (Koyré 1973, p. 24).

Paris Lectures and may have even played some role in arranging them.[38] In the audience along with Koyré were Hering and Hering's students from Strasbourg – who had just spent two semesters with Husserl in Freiburg and who had accompanied Husserl and his wife on their trip to France – Emmanuel Levinas (Moran 2005, p. 38). Husserl asked Koyré to review and edit the translation of the expanded version of these lectures being carried out by Levinas and Gabrielle Peiffer – the now famous *Méditations cartésiennes* (Husserl 1931). These arrangements may have been made when Koyré visited Husserl on 8 April 1929, for Husserl's 70th birthday (Husserl 1968, p. 52, p. 161). However, details of Koyré's involvement in the project are limited since little of whatever correspondence there was between Husserl, Levinas, and Koyré survives.

After the Paris Lectures, Husserl attended Koyré's thesis defence before leaving with Hering to lecture at Strasbourg. Alexandre Kojève recounts that the defence took place on the first day of March at the Sorbonne. Two theses were presented: *The Philosophy of Jacob Boehme* (Koyré 1929b) – a portion of which would appear in Husserl's *Jahrbuch* that same year (Koyré 1929a) – and *Philosophy and the National problem in Russia at the beginning of the nineteenth century* (Koyré 1929c). Koyré's defence committee for the primary thesis on Böhme was comprised of Étienne Gilson, Léon Brunschvicg, and Émile Brehier. Gilson went so far as to say that the thesis on Böhme was "one of the greatest books on the history of philosophy and was exactly what everyone had expected from the candidate" (Kozhevnikov 1929, p. 8). In 1930, Koyré took a position in Montpellier but returned to teach at the EPHE in Paris in 1931.

Upon returning to Paris, Koyré founded the journal *Recherches philosophiques*, which Ethan Kleinberg suggests might be seen as "the continuation of Husserl's *Jahrbuch*, which stopped publication in 1930", though this new publication "wanted to move beyond the scope of phenomenology as presented by Husserl" (Kleinberg 2005, 58). This would certainly fit with aims of the Bergzabern Circle noted earlier. Kleinberg further notes that many of the figures from the early phenomenological circles were featured in *Recherches philosophiques*, which included an entire section devoted to presenting and reviewing phenomenology. The journal also featured the early work of Levinas, including "De l'evasion" (Levinas 1936), who had started following Koyré's lectures at the EPHE.[39] Along with the publication of the *Méditations cartésiennes*, and Levinas' dissertation, *Théorie de l'intuition dans la phénoménologie de Husserl* (Levinas 1930), Koyré was instrumental to the reception of phenomenology in France.

[38] The invitation to speak in Paris was largely due to Lev Shestov. After Shestov and Hering engaged in a debate over phenomenology, Shestov met Husserl in Amsterdam in 1928, and the two became friends. Shestov arranged for Husserl to come and speak in Paris. However, Shestov did not work at the Sorbonne, where the lectures were held. But Brunschvicg, who was working with Koyré and had an interest in phenomenology, did. The exact role that Koyré played in the entire affair is rather unclear, particularly in light of the fact that he and Shestov had a less than cordial relationship. See "Entretiens avec Léon Chestov" (Fondane 1982).

[39] Koyré and Levinas also authored a set of reviews of German philosophy together for the *Revue Philosophique de la France et de l'Étranger*. See Koyré and Levinas 1933, pp. 285–295.

Koyré indirectly contributed to the Hegelian and Heideggerian influences on the development of phenomenology in France. In 1933, Kojève took over Koyré's course on Hegel and "mistakenly spread the notion that Hegel's phenomenology was essentially the same as Husserl's" (Dupont 2014, p. 162). However, the early phenomenology which Koyré had been trained in borrowed nothing from Hegel aside from its title. Husserl, it seems, had very little interest in Hegel, and nowhere does he associate his phenomenology with the philosophical analyses presented in the *Phenomenology of Spirit*. The conflation of the two philosophies seems to have been more the doing of Kojève than of Koyré. Concerning Heidegger, Koyré was initially drawn to his work, but like the other members of the Bergzabern Circle became a critic of Heidegger's version of phenomenology, even before the war. Levinas, who had also written early articles on Heidegger for the French philosophical audience, recalls that it was Koyré who had warned him about Heidegger's political views before Hitler came to power (Levinas 1989, p. 486).[40] Following a visit with Koyré in August of 1930, von Sybel writes:

> [Koyré and I] were talking a lot about Heidegger, from whom on the one hand comes such a strong call to self-reflection and authenticity, while on the other hand he has ruined all that phenomenology actually strived to be. By now none of the old phenomenologists can keep up with the force of his appearance. (Feldes 2013, pp. 217–218)

However, Koyré had been instrumental in the publication of Heidegger's 1929 essays *Was ist Metaphysik?* and *Vom Wesen des Grundes* in French.[41] The appropriation of Hegelian philosophy and also of Heidegger's *Being and Time*, along with the works of nineteenth- and early-twentieth-century French thinkers, resulted in the development of a distinct branch of phenomenology in France, one which is often difficult to reconcile with Husserlian phenomenology.[42]

14.6 Husserl and Koyré on Galileo and the Mathematization of Nature

The above biographical details concerning Koyré's student years and early career provide us with information that can help us to track the influence that Husserl might have had on his work. Another point of interest has been on the possible influence that Koyré may have had on his former teacher with respect to his

[40] In 1955, Koyré and Levinas both refused to attend a conference organized in Cerisy in honour of Heidegger because of Heidegger's Nazism. It is worth noting that Koyré's mother, Catherine, died in 1940 fleeing from Paris after it had been conquered by the Germans.

[41] For Koyré's later remarks on Heidegger, see Koyré (1946), his preface in Heidegger (Heidegger 1931, pp. 5–8), and his remarks in Wahl (Wahl 1947, pp. 71–79). See also Zambelli's introduction to Koyré (Koyré 1998, pp. 523–530) and the discussions (both philosophical and historical) of the relationship between Koyré and Heidegger in Geroulanos (Geroulanos 2010).

[42] One might note here that while Husserl paid little attention to Hegel, Heidegger lectured on Hegel in the SS 1929 and WS 1930/1931 in Freiburg.

understanding of Galileo and the mathematization of nature. Koyré began writing articles on Galileo as early as 1935 and in rapid succession published "Galilée et l'expérience de Pise" (Koyré 1937a), "La loi de la chute des corps Galilée et Descartes" (Koyré 1937b), and his *Études galiléennes* (Koyré 1939). These works on sixteenth- and seventeenth-century physics were by no means the first works in the history of science, but they initiated a new important phase in the discipline. As Bernard Cohen writes:

> Koyré was able to show us that the central concepts of philosophy at any given time may be a determining element of the nature of the scientific thought of the age, and vice versa. Some examples are the effect of the geometrization of space in the Renaissance, the concept of the infinite universe, matter and spirit. (Cohen 1987, p. 159)

In his writings, Koyré argues that Descartes and Galileo championed a new view of nature and of science. These two thinkers found themselves at the end of a long tradition that was moving towards a new physics and a "conversion of the human mind from *theoria* to *praxis*, from the *scientia contemplativia* to the *scientia active et operative*, which transforms man from a spectator into an owner and master of nature" (Koyré 1957, p. vii). Central to this change in how science viewed and understood the world was the mathematization of nature.

David Carr is largely responsible for spreading the notion that a visit from Koyré led Husserl to add *Crisis* §9: Galileo's Mathematization of Nature, a section that was not present in the original Prague Lecture. In a footnote to the "Translator's Introduction" to the *Crisis*, Carr writes:

> It is interesting to speculate that the Galileo section might have resulted from a reported visit during this period by Husserl's friend and former student Alexandre Koyré, who published his monumental *Études Galiléenes* in 1940. The striking similarity between Husserl's and Koyré's interpretation of the significance of Renaissance science is best seen in the latter's *From the Closed World to the Infinite Universe*. (Husserl 1970, p. xix)

However, this speculation is now believed to be inaccurate. While Koyré was in agreement with the interpretation of Galileo in the *Crisis*, there is little to suggest that Koyré influenced Husserl in this regard. Reinhold N. Smid has shown that Koyré's final visit with Husserl was in July of 1932. This was 3 years before Koyré's works on Galileo would appear, as noted above (Husserl 1993, p. il, n. 2).

In his recent study of Husserl's *Crisis*, Dermot Moran sketches the bizarre history of the Galileo section. This piece of the text was a late addition to the manuscript published in *Philosophia* in 1936. Before the article went to print, Husserl replaced the original three or four pages with an expanded discussion roughly ten times the length of the original. These were not his first ruminations on Galileo and the shift towards the mathematization of nature which he felt characterized modern science. Husserl's work on this theme dates back to at least 1928, as is evident from the text "Idealization and the Science of Reality – The Mathematization of Nature", the first supplementary text found in Carr's translation of the *Crisis*. Interestingly, Koyré had visited Husserl in mid-October 1928 (Schuhmann 1977, p. 337; Husserl 1968, p. 48). However, this was not the visit that Carr had in mind. "Furthermore", Moran writes

> [...] in 1937, Koyré himself remarked to Ludwig Landgrebe that he agreed with Husserl's Galileo interpretation (see Hua XXIX, p. il), suggesting that Koyre was influenced by Husserl, rather than the other way round. (Moran 2012, pp. 72–73)[43]

Moran goes on to claim that the characterization of Galileo that we find in Koyré's own works essentially conforms with Husserl's – as an a priori mathematical physicist.[44]

Moran's suggestion that it may have been Husserl that influenced Koyré with respect to Galileo, rather than the other way around, makes sense. Recall that in what would likely have been Koyré's first course with Husserl, Husserl made a passing reference to Galileo. Such references are not uncommon in Husserl's *Nachlass*. It is true that many of Husserl's references to Galileo in his unpublished writings were written in the mid-1930s. However, Gurwitsch points out that some of the preparatory studies for the *Crisis* date from the late 1920s, perhaps referring to texts dealing with the "Mathematisierung der Natur" written in 1926,[45] and notes also that "some of the relevant ideas can be found, at least in germinal form, as early as 1913" (Gurwitsch 1974, p. 33). In fact, there are numerous, albeit brief, references to Galileo from the period of 1909–1920. They tend to single out Galileo as ushering in a new era in the natural sciences. In the Prague Lecture delivered in November 1935, Husserl connects Galileo and Descartes a number of times and explicitly talks about the Galilean mathematization of nature (Husserl 1993, p. 110, p. 123, p. 131). It may have been that the meeting between Koyré and Husserl in 1932 was responsible for informing the way that Koyré would tackle Galileo and Descartes in his work on the history and philosophy of science.

Tracing Husserl's engagement with Galileo is in no way meant to downplay the novelty and importance of Koyré's work. It is undeniable that Koyré is more than a mere historian attempting to demonstrate the merits of his teacher's reading of Galileo and modern science. Koyré is also pushing towards a phenomenological treatment of the history and philosophy of science towards a historical–teleological critique of reason, using the scientific revolution as a case study. Reconstructing Galileo's thought we witness a change in the foundation of meaning in modern science. This is not only a technical shift but an epistemological one. In making these points salient features of historical analysis, Koyré's work goes far beyond what Husserl was able to do in his *Crisis* writings.

Husserl claims that from Descartes onwards, we find a new ideal of scientific theories and of scientific explanation. Science in the "modern" sense is, according to Husserl, "rational and all-inclusive, or rather the idea that the infinite totality of what is in general is intrinsically a rational all-encompassing unity that can be mastered, without anything left over, by a corresponding universal science...Its rationalism soon overtakes natural science and creates for it the completely new

---

[43] Moran goes on to suggest that Husserl's interest in Galileo may have been influenced by Jacob Klein. See Klein 1992, pp. 118–121.

[44] For a detailed analysis of Husserl's position with respect to Galileo, see Moran 2012, pp. 66–71.

[45] See Husserl (Husserl 2012, p. 324) and *Realitätswissenschaft und Idealisierung. Die Mathematisierung der Natur* (1926–1928) in Husserl 1976, pp. 279–293.

idea of *mathematical natural science* – Galilean science, as it was rightly called for a long time" (Husserl 1970, pp. 22–23). In the *Crisis* writings, Husserl intends to offer a critique of a style of thought which he refers to as *Galilean reasoning*, which he takes to be representative of "modern" rationality. Gurwitsch defines Galilean reasoning as:

> The cleavage between the world as it presents itself in the perceptual experience of everyday life and the world as it is in scientific truth and 'in reality.' [...]. Modern science of the Galilean style starts by refusing to accept the perceptual world at face value. Instead, reality is believed to contain, embody, and conceal a mathematical structure. As to its true and real condition, in contradistinction to its perceptual appearances, the world (to express it in more modern terms) is a mathematical manifold. To pierce through the veil of appearances, to discover the mathematical structure of the universe, and to disclose reality as a mathematical manifold constitute precisely the task of Galilean science. (Gurwitsch 1974, pp. 34–35)

While Galileo stands out as the primary target of Husserl's remarks on the mathematization of nature at §§ 8–9a of the *Crisis*, the focus begins with Platonism.

Following the dawn of Galilean reasoning, we see the subordination of philosophy to science. Science becomes a fact which philosophy must accommodate, rather than justify. According to Gurwitsch, the historical significance of Husserl's analysis of Galilean reasoning is that it challenges and even abandons the acceptance of science as fact and views it instead as a problem. By this we do not mean that Husserl thinks that scientific theories and practice are internally inconsistent or corrupt. Husserl does not challenge the validity of science. The problem lies in the presupposition that "real reality" is something divorced from experience which possesses a mathematical structure. Gurwitsch writes:

> Husserl's analysis of Galilean science [...] in calling attention to the presuppositions of Galilean science, raises the problem of its *sense*, and of the sense of the limits of its validity [...]. Husserl was not a historian of science and never made any claim to be one, even though he presented his views on 'science as a problem' in the form of an analysis of Galileo's work. (Gurwitsch 1974, p. 38)

How then might Husserl's work have inspired Koyré? "It is now generally accepted", writes Zambelli, "that Alexandre Koyré was a 'phenomenological' historian of philosophical and scientific thought" (Zambelli 1999, p. 209). Hering gives us a hint of this phenomenological approach to history that we find in Koyré's work when he writes that Koyré "never imitated the many scholars of that time who tried to understand a philosopher only by way of showing contemporary influences upon his thought. Koyré always went deeper and tried to show the fundamental intuitions of the author" (Hering 1965, p. 453). The relationship between scientific theories and practice and the historical contexts in which these theories and practices emerge, endure, and are overturned was of interest to Koyré. It should go without saying that science as a historical reality is deeply connected to the economic, political, social, and cultural milieu in which it exists. It is also informed by and evaluated against the philosophical standards of the time, namely, the accepted metaphysical systems and parlances, and the accepted theories of truth and knowledge. The interaction between science and history results in a dynamic and mutually affective entity, often considered to be progressing, but towards what no one seems to know with any certainty or precision. But what is of interest to the phenomenologist in all of this is what we can learn about the relationship between consciousness and its object through the study of the history of science. The historical contextualization of science is an important and perhaps necessary first step in such analyses, but not the end goal. The phenomenologist is not concerned with accurately portraying the minute details of history but with uncovering the life of the mind as it plays out in time in order to determine the conditions of the possibility of such a life.

Koyré argues that a shift in the natural attitude occurred as a result of a decisive mutation in human thought during the scientific revolution. He writes that it is this mutation which:

> [...] explains why the discovery of things which seem to us nowadays childishly simple required such prolonged efforts – and not always crowned with success – by the greatest of geniuses, by Galileo and Descartes. This is because it was not a matter of battling against theories which were simply inadequate or erroneous, but of changing the very intellectual framework itself, of overthrowing an intellectual attitude, one which was when all is said and done a perfectly natural one, and substituting it for another, one which was not natural at all. (Koyré 1978, p. 3)

According to Koyré, it was not simply experimentation and observation that led to the scientific revolution. It required a transformation of thought and the adoption of new metaphysical and epistemic frameworks in order to account for phenomena in a different way. Experimentation and observation presuppose a language in which questions are to be posed and data interpreted. They presuppose a conceptual framework in which certain types of entities exist and where certain forms of explanation are accepted. This is what we learn through Koyré's phenomenological analysis of Galilean reasoning.

14.7 Conclusion: Koyré's Debt to Phenomenology

Nothing that has been argued herein is meant to suggest that Husserl was Koyré's primary intellectual influence. His epistemological studies are clearly linked to the work of his mentors Meyerson and Brunschvicg, as well as the work of Lucien Lévy-Bruhl.[46] Insofar as Koyré emphasizes the historical nature of reason, Jorland argues that his position is most closely related to that of Brunschvicg.[47] However,

[46] In 1930, Koyré published a review of Lévy-Bruhl's *The Soul of the Primitive*. Husserl had a copy of this review, and it may be the reason Husserl read Lévy-Bruhl when preparing the *Crisis* writings. For more on Koyré and Lévy-Bruhl, see Zambelli (1995).

[47] "En ce sens, Alexandre Koyré sera plus proche de Brunschvicg que de Meyerson: il cherchera à montrer, de manière régressive, comment une certaine image du monde présuppose une certaine forme de la pensée, une certaine ontologie, puis, de manière progressive, par quelle série de concepts et d'opérations cette ontologie donne accès à certains objets. Mais, fidèle à sa formation phénoménologique initiale, Alexandre Koyré se démarquera aussi bien du psychologisme de l'un que de la philosophie de la conscience de l'autre. D'ailleurs il réunit dans une même inspiration, celle du retour aux choses, Meyerson et Husserl. Il en retiendra des éléments pour une histoire de

Koyré's position is not one of relativism. Following Husserl, it is decidedly anti-relativistic. Koyré also strives to distance himself from the psychologistic interpretation of Meyerson's critical philosophy as presented in *Identité et Réalité* (Meyerson 1908). Again, following Husserl, Koyré is anti-psychologistic. The middle view between these two positions which Koyré attempts to carve out is a phenomenological one or at least one that is quasi-phenomenological. I will leave a more in-depth analysis of this to scholars more versed in nineteenth- and early-twentieth-century French epistemology and neo-Kantianism than I am.

Koyré's understanding of the theory-ladenness of observation is perhaps based on a reading of Meyerson in light of Husserl. Koyré is, like Meyerson, an anti-positivist. The problems in epistemology and sociology that French philosophers in the early twentieth century were dealing with provide part of the backdrop against which Koyré ought to be read. But the influence of phenomenology is also important, if not essential, for understanding Koyré's writings. When tracing this influence, we should be careful not to focus solely on the influence of Reinach and the realist phenomenologists. Koyré remained faithful to much of Husserl's teachings and the spirit of his project and extended the phenomenological method into areas which Husserl had only started to explore near the end of his career. The two remained close throughout Husserl's life, despite Husserl's decision to reject Koyré's draft–dissertation.

Consciousness not only constitutes its objects in terms of a set of a priori laws of thought (which are not reducible to laws of the brain), it also constitutes them within a metaphysical or conceptual framework – what Meyerson refers to as an *ontology*. As Husserl explains, consciousness bestows meaning on its objects and constitutes them as having a particular being–sense. So the laws of thought themselves are not relative, but how we conceptualize and understand the objects to which these laws are applied is dependent upon theoretical frameworks which are, at least in part, socially constructed. Hence Koyré's inheritance of Husserl's anti-relativism and anti-psychologism. However, there is also a certain Platonism that underlies Koyré's thought. While he may not have rejected Husserl's transcendental phenomenology entirely, he seems committed to the idea that there is a "real reality" behind the veil of appearances and that it is not nonsense to speak of such a reality, even if we can have no knowledge of it.

As a final note on the history between Koyré and Husserl, we should point out that after the Second World War, Koyré was involved with Father von Breda's *Colloque international de phénoménologie*. In April 1951, Koyré attended the inaugural meeting of the *Colloque*, held in Brussels. The programme for the conference had included a paper by Hendrik Pos entitled *La phénoménologie et la connaissance scientifique*; however, Pos was unable to attend. In his absence, Koyré offered some thoughts on the same theme. He is reported to have argued that while phenom-

---

la pensée et ce thème platonicien. En effet pour Koyré, l'idéal de la science, l'idéal de la déductibilité totale du réel, est. parfaitement légitime, mais il ne l'est. que pour un monde des essences, pas pour « les choses temporelles et changeantes »". (Jorland 1981, p. 68).

enology had taken a backseat to positivism in the philosophy of science, phenomenology could be of use in answering questions concerning the foundations and philosophical implications of the sciences. He also noted the usefulness of phenomenology for cognitive science, which tends to replace causal explanations with descriptions of structures. However, Father Van Breda was critical of the so-called phenomenology that Koyré spoke of (Van Haecht 1951, pp. 438–445).

What remained of phenomenology after the Second World War was much different from either the realist or transcendental phenomenologies that Koyré had been schooled in. The new phenomenology which came into existence in both France and Germany had been irreversibly altered by Heidegger, both by his philosophy and by his person. This is perhaps a part of why Koyré hesitated in answering Spiegelberg concerning the extent to which he still considered himself a phenomenologist. The other reason may have been that Koyré accepted Husserl's claim that his students had misunderstood the reduction and his transcendental idealism and worried that he himself was guilty of this. In a letter to Hedwig Conrad-Martius on 10 January 1957, Koyré intimates that the phenomenology of the *alte Garde*, the Göttingen Circle, is just as much a part of phenomenology as transcendental phenomenology or existential phenomenology (Koyré 1998, p. 530, n. 39). But it is unclear if he is disparaging of Husserl's later work in saying so. There is good reason to think that he is not.

More work needs to be done on the relationship between Koyré and Husserl and between Koyré and phenomenology more broadly understood. Such work promises to shed new light on both Koyré's philosophical project and on the subtle theoretical disputes between the earliest members of the phenomenological movement and their progenitor. I hope that this paper – which outlines the history between Koyré and Husserl fully accepting that "histories, even the most truthful, never present us with things as they really were" (Koyré 1950, p. xvi) – might serve as a foundation for such work.

Acknowledgements I would like to thank Christian Dupont for his many helpful suggestions and corrections – both historical and philosophical – on an earlier version of this paper and to Anna Yampolskaya and Joachim Feldes for assisting me with my research.

References

Baron L (1983) Discipleship and Dissent: Theodor Lessing and Edmund Husserl. Proceedings of the American Philosophical Society 127/1:32–49.
Bergson H (2002) Key Writings. Continuum, New York.
Chimisso C (2008) Writing the History of the Mind: Philosophy and Science in France 1900–60s. Ashgate, Burlington.
Cohen IB (1987) Alexandre Koyré in America: Some Personal Reminiscences. History and Technology 4:55–70.
Conrad–Martius H (1921) Einleitung. In Stein 1921, pp. v–xxxvii.
Couturat L (1896) De l'infini mathématique. Alcan, Paris.

Delbos V (1911) Husserl : Sa critique du psychologisme et sa conception d'une Logique pure. Revue de Métaphysique et de Morale 19/5:685–698.
Dupont C (2014) Phenomenology in French Philosophy: Early Encounters. Springer, Dordrecht.
Feldes J (2013) Alfred von Sybel – A Life between the Lines. Symposium 17/2:204–223.
Feuling D (1932) La Phénoménologie. Journées d'Études de la Société thomiste [12 septembre 1932]. Cerf, Juvisy.
Fondane B (1982) Rencontres avec Léon Chestov. Plasma, Paris.
Foucault M (1988) Politics, Philosophy, Culture: Interviews and Other Writings, 1977–1984. Routledge, New York.
Foucault M (1998) Aesthetics, Method, and Epistemology. The New Press, New York.
Geroulanos S (2010) An Atheism that Is Not Humanist Emerges in French Thought. The Stanford University Press, Stanford.
Gillispie CC (1973) Entry: Koyré. Dictionary of Scientific Biography. Vol. VII. Scribner's Sons, New York, pp. 482–487.
Gillispie CC (2006) Alexandre Koyré. Transactions of The American Philosophical Society 96/5:283–299.
Gurwitsch A (1974) Phenomenology and the Theory of Science. The Northwestern University Press, Evanston.
Heidegger M (1931) Qu'est ce-que la métaphysique ? Bifur 8:5–27.
Hering J (1927) Sub specie aeterni. Réponse à une critique de la philosophie de Husserl. Revue d'histoire et de philosophie religieuse 7:351–364.
Hering J (1939) La phénoménologie d'Edmund Husserl il y a trente ans souvenirs et réflexions d'un étudiant de 1909. Revue Internationale de philosophie 1/2:366–373.
Hering J (1950) Phenomenology in France. In Farber M (ed). Philosophic Thought in France and the United States: Essays representing major trends in contemporary French and American philosophy. SUNY Press, Albany, pp. 67–85.
Hering J (1959) Edmund Husserl. Souvenirs et réflexions. In Taminiaux J (ed). Edmund Husserl 1859–1959 : Recueil commémorative publié à l'occasion du centenaire de la naissance du philosophe. Martinus Nijhoff, The Hague, pp. 26–29.
Hering J (1965) In Memoriam – Alexander Koyré. Philosophy and Phenomenological Research 25/3:453–454.
Hering J (2015) Phänomenologie als Grundlage der Metaphysik? Studia Phaenomenologica XV:37–52.
Husserl E (1900–1901) Logische Untersuchungen. Niemeyer, Halle.
Husserl E (1913) Ideen zu einer reinen Phänomenologie und phänomenologischen Philosophie: Allgemeine Einführung in die reine Phänomenologie. Jahrbuch für Philosophie und phänomenologische Forschung 1/1:1–323.
Husserl E (1931) Méditations cartésiennes : Introduction à la phénoménologie. Traduit de l'allemand par Melle Gabrielle Peiffer et M. Emmanuel Levinas. Armand Colin. Paris.
Husserl E (1939) Die Frage nach dem Ursprung der Geometrie als intentional–historisches Problem. Revue internationale de philosophie 1/2:203–225.
Husserl E (1968) Briefe an Roman Ingarden. Martinus Nijhoff, The Hague.
Husserl E (1970) The Crisis of European Sciences and Transcendental Phenomenology. Trans. David Carr. The Northwestern University Press, Evanston.
Husserl E (1976) Husserliana VI: Die Krisis der europäischen Wissenschaften und die transzendentale Phänomenologie. Eine Einleitung in die phänomenologische Philosophie, 2 Auflage. Martinus Nijhoff, The Hague.
Husserl E (1993) Husserliana XXIX: Die Krisis der europäischen Wissenschaften und die transzendentale Phänomenologie. Ergänzungsband. Texte aus dem Nachlass 1934–1937. Kluwer, Dordrecht.
Husserl E (1994a) Husserliana Dokumente 3: Briefwechsel, Band III: Die Göttinger Schule. Kluwer, Dordrecht.

Husserl E (1994b) The Idea of Phenomenology. Trans. Lee Hardy. Kluwer, Dordrecht.
Husserl E (1998) The Paris Lectures. Trans. Peter Koestenbaum. Kluwer, Dordrecht.
Husserl E (1999) Cartesian Meditations: An Introduction to Phenomenology. Translated by Dorion Cairns. Kluwer, Dordrecht.
Husserl E (2005) Husserliana Materialien: Einführung in die Phänomenologie der Erkenntnis. Vorlesung 1909. Springer, Dordrecht.
Husserl E (2008) Introduction to Logic and Theory of Knowledge. Lectures 1906/07. Translated by Claire Ortiz Hill. Springer, Dordrecht.
Husserl E (2012) Husserliana XLI: Zur Lehre Vom Wesen und zur Methode der Eidetischen Variation. Texte aus dem Nachlass 1891–1935. Kluwer, Dordrecht.
Ingarden R (1962) Edith Stein on her Activity as an Assistant of Edmund Husserl. Philosophy and Phenomenological Research 23/2:155–175.
Jorland G (1981) La science dans la philosophie : les recherches épistémologiques d'Alexandre Koyré. Gallimard, Paris.
Jorland G (1994) Koyré phénoménologue ? In Vinti C (ed). Alexandre Koyre – L'avventura intellettuale. Edizioni Scientifiche Italiane, Napoli, pp. 105–126.
Klein J (1992) Greek Mathematical Thought and the Origin of Algebra. Translated by Eva Brann. Dover, New York.
Kleinberg E (2005) Generation Existential. Heidegger's Philosophy in France, 1927–1961. The Cornell University Press, Ithaca.
Koyré A, Levinas E (1933) Philosophie allemande. Revue Philosophique de la France et de l'Étranger 116:285–295.
Koyré A (1911–1922) Briefe an Max Scheler, Theodor Conrad, und Hedwig Conrad–Martius. Bayerische Staatsbibliothek München. Ana 315 E II, 373 B II, 378 B II, Conrad-Martiusiana C II. [Unpublished manuscript].
Koyré A (1912) Sur les nombres de M. Russell. Revue de métaphysique et de morale 20/5:722–724.
Koyré A (1922a) Bemerkungen zu den Zenonischen Paradoxen. Jahrbuch für Philosophie und phänomenologische Forschung 5:603–628.
Koyré A (1922b) Essai sur l'idée de Dieu et les preuves de son existence chez Descartes. Ernst Leroux, Paris.
Koyré A (1923a) Descartes und die Scholastik. Friedrich Cohen, Bonn.
Koyré A (1923b) L'idée de Dieu dans la philosophie de saint Anselme. Ernst Leroux, Paris.
Koyré A (1925) Max Scheler, *Wesen und Formen der Sympathie*. Revue Philosophique de la France et de l'Étranger 100:456–457.
Koyré A (1929a) Die Gotteslehre Jakob Boehmes. Jahrbuch für Philosophie und phänomenologische Forschung. E. Husserl zum 70. Geburtstag gewidmet. Festschrift: 225–281.
Koyré A (1929b) La philosophie de Jacob Boehme. Étude sur les origines de la métaphysique allemande. Paris, Vrin.
Koyré A (1929c) La philosophie et le problème national en Russie au début du XIXe siècle. Paris, Champion.
Koyré A (1937a) Galilée et l'expérience de Pise : à propos d'une légende. Annales de l'Université de Paris 12:422–53.
Koyré A (1937b) La loi de la chute des corps Galilée et Descartes. Revue Philosophique de la France et de l'Étranger 123/5–8:149–204.
Koyré A (1939) Études galiléennes. Hermann, Paris.
Koyré A (1950) Introduction. In Anscombe E (ed) Descartes. Philosophical Writings. Bobbs-Merrill, Indianapolis, pp. vii–xliv.
Koyré A (1955) Influence of Philosophic Trends on the Formulation of Scientific Theories. The Scientific Monthly 80/2:107–111.
Koyré A (1957) From the Closed World to the Infinite Universe. The Johns Hopkins Press, Baltimore.

Koyré A (1973) The Astronomical Revolution. Copernicus, Kepler, Borelli. Translated by REW Maddison. The Cornell University Press, Ithaca.
Koyré A (1978) Galileo Studies. Trans. John Mepham. Sussex, Harvester Press.
Koyré A (1998) Present Trends in French Philosophical Thought. Journal of the History of Ideas 59/3:521–548.
Koyré A (1999) Insolubilia. Eine Logische Studie über die Grundlagen der Mengenlehre. Giornale critico della filosofia italiana anno LXXVII 6/18:323–354.
Koyré A (1946) L'évolution philosophique de Martin Heidegger. Critique 1:73–84; 2:161–183.
Kozhevnikov A (1929) Zashchita dissertatsii A. V. Koyré. Eurasia 16:8.
Kuhn T (1970) Alexander Koyré and the History of Science. Encounter 34:67–69.
Kuhn T (1996) The Structure of Scientific Revolutions. 3rd edition. The University of Chicago Press, Chicago.
Levinas E (1930) Théorie de l'intuition dans la phénoménologie de Husserl. Alcan. Paris.
Levinas E (1936) De l'évasion. Recherches philosophiques 5:373–392.
Levinas E (1989) As If Consenting to Horror. Critical Inquiry 15/2:485–488.
Luquet GH (1901) Edmund Husserl – *Logische Untersuchungen. 1te Theil*, et Max Scheler – *Die transzendentale und die psychologische Methode.* Revue philosophique de la France et de l'étranger 51:414–419.
Majer U (1997) Husserl and Hilbert on Completeness: A Neglected Chapter in Early Twentieth Century Foundations of Mathematics. Synthese 110:37–56.
Meyerson É (1908) Identité et Réalité. Alcan. Paris.
Moran D (2005) Edmund Husserl: Founder of Phenomenology. Polity, Cambridge.
Moran D (2012) Husserl's Crisis of the European Sciences and Transcendental Phenomenology: An Introduction. The Cambridge University Press, Cambridge.
Mulligan K (1987) Speech Act and Sachverhalt. Reinach and the Foundations of Realist Phenomenology. Kluwer, Dordrecht.
Reinach A (1902–17) Briefe von Reinach an Conrad. Bayerische Staatsbibliothek München. Ana 378 B II. Transcribed by Karl Schuhmann [Unpublished manuscript].
Reinach A (1911) Zur Theorie des negativen Urteils. In Pfänder A (ed). Münchener Philosophische Abhandlungen: Theodor Lipps zu seinem sechzigsten Geburtstag gewidmet von früheren Schülern. Barth, Leipzig, pp. 196–254.
Reinach A (1969) Concerning Phenomenology. The Personalist 50/2:194–221.
Russell B (1992) The Collected Papers of Bertrand Russell. Volume 6: Logical and Philosophical Papers 1909–1913. Routledge, New York.
Schuhmann E, Schuhmann K (2001) Husserls Manuskripte zu seinem Göttinger Doppelvortrag von 1901. Husserl Studies 17:87–123.
Schuhmann K (1977) Husserl-Chronik. Martinus Nijhoff, The Hague.
Schuhmann K (1987) Koyré et les phénoménologues allemands. History and Technology 4:149–167.
Spiegelberg H (1953) Notes about Interviews with various Philosophers, Psychologists, Psychiatrists etc. Herbert Spiegelberg Papers, WUA00070, Box 2, Folder 9, 070-NBK1953 [Unpublished manuscript].
Spiegelberg H (1969) The Phenomenological Movement: A Historical Introduction, 2nd edition. Martinus Nijhoff, The Hague.
Stein E (2005) Übersetzungen V: Übersetzung von Alexandre Koyré Descartes und die Scholastik. Herder. Freiburg im Breisgau.
Stein E (1921) (ed) Adolf Reinach. Gesammelte Schriften. Halle, Niemeyer.
Van Haecht L (1951) Le colloque international de phénoménologie. Bruxelles, 12 avril 1951. Revue Philosophique de Louvain. Troisième série 49/23:438–445.
Wahl J (1947) Petite histoire de « l'existentialisme ». Club Maintenant, Paris.
Webb D (2013) Foucault's Archaeology: Science and Transformation. The Edinburgh University Press, Edinburgh.

Zambelli P (1995) Alexandre Koyré versus Lucien Lévy–Bruhl: From Collective Representations to Paradigms of Scientific Thought. Science in Context 8/3:531–555.
Zambelli P (1999) Alexandre Koyré im 'Mekka der Mathematik': Koyré's Göttinger Dissertationsentwurf. NTM Zeitschrift für Geschichte der Wissenschaften, Technik und Medizin 7:208–230.
Zambelli P (2007) Segreti di Gioventù. Koyré da SR a S.R: Da Mikhailovsky a Rakovsky? Giornale Critico della Filosofia Italiana 3/1:109–151.
Zambelli P (2009) Alexandre Koyré : la fondation du « Centre » et l'histoire des sciences des deux côtés de l'Atlantique. Traduction par Irène Molina. Paper presented at the Conférence à l'occasion du cinquantenaire du Centre Alexandre Koyré, Université de Florence, en février 2009 [Unpublished manuscript]

Chapter 15
Kuhn, Sarton, and the History of Science

J.C. Pinto de Oliveira and Amelia J. Oliveira

Abstract This may possibly be the first work on Kuhn and Sarton despite the importance of the subject, since Sarton was one of the main representatives of the "old" historiography of science, essential to understanding Kuhn's critical proposal of a "new historiography of science" and the "historiographic revolution". The theme has not been explored up until now by Kuhn scholars perhaps due to the vastness of Sarton's work and the inadequate resource of merely citing excerpts out of context. Seeking to circumvent these problems, in a Koyréan way, we focus here on a case study. It is a peculiar case study and convenient for our purposes because, while it is of great importance for Sarton's conception regarding the history of science and is in fact present throughout his academic career, it is not very extensive and can therefore be addressed in a more complete manner. We refer to Sarton's treatment of Leonardo da Vinci and a discovery that he did *not* make: the discovery of blood circulation in the seventeenth century, attributed to William Harvey. Although Kuhn did not write about Leonardo or Harvey, we aim to show that he clearly positioned himself contrary to Sarton, albeit indirectly, with respect to this particular historical episode, as well.

Keywords Kuhn • Sarton • Koyré • Butterfield • History of science • Historiographic revolution • Leonardo da Vinci • William Harvey • Galen • Blood circulation

15.1 Introduction

In the mid-twentieth century, the history of science underwent changes significant enough to speak, as did Thomas Kuhn, in terms of a "historiographic revolution" (Kuhn 1970, p. 3). In *The Structure of Scientific Revolutions*, he presents a new historical perspective and cites some authors, especially Alexandre Koyré, who fomented the change, but no traditional historians, responsible for the books, which

J.C. Pinto de Oliveira (✉) • A.J. Oliveira
Department of Philosophy, IFCH, State University of Campinas,
Rua Cora Coralina, 100, 13083-896 Campinas, SP, Brazil
e-mail: jcpinto@unicamp.br

R. Pisano et al. (eds.), *Hypotheses and Perspectives in the History and Philosophy of Science*, DOI 10.1007/978-3-319-61712-1_15

he argues have misled us about science "in fundamental ways" (Kuhn 1970, p. vi, pp. 1–3).

It is only in his later writings that Kuhn's criticism is clearer with respect to the identification of followers of traditional historiography. In two of the few passages where names are cited, Kuhn refers to Condorcet, Comte, Dampier, and Sarton (Kuhn 1977, p. 148, pp. 106–107). George Sarton (1884–1956) is criticized despite his significant role at the time in establishing the history of science as a discipline of study, as well as his "monumental researches" (Kuhn 1977, p. 109, p. 148).[1]

Notwithstanding, Kuhn makes no critical reference to Sarton as representing the older history of science that is comparable to his complimentary references to Koyré as a typical representative of the new historiography. But we believe Sarton could play this role very well. We argue that he is a typical representative of traditional historiography, which allows us, through contrast, to better understand Kuhn's "new historiography of science" and the "historiographic revolution", and we try to show this by taking into account some of Koyré's concerns.

Koyré seems to avoid direct criticism to Sarton. His references to him are rare and when they occur are positive (and trivial). In *Études d'Histoire de la Pensée Scientifique*, for example, which is a collection of writings from different periods, Koyré only makes two references to Sarton. They occur in the texts about the origins of modern science (in which he criticizes positivism) and about Leonardo. And they are limited to a reference to the publication of the "great works of Thorndike and Sarton" (Koyré 1973, p. 61) and a mention of Sarton's comment that Verrocchio was Leonardo's St. John the Baptist (Koyré 1973, p. 108). In addition to this scant presence, it is also worth noting the conspicuous absence of Sarton's name in the final text of the book. There Koyré comments precisely on a brief history of the historiography of science outlined by Guerlac which, albeit in passing, cites Sarton in opposition to the "enlarged and deepened conception of the history of science", represented by Koyré (Guerlac 1963, pp. 808–809. See also Bernard Cohen 1987, p. 61).[2]

It can be said that Koyré has regarding Sarton the same concerns with tolerance and contextualization he demonstrates with respect to the Enlightenment

[1] In his 1963 book, Joseph Agassi had already criticized Sarton directly as being an inductivist historian. (Cf. Agassi 2008, p. 198, note 4; pp. 201–202, note 20; p. 211, note 58; p. 226, note 105). See Kuhn's review (Kuhn 1966, p. 257).

[2] Sarton's name is also missing from a list that appears at the beginning of *Études Galiléennes* in which Koyré includes the names of Duhem, Meyerson, Cassirer, Brunschvicg, and Enriques to justify "interest in the study of the history of science" and the "philosophical fertility of this study" (Koyré 1966, p. 1). In fact, Sarton's name is not mentioned anywhere in the book (nor in Jorland's book about Koyré (Jorland 1981)). Perhaps Koyré availed himself of the fact that Sarton's research had not extended much beyond the fifteenth century (and thus referred to a different historical period) to avoid entering into conflict with Sarton's historiography. However, even in the 1952 *Colloque* about Leonardo, in which both were present, Koyré, in his final synthesis of the event, does not refer critically to Sarton and the "positivist" passage of his communication cited below. On the contrary, the references are complimentary (Koyré 1953, pp. 237–238, p. 240, p. 242), but Koyré omits them in the similar text of a lecture about Leonardo that he delivered the following year (Koyré 1973, pp. 99–116).

historiography, especially in a text on Condorcet (Koyré 1948, pp. 132–133; Pinto de Oliveira 2012, p. 121). It is this same concern that mobilizes us here.

Thus, this may possibly be the first work on Kuhn and Sarton despite the importance of the subject. The theme has not been explored up until now by Kuhn scholars perhaps due to the vastness of Sarton's work and the inadequate resource of merely citing excerpts out of context. Seeking to circumvent these problems, we focus here on a case study. It is a peculiar case study and convenient for our purposes because, while it is of great importance for Sarton's conception regarding the history of science and is in fact present throughout his academic career, it is not very extensive and can therefore be addressed in a more complete manner. We refer to Sarton's treatment of Leonardo da Vinci and a discovery that he did *not* make: the discovery of blood circulation in the seventeenth century, attributed to William Harvey.

The scientific work of Leonardo da Vinci may have served as the main inspiration for Sarton's historical research.[3] In a comparison of studies of manuscripts written by Leonardo, Sarton describes his engagement with the works of the Renaissance author:

> During the years 1916–1919, I was myself engaged in a very searching analysis of the Leonardo MSS [manuscripts], delivered a series of lectures at the Lowell Institute of Boston, and then apparently dropped the subject. I did not drop it but having realized that it was impossible to appreciate correctly Leonardo's scientific thought without a deeper understanding of mediaeval thought than I then possessed, I undertook a systematic investigation of all the mediaeval writings. As my readers know, I have been engaged in that work for the last twenty-five years, and I am still a century short of Leonardo! (Sarton 1944, p. 185)

The twenty-five years work to which Sarton refers is the "monumental" *Introduction to the History of Science*. As Dorothy Stimson stated, this work, published in five volumes from 1927 to 1948, was the result of Sarton's effort "to make a thorough study of Leonardo da Vinci's background" (Stimson 1962, p. xv).[4] But the fact is that Sarton never achieved his goal of producing a work on Leonardo that fulfilled his own expectations. Arnold Thackray and Robert Merton commented on his abandonment of the project aimed at the comprehensive study of Leonardo:

> By 1918 [...] Sarton mentioned a study of Leonardo da Vinci's scientific manuscripts, which would take "about six months" [...]. As early as July 1918 he wrote in tones of mingled consternation and delight that "Leonardo was interested in almost everything…[My book]

[3] Sarton was certainly under the influence of Duhem's book on Leonardo (Duhem 1984 [1906–1913]).

[4] It is interesting to note that in 1949, the year following the publication of the last volume, Sarton speaks again of studying Leonardo and again relates it to the *Introduction:* "Leonardo, sometimes called the father of modern science, was the child of the Middle Ages". Therefore, to assess his thinking adequately, Sarton believed a systematic knowledge of medieval science, "century by century", would be necessary (Sarton 1962, pp. 367–368). In the 1950s, he updated this note, referring then to his past *thirty* or *thirty-five years* of work (Sarton 1953a, p. 11, note 3; Sarton 1957, p. 306, note 4; Sarton 1955, p. x).

> will be in fact an encyclopedia of the positive knowledge attained at the end of the Fifteenth Century". Soon he was seeking specialist help [...]. Reference to the study of Leonardo – his main concern – continues in Sarton's correspondence and annual reports for several more years, though the promised book was never to materialize. (Thackray and Merton 1972, pp. 486–487)

Thackray and Merton (1972, p. 487) assert that Sarton published only two studies about Leonardo, "two popular studies", in 1919 and 1952 (cited here as they were republished, respectively, in Sarton 1948 and 1962).[5] However, despite Sarton having written little about his object of inspiration, what he did write points to the importance he attributed to Leonardo in the history of science, reiterated particularly in his references to William Harvey's discovery of blood circulation in the seventeenth century.

In the present chapter, we seek to show how George Sarton's conception of history of science aligns itself with the traditional perspective, targeted by Kuhn in his criticisms. In the next section, we restricted ourselves to Sarton's treatment of Leonardo and the discovery of blood circulation as a case study.[6] In section three, we argue that Sarton's approach illustrates very well the "old" historiography of science, as opposed to Kuhn's "new historiography of science". Although Kuhn did not write about Leonardo or Harvey, we aim to show that he clearly positioned himself contrary to Sarton, albeit indirectly (in Kuhn's assessment of the work of Butterfield), with respect to this particular historical episode, as well.

15.2 Sarton, Leonardo, and Blood Circulation

In the opening lines of "The Quest for Truth: Scientific Progress during the Renaissance", Sarton writes:

> Many people misunderstand science, and hence one can hardly expect them to have a fair idea of its history. The history of science might be defined as the history of the discovery of objective truth, of the gradual conquest of matter by the human mind; it describes the age long and endless struggle for the freedom of thought – its freedom from violence, intolerance, error, and superstition.

[5] In addition to the studies identified by Thackray and Merton, we consider here other works by Sarton about Leonardo: the articles "Une encyclopédie léonardesque" (Sarton 1919) and "Léonard de Vinci, ingénieur et savant" (Sarton 1953a); two reviews of works by Leonardo organized by different editors (Sarton 1944, 1953b); the preface to McMurrich's book *Leonardo da Vinci, the Anatomist: 1452–1519* (Sarton 1930) and Chapter 6 in *Six Wings: Men of Science in the Renaissance* (his last book, which he did not live to see published in 1957). In the 1919 article, intended to disseminate his future work on Leonardo, Sarton hints that it would be a great work composed of three parts (possibly three volumes). He said it would be entitled "Leonardo: An encyclopedic survey of artistic, scientific and technical thought at the height of the Italian Renaissance" and would be finished by the end of 1920 (Sarton 1919, p. 237).

[6] In a study that is underway, we seek to understand Sarton's history of science taking into consideration a more extensive content of his work.

> The history of science is one of the essential parts of the spiritual history of mankind; the other main parts are the history of art and the history of religion. It differs from these other parts in that the development of knowledge is the only development which is truly cumulative and progressive. Hence, if we try to explain the progress of mankind, the history of science should be the very axis of our explanation. (Sarton 1962, p. 102)[7]

The history of the discovery of blood circulation is one of the accounts Sarton presents synthetically at the beginning of *The Life of Science* as an example of this truly progressive development of knowledge. According to Harvey, blood was pumped by the heart and carried to other parts of the body by the arteries and returned to the heart through the veins. Upon its return, it passed from one side of the heart to the other "through the lungs" where it was resupplied with air before being recirculated again, "in a state of ceaseless motion" (Harvey 1993 [1628], p. 73).

Sarton presented Harvey's discovery as a triumph over scientists who had failed before him to discover the truth, which according to Sarton, had been within reach for a long time. This can be easily observed in Sarton's synthesis of ideas presented by Galen fourteen centuries before Harvey. Among other differences, Galen believed that blood passed from one side of the heart to the other through invisible pores directly, not via the lungs, as Harvey theorized. Sarton writes:

> To explain the impossible, Galen had been obliged to assume that it [the blood] passed through innumerable *invisible* pores in the solid wall which divides the right heart from the left. Nobody ever detected these pores for they are not simply invisible but nonexistent. Yet Galen, supreme pontiff of Greek medicine, and nine centuries later Avicenna, the infallible medical pope of the middle ages, had spoken *ex cathedra* with such indisputable authority that this gratuitous assumption was generally taken for gospel. (Sarton 1948, p. 8)

Sarton argues that ancient and medieval science lacked the parameters of modern science that resulted from the scientific revolution: the experimental method, the skeptical or critical spirit, and rationalism (Cf. Sarton 1950, p. 155). That is why, in his view, men like Galen and Avicenna were unable to present anything other than a "gratuitous assumption" about the circulation of blood.

In the introductory chapter to his major work, Sarton declares his greater interest in modern science, being that which, according to him, accumulated more truths and presented "tremendous" progress (Sarton 1927, p. 14). It is with this enthusiasm for modern science that he looks to the past:

> The historian of science can not devote much attention to the study of superstition and magic, that is, of unreason, because this does not help him very much to understand human progress. Magic is essentially unprogressive and conservative; science is essentially progressive. The former goes backward; the latter, forward. (Sarton 1927, p. 19)

And further on:

> The amount of positive knowledge available in the Middle Ages was exceedingly small [...] There was little opportunity for induction, and knowledge, under scholastic influence, took almost exclusively a deductive form. There was not much choice; granted the arbitrary

[7] He presents his cumulative view in a very similar way in Sarton 1927, pp. 3–4, and 1936, p. 5, passages cited often to illustrate Sarton's conception of the progress of science. See also Sarton 1937, p. 10.

> premises of the theologians and the scarcity of positive knowledge, scholasticism was almost an unavoidable consequence [...]. In spite of political crises and of obscurantist tendencies, positive knowledge must increase and accumulate; every progress in that direction, be it ever so small, was final and irrevocable. Thus we may say that the cure of scholasticism was simply the progress of positive knowledge, and this means the progress of the experimental method. (Sarton 1927, pp. 23–24)

In general, the view presented in the introductory chapter of his extensive work is organized around a comparison between the "relative sterility of scholasticism" and the "immense, almost unconceivable, fertility" of modern science (Sarton 1927, p. 28). The opposition between the two periods is emphasized by his repeated prescription of the experimental method for the "cure of scholasticism".[8] Sarton writes:

> The history of science may always be considered under two aspects, either positively as the gradual unfolding of truth, the increase of light, or negatively as the progressive triumph over error and superstition, the decrease of darkness. The modern scientist studying mediaeval science gets a little impatient, because he is accustomed to a much faster pace; he would like to be able to watch the progress of science, and for many centuries the pace was often so slow, with so many stops and regressions, that one has the feeling that there was no progress at all. (Sarton 1927, p. 25)

This consideration by Sarton leads us back to his account of the discovery of blood circulation, in which the two aspects he cites as being components of science are clearly identifiable. Harvey's contribution was a giant step and led to rapid progress compared to the conceptions of Galen (and Avicenna) which, for many years, were, according to Sarton, responsible for hiding the truth.

In his book about Galen, Sarton (1954) then appears to complete what he believes to be the task of the science historian: on the one hand, to unveil the gradual search for the truth; on the other, present the progressive triumph over error and superstition. When he justifies the importance of Galen for modern science, he presents the contributions of the ancient anatomist, physiologist, physician, surgeon, and pharmacist.[9] However, according to Sarton, Galen never fully realized his potential as a scientist because he

> [...] began his life as a lover of scientific truth, an honest investigator, one of the very few ancients who understood and illustrated the experimental method, yet he ended it as a theologian. He had been carefully trained to be open-minded, impartial, and tolerant, and he preserved some kind of eclecticism in philosophy as well as in medicine, yet he created a scientific doctrine, teleology, which was as dogmatic as anything could be. (Sarton 1954, p. 59)

It is interesting to note how value-laden his historical description of this "scientific doctrine" is, which lasted for many centuries, keeping alive what Sarton refers to as "Galenic aberrations" (Sarton 1954, p. 51). The following passage illustrates well this point of view regarding Galen:

---

[8] In addition to this passage, the expression "cure" also appears on pp. 25 and 29.

[9] It is interesting to note Sarton's search for Galen's contributions to areas of specialization that came to exist only after the so-called scientific revolution.

> His description of the blood vessels was very insufficient and confusing, but who would dare blame him for that? He might have discovered the pulmonary circulation, but he did not, and his influence blocked the way for the discovery of the real circulation. Harvey himself, as late as 1628, had to be careful not to offend the prejudices and not to wound the feelings of his Galenic readers. (Sarton 1954, p. 47)

It was because of this that David Lindberg refers directly to Sarton and his *Galen of Pergamon* to illustrate how Galen had been the target of abuses by historians who were "angry at him for not being modern" (Lindberg 2007, p. 130).

Modern science, according to Sarton, was a child of the Renaissance, born of those who, imbibed with the spirit of experimentation and the search for the truth, promoted a rapid expansion of knowledge (Sarton 1948, p. 78). It is worth noting again here the article "The Quest for Truth", where he presents the Renaissance as the period of transition between the Middle and the Modern Ages in which the innovations were huge and revolutionary. The novelties were so numerous, according to Sarton, that it is not possible to speak of a Renaissance, since the period was truly a "real birth, a new beginning" (Sarton 1962, p. 104).

Leonardo appears to him to be an example of the true scientific spirit: rather than indebting himself to past authorities, he observes nature; he has a critical spirit and conducts experiments. He is able to overcome the fundamental vice of the scholastics, anticipating by a century and a half, in practice, the method propagated by the philosopher Francis Bacon. These are some of the considerations presented by Sarton in the text which bears the significant title "Leonardo and the Birth of Modern Science".[10]

In "Leonardo da Vinci",[11] Sarton once again provides evidence of the revolutionary aspects of the work of the great genius, pioneer of so many branches of scientific knowledge, and asserts that "the historian of science is impressed by Leonardo's gadgets" (Sarton 1962, p. 134). And from an anachronistic perspective, as pointed out by Lindberg with respect to Galen, Sarton examines the difficulties faced by the great Renaissance man, asserting that while he had no superstitions, he was unable to completely free himself of old prejudices.

In fact, Sarton presents the history of blood circulation in a way that reveals the difficulties encountered by those who might have discovered "the truth" earlier. The list of observers, whose paths were blocked by the "gospel" according to Galen (and later Avicenna), includes names of scientists who lived the golden age of the search for truth, the Renaissance:

> When I shut my eyes and evoke the past, I imagine that this great discovery was enclosed in a chest of which intelligent observers like Leonardo, Vesalius, Servetus or Columbus

[10]The 1919 text is the first of two studies by Sarton about Leonardo cited by Thackray and Merton (and republished in *The Life of Science*). The original title is "The Message of Leonardo: his Relation to the Birth of Modern Science" (Sarton 1948, p. 188). See also Strelsky 1957, item 69. In Sarton 1937, pp. 99–100, he refers to what he calls the "experimental spirit" and says that "we may consider Leonardo da Vinci its first deliberate vindicator".

[11]The second of Sarton's two texts about Leonardo cited by Thackray and Merton. Published in 1952, it corresponds to item 8 of Katharine Strelsky's catalogue, cited in the preceding footnote. See also footnote 5, above.

> could have easily found the secret if they had set their hearts upon it, but they did not dare approach near enough because Prejudice sat on the lid. I can see those great men standing shyly around the coffer, mysteriously attracted by it, yet awed into impotence, while Truth was prisoner inside. (Sarton 1948, p. 9)[12]

Regarding the "intelligent observers" who could have found a definitive explanation for blood circulation, Sarton appears to regret in particular the fact that *even* Leonardo da Vinci

> […] endowed with so much genius and originality, and had himself dissected a large number of bodies and examined very minutely many a heart [...] was subjugated by this intangible dogma. (Sarton 1948, p. 9)

Sarton goes so far as to assert that Leonardo had perceived the true explanation but that the "invisible pores" imagined by Galen and Avicenna were too sacred to be contested. Leonardo, he believed, was

> […] so completely dominated by the Galenic prejudice that he was not only able to see the invisible pores, but even to draw them. It would be difficult to think of a better example of the limitations of the genius (Sarton 1953a, p. 17).[13]

According to Sarton, it was prejudice that kept Leonardo from making the discovery about blood circulation:

> Galen's triumphant dogmatism made even a Leonardo see the inexistent. But for this illusion which sidetracked him hopelessly, Leonardo might conceivably have discovered the circulation of the blood before Harvey, for he had as much anatomical and mechanical knowledge as was needed. He had all that was necessary to see the truth, except that in this particular case he was blinded by an overpowering prejudice. (Sarton 1930, p. xix)[14]

Thus, the way Sarton refers to Leonardo because of the discovery he did *not* make is illustrative of the importance Sarton attributed to him in the history of science. In

[12] Curiously, Koyré writes that with Leonardo, science was associated with a "[…] substitution of *fides* and *traditio*, of the knowledge of others, for personal *vision* and *intuition*, free and unembarrassed" and that in light of this, "Galileo and his friends, the members of the *Accademia dei Lincei* [...] rejected authority and tradition and wanted to *see* things as they were" (Koyré 1973, pp. 115–116. See also p. 399, where he speaks of an *itinerarium mentis in veritatem*). It is perhaps for this reason that Kuhn says that what he values in Koyré (as in Brunschvicg and Meyerson) is the historian and not the philosopher (Kuhn 1994, pp. 158–159 and Kuhn 1995b, p. 13). For Kuhn's restrictions regarding Koyré as a historian, as well, see Pinto de Oliveira 2012. As one can see on pp. 118–119 of this article, these restrictions appear to include the idea that Koyré still has one foot in the "old" historiography. This could explain, in part, Koyré's sympathy for Sarton, irrespective of specific reasons or circumstances, such as the complimentary letter of recommendation written by Sarton for Koyré to the New School for Social Research of New York in 1940 (Koyré 1986, p. 59, p. 63).

[13] In the 1952 *Colloque* about Leonardo, Koyré does not refer critically to Sarton and to the passage cited here. He limits himself again to positive and trivial references. In turn, although Sarton surprisingly does not mention Koyré in Sarton 1952, he did not represent effective opposition to Koyré, unlike Aldo Mieli, for example (Koyré 1986, pp. 35–37, p. 63). See also Bernard Cohen (1987, p. 68, note 1) and Chimisso (2008, pp. 131–132).

[14] See also Sarton (1959 [1929], p. 87). In addition to the Galenic prejudice, Sarton also identifies "Platonic prejudices" (Sarton 1962, p. 137; Sarton 1953a, p. 17).

the reviews written by Sarton regarding Leonardo, his assessment of the works is guided precisely by the interpretation that the organizers present with respect to Leonardo's "non-discovery". In his 1944 review, we read:

> In order to measure the value of the Richter and MacCurdy collections for the historian of science, let us consider a crucial example, Leonardo's views on the circulation of the blood. Richter does not hesitate to say "Leonard had a clear conception of it" (his vol. 2, 105, note). That statement is as preposterous as it is dogmatic. The few extracts quoted by himself do not in the least justify it. MacCurdy does not dogmatize, but he gives us a much larger selection of anatomical and physiological items and enables us to reach truer conclusions. (Sarton 1944, pp. 185–186)

In the review of Leonardo's works compiled by O'Malley and Saunders, Sarton highlights precisely Leonardo's anatomy notebook, and asserts:

> The main advantage of this publication for English-reading historians of science is the fact that all the drawings and the texts concerning them are grouped in systematic order [...] An introduction contains all the information relative to Leonardo's life, anatomical illustrations anterior to him, his anatomical achievements, his MSS [manuscripts], etc. Let us give an example; the best that one could choose is the conclusions on Leonardo's knowledge of the heart and the movement of the blood. (Sarton 1953b, p. 65)

Sarton then cites a passage in which O'Malley and Saunders argue that Leonardo had no knowledge of blood circulation and that, until 1500, his opinion was derived from the degraded view of the ancient wise man, Galen. The great praise of the book is due, above all, to his having shown Leonardo's limitations with respect to anticipating Harvey's discovery.[15]

15.3 Kuhn, Butterfield, and Sarton

As we saw, in one of the few passages where he cited names responsible for the traditional historiography of science, Kuhn (1977, p. 148) refers to a tradition that extends "from Condorcet and Comte to Dampier and Sarton". According to him, this tradition

> [...] viewed scientific advance as the triumph of reason over primitive superstition, the unique example of humanity operating in its highest mode. [...] the chronicles which this tradition produced were ultimately hortatory in intent, and they included remarkably little information about the content of science beyond who first made which positive discovery when. [...]. Though I know it will give offense to some people whose feelings I value, I see no alternative to underscoring the point. Historians of science owe the late George Sarton an immense debt for his role in establishing their profession, but the image of their specialty which he propagated continues to do much damage even though it has long since been rejected. (Kuhn 1977, p. 148. See also p. 106)

[15] He confirms this appreciation in Sarton (1957, pp. 174–175, p. 294 note 4).

In a long interview in 1995, when asked why he had not associated himself directly with Sarton at Harvard,[16] Kuhn cited a significant divergence between them:

> He [Sarton] certainly was a Whig historian and he certainly saw science as the greatest human achievement and the model for everything else. And it wasn't that I thought that it was *not* a great human achievement, but I saw it as one among several. I could have learned a lot of data from Sarton but I wouldn't have learned any of the sorts of things I wanted to explore. (Kuhn 2000, p. 282)[17]

Immediately following that comment, in a rapid assessment of the American academic environment in which he began his studies of the history of science, Kuhn highlights the discrepancy between his point of view and that of Sarton and a few other contemporaries. According to him, what they did was not "quite history; it was textbook history" (Kuhn 2000, p. 282. See also Kuhn 1970, p. 1, pp. 137–138).

In addition, in the *Structure*, Kuhn writes that textbooks

> begin by truncating the scientist's sense of his discipline's history and then proceed to supply a substitute for what they have eliminated. Characteristically, textbooks of science contain just a bit of history, either in an introductory chapter or, more often, in scattered references to the great heroes of an earlier age. From such references both students and professionals come to feel like participants in a long-standing historical tradition. Yet the textbook-derived tradition in which scientists come to sense their participation is one that, in fact, never existed. For reasons that are both obvious and highly functional, science textbooks (and too many of the older histories of science) refer only to that part of the work of past scientists that can easily be viewed as contributions to the statement and solution of the texts' paradigm problems. Partly by selection and partly by distortion, the scientists of earlier ages are implicitly represented as having worked upon the same set of fixed problems and in accordance with the same set of fixed canons that the most recent revolution in scientific theory and method has made seem scientific. (Kuhn 1970, pp. 137–138)

In his criticism of the older historiography in *The Essential Tension*, Kuhn points to examples of works that he believed were changing the history of the scientific development in a significant way. Of particular interest to us here are his considerations regarding the work of Herbert Butterfield. For Kuhn, this (general) historian's book about the history of science, *The Origins of Modern Science,* first published in 1949, contributed significantly to break from the misleading view of modern science as

[16] Both were at Harvard from 1940 to 1951. Kuhn also comments in another excerpt from the interview: "I had sat in on some lectures of Sarton's as an undergraduate and found them turgid and dull" (Kuhn 2000, p. 275). It is worth noting that, despite the many references he made to Sarton in *The Copernican Revolution*, it should be taken into account what Kuhn wrote in *The Essential Tension* about the whiggish histories: they are "sometimes invaluable as reference works, but otherwise virtually useless to the man whose interests include the development of ideas" (Kuhn 1977, p. 135; see also Koyré 1986, p. 136).

[17] The expression "Whig" was applied to the historiography by Butterfield in 1931. As summarized by the author, Whig history, or "whiggism" is "the tendency in many historians to write on the side of Protestants and Whigs, to praise revolutions provided they have been successful, to emphasise certain principles of progress in the past and to produce a story which is the ratification if not the glorification of the present" (Butterfield 1973 [1931], p. 9). Kuhn uses this expression frequently (as well as "textbook history") to refer critically to traditional historiography of science.

having emerged due to the new observations and methods. And when he highlights the importance of examining medieval science to understand the essential novelties of the seventeenth century, Kuhn points to Butterfield's "pioneering synthesis" as a reference on a path worth following (Kuhn 1977, pp. 108–109).

According to Kuhn (Kuhn 1977, p. 35, note 3), Butterfield, in his studies regarding the origins of modern science, "plausibly explained the main conceptual transformations of early modern science as 'brought about, not by new observations or additional evidence in the first instance, but by transpositions that were taking place inside the minds of the scientists themselves'".[18] However, while Butterfield may have written an "admirable" work and traced a better path for the historical analysis of science, he did not, in Kuhn's judgment, remain entirely faithful to this path throughout the work. Kuhn writes:

> Butterfield's first four chapters plausibly explained the main conceptual transformations of early modern science [...] The next two chapters, "The Experimental Method in the Seventeenth Century" and "Bacon and Descartes", provided more traditional accounts of their subjects. Although they seemed obviously relevant to scientific development, the chapters which dealt with them contained little material actually put to work elsewhere in the book. (Kuhn 1977, p. 35, note 3)

In another reference, Kuhn's caveat is more specific. The object of his praise is no longer all of the first four chapters. Despite referring again to the *Origins of Modern Science* as "admirable", Kuhn says that

> One aspect of Butterfield's discussion has, in fact, helped to preserve the myths. The historiographic novelties accessible through his book are concentrated in chaps. 1, 2, and 4, which deal with the development of astronomy and mechanics. These are, however, juxtaposed with essentially traditional accounts of the methodological views of Bacon and Descartes, illustrated in application by a chapter on William Harvey. (Kuhn 1977, p. 131, note 2)[19]

The chapter about William Harvey to which Kuhn refers to is Chapter 3. Kuhn claims it fails to provide a plausible explanation of the conceptual transformations at the beginning of modern science and supports "essentially traditional accounts of the methodological views". It is worth asking at this point: What characteristics of Chapter 3 of the *Origins of Modern Science* render it an "essentially" traditional account? Not surprisingly, they are the same ones that are present in Sarton's analysis regarding the discovery of blood circulation.

The first note in Butterfield's chapter on Harvey that calls our attention as being characteristic of a traditional history is the comparison between the development of knowledge beginning with the Renaissance and that which preceded it. In this

[18] The reference to Butterfield as an example of a "perceptive historian" appears already in *Structure* (Kuhn 1970, p. 85). See also Kuhn 1995a [1957], p. 283.

[19] In his book about Butterfield, after referring to the above passage, Sewell wrote: "In making these observations Kuhn was apparently unaware of their relevance to the problem of reconciling Butterfield's concept of technical history and an expository historiography based on his belief in providence. Butterfield's three ways or levels formulation had sought to distinguish these, but in the course of doing so he effectively integrated them" (Sewell 2005, p. 163). A discussion of this point is beyond the scope of this work, however.

aspect, Butterfield exalts the great advances in the art of observation, remembering that artists (Leonardo da Vinci, in particular) "were the first to cry out against mere subservience to authority – the first to say that one must observe the nature for oneself" (Butterfield 1966, p. 51).

Much like Sarton, Butterfield exalted the discovery of blood circulation as the result of merit and the individual capacity of the discoverer to extricate himself from past tradition, remembering that Harvey would have declared that he "learned and taught anatomy, 'not from books but from dissection'" (Butterfield 1966, p. 61). In this evaluation, the historian emphasizes the role of Harvey's experimental method, which reveals an "extraordinarily modern flavour" (Butterfield 1966, p. 62) and breaks definitively from the Galenic view. After presenting a synthesis of this view, he asserts:

> Here we have a complex system of errors concerning which it has to be noted that the doctrine was not only wrong in itself, but, until it was put right, it stood as a permanent barrier against physiological advance – for, indeed, nothing else could be right. It is another of those cases in which we can say that once this matter was rectified the way lay open to a tremendous flood of further change elsewhere. (Butterfield 1966, pp. 54–55)

According to Butterfield (1966, p. 55), Galen's dominance endured for a long time and extended over many thinkers – "even Leonardo da Vinci". In his analysis of the evolution leading to Harvey's discovery, Butterfield discusses the contributions of scientists who could have made the discovery but did not because they were overly influenced by Galen:

> Until the seventeenth century, therefore, *a curious mental rigidity* prevented even the leading students of science from realising essential truths concerning the circulation of the blood, though we might say with considerable justice that they already held some the most significant evidence in their hands. (Butterfield 1966, p. 58, our *italics*)

Butterfield commented that the consolidated reception of Harvey's work took 30–50 years, "though his arguments would perhaps seem more cogent to us today than those of any other treatise that had been written up to this period" (Butterfield 1966, p. 65). For Butterfield, what made the big difference was the experimental method:

> Only now could one begin to understand respiration itself properly, or even the digestive and other functions. Given the circulation of blood running through the arteries and then back by the veins, one could begin to ask "what it carries, and why, how and where it takes up its loads" […] Both in regard to methods and results, therefore, we seem to have touched something like the genuine scientific revolution at last. (Butterfield 1966, pp. 65–66)

It is interesting to note that Butterfield's approach to the so-called chemical revolution is also traditional. See, for example, the passage (Butterfield 1966, p. 211) where he describes how researchers, "incapacitated by the phlogiston theory", failed to see "the truth under their very noses".[20]

---

[20] Compare this to Sarton's abovementioned text about the truth hidden in a chest (Sarton 1948, p. 9).

15.4 Concluding Remarks

"For forty years the name of George Sarton has been practically synonymous with the history of science". The opening statement of Dorothy Stimson's preface to *Sarton on the History of Science*, published in 1962, is still very expressive 50 years later but is not lacking in ambiguity. It was also in 1962 that Kuhn, with *The Structure of Scientific Revolutions,* announced the emergence of a new history of science in response to an older history of science, which existed as an autonomous academic discipline thanks to, above all, the untiring work of Sarton.

Thus, if Stimson's recognition of Sarton's contribution to the establishment of the field still carries weight,[21] it is also true that such an assertion can now be challenged in light of the new conception of the history of science, which scarcely mentions the historical work of Sarton.[22]

As we have seen above, Butterfield maintains some traces of the history that Kuhn assessed as being traditional because of Butterfield's distorted view of the role of the experimental method in the scientific revolution (especially in the case of physiology). We should remember that Kuhn refers to physiology as being a field in which experimentation did not follow the Baconian model but rather the classic model of Galen (Kuhn 1977, p. 136). It is also worth noting that, in one of his rare mentions of Leonardo da Vinci in his work, Kuhn writes:

> [...] as Leonardo's career also indicates, instrumental and engineering concerns do not make a man an experimentalist" (*Kuhn* 1977, p. 49; see also Koyré 1973, p. 110).

The aspects that approximate Butterfield to Sarton are the same ones that distance him from Kuhn. *The Origins of Modern Science* represents a step toward the new historiography but, according to Kuhn, still has a foot in the traditional historiography. The characteristics of this historiography which Kuhn identifies in Butterfield can also be found in Sarton's historical account of the discovery of blood circulation. Thus, this episode seems to us to be a concrete example of Whig history not explicitly pointed out by Kuhn, an example that attests to the substantial difference between the "old" and the "new" historian of science.[23]

It is worth remembering at this point a passage by Sarton cited earlier in this chapter: "Many people misunderstand science, and hence one can hardly expect them to have a fair idea of its history" (Sarton 1962, p. 102). Kuhn, on the other hand, throughout *Structure*, suggests the opposite: that if one does not have a good

---

[21] See, for example, Kuhn (1977, p. 148), Fichant (1969, p. 67), and Kragh (1989, p. 19).

[22] Helge Kragh (1989, pp. 18 and 198, note 43), for example, states that Sarton's view is, according to modern standards, "somewhat naive and surprisingly ahistorical", with reference to the judgment of Rupert Hall (1969) who, despite considering Sarton a man of great knowledge, admits that one must ask if he was in fact a historian. Kragh also cites Kuhn (1977, p. 148) for evidence of Sarton's supposedly ahistorical view (for a discussion, see Pinto de Oliveira 2012, pp. 119–121). Sayili (2005) and Pyenson (2007) are among the few interpreters who recently sought to reaffirm the importance of Sarton's work as a historian.

[23] In Pinto de Oliveira 2015, the author relates Carnap (and Reichenbach) to Sarton's "old" historiography.

understanding of the history of science, one will not understand science itself (see, for example, Section I: Introduction).

To conclude, we cite two passages by Sarton which, like other more diffuse passages cited in the text, can be seen as "negative images" of Kuhn's normal science (or Sarton's "abnormal science"):

> Neither do I mean to imply that all the schoolmen were dunces. Far from that, not a few were men of amazing genius, but their point of view was never free from prejudice; it was always the theological or legal point of view; they were always like lawyers pleading a cause; they were constitutionally unable to investigate a problem without reservation and without fear. Moreover, they were so cocksure, so dogmatic. (Sarton 1948, p. 77)
>
> It has often been repeated that anatomy was neglected in the Middle Ages because of religious prejudices. Anatomy was not completely neglected, and dissections, even human dissections, were made from time to time, but these dissections were few and they were not made with sufficient application nor with sufficient freedom of thought. The shackles of the medieval anatomists were less religious than scholastic. Medical men had not acquired the habit of seeing with their eyes open without prejudices. Indeed, they were so much dominated by older masters such as Galen and Avicenna that they were not only blind to reality but able to see things which were not there at all; Galen's words were more convincing to them than reality itself! It is a bit difficult for us to imagine such a state of mind, though it has not yet completely disappeared. The renovation of anatomy was finally accomplished by men who were good observers, had dexterous hands and sharp eyes, and were not inhibited by prejudices. (Sarton 1962, p. 134)

Kuhn positions himself against negative assessments of this type when he speaks of normal science. Long before *Structure* and the concept of paradigm, as John Preston points out (Preston 2008, p. 5), Kuhn already viewed these "prejudices" or "points of view" and "principles" or "conceptual frameworks", not as impediments but as essential to the development of science.[24]

Moreover, Kuhn speaks of "dogmas", as well. In "The Function of Dogma in Scientific Research",[25] he writes:

> At some point in his or her career every member of this Symposium [Oxford, 1961] has, I feel sure, been exposed to the image of the scientist as the uncommitted searcher after truth. He is the explorer of nature – the man who rejects prejudice at the threshold of his laboratory, who collects and examines the bare and objective facts, and whose allegiance is to such facts and to them alone. […] To be scientific is, among other things, to be objective and open-minded. (Kuhn 1963, p. 347)

And:

> Almost no one, perhaps no one at all, needs to be told that the vitality of science depends upon the continuation of occasional tradition-shattering innovations. But the apparently contrary dependence of research upon a deep commitment to established tools and beliefs receives the very minimum of attention. I urge that it be given more. Until that is done,

[24] The same is true of Koyré and other authors, such as Metzger and Febvre (Redondi 1986, pp. xvi–xvii).

[25] Kuhn expressed dissatisfaction with this text and did not want it included in the collections of his essays (Kuhn 2000, p. 2, note 1). However, he certainly had no objections to the first and last paragraphs, which we cite here to conclude our chapter.

some of the most striking characteristics of scientific education and development will remain extraordinarily difficult to understand. (Kuhn 1963, p. 369)

We can thus conclude that Sarton fits Kuhn's image of the traditional historian of science very well. He accepted and propagated the image of the scientist as "the uncommitted searcher after truth" and viewed negatively, as mere prejudice that obstructs the pure and objective view of the facts, that which Kuhn understood positively to be the paradigm that necessarily guides (normal) scientific research. For Sarton, as Kuhn said regarding the traditional conception of science, it is as if, since the beginning, "scientists have striven for the particular objectives that are embodied in today's paradigms" (Kuhn 1970, p. 140). Or, to be more precise, for Sarton, there are no paradigms in science. There are only pure and objective views of the facts "that were there all the time" (Kuhn 1970, p. 141) or Sarton's "abnormal science": deformed views of the facts resulting from error, superstition, and prejudices.

These conceptions are entirely consistent with the idea of strictly cumulative progress in science, which Kuhn criticizes and Sarton explicitly defends (see Pinto de Oliveira forthcoming).

Acknowledgments We are grateful to Paul Hoyningen-Huene for his comments on an earlier draft of this work and to Anne Kepple for translations and revisions. Research supported by FAPESP – Proc. 2013/20172-0.

References

Agassi J ([1963] 2008) Towards an historiography of science. In Science and its history. A reassessment of the historiography of science. Boston Studies in the Philosophy of Science. Vol. 253. Springer–Verlag, New York, pp. 119–242.

Bernard Cohen I (1987) Alexandre Koyré in America: some personal reminiscences. In Redondi P (ed). Science: The renaissance of a history. History and Technology 4:1–4. Proceedings of the international conference Alexandre Koyré, Paris, Collège de France, 10–14 June 1986. Harwood Academic Publishers, London, pp. 55–70.

Butterfield H ([1949] 1966) The origins of modern science 1300–1800. The Free Press, New York.

Butterfield H ([1931] 1973) The Whig interpretation of history. Penguin, Harmondsworth.

Chimisso C (2008) Writing the history of the mind: Philosophy and science in France, 1900 to 1960s. Ashgate, Aldershot.

Duhem P ([1906–1913] 1984) Études sur Léonard de Vinci : ceux qu'il a lus et ceux qui l'ont lu. Archives Contemporaines, Paris.

Fichant M (1969) Sur l'histoire des sciences. In Pécheux M, Fichant M (eds). Sur l'histoire des sciences. Maspéro, Paris, pp. 13–57.

Guerlac H (1963) Some historical assumptions of the history of science. In Crombie AC (ed). Scientific change: historical studies in the intellectual, social and technical conditions for scientific discovery and technical invention, from Antiquity to the present – Symposium on the history of science, University of Oxford, 9–15 July, 1961. Heinemann Educational Books, London, pp. 797–812.

Harvey W ([1628] 1993) On the motion of the heart and blood in animals. Translated by Robert Willis. Great Minds Series. Prometheus Books, Buffalo.

Jorland G (1981) La science dans la philosophie : les recherches épistémologiques d'Alexandre Koyré. Gallimard, Paris.

Koyré A (1948) Condorcet. Journal of the History of Ideas 9/2:131–152.
Koyré A (1953) Rapport final. In Febvre L (ed). Léonard de Vinci et l'expérience scientifique au seizième siècle. Presses Universitaires de France, Paris, pp. 237–246.
Koyré A ([1939] 1966) Études galiléennes. Hermann, Paris.
Koyré A ([1966] 1973) Études d'histoire de la pensée scientifique. Gallimard, Paris.
Koyré A (1986) De la mystique à la science : cours, conférences et documents 1922–1962. Redondi P (ed). EHESS Éditions, Paris.
Kragh H (1989) An introduction to the historiography of science. The Cambridge University Press, Cambridge.
Kuhn TS (1963) The function of dogma in scientific research. In Crombie AC (ed). Scientific change: historical studies in the intellectual, social and technical conditions for scientific discovery and technical invention, from Antiquity to the present – Symposium on the history of science, University of Oxford, 9–15 July, 1961. Heinemann Educational Books, London, pp. 347–369.
Kuhn TS (1966) Review of Agassi J. Towards an historiography of science. The British Journal for the Philosophy of Science 17/3:256–258.
Kuhn TS ([1962] 1970) The structure of scientific revolutions. The University of Chicago Press, Chicago.
Kuhn TS (1977) The essential tension. The University of Chicago Press, Chicago.
Kuhn TS ([1991] 1994) Paradigms of scientific evolution. In Borradori G (ed). The American philosopher: conversations with Quine, Davidson, Putnam, Nozick, Danto, Rorty, Cavell, Macintyre, and Kuhn. Translated by Crocitto R. The University of Chicago Press, Chicago, pp. 153–167.
Kuhn TS ([1957] 1995a) The Copernican revolution: Planetary astronomy in the development of western thought. The Harvard University Press, Cambridge.
Kuhn TS (1995b) Un entretien avec Thomas S. Kuhn. Translated and edited by Christian Delacampagne. Le Monde, LI année, no. 15.561, dimanche 5 – lundi 6 février, p. 13.
Kuhn TS (2000) The road since Structure. The University of Chicago Press, Chicago.
Lindberg DC (2007) The beginnings of western science: The European scientific tradition in philosophical, religious, and institutional context, prehistory to A.D. 1450. The University of Chicago Press, Chicago.
Pinto de Oliveira JC (2012) Kuhn and the genesis of the "new historiography of science". Studies in History and Philosophy of Science 43:115–121.
Pinto de Oliveira JC (2015) Carnap, Kuhn, and the history of science: A reply to Thomas Uebel. Journal for General Philosophy of Science 46: 215–223.
Pinto de Oliveira JC (forthcoming) Thomas Kuhn, the image of science and the image of art: The first manuscript of Structure. Perspectives on Science.
Preston J (2008) Kuhn's The structure of scientific revolutions: A reader's guide. Continuum, London.
Pyenson L (2007) The passion of George Sarton. A modern marriage and its discipline. The American Philosophical Society, Philadelphia.
Redondi P (1986) Préface. De l'histoire des sciences à l'histoire de la pensée scientifique : le combat d'Alexandre Koyré. In Koyré 1986, pp. ix–xxvii.
Rupert Hall A (1969) Can the history of science be history? The British Journal for the History of Science 4:207–220.
Sarton G (1919) Une encyclopédie léonardesque. Raccolta Vinciana 10:235–242.
Sarton G (1927) Introduction to the history of science. Vol. 1. Williams and Wilkins, Baltimore.
Sarton G (1930) Preface. In McMurrich JP. Leonardo da Vinci, the anatomist: 1452–1519. The Williams & Wilkins Co., Baltimore, pp. xv-xx.
Sarton G (1936) The study of the history of science. The Harvard University Press, Cambridge.
Sarton G ([1931] 1937) The history of science and the new humanism. Harvard University Press, Cambridge.

Sarton G (1944) Review of Richter JP, Richter IA The literary works of Leonardo da Vinci; MacCurdy E. The notebooks of Leonardo da Vinci. Isis 35/2:184–187.
Sarton G (1948) The life of science. Essays in the history of civilization. Henry Schuman, New York.
Sarton G (1950) Boyle and Bayle. The sceptical chemist and the sceptical historian. Chymia 3:155–189.
Sarton G (1952) A guide to the history of science. Chronica Botanica Company, Waltham, Massachusetts.
Sarton G (1953a) Léonard de Vinci, ingénieur et savant. In Febvre L (ed). Léonard de Vinci et l'expérience scientifique au seizième siècle. Presses Universitaires de France, Paris, pp. 11–22.
Sarton G (1953b) Review of O'Malley CD, Saunders JB. Leonardo da Vinci on the human body. Isis 44/1–2: 65–66.
Sarton G (1954) Galen of Pergamon. The University of Kansas Press, Kansas.
Sarton G (1955) Appreciation of ancient and medieval science during the Renaissance (1450–1600). University of Pennsylvania Press, Philadelphia.
Sarton G (1957) Six Wings: men of science in the Renaissance. The Indiana University Press, Bloomington.
Sarton G ([1929] 1959) Science in the Renaissance. In Thompson JW et al. The civilization of the Renaissance. Frederick Ungar Publishing Co., New York, pp. 75–95.
Sarton G (1962) Sarton on the history of science. Essays by George Sarton. Stimson D (ed). The Harvard University Press, Cambridge.
Sayili A (2005) George Sarton and the history of science. Foundation for Science, Technology and Civilization. Retrieved: http://www.muslimheritage.com/uploads/Sarton.pdf
Sewell KC (2005) Herbert Butterfield and the interpretation of history. Palgrave Macmillan, New York.
Stimson D (1962) Preface. In Sarton 1962, pp. v–x.
Strelsky K (1957) Bibliography of the publications of George Sarton. Isis 48/3:336–350.
Thackray A, Merton R (1972) On discipline building: The paradoxes of George Sarton. Isis 63/4:472–495.

Chapter 16
On the Conceptualization of Force in Johannes Kepler's *Corpus*: An Interplay Between Physics/Mathematics and Metaphysics

Raffaele Pisano and Paolo Bussotti

Abstract In this chapter, we present the concept of force in Kepler. We follow the development of this concept during Kepler's scientific career, starting from his early considerations in the *Mysterium Cosmographicum* (1596) until his ripest conceptions expounded in the *Epitome Astronomiae Copernicanae* (1618–1621). Kepler tried to supply a dynamical explanation to the planetary movements. This is an important novelty because astronomy was traditionally a kinematical science. Based on the main accredited literature we present a historical account and theoretical/nature of science (NoS) developments: (1) Keplerian forces; (2) physical astronomy; and (3) orbits, force, and the relation between distances and forces. Koyré had a prominent role in the studies on astronomical revolution and on Kepler. He was also important in fully clarifying the pivotal function that physical astronomy—and hence the concept of force—had in Kepler's system of the world. Indeed, Koyré is still nowadays an almost unavoidable reference point for Kepler's *Forschung*. We refer to several interpretations of his. This is the reason why we think appropriate to present this work in homage to Alexandre Koyré.

Keywords Kepler • Newton • Koyré • Concept of force • Gravity • *Virtus Tractoria* • *Virtus Promotoria* • Relationships Physics–Mathematics • Science in context • Nature of science • Comparative history of physics • Historical epistemology of science • *Mysterium Cosmographicum* • *Astronomia Nova* • *Epitome Astronomiae Copernicanae*

R. Pisano (✉)
Lille University, Villeneuve d'Ascq, France

Archive Poincaré, Lorraine University, Nancy, France

Unit HPS, Sydney University, Sydney, NSW, Australia
e-mail: raffaele.pisano@univ-lille3.fr

P. Bussotti
Department of Human Studies, Udine University, Udine, Italy
e-mail: paolobussotti66@gmail.com

R. Pisano et al. (eds.), *Hypotheses and Perspectives in the History and Philosophy of Science*, DOI 10.1007/978-3-319-61712-1_16

16.1 A Specification on the Content of This Chapter

While referring to the concept of force[1] in Kepler, it is possible to deal with four kinds of problems:

(a) Analysis of the mechanisms by means of which the forces act. For example: is the force exerted by a body on another body a function of the distance between the two bodies?
(b) Analysis of the way in which the force's action is spread. For example: is the force spread as an action at a distance or does the force act by contact or by mechanisms that cannot be reduced to these two hypotheses?
(c) Analysis pertaining to the features of the bodies emanating the forces. For example: does a body exert a force because such a body has a soul or an intellect or only because of its corporeal nature?
(d) Analysis of the concept of force inside the general conception of an author. In the specific case of Kepler, analysis of his concept of force inside his geometrical-harmonic conception of the world, where astrology and religion also play a role.

In this chapter, we deal with items (a) and (b) and with the connected problems. We sometimes refer to item (c), and we do not deal with the general problem (d). In our discussion, we also address the following three questions, which are significant in the interpretation of Kepler's thought:

(a) What exactly did Kepler mean with *virtus tractoria*, *virtus motrix,* and *vis* and did their meaning change in the course of his scientific career?
(b) What is the link between Kepler's kinematical astronomy and his physical astronomy? Thus: what are the relationships between the first and second of Kepler's laws and his concept of force?
(c) Can Kepler's concept of force work—at least in part—as a coherent physical theory?

We first analyze the notion of force in the *Mysterium Cosmographicum*. Afterwards, we address Kepler's conception of gravity. Finally, we face our main subject: the forces responsible for the planetary motions in *Astronomia Nova* and *Epitome Astronomiae Copernicanae* (KGW, VII). We remark, that, as usual in our historical investigations, we worked on Latin texts. Although some English versions were of help, we provided our translations. Some images are in the public domain (i.e., *Google Books*); others are incorporated such as "with permission of." Therefore, all cases are cited below.

[1] A terminological specification: while dealing with the term "force" in Kepler, we refer to Latin words used by Kepler as *vis* and *virtus*. Kepler spoke of *vis* and *virtus* and of their actions, but did not define them. This means, we are free to use the English term "force" to indicate these concepts, without being compromised with Newton's concept of force. For the sake of brevity, we avoid an introduction about *de motu locali* related to the genesis of the theory of *impetus* from Aristotle (384–322 BC)/John Philoponus (490–570) to Jean Buridan (ca. 1300–ca. 1360; see Buridan 1509), Nicolas d'Oresme (1320? 1325?–1382), and so on, until to Niccolò Tartaglia (1499?–1557). Tartaglia also presented contributions to the art of warfare in *Nova scientia* (Tartaglia 1537, Books I–II) and *Quesiti et inventioni diverse* (Tartaglia 1554, Books I–III; Pisano and Capecchi 2015).

16.2 A Concise Overview of the Literature Concerning the Concept of Force in Kepler

The literature on the concept of force in Kepler is rather conspicuous[2]: after the admirable technical explanations of Kepler's astronomy (of which his ideas on forces are a part) by Small (1804) and Delambre (1817, 1819, 1821), Goldbeck (1896) published a contribution on Kepler's concept of gravity which is dated but still important. The classical work by Dreyer (1906) dedicates some pages to the problem of the forces responsible for the planetary motions in Kepler. Caspar gave remarkable contributions in his biography of Kepler (1948) and in his *Nachbericht* to *Astronomia Nova* (KGW III, pp. 426–484) and *Epitome Astronomiae Copernicanae* (KGW VII, pp. 541–610). Caspar's work is fundamental to understand a series of technical aspects concerning the way in which the various forces conceived by Kepler act.

After Caspar's studies, Koyré[3] in *La Revolution Astronomique* (1961c) dedicates many parts of the chapter on Kepler (by far the longest one of his masterpiece) to the concept of force. Koyré provides a picture in which all four aspects (a)–(d) are analyzed. The work by Koyré is fundamental because, first of all, a translation into modern languages of many passages by Kepler, which were available only in Latin, is offered. Furthermore, Koyré provides the reader with a series of commentaries that strictly connect the texts to which he is referring. We think that this is a quite good and correct historiographical praxis, because the reader can achieve a personal idea of the texts on which the historian's interpretation is based. Moreover, Koyré also develops a series of original ideas on Kepler's astronomy, specifying the details in the running text and in a huge and detailed series of notes. Because of these reasons, he is a turning point for the studies on Kepler. A further reference point is Stephenson's *Kepler's Physical Astronomy*, published in 1987, without any doubt the most complete work published up to now on items (a) and (b). After that, the series of papers by Davis published in 1992 in *Centaurus* (and especially Davis 1992e)

[2] Here, we mention the works to which we have directly referred in the running text and other important contributions which are, at least in part, dedicated to the concept of force in Kepler: Baigre (1990); Baker and Goldstein (1994); Beer and Beer (1975); Blum and Helmchen (1987); Bussotti (2011); Boner (2013); Caspar ([1948] 1962); Cohen (1994, 2011, pp. 161–177); Davis (1981, 1992e), Delambre ([1817] 1965, [1819] 1965, [1821] 1969); Dijksterhuis ([1950] 1961, part IV, C, §§ 36–50, 53–55); Donahue (1981, 1994, 1993, 1996); Dreyer ([1906] 1953, chap. XV); Elena (1983); Gingerich (1975); Goldbeck (1896); Granada (2010); Guidi Itokazu (2006, 2007); Holton (1956); Hoyer (1979); Ihmig (1990); Jaballah (1999. Two volumes. First volume: pp. 60–80; second volume: pp. 39–43, pp. 78–85); Jammer (1957, chap. V); Jardine (1984, 2000); Koyré (1934, 1957, 1961a, b, c); Knobloch (1997, 2012); Krafft (1975, 1991); Martens (2000); Mittelstrass (1972); Petroni (1989); Pisano and Bussotti (2013); Rabin (2005); Radelet–de Grave (1996, 2007, 2009); Schuster (2000, 2013); Small (1804); Stephenson ([1987] 1994a); Treder (1975); Voelkel (1999); and Westfall (1971).

[3] Koyré also wrote about the way to develop the history of science. This aspect is significant in his research on astronomical revolution. See Koyré 1963, 1961a; see also 1961b, 1966, 1971, 1973, 1986. His reply to Henry Guerlac's (1910–1985) talk (Guerlac 1963; see also Crombie 1963) was superb. Recently, on the subject, see Pisano and Bussotti 2015a, b.

expresses new ideas on Keplerian forces. The contributions of Jardine (1984, 2000), Voelkel (1999), Martens (2000), and Boner (2013) tend, with different ideas and methods, to basically analyze the items (c) and (d). Rabin (2005) and Guidi Itokazu (2006, 2007) deal, in substance, with (a) and (b), also discussing the nature of *species immateriata*. Therefore, given this abundance of literature, why to write a further contribution on Kepler's concept of force? There are two main reasons:

(a) We wish to offer the reader a contribution to the items (a) and (b), which is useful to get a precise, but relatively concise idea of Kepler's concept of force. This is a novelty because the contributions to this problem are entire volumes (such as Stephenson's), a conspicuous part of volumes (such as Koyré's), papers concerning specific aspects of Keplerian forces (such as Davis 1992e), or articles in which the technical–astronomical–mathematical aspects are not dealt with in detail.
(b) In 1994 Barker and Goldstein wrote: "The nature of Kepler's *virtus motrix* remain[s] controversial" (Barker and Goldstein 1994, p. 67). Twenty years later this sentence is still true. We add that, more in general, the nature of Kepler's concept of force and the role of his dynamical concepts inside his astronomy remain controversial, despite the numerous and profound contributions published on this subject. In this chapter we wish hence to supply a contribution to Kepler's *Forschung* as far as the concept of force is concerned.

16.3 The Conceptualization of Force in *Mysterium Cosmographicum* ([1596] 1621)

The *Mysterium Cosmographicum*[4] (KGW, I) is almost entirely dedicated to the theory of the regular polyhedron applied to the disposition and orbits of the planets in the solar system. In his preface Kepler pointed out that his research aimed at discovering "[…] Numbers, quantities and movements of the celestial bodies."[5] Copernicus' arguments had been mathematical, whereas his ones were physical and metaphysical. It is necessary to stress that *physics* and *metaphysics* are correlated in Kepler because both of them refer directly to the universe: metaphysics deals with the archetypical structures (*regular convex polyhedron* and *harmonic relationships*), whereas physics has to do with the laws of planetary movements. Kepler tried to provide a global theory starting from the *Mysterium Cosmographicum*. In this early work, Kepler dealt with metaphysical arguments. Only in the chapters XX–XXII (the whole book contains 23 chapters), he tried to characterize mathematically a force responsible for the movements of the planets around the Sun. This is an

[4] On *Mysterium Cosmographicum* see: Aiton (1977) and Field (1979, 1988). Recently, see also Pisano and Bussotti 2012.

[5] "[…] Numerus, Quantitas, et Motus Orbium" (KGW, I, p. 9, line 34; see Fig. 3; authors' translation).

important fact in the history of science (particularly of astronomy and physics) because it is a clear attempt to offer a precise characterization of physics inside astronomy. This attempt is naïve and unsatisfactory, and Kepler was aware of this; nevertheless it represents a starting point of a new conception (KGW, I, Chapters XX–XXII, pp. 68–72, pp. 75–77) (Figs. 16.1 and 16.2).

As to the mathematical aspects of the planetary movements within his physical analysis, Kepler proposed the following reasoning (KGW I, Chapter XX, pp. 70–72). Given two planets A and B, let A indicate the planet immediately superior to B, whereas D and d are their respective distances from the Sun, with $D > d$, and T and t the periodical times of A and B, respectively. The distance has a double effect on the periodical times. Particularly, because:

1) If $D > d$, the orbit of A is larger;
2) A is slower.

Item (1) needs no consideration. Item (2) is more problematic because it implies the existence of a functional link between speed and distance of a planet from the Sun. In Chapter 20 Kepler wrote:

> First of all, everyone wants the motion of each planet to be the slower the further the planet is from the centre. Indeed, nothing is more reasonable, witness Aristotle, *De Coelo*, book II, chapter 10 that "the motion of each should be in proportion to the distance."[6]

Kepler clarifies that he agrees with this idea by Aristotle, although it is expressed in a context whose physical reality Kepler did not recognize. Thence, speed and distance from the Sun are inversely proportional. On the other hand, as we show, the *anima motrix* (KGW I, p. 70, line 21) responsible for the force[7] moving the planets is in the Sun and "[...] impels each body [...] strongly in proportion to how near it is."[8] This means that the force emanating from the Sun decreases as the distance. The logical conclusion is that the speed and the Sun force are directly proportional and the force produces velocities. Kepler can hence claim, at the beginning of Chapter 21:

> It was demonstrated by it on the authority of Aristotle that the new hypotheses are preferable. For, by them the motions are under two headings, both from the extent of the power, and the speed of revolution, made proportional to the Copernican distance, which could be obtained in no way, if one adheres to the tradition of the ancients on the universe.[9]

---

[6] KGW, I, p. 69, lines 1–4. See also Kepler 1981, p. 197.

[7] KGW, I, p. 70, lines 21–22. See also Kepler 1981, p. 199.

[8] With regard to the force of the Sun or, anyway, to the force acting between the Sun and planets (also considering the occurrences where Kepler speaks of the hypothesis, which he refuses, in which the planets have a soul producing a motive force, too), Kepler uses two words: *vis* and *virtus*. The latter is used far more than the former. In particular: *vis* is used at p. 11, line 11 and at p. 71, line 13, whereas *virtus* at: p. 11, lines 5 and 21; p. 70, lines 23 and 29; p. 71, line 9; pp. 71–72, lines 38–1; p. 72, lines 1, 11, 12, 27; p. 76, lines 25 and 26; p. 77, lines 4 and 22. The two words are also used with different meanings with which we do not deal.

[9] KGW, I, p. 72, lines 1–6. See also Kepler 1981, p. 209.

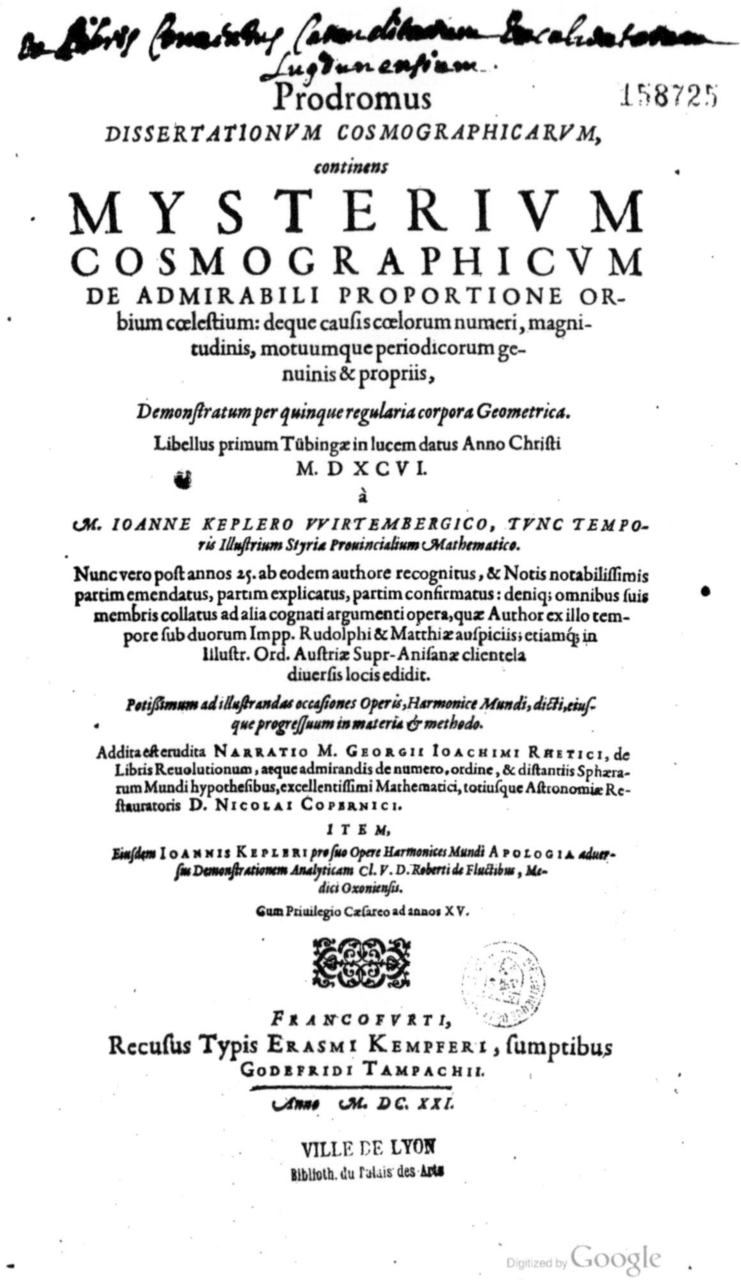

Prodromus
DISSERTATIONVM COSMOGRAPHICARVM,
continens
MYSTERIVM
COSMOGRAPHICVM
DE ADMIRABILI PROPORTIONE ORbium cœleſtium: deque cauſis cœlorum numeri, magnitudinis, motuumque periodicorum genuinis & propriis,
Demonſtratum per quinque regularia corpora Geometrica.
Libellus primum Tübingæ in lucem datus Anno Chriſti M. D XCVI.
à
M. IOANNE KEPLERO VVIRTEMBERGICO, TVNC TEMPOris Illuſtrium Styriæ Prouincialium Mathematico.
Nunc vero poſt annos 25. ab eodem authore recognitus, & Notis notabiliſſimis partim emendatus, partim explicatus, partim confirmatus: deniq; omnibus ſuis membris collatus ad alia cognati argumenti opera, quæ Author ex illo tempore ſub duorum Impp. Rudolphi & Matthiæ auſpiciis; etiamq; in Illuſtr. Ord. Auſtriæ Supr-Aniſanæ clientela diuerſis locis edidit.
Potiſſimum ad illuſtrandas occaſiones Operis, Harmonice Mundi, dicti, eiuſque progreſſuum in materia & methodo.
Addita eſt erudita NARRATIO M. GEORGII IOACHIMI RHETICI, de Libris Reuolutionum, atque admirandis de numero, ordine, & diſtantiis Sphærarum Mundi hypotheſibus, excellentiſſimi Mathematici, totiuſque Aſtronomiæ Reſtauratoris D. NICOLAI COPERNICI.
ITEM,
Eiuſdem IOANNIS KEPLERI pro ſuo Opere Harmonices Mundi APOLOGIA aduerſus Demonſtrationem Analyticam Cl. V. D. Roberti de Fluctibus, Medici Oxonienſis.
Cum Priuilegio Cæſareo ad annos XV.
FRANCOFVRTI,
Recuſus Typis ERASMI KEMPFERI, ſumptibus GODEFRIDI TAMPACHII.
Anno M. DC. XXI.

Fig. 16.1 Frontispiece of *Mysterium Cosmographicum* (Kepler [1596] 1621. Image source: Google Books, public domain; see also Kepler 1993)

MYSTERIVM COSMOGRAPHICVM / CAPVT XX 71

orbis Mercurialis ad Veneriam. Iam verò commiscet se huic motuum proportioni debilitas motricis animae in remotiori. Dispiciendum igitur, cum hac debilitate vt comparatum sit. Ponamus igitur, id quod valde verisimile est, eadem ratione motum à Sole dispensari, qua lucem. Lucis autem ex centro prorogatae debilitatio qua proportione fiat, docent Optici. Nam quantum lucis est in paruo circulo, tantundem etiam lucis siue radiorum solarium est in magno. Hinc cum sit in paruo stipatior, in magno tenuior, mensura huius attenuationis ex ipsa circulorum proportione petenda erit, idque tàm in luce, quàm in motrice virtute. Quare quantò amplior Venus Mercurio, tantò istius, quàm illius motus fortior, siue citatior, siue pernicior, siue vigentior, seu quocunque verbo rem exprimere placet. At quantò orbis orbe amplior, tantò plus temporis etiam requirit ad ambitum, etsi vtrinque sit aequalis vis motus. Ergo hinc sequitur, vnam elongationem Planetae à Sole maiorem bis facere ad augendam periodum: et contra, incrementum periodi duplum esse ad ἀποστημάτων differentiam.

Dimidium igitur incrementi additum periodo minori, exhibere debet proportionem veram distantiarum: sic vt aggregatum sit, vt distantia superioris, et, simplex minor periodus repraesentet inferioris, sc. Planetae sui distantiam in eadem quantitate. Exemplum. ☿ motus periodicus est 88. ferè dierum, Veneris 224 cum besse fermè, differentia 136. et bes, dimidium 68. et pars tertia. Hoc iunctum cum 88. efficit 156, et
73 trientem. Ergo vt 88. ad 156.[1] cum tertia, sic semidiameter circuli Mercurialis medij ad mediam Veneris. Hoc modo si in singulis opereris, atque prouenientes binas distantias per numeros sinuum explices, sic vt semper superioris semidiameter sit sinus totus:

	♃	574		572
prouenient	♂	274	At est	290
semidiame-	terrae	694	in	658
ter orbis	♀	762	COPERNICO	719
	☿	563		500

Propiùs, vt vides, ad veritatem accessimus. Etsi verò dubito, an demonstratiua methodo, quod theorema instituerat, praxis ista diuisae differentiae assecuta fuerit per omnia: tamen non omnino nihil in hisce numeris latêre, credere me iubet alia numerandi methodus, qua ad eosdem numeros reuoluar. Quia enim probabile est, fortitudinem motus cum distantijs esse in proportione: erit et hoc probabile, quòd quilibet Planeta, quantum superat superiorem fortitudine motus, tantùm superetur in distantia. Esto igitur, exempli gratia, Martis et distantia et vir-

27) 275 *statt* 572

Fig. 16.2 Dynamical arguments in the *Mysterium Cosmographicum* (KGW, I, Chapter XX, p. 71. With permission of *Bayerische Akademie der Wissenschaften*)

Therefore, the direct proportionality between force and speed and the inverse proportionality between these two quantities and the distance are established. Kepler underlines the novelty of this new acquisition concerning the dynamics of the skies, an acquisition he thought impossible in the context of ancient astronomy. Once this general reference frame is determined, how do things work from a mathematical point of view? Kepler writes what follows:

> As a consequence, one excess in the distance of a planet from the sun acts twice in the period's increment. Conversely, the increase of the period is double the difference in the distances. Thus, if one adds half the increase to the smaller period, he should show the true ratio of the distances: the sum is as the distance of the superior planet, and the simple lesser period represents the distance of the inferior one, that is, of its own planet in the same proportion.[10]

Using the symbolism we have introduced, Aiton claims[11] that the part of the quotation beginning with "As a consequence" and finishing with "distances" can be transcribed like this:

$$\frac{T-t}{t}=\frac{2(D-d)}{d} \tag{16.1}$$

which can be easily transformed into

$$\frac{D}{d}=\frac{t+1/2(T-t)}{t}=\frac{t+T}{2t} \tag{16.2}$$

This is, of course, true. Nevertheless, we think it is really hard to interpret the meaning of the first part of Kepler's quotation, without reading the second one. However, this can be interpreted in a relatively easy manner: half the increment of the period is half the difference between the two periodical times, which has to be added to the periodical time of the inferior planet. This sum is to *D* as *t* is to *d*. From Eq. (16.2) it is quite easy to deduce Eq. (16.1) and hence to recognize the equivalence of Eqs. 16.1 and 16.2. But, relying only upon Eq. 16.1, it seems to us rather complicated to understand what Kepler meant. Because the periodical times of the planets were known, the ratio of the distances could be directly calculated. These calculations reasonably fit with the calculations derived from the polyhedral theory. However, we remark:

1. The relationship expressed by Eq. 16.2 is not exactly a general law of the solar system because it does not connect two arbitrary planets, but a planet to the one immediately superior.[12] Thus, if a planet were discovered between the two considered, the consequence would have been a variation of the ratio *D/d* with respect to that calculated before the discovery of the new planet. This means

---

[10] KGW, I, p. 71, lines 13–20. See also Kepler 1981, p. 201.

[11] Aiton's note 7 and 8 to Chapter XX of the *Mysterium*. See Kepler 1981, p. 249; Aiton 1977, 1978.

[12] Cfr. Stephenson [1987] 1994a, pp. 13–14.

that Eq. 16.2, apart from the fact it is wrong, can be applied only as an empirical rule and does not reach the status of a general law.
2. Kepler claimed (KGW, I, Chapter XX, p. 71) that the relationship between periodical times and Sun–planet distance are influenced by two facts: the distance itself and the orbital speed that, according to Kepler, decreases in inverse proportion to the distance. This means that the distance from the Sun has a double effect (*duplum*: KGW, I, Chapter XX, p. 71, line 15) on the periodical time. Given the explained reasoning and also considering that from a linguistic point of view, the expression *duplicata proportio* was commonly used by mathematicians to indicate two quantities that are as the squares of two other quantities, one could expect the conclusion that the periodical time varied as the square of the Sun–planet distance. This is the reason why we think it would have been almost impossible to understand Kepler's thought relying upon Eq. (16.1). Only Eq. (16.2) clarifies. If Kepler had followed this line of reasoning, he would have obtained the relation $T/D^2 = k$, a general law connecting time and distances as, in fact, the third law $T^2/D^3 = k$, will do. Instead of this argumentation, Kepler followed the reasoning that led him to Eq. (16.2).

This is a clear indication Kepler was working at the idea of a physical-dynamical astronomy, but that his conceptions on the planetary laws were in a gestational phase.

With regard to the physical nature of the force on which the movements of the planets depend, Kepler essentially proposed these hypotheses (KGW, I, p. 70):

1. Each planet has a moving soul (*anima motrix*) whose power decreases in proportion to its distance from the Sun.
2. A sole *anima motrix exists*. It is in the Sun and acts on each planet in inverse proportion to the Sun–planet distance.

Kepler is favorable to this second hypothesis[13] because:

> [...] as the source of light is in the Sun and the principle of the circle in Sun's place, that is in the centre, so the life, the motion and soul of the world are posed in the Sun itself, in the same way as the rest characterises the fixed stars and the movement is typical of the planets.[14]

As to the nature of *anima motrix*, Kepler in the *Mysterium Cosmographicum* used a language typical of a vitalistic conception of the universe. Nevertheless, he treats the *anima motrix* with mathematical means. This represents an important novelty. We remind the reader of the beginning of the second edition of *Mysterium Cosmographicum* (*In Titulum libri Notae Auctoris*), where Kepler claimed (KGW, VIII, p. 15, lines 1–7) that he had been deeply influenced by Scaligero's conception (Cfr. Scaliger 1577). Certainly, Kepler took the expression *anima motrix* from the

[13] For this problem, see KGW, I, p. 70, lines 18–34. Quotation, p. 70, lines 25–26.

[14] "Sicut igitur fons Lucis in Sole est., et principium circuli in loco Solis, scilicet in centro: ita nunc vita, motus et anima mundi in eundem Solem recidit: vt ita fixarum sit quies, Planetarum actus secundi motuum" (KGW, I, p. 70, lines 24–26).

philosophical tradition and from vitalistic Renaissance conceptions, but we think that in *Mysterium Cosmographicum*, he is thinking of *anima motrix* as a physical entity that exerts a mechanical action independently of its ontological status. The action of the *anima motrix* can, hence, be translated with the term *force*, although Kepler gives no definition. Finally, two main features of Keplerian forces will remain unmodified in his successive assumptions:

- The action of the forces responsible for the movements of the planets around the Sun has an intensity that is inversely proportional to the Sun–planet distance.
- Forces produce velocities and not modifications of velocities (accelerations). This is why some authors have spoken of "Aristotelian inheritance" (Davis 1992e, pp. 165–168) in Kepler.

16.4 The Conceptualization of Gravity in *Astronomia Nova* (1609)

With regard to gravity, a short premise is necessary: as we show, in Kepler gravity has a completely different character with respect to the forces responsible for the planetary motions. Thus, one could object that it is not appropriate to deal with gravity in a chapter dedicated to the concept of force in Kepler. In contrast to this, we think Kepler's conception of gravity is a significant, though not prominent, element of his dynamical conception. This is why we address such a subject.

Isaac Newton (1642–1727)[15] showed that gravity is the force responsible both for the fall of the bodies on the Earth and for the planetary movements.[16] Nevertheless, in Kepler's lifetime no scientist had such a conception. There were various ideas about how gravity acted, but no precise quantitative treatment. Galileo Galilei (1564–1642) in *Discorsi e dimostrazioni intorno a due nuove scienze* (1638) was probably the first one who provided a relatively satisfying set of ideas on gravity[17] expressed in a mathematical form. On the other hand, if, before Kepler no dynamic explanation of the planetary motions was available, the idea of a possible connection

[15] January 4, 1643 according to the Gregorian calendar; December 25, 1642 according to the Julian calendar. Newton's country switched from the Julian calendar to the Gregorian calendar in September 1752.

[16] See, in particular, Propositions I–IX of the third book of the *Principia*, Newton [1726] [1739–1742] 1822, III, vol. III, pp. 17–40; see also Newton 1687; [1713] 1729. Particularly in Newton's Geneva edition, our work in progress proposes a critical translation from Latin into English of the whole *Newton Geneva Edition* (1822) of the *Principia* (Oxford University Press, 2020, 5 Vols.; see recently Bussotti and Pisano 2014a, b; Pisano and Bussotti 2016).

[17] We mainly refer to his studies within *Theoremata* circa *centrum gravitatis solidorum* as an appendix of the *Discorsi e dimostrazioni matematiche intorno a due nuove scienze attinenti alla meccanica e ai moti locali* (1638). On these studies and their relationships with Kepler's science see Pisano and Bussotti 2012.

between gravity and physical causes of these motions was even more extraneous to the scientists.

In the introduction to the *Astronomia Nova*, Kepler explained the principles on which the *doctrina de gravitate* is based (KGW, III, pp. 24–28).[18]

1. Each body stays at rest, unless it is not in the influence sphere of a like body (*cognatum corpus*). On *cognata corpora* Kepler wrote:

   > Gravity is a reciprocal corporeal disposition among cognate bodies to unite or join together (the magnetic faculty is another example of this kind).[19]

2. A force (*virtus tractoria*) exists between like bodies and it depends on the mass (complex concept that we try to clarify) of each body, as can be deduced from the following interesting quotation:

   > If the moon and the earth were not maintained each in its own circuit by an animate force or something equivalent to it, the earth would ascend towards the moon by [the] fifty-fourth part of the interval, and the moon would descend towards the earth about fifty-three parts of the interval. In that point, they would be joined together, assumed that their substance is of the same density.[20]

After three lines, Kepler used the expression *virtus tractoria* to indicate such an attraction.

These are the two basic properties of gravity, which can be considered almost a definition of the concept. However, when is it possible to consider two bodies as *cognata*? Kepler does not supply any precise answer. Notwithstanding, it is possible to infer from the development of his reasoning that two bodies are *cognata* if they have the same material composition. The problem of the composition of matter was a complicated subject at that time. As cited, Kepler claimed that a stone and the whole Earth exert a mutual attraction because they are *cognata* bodies (KGW, III, pp. 25–26). However, Kepler was certainly aware that the material composing the Earth is not exactly the same as that composing the stone. Furthermore, without a precise chemical theory, it is difficult to deal with matter's composition. Our interpretation is that Kepler used the word *cognata* in an intuitive sense, as to say that the Earth and the stone are composed more or less of similar materials, although, for example, the matter composing the Sun is different. On the other hand, the force exerted by the Sun is a *virtus promotoria* (that induces movement and not

---

[18] As to the concept of gravity in Kepler, see Elena (1983), the fundamental Goldbeck (1896), Ihmig (1990) and Trader (1975).

[19] "Gravitas est. affectio corporea, mutua inter cognata corpora ad unitionem seu conjunctionem (quo rerum ordine est. et facultas Magnetica)." (KGW, III, p. 25, lines 21–23, author's parentheses. See also Kepler 1992, p. 55.)

[20] "Si Luna et Terra non retinerentur vi animali, aut alia aliqua aequipollenti, quaelibet in suo circuitu; Terra ascenderet ad Lunam quinquagesimaquarta parte intervalli, Luna descenderet ad Terram quinquaginta tribus circiter partibus intervalli: ibique jungerentur: posito tamen, quod substantia utriusque sit unius et ejusdem densitatis." (KGW, III, p. 25, lines 37–41; see also p. 27, 11–15; Kepler 1992, p. 55.)

that attracts)[21] and not *tractoria*. But what about the relationships between the Earth and the other planets? Are they cognate bodies of the Earth, so that they exert a mutual *virtus tractoria*? In this case, too, Kepler does not offer a direct answer. Nevertheless, there are some interesting indications: in the *Epitome*, Kepler wrote:

> Certainly, Jupiter's body casts a shadow as the Earth and the Moon; Venus' body is missing of light in the part, which is opposite to the sun, as the Earth and the Moon.[22]

In the previous passage in the same section, Kepler had also taken into account the possibility of comparing the Earth with the fixed stars. The brief mentioned quotation seems to indicate Kepler is favorable to the idea that the Earth and the planets are *cognata corpora*, even though the property of casting a shadow is, admittedly, too vague for two bodies to be considered *cognata* only based on it. On the other hand, the planets have different densities and according to Kepler they are the denser the nearer to the Sun, although density does not decrease in linear proportion with the distance from the Sun.[23] However, this does not seem an objection against the possible *cognatio* Earth–planets because Kepler considered *cognata* a stone and the whole Earth and, certainly, not all stones have the same density. A further problem concerns the different roles the planets hold in Kepler's ideas concerning the relationship between harmony, geometry, and astrology inside the archetypical structure of the world.[24] However, these different roles rely upon the aspects, upon the different velocities of the planets, upon their "character," and upon the polyhedron with which they are associated, rather than upon their material composition. Furthermore, both the Earth and the planets share certain properties:

1. They are magnetic bodies with a dipolar magnetic axis, whereas the Sun has a different magnetic property.
2. Kepler saw a similarity between magnetic force and gravity, although he did not identify these two interactions.
3. The planets and the Earth belong to the moving part of the universe, whereas the Sun and the fixed stars do not (the Sun moves only *in loco* around its center).

---

[21] We remark that in a letter to Maestlin on March 5, 1605 (KGW, XV, pp. 170–176) Kepler defined the force of the Sun that determines the movements of the planets as *virtus promotoria* because it induces the movement through a mechanism different from gravitational attraction (*virtus tractoria*). In the *Astronomia nova* Kepler gave such *virtus promotoria* the name of *virtus motrix*.

[22] KGW, VII, p. 79. Original Latin text: "Iovis certe corpus umbram jacit, ut Terra et Luna, Veneris corpus parte a Sole aversa lumine caret, ut Terra et Luna."

[23] KGW, VII, Book IV, Part I, chapter entitled "De raritate et densitate horum sex globorum. Quid tenendum?" pp. 283–284. See, for example, Koyré 1961a, b, c, pp. 354–356.

[24] With regard to the difficult problems concerning the relations among these three aspects of Kepler's thought, we mention, without any claim to be exhaustive: Barker-Goldstein (2001); Bialas (2003); Boner (2006, 2008, 2009, 2011, 2013); Bruhn (2005); Escobar (2008); Fabbri (2009); Field (2009); Gingerich (2011); Granada (2009); Grössing (2003, 2005); Haase (1998); Juste (2010); Menschl (2003); Rabin (1997); Schwaetzer (1997); Stephenson (1994b); Voltmer (1998); and Westman (2001). On the structure of science Nagel (1961) is always of interest (for our aims).

This distinction is important in Kepler's theory starting from the *Mysterium Cosmographicum.*[25] This happens exactly because of archetypical reasons connected to the Holy Trinity.

Therefore, we propend to think that, according to Kepler, the Earth and the planets are cognate bodies. On the other hand, gravity can be interpreted as a local force acting in systems as Earth–Moon, whose radius is short in comparison to the distances among the planets.[26] Because of this, gravity does not play a role for the planetary movements. Given the general frame of Kepler's theory, the Earth and the fixed stars cannot be considered *cognate bodies*. As to the Moon, it is a cognate body of the Earth, but the Sun is not. Once the basic concepts and problems connected to the Keplerian conception of gravity are introduced, it is necessary to mention gravity's fundamental properties:

1. Against Aristotle, Kepler thought that no natural place exists in the universe to which a certain body tends. Strictly connected with this conception, Kepler claimed that no mathematical point can attract bodies. The mutual attraction exists only between like bodies.

   A mathematical point, independently of being or not the centre of the world, can neither exert any effect on the motion of heavy bodies nor act as an object to which such bodies tend.[27]

2. The heaviness and the lightness are relative concepts.

   Actually, nothing consisting of corporeal matter is absolutely light. It is only comparatively lighter, since it is less dense, either because of its own nature or by means of heat's influence.[28]

This is a further conception in which Kepler did not agree with Aristotle and Aristotelian tradition.

3. The single parts of a body attract another like body. This is an important passage because it implies that a body, however small and light it is, has its own power to attract like bodies. Therefore, gravity in itself is a property of each body. Its intensity is variable, not its essence. In a letter to Fabricius on October 11, 1605 Kepler wrote:

   I define gravity in a different manner, namely, gravity is that force that intrinsically moves a stone, a magnetic force that unifies similar bodies, which is present both in a big and in a little body, and is divided by the volumes (*moles*) of the bodies and gets the same dimen-

---

[25] To the relations between Sun–planets–fixed stars and the Holy Trinity is dedicated the *Primariae Demonstrationis Delineatio*, second chapter of the *Mysterium Cosmographicum* (KGW, I, pp. 23–27). See also Pisano Bussotti 2012, pp. 126–127; Peuerbach 1473; Pichler 2003.

[26] Cfr.: Davis 1992e, p. 181; Stephenson [1987] 1994a, pp. 4–7.

[27] "Punctum mathematicum, sive centrum mundi sit sive non, nequit movere gravia neque effective neque objective." (KGW, III, p. 24, lines 37–38. See also Kepler 1992, p. 54.)

[28] "Leve vero nihil est. absolute, quod corporea materia constat, sed comparate levius est., quod rarius est. sive natura sua, sive ex accidente calore." (KGW, III, p. 27, lines 16–17. See also Kepler 1992, p. 57.)

sions with the body. Thus, if a stone were posed behind the earth and its volume had a well determined proportion to that of the earth and if both the stone and the earth were free from every other movement, then, not only the stone would go toward the earth, but the earth would go toward the stone, too. They would meet in a point of their unifying straight line such that the distance of this point from the two bodies is in inverse proportion to their weight.[29]

The word *molem* (*moles*) introduces the fundamental question of the mass. Namely: how does gravity between two bodies act? Here the word *moles* is used as a synonym of *magnitudo*, size, namely volume. The final part of the quotation is interesting as well: if we have two like bodies, separated by a distance d and if their respective weights (*pondera*) are P_1 and P_2 they will meet in a point of the segment connecting the two bodies characterized by the relation:

$$d_1 : d_2 = P_2 : P_1$$

where d_i are the distances between the initial position of each body and the impact-point.[30]

The most complete exposition of Kepler's concept of mass is expressed in the *Epitome Astronomiae Copernicane* through the notion of *copia materiae* (Ihmig 1990, pp. 182–183). In this regard, what Kepler wrote in the *Epitome* is particularly important, where we read that the *copia materiae* of Saturn coincides with the product of its volume (*moles*) by its density. Indeed, we read:

For, if someone followed this, he would make a mistake against another law of truth, as far as he would introduce a non-different *copia materiae*, but an equal one for every planet. For, if Saturn's *moles*, 10, is multiplied by its density, 5, the *copia materiae* would result equal to 50, that is the same as if you multiplied Jupiter's *moles*, 5, by its density, 10.[31]

Through the concept of *copia materiae*, Kepler was trying to define a quantity that is valid for each body. Saturn is clearly only an example that can be generalized to each body.[32]

---

[29] "Aliter ego definio gravitatem, seu illam vim, quae intrinsece movet lapidem, vim magneticam coagmentantem similia, quae eadem numero est. in magno et parvo corpore, et dividitur per moles corporum accipitque dimensiones easdem cum corpo re. Itaque si lapis aliquis esset pone Terram positus in notabili aliqua proportione magnitudinis ad molem Telluris, et casus daretur, utrumque liberum esse ab omni alio motu: tunc ego dico futurum, ut non tantum lapis ad Terram eat, sed etiam Terra ad lapidem, dividantque spatium interjectum in eversa proportione ponderum." (KGW, III, p. 456, Nachbericht.)

[30] We have no room here to deal with the difference between masses and weights (namely weight-force). As to this problem, also in connection with the relationship between physics and mathematics, see: Pisano 2011, 2013a; Pisano and Capecchi 2013; Gillispie and Pisano 2014; Dhombres 1978, 2013; Alvarez and Dhombres 2011. On foundations of mathematics, see an interesting essay by Heinzmann (2009).

[31] "Nam si quis hoc sequeretur, is peccaret jam in aliam varietatis legem, introducens copiam materiae non inaequalem, sed eandem per omnes planetas. Multiplicata enim mole Saturni 10, in densitatem 5, prodiret copia materiae 50, tantundem scilicet, quantum, si molem Iovis 5 in densitatem ejus 10 multiplicasses." (KGW, VII, p. 283, lines 31–34.)

[32] Kepler speaks of Jupiter in the mentioned lines (KGW, VII, pp. 283–285).

4. The further Kepler's significant assertion is that gravity and magnetism are similar forces. It is well known that Kepler was profoundly influenced by *De Magnete, Magneticisque Corporibus, et de Magno Magnete Tellure* (1600) by William Gilbert (1544–1603). In the specific case of gravity, the comparison between magnetism and gravity makes it clear that, although Kepler tried a pure quantification of the concept of mass, he was still partially tied to the conception that an attraction can arise only because of some qualities held by a body and not for the mere fact that a body is a body. This is perhaps the most difficult aspect of the Newtonian concept of mass. Such a thesis is connected to the fact that gravity acts only between *cognata corpora*: an attraction exists because there are some qualities shared by two bodies (this is the sense of *cognata*) as the magnetic attraction exists between two magnetic bodies, not between any two bodies.
5. One of the most important consequences of Kepler's ideas on gravitation is given by his tides theory. Kepler does not develop this theory in detail, but his general ideas are clear and clearly expressed: the tides are due to the attraction of the Moon on the Earth. The following quotation needs no further explanations:

> If the earth would cease to attract its waters to itself, the whole water of the sea would be lifted up, and would flow to the body of the moon. The sphere of influence of the moon's attractive power is extended until the earth, and in the torrid zone calls the waters forth, particularly when it comes to be overhead in one or another of its passages. This effect is insensible in enclosed seas, but remarkable where the bed of the ocean are widest and there is much space for waters' reciprocation.[33]

The problem of the tides was discussed in the seventeenth century (Cohen 1994, 2011) by important scientists such as Galileo (Pisano 2009; Pisano and Bussotti 2014a, 2017; Naylor 1976; see also Abattouy 2006; Abattouy, Renn and Weinig 2001) and René Descartes (1596–1650) who gave wrong explanations, whereas Kepler caught the real nature of this phenomenon and provided a substantially correct explanation. Kepler was favored in this important discovery by his profound knowledge of Renaissance literature, in this case by Patrizi's *Nova de universis philosophia* (Goldbeck 1896, pp. 15–19). For, Patrizi refers to the opinions of the ancients on the tides. These opinions were, in many cases, more advanced than those of Kepler's contemporaries. Actually, Kepler quotes Patrizi in *Astronomia Nova*, however, in a slightly different context (KGW, III, p. 62, lines 27–28).

The conceptualization of gravity also represents an access key to Kepler's ideas on *inertia*: in his opinion, if there is no external action, each body stays at rest. We read in the introduction to *Astronomia Nova*:

---

[33] "Si Terra cessaret attrahere ad se aquas suas; aquae marinae omnes eleverentur, et in corpus Lunae influerent. Orbis virtutis tractoriae, quae est. in Luna, porrigitur usque ad Terras, et prolectat aquas sub Zonam Torridam, quippe in occursum suum quacunque in verticem loci incidit, insensibiliter in maribus inclusis, sensibiliter ibi ubi sunt latissimi alvei Oceani, aquisque spaciosa reciprocationis libertas." (KGW, III, p. 26, lines 1–7; see also Kepler 1992, p. 56.)

> Each corporeal substance, as far as it is corporeal, has been created so to be suitable to stay rest in every place in which it is located by itself, outside the sphere of influence of a cognate body.[34]

Thus, according to Kepler, each body has its own inertia: it naturally stays at rest unless a force moves it (Pisano and Capecchi 2013). The movement is a physical condition opposed to the lack of movement. This confirms that, in Kepler's conception, the forces have to explain not only the change of movement, but the origin of movement itself.[35]

As to gravity's action ray, we have seen that the accepted interpretation, with which we agree, is that gravity–action is confined to systems as planet–satellite. Nevertheless, in the introduction to *Astronomia Nova* Kepler gave the impression of thinking that the action-ray of gravity might be potentially infinite. This can be deduced by the following quotation:

> Therefore, if the moon's power [moon's *virtus tractoria*] of attraction extends to the earth, it will be more probable that the earth's power [earth's *virtus tractoria*] of attraction extends to the moon and far beyond. Accordingly, nothing, which consists, in some way, of terrestrial material, carried up on high, ever escapes the grasp of this mighty power of attraction [*virtus tractoria*].[36]

Therefore, the *virtus tractoria* of the Earth is extended beyond the Moon. However, gravity is not responsible for the planetary movements because the Sun and the planets are not *cognate* and the gravitational fields of the planets do not mutually interact because of their distance, independently of being the planets *cognata corpora.*

In the perspective of the history of scientific ideas, Kepler's conception of gravity is quite interesting, basically because of the concept of mass: Kepler, as we have seen, first considered the mass as the product of the volume by density. Furthermore, he thought that the more massive a body is, the stronger its attraction is. Therefore, without dealing with the absurd concept of scientific predecessor, these ideas by Kepler were advanced in comparison to his contemporaries'. On the other hand, gravity is still tied to qualitative properties of the bodies—the *cognatio*—and mass is not distinguished by weight because this implies that a body attracts another body producing an acceleration and the weight is exactly the *copia materiae* multiplied

---

[34] "Omnis substantia corporea, quatenus corporea, apta nata est. quiescere omni loco, in quo solitaria ponitur, extra orbem virtutis cognate corporis." (KGW, III, p. 25, lines 9–10. See also Kepler 1992, p. 55.) Several references to Kepler's concept of inertia are available in a recent Leibnizian book of one of us (Bussotti 2015, Chap. 3). Particularly on Leibniz, on the occasion of his anniversary, see *Leibniz and the Dialogue between Sciences, Philosophy and Engineering, 1646–2016* (Pisano, Fichant, Bussotti and Oliveira 2017; Bussotti and Pisano 2017).

[35] Clearly, this conception of inertia is different from Newton's, but also from Galilei's and Descartes'. See also De Gandt 1995; Pute and Mandelbrote 2011; Bussotti and Pisano 2013; Shank 2008.

[36] "Sequitur enim, si virtus tractoria Lunae porrigitur in Terras usque, multo magis virtutem tractoriam Telluris porrigi in Lunam et longe altius, ac proinde nihil eorum quod ex terrena materia quomodocunque constat, inque altum subvehitur, complexum hunc fortissimum virtutis tractoriae unquam effugere." (KGW, III, p. 27, lines 11–15. See also Kepler 1992, p. 57.)

by acceleration, which is extraneous to Kepler. Furthermore, considerations on the product density by volume are inserted, as seen, inside a chapter of *Epitome* entitled *De raritate et densitate horum sex globorum* (KGW, VII, pp. 283–284) which is deeply involved with the harmonic relations concerning the planets, their distances from the Sun, and their *copiae materiae*. Thence, the general conception in which these considerations are explained is far from a merely quantitative context and from the trains of thought that took place with Newtonian science. Notwithstanding, Kepler's concept of *copia materiae* seems to us a clear attempt to give a quantitative definition of a notion to which Kepler ascribed a remarkable importance inside his theory, certainly because of harmonic reasons, but also because of physical reasons, for example, Kepler's ideas on tides.

Gravity, as *virtus tractoria*, has no effect on the mean motion of the planets because, as we show in the prosecution of this chapter, such motion is due to the Sun's *virtus promotoria*. Actually, the libratory[37] motions of the planets do not depend on the *virtus promotoria*, but on the magnetic attraction the Sun exerts on the planets. It is clarified that, concerning the libration of the planets, magnetism is not used by Kepler as an analogy, but referring exactly to magnetic attraction. Kepler thought of a similarity between gravity and magnetism. This means that, from a merely theoretical point of view, the force responsible for the libration of the planets could be gravity rather than magnetism. This is not the case because: (1) the Sun and the planets are not cognate bodies, although they share magnetic properties; (2) the action-ray of gravity is short. But, if, in a conceptualization of Keplerian forces, magnetism is included—and it is by most of the literature—gravity cannot be excluded. On the contrary, gravity has to be considered a force because it is *potentially* a cause of movement. Such a potentiality can become actual in the circumstances we have explained, whereas the *virtus promotoria* always exerts an actual action.

16.5 Physical Astronomy and Forces

In his complex intellectual itinerary, Kepler reached to determine three forces responsible for the movement of the planets around the Sun. The movement of each planet can be decomposed (as we have seen at the beginning of our chapter) into four main motions:

1. The mean motion around the Sun.
2. The elliptical orbit, which can be interpreted as a perturbation of the mean movement.
3. The movement of the planetary apses-line, which is a movement of the reference frame.
4. The movement in latitude.

[37] Kepler called "libration" the approaching and moving away of a planet to/from the Sun. The libratory or librating force is the force responsible for the radial motion.

Kepler thought of three forces to explain these four movements. In what follows, we deal with this fundamental part of Kepler's physical and dynamical theory.

16.5.1 The Movements of the Planets Around the Sun

An astronomical model is dynamical if it provides the mechanism through which the forces act to determine the motions or, in the modern conception, the change of motion. From a theoretical point of view, it would be possible to define a force (or whatever other word one uses to indicate the causes of motion) in a manner different from Newton's and to reach different conclusions. However, it is certain that, for a model to be considered dynamical, it is necessary to determine:

1. The features of the forces
2. A functional link between the forces and the way in which the movements or the changes of the movements occur

The observations and the experiments will decide if the theoretical conception is correct to explain the external world. However, in an initial phase, the problem of physical astronomy is a problem of rational mechanics, not of observational astronomy or experimental physics. The mathematical apparatus has to provide a certain amount of consequences independently of the observations. The observations and the experiments can play an important heuristic role, but they cannot be part of the theoretical apparatus. It is hence clear that Ptolemy's (90–168 AC), Copernicus', and Tycho Brahe's (1549–1601) astronomies are not dynamical. Newtonian astronomy was different. As to Kepler, the situation is far more complicated. In the conceptual odyssey that brought Kepler to determine the area law and the elliptical form of the orbits, there are:

1. Direct references to forces and to their characteristics
2. A series of deductions whose justifications rely apparently upon the fact that some physical phenomena are impossible according to the nature of forces

Nevertheless, Kepler did not deduce his correct kinematical astronomical models from his conception of the forces, because, according to his opinion, forces produce velocities and not accelerations and they act as the inverse of the distance and not of the distance–square. In fact, if a force acts according to the inverse-distance law the orbit cannot be a conic section. As to the fact that a force produces velocities and not accelerations, it is not conceivable which consequences this conception should have as to rational mechanics. With regard to possible interpretations of the link between Kepler's dynamical astronomy and his concept of force, Stephenson stresses that a correct dynamics cannot be drawn from such a concept (Stephenson [1987] 1994a, p. 149), but that, anyway, Kepler deduced some significant *quantitative* relations from his general *qualitative* conception with regard to important angular quantities connected to the planet's path (Ibidem, pp. 146–172). They are the true anomaly, the eccentric anomaly, the mean anomaly, and the equation of center, divided into its physical and optical components. Apart from other problems dealt with by Stephenson and that we analyze, these quantities and their relations to the variable distance Sun–planet are not deduced from quantitative assumptions on the forces, as the one that a force acts as the inverse of the distance or as the inverse of the square distance.

Our idea is that the physical and dynamical aspect of Keplerian astronomy had a heuristic role, where the meaning of the term *heuristic* has to be clarified (Petroni 1989, pp. 113–119, pp. 179–196). This does not mean that Kepler was not thinking he was constructing a physical and dynamical astronomy. However, Kepler's astronomy was still kinematical as far as it is correct. Kepler had the enormous merit to catch the importance of the physical and dynamical astronomy and of the concept of force, but he did not develop this part of his conception in the necessary quantitative manner, whereas only a transcription into mathematical terms can determine if a certain idea is in itself coherent and applicable to the external world.

16.5.2 *The Conceptualization of Force in* Astronomia Nova *(1609) and in Epitome Astronomiae Copernicanae (1618–1621)*

Kepler developed his ideas on the concept of force in the *Astronomia Nova* (KGW, III) and provided many specifications in the *Epitome Astronomiae Copernicanae* (1618–1621; KGW, VII). In the following we provide Kepler's explanations and our historical and epistemological interpretations.

16.5.2.1 What Force Is Responsible for the Mean Motion of the Planets Around the Sun?

Kepler thought that the paths of the planets are determined by the combined effect of three forces. Each of them determines one or more than one movement whose composition provides the global motion of the planet. In this section, we deal with the problem of the force that determines the mean motion. This section is divided into two parts. In the first we explain and clarify Kepler's conception, and in the second, interpretations of Kepler's work are presented.

Kepler's Accounts

It is known that Copernicus did not supply a dynamical explanation of the movements in the solar system (Kepler 2010). However, Kepler was going to determine the laws of the planetary motions and Copernicus' theory was without doubt more favorable, at least from a psychological point of view, than the Ptolemaic one: the Sun is the celestial body that provides the light (Lindberg 1986; Malet 1990), it is the biggest body of the whole system, and, if it is also posed at the center (even though not the geometrical center) of the system, the question whether the Sun itself is responsible for the planetary movements can arise in the mind of a scientist. We have already seen that Kepler in his *Mysterium Cosmographicum* was favorable to the idea that the force responsible for the motions of the planets was centered in the Sun. In the *Astronomia Nova*, at least as far as the mean motion of the planets is concerned, Kepler adopted this idea definitively. Particularly,

- The force, the *virtus motrix* that determines the movements of the planets is in the center of the system (KGW, III, p. 237, lines 14–24).
- The Sun is in the center of the system; the Sun rotates around its axis (KGW, III, p. 243, lines 22–25).
- The *virtus motrix* depends on the rotation of the Sun. It is spread through a *species immateriata* (KGW, III, pp. 240–242, p. 350) radiating from the Sun.
- The *virtus* is *promotoria*[38] (induces movements), not *tractoria* (attracts).
- The *virtus* is not a geometrical body, but like a kind of surface (KGW, III, p. 240, lines 24–39).

These are the general features of the *virtus motrix*, but the nature of its action needs a series of specifications.

1. First: how does the intensity of the *virtus motrix* vary in function of the Sun–planet distance? Kepler is clear on this question: the intensity of the *virtus motrix* is inversely proportional to such a distance (see the whole chapter XXXII, KGW, III, pp. 233–236). Kepler reaches this conclusion by means of the following reasoning:

> […] Insofar as a planet is farther from the point, which is assumed as the centre of the world, it is less strongly constrained to move about that point. Thence, it is necessary that the cause of the weakening is either in the body itself of the planet, in a motive force placed therein, or exactly in the supposed centre of the world.[39]

From this quotation, it follows that the *virtus motrix* decreases inversely as the distance from the center. However, before answering the question whether the *virtus motrix* resides in the Sun or in the planets, Kepler poses another question, based on a presupposition, which is a fundamental assumption of his physics:

> In this condition, the intension and remission of motion is always in the same ratio as the approach and recession from the centre of the world.[40]

---

[38] This adjective is not used in *Astronomia Nova*, but in the letter to Fabricius mentioned in Footnote 31.

[39] KGW, III, Chapter XXXIII, p. 236, lines 12–16. Original Latin text: "quo longius abest Planeta a puncto illo, quod pro centro mundi assumitur, hoc debilius illum incitari circa illud punctum: necessarium est. igitur, ut causa hujus debilitationis insit aut in ipso Planetae corpore, eique insita vi motrice, aut in ipso suscepto mundi centro." See also Kepler 1992, p. 376.

[40] KGW, III, Chapter XXXIII, p. 236, lines 20–21. Original Latin text: "Ut hic intentio et remissio motus, cum accessu et recessu a centro mundi, in proportione perpetuo coincidit." See also Kepler 1992, p. 376.

That is, the speed[41] decreases inversely as the distance from the center of the world. This means that solar virtue and speed are directly proportional and both of them inversely as the distance from the Sun. But, Kepler continues, either the weakening of the virtue is the cause of the distance of a planet from the Sun, or vice versa, or both quantities depend on a third cause (KGW, III, p. 236, lines 21–24). Because in this case there is no reason to think of a third cause, the distance is a relation, a relation cannot be the cause of a physical condition; and thus, the cause is the force (*ivi*, lines 34–36). To conclude the reasoning Kepler excludes that the force moving a planet is an internal force because it could only be produced by the soul of the planet. However, Kepler cannot understand either how the soul can regulate the planet's speed, according to the distance, without suffering any exhaustion or moving the planet in a sky from which the celestial spheres have been eliminated, given that a planet has neither wings nor feet (*Ivi*, p. 237, lines 3–10). Therefore, the conclusion is that the cause of the mean motion of the planets is a force, a virtue emanating from the Sun. Kepler concludes this part of his reasoning with three lines that cannot be equivocated: the force is in the Sun; force and speed decrease as the distance from the Sun. Here the word *velocitas* (*velocitatem*) is used. Kepler wrote:

> Thus, if the distance of the world's centre from the body of a planet determines its slowness, and approach determines its speeding up, it follows as a necessary consequence that the source of motive power is at the supposed centre of the world.[42]

Two considerations concerning the speed of the planets are necessary:

(a) Because the *virtus motrix* induces a merely uniform circular motion, if all the planets had the same mass and if no other force existed among sun and planets, then every orbit would be circular with the Sun in the center, and would be traversed with a uniform speed. The velocity would have only the transversal component, not the radial one.
(b) However, the planets do not have the same mass. Given their natural inertia, the more massive a planet, the more the resistance it opposes to the motion. Hence, the periodical time of a planet is a function of the solar force, of the mass, and of the orbit–length.[43] Furthermore, beyond the solar virtue, there are other forces determining the planetary path, as we show.

This is why the planets that are the farthest from the Sun are the slowest and why a planet, in its orbit, moves quicker at perihelion and slower at aphelion, albeit the non-circularity of the orbit depends, according to Kepler, on another force.

---

[41] We use here the word speed, high or low speed, to indicate what Kepler calls "intentio et remissio motus." This is the almost universally accepted translation. However, in the next section we show that not all scholars agree this is the right translation.

[42] KGW, III, Chapter XXXIII, p. 237, lines 14–16. Original Latin text: "Quod si itaque elongatio centri mundi a corpore Planetae, praestat Planetae tarditatem, appropinquatio velocitatem; fons itaque virtutis motricis in illo suscepto mundi centro insit necesse est." See also Kepler 1992, p. 377.

[43] On this question Koyré 1961a, b, c, pp. 202–205 and the connected note 18, p. 408, is exhaustive.

2. The action of the *virtus motrix* is spread by means of a *species immateriata* that radiates from the Sun. Each point of the Sun's surface emanates a ray of the *species immateriata.*[44] The Sun rotates around its axis. When the ray of the *species immateriata* touches the planet, it induces it to move in the same direction as the Sun's rotation. With regard to the features of the *species*, the Chapter XXXIII of *Astronomia Nova*, entitled *Virtutem quae planetas movet, residere in corpore solis*, is fundamental: Kepler develops a comparison and an analogy between *virtus motrix* and light. He used the expression *Cognatio virtutis Solaris motricis* cum *luce* (KGW, III, p. 239, lines 7–8). We have seen, with regard to gravity, that the word *cognatio* indicates, for Kepler, a strong analogy. However, it is evident that light in itself cannot move the planet; but Kepler poses a refined question: is it possible that light does not move the planets, but is the means by which the *virtus motrix* is spread (*Ivi*, p. 240, lines 3–5)? Kepler answers this is not possible because of the reasons clarified in the following Table 16.1.

Because of this, the immateriality connotes both *species immateriata* and light, but the items (A)–(C) imply that the two emanations are different, beyond the analogy. After having identified the main properties of his *species immateriata*, Kepler poses a profound problem: the *species immateriata* acts on the planets. Furthermore, its action depends on its distance from the planets and on the mass of the planets. But distance and mass are two quantifiable magnitudes; Kepler uses the general adjective *geometricus*. How is it possible that something immaterial is the subject of a quantification? Kepler explicitly writes:

> However, it seems contradictory that [the *species immateriata*] lacks of matter, but is, nonetheless, subject to geometrical dimensions. This has the consequence that it is poured out throughout the world, and, nonetheless, does not exist anywhere but where there is something movable.[45]

In this case, Kepler provides an answer that is, in fact, a truism:

> The answer is this: In spite of the fact that the motive power is not anything material, nonetheless, as it is destined to carry matter (that is, the body of a planet) it is not free from geometrical laws, at least, insofar as this material action of carrying things about is concerned.[46]

But this truism is revealing: Kepler recognizes the existence of a nonmaterial entity, which nevertheless is a corporeal entity, in the sense that it does not depend

[44] We deal with the exact meaning of the word *immateriata* in the next section while providing the interpretations of Kepler's thought. In this section we wish only to present Kepler's theory without interpretation as far as this is possible.

[45] Ibidem, Chapter XXXIII, p. 241, lines 4–6. Original Latin text: "Videntur autem pugnantia, materia carere, et tamen dimensionibus Geometricis subjacere: diffundi per mundi amplitudinem, et tamen nuspiam esse nisi ubi est mobile." See also Kepler 1992, p. 382.

[46] Ibidem, Chapter XXXIII, p. 241, lines 7–10. Original Latin text: "Respondetur autem sic: quamvis virtus motrix non sit materiale quippiam, quia tamen materiae hoc est. corpori Planetae vehendo destinatur, non liberam esse a legibus Geometricis, saltem ob hanc materialem actionem transvectionis" (*Ivi*, p. 241, lines 7–10). See also Kepler 1992, p. 383.

Table 16.1 A comparison between light and *virtus motrix*

Light	*Virtus motrix–Species immateriata*
(A) Light is blocked by the opaque	This is not the case with *species immateriata*. The planets also move when they cannot perceive solar light (for example, occultations) (*Ivi*, p. 240, lines 6–8).
(B) Light flows out in straight lines spherically (*orbiculariter*). Light is spread hence in every direction.	"The moving virtue acts in straight lines also but circularly (*circulariter*); that is it presses in only one direction of the heaven, from west to east, not backwards nor toward the poles, etc." (Rursum lux rectis effluit orbiculariter, virtus movens rectis quidem sed circulariter; hoc est. in unam tantum plagam mundi ab occasu in ortum nititur, non contra, non ad polos etc. (KGW, III, p. 240, lines 8–12).) This means that, assuming as direction North–South that of the Sun's poles, the *virtus* propagates parallel to the plane of the solar equator. In this manner, a series of mutually parallel planes (each parallel to the Sun's equator) exists, in which the *virtus motrix* is spread. By the way, this is coherent with Kepler's idea that the *virtus motrix* acts because the Sun rotates around an axis perpendicular to its equator.
(C) The intensity of a light-ray is not reduced while light traverses the skies. Rather the global intensity of light is the same for every *sphere* concentric to the sun. This means that its effect is less the farther a sphere is from the emanating center because light is less dense in these spheres.	For the *virtus* the same property is valid, but the word *sphere* has to be replaced with the word circle (*circulum*) because the virtue is spread *circulariter* and not *orbiculariter*. (*Ivi*, p. 240, lines 24–27). This means that the virtue is spread like a surface.

on the action of a soul or of an intelligence. In the following section, we comment on this important question.

3. The effect of the *virtus* attenuates as $1/r$ (where r is the distance Sun–planet). However, in Chapter XXXVI, which is fundamental to understanding Kepler's concept of *virtus motrix*, he reasons like this (see, in particular, KGW, III, p. 248, line 20, p. 250 line 2): the *virtus motrix* should attenuate as the square or as the cube of the distance from the Sun. However, through a series of argumentations, many of which are rather unclear, Kepler arrives at the conclusion this is not the case. The whole reasoning is based on an analogy between light and *virtus motrix*. Notwithstanding, at the end of his argumentation (KGW, III, p. 250, lines 37–40, p. 251, line 1) Kepler claims that, although the reasoning he has expounded could work for light, it does not work for the *virtus motrix* because the *virtus* is not spread in a sphere *orbiculariter*, but rather in the circle in which the planet moves. To this subject is dedicated the final page of the XXXVI. We comment on this in the following section.

Let us summarize the mechanism by which the solar virtue produces the mean movement of the planets:

(a) The virtue is spread by immaterial rays emanating from every point of the Sun's surface.
(b) The intensity of the rays does not decrease with the distance, but the intensity of the virtue decreases (as $1/r$) because the quantity of rays is the same for every arc of circumferences concentric to the Sun, hence the rays' density is less the farther a circumference is from the Sun.
(c) The sun rotates and the rays of the virtue (imagined by Kepler rigidly attached to the Sun) rotate with the same angular velocity as the Sun.
(d) When a ray touches a planet it induces a movement in the same direction as the rotating direction of the Sun.
(e) The farther a planet is from the Sun, the slower its motion is because, in a given range of time t, a superior planet is touched by fewer rays than an inferior one. The mass also has an influence on the mean motion because the more massive a planet is, the bigger its resistance to the action of the solar *virtus* is, due to the natural inertia of the masses.

Historiographic Accounts and Interpretations

1. With regard to *virtus motrix*, the most problematic question concerns the way in which the virtue is spread: the traditional interpretation consists in identifying Kepler's idea that the *virtus* is spread on a surface with the one that the *virtus* is spread on a *plane* surface. Because *virtus* acts *circulariter*, most scholars have interpreted Kepler's thought as if the *virtus* were confined near the ecliptic and have referred to the term *surface* as if the solar virtue were spread in circumferences on the plane surface of the ecliptic.[47] In this manner, Kepler's idea that force attenuates as the inverse of distance was saved because a circumference increases as the ray. Hence, it can be comprehensible that, if an action is spread in a circumference, its attenuation is proportional to the distance from its center. If an action is spread as the external part of a body (as the surface a sphere), however, its attenuation would be proportional to the square of the distance from the center, as Kepler knew with regard to light (Stephenson [1987] 1994a, pp. 68–69). Stephenson's work has been a turning point on this problem, too: he does not agree with this traditional interpretation and proposes a different explanation (Ibidem, pp. 70–75) quoting a passage by Kepler in Chapter XXXVI of the *Astronomia Nova* (KGW, III, p. 251, lines 3–30). In the chapter preceding the one analyzed by Stephenson, Kepler had established an analogy between the *virtus motrix* and the magnetic force, but this is not

[47] See, for example, Dreyer [1906] 1951, pp. 387–388; Caspar [1948] 1962, p. 143; and Dijksterhuis [1950] 1961, IV, 44. Koyré is quite prudent on this aspect and, although he seems to agree with Dreyer's, Caspar's, and Dijksterhuis's interpretation, he does not underestimate the difficulty of an exegesis of Kepler's assertions (Koyré 1961a, b, c, pp. 210–214).

necessary for our aims and to understand Stephenson's interpretation. Rather, it is important to summarize the main characteristics of the *virtus motrix*:

(a) Some *filamenta* exist in the solar body—Kepler calls them magnetic (KGW, III, p. 246, line 4), but, as already mentioned, this is not important in our context—from every point of which the *species immateriata* of the solar virtue (KGW, pp. 239–240) is spread.
(b) The *filamenta* are parallel to the Sun's equator (KGW, III, p. 246, lines 4–11) and are disposed in circles, which become smaller from the equator to the pole until reducing to a point in the poles of the Sun.
(c) A single great circle of the Sun's body lies under the Zodiac or ecliptic.
(d) Despite this, the rays of the *species immateriata* converge not only to the single points of the planetary paths, but also to the poles, which are over the poles of the body of the Sun (KGW, III, p. 251, lines 16–19).
(e) This does not imply that the planets move in all directions because the *filamenta* do not move the planet if they are considered in isolation, but only because the Sun, by rotating, carries around the *filamenta* and the moving image (*speciem moventem*) spreading from them (KGW, III, p. 251, lines 21–30).
(f) The planet will not go to the poles because the *filamenta* of the Sun are not directed towards the poles and because the Sun does not rotate in that direction.

The problems are hence: (1) to explain the exact way in which the virtue is spread; (2) to explain by means of this mechanism why the planets rotate around the Sun in orbits that are near the Sun's equator, even though the orbital planes do not coincide with the plane of Sun's equator. Stephenson's interpretation is essentially based on points (d) and (e) and on the translation of *species movens* as *moving image*. Stephenson explains the situation in this manner:

> This passage establishes conclusively that the solar virtue was not confined to the ecliptic. It propagated in all directions; only its motion was circular. Kepler went on to explain how it came about that the planets' motion remained near the ecliptic [Stepehenson refers to the final part of Chapter XXXVI. KGW, III, p. 251, line 31–252, line 3] [...]. He [Kepler] had always emphasized that it was not the mere presence of the species, but its motion, which impelled the planets. This motion derived entirely from the rotation of the solar body. So far as concerned a planet in the plane of the solar equator, the sun's motion was quite unidirectional. That is, the appearance of the visible manifestation of the motion was altogether in the direction designated by "east". From a higher latitude the motion appeared more complex. Extrapolating to the view from above the pole, the appearance of the sun's motion would have tended equally in all directions. The planet, which was carried by the "whole image, composed out all the filaments", would thus have been impelled in all directions equally, and would not have moved.[48]

Afterwards, Stephenson is able to prove that, if Keplerian virtue holds the features he thinks to have clarified, then it attenuates as the distance (Stephenson [1987] 1994a, pp. 74–75).

---

[48] Stephenson [1987] 1994a, pp. 73–74 and see the figure on p. 73.

Stephenson's argumentation that the solar virtue is not confined to the ecliptic is convincing. His general interpretation is compatible with item (d) and with the final explanation given by Kepler at the end of Chapter XXXVI. However, it is a visual explanation, because it depends on the angle with which a possible celestial body sees, from different latitudes, the *filamenta* of the sun. However, Kepler had claimed that the species *immateriata* is spread in the opaque, too, therefore the visibility of the Sun cannot play a role in the explanation of this force. This is a conceptual problem. Moreover, Guidi Itokazu argues, by means of arguments we think sound, that the translation of *species* by imagine is not convincing (Guidi Itokazu 2006, pp. 223–225), and the great part of Stephenson's interpretation relies upon this translation. Furthermore, in Stephenson's explanation the fibers in the Sun's body move *circulariter*, but the *species immateriata* is spread in every direction; this means *orbiculariter*, whereas, as we have seen, Kepler had claimed that the *species immateriata* itself is spread *circulariter*. Finally, a psychological mechanism of the celestial bodies seems to play a role as to their motion: if a body were in the Sun's pole lines, it would see part of the filaments rotate in a direction, part in the opposite direction, and hence, it could not decide in what direction to rotate. These kinds of psychological mechanisms appear extraneous to the way in which Kepler dealt with the *species immateriata*. Thus, Stephenson's interpretation catches some important aspects of Kepler's solar virtue, but neglects others.

On our side, we think that an advanced interpretation, with respect to the cited secondary literature, is possible: as we have outlined above, the *species immateriata* is spread from every point of each circumference of the solar body parallel to the equator, not in every direction, but in the direction parallel to the equator itself. In this manner, the *species immateriata* is really diffused *circulariter*, it is not confined to the ecliptic, and it is stronger at the Sun's equator because it is the greatest parallel. Furthermore, the species deriving from the Sun's equator tends to make a planet move around the Zodiac, whereas the species deriving from a parallel with latitude λ tends to make a planet move at latitude λ. However, because the latitude λ parallel is smaller than the Sun's equator, the global effect of the *species* spread from this parallel is less than the effect of the *species* spread from the Sun's equator. Therefore, the *species* induces different speeds in the planets, a speed, which is null at the poles and maximal at the equator. The difference of speed in the different planes parallel to the Sun's equator tends to push the planets towards the plane where the speed has a maximum, namely towards the equatorial plane. This effect is not complete; hence, the planets rotate in proximity of the Sun's equator, but not exactly in this plane. Following this interpretation, the meaning of the Sun's virtue as *virtus promotoria* acquires a new pregnant meaning: it is *promotoria* also because it induces different speeds (motions) at different latitudes and this explains why the planets are confined near the Sun's equator.

This is our interpretation with which we try to make most of Kepler's assertions coherent. We think that it is possible to provide a sufficiently founded historical explanation and epistemological interpretation of the *virtus motrix*, but not a complete explanation of all its characteristics. This depends on:

- The continuous analogies made by Kepler with light and with magnetic force. These analogies tend to give a general idea of the *virtus motrix*. They are important in Kepler's narrative style. However, they indicate that this concept probably was not completely clear in Kepler's own mind. Otherwise why resort continuously to analogies?
- The fact that some explanations are given in a certain chapter of *Astronomia Nova* and are provided with slightly different terms in other chapters.
- The lack of a definition of the *virtus motrix*. If we indicate such a force as F, there is no definition of the form F = something. A series of its features is provided, from which it is problematic to deduce a definition.

2. With regard to *species immateriata*'s composition, and in particular as to the translation of the word *immateriata,* there was a general agreement among the scholars: the right translation of *immateriatus* is immaterial = nonmaterial. However, in 2005 Rabin proposed a different interpretation: her analysis starts from the consideration that the word *immaterialis* existed. Then why to use the word *immateriatus*? Rabin stresses that Kepler did not seem comfortable while admitting that something immaterial is the subject of geometrical quantification (Rabin 2005, p. 50). After having highlighted several contexts in which Kepler employed this word, she concludes, "That Kepler came to use the word *immateriatus* to indicate something material fits into the context of his ideas about the universe" (*Ivi*, p. 52). Guidi Itokazu argues against Rabin's interpretation and proposes the thesis that *immateriatus* is referred to an emanation having no mass (Guidi Itokazu 2007, p. 306).

A linguistic problem becomes interesting when a conceptual problem is connected to it: as Rabin points out, the conceptual problem is, "So, how can something immaterial be the cause for moving the planets?" (Rabin, 2005, p. 50). In our opinion, here, Kepler moved a fundamental step towards conception that has become typical of modern science:

(a) The *species immateriata* connoting the *virtus motrix* (as we have already stressed) has a material origin in the Sun's body.
(b) It moves the planets.
(c) The *virtus motrix* is not material.[49] Kepler is explicit on this. Probably, as Rabin claims, Kepler felt uncomfortable with this assertion because of its novelty. Nevertheless, he is clear.
(d) We interpret that Kepler thought of a new kind of interaction: something immaterial, but not tied to *animae* or *intelligentiae*, that acts on the matter. The physical world can be divided into material and immaterial components. The immaterial can have a profound influence on the material ones. Let us think—in modern physics—of electromagnetic radiation. We are not comparing *virtus motrix* and electromagnetic waves: this is wrong, due to its anachronism, and to the fact that the *virtus motrix* does not have the features of a wave. However, it has those of an immaterial entity acting on the bodies. Its action is interesting:

[49] See our quotation from *Astronomia Nova* in the previous section (KGW, III, p. 241, lines 7–10).

every ray of the virtue can be imagined as an immaterial ray which, rotating with the Sun, hits a planet and moves it. Therefore, this conception is different from that of Newton's central forces: in Kepler there is no action at a distance: the ray of the *virtus* is like a lace that moves the planet by its mechanical action. This is the meaning of the locution *vis promotoria*, whereas, in Keplerian language, the central forces would be *vires tractoriae*. Therefore, mechanical action is not in contradiction with immateriality. Finally, as Guidi Itokazu claims, it makes sense to translate *immateriatus* as "lacking of mass". What is important is to stress the features of this immateriality, which we have tried to do in this part.

Once given a general explanation of why the planets rotate around the Sun, Kepler faced a further problem: the orbits of the planets are eccentric and they are ellipses. The described mechanism can explain why the planets rotate around the Sun, but not why their distance from the Sun is variable. As to this problem, Kepler provided an ingenious explanation, which is the subject of the next section.

16.5.2.2 What Force Makes an Orbit Elliptical?

We follow the same procedure used in the previous section to explain the features of this force: Kepler's ideas are expounded in the first section; a section dedicated to the interpretations follows.

Kepler's Accounts

The problem of determining the force that can determine the approaching and the moving away of the planets from the Sun and, in particular, makes an orbit elliptical is dealt with in *Astronomia Nov*a (KGW, III, pp. 348–364), but explained more clearly in *Epitome* (KGW, VII, pp. 337–342). For this reason, we follow Kepler's argumentations expounded in *Epitome*.

Kepler thought that both the Sun and the planets are magnets, therefore beyond the *virtus motrix*, there is a magnetic force, too. In modern terms: Kepler believed that the Sun was an enormous magnet, of which one pole is on the surface and the other is internal.

Let us consider the planet in position *A* (Figs. 16.3a). The magnetic poles of the planet, indicated by the point and the tail of the arrow, are equidistant from the Sun. There is no magnetic attraction here. On the basis of the mechanism seen in Sect. 16.5.2.1. the planet proceeds in its orbit. In positions B, C, D the planet has the pole, called by Kepler *amicus*, turned towards the Sun, hence the Sun attracts the planet. After having reached and passed the perihelion in E, the *discors* pole (KGW, VII, p. 337, lines 20–29) of the planet is turned towards the Sun, hence the magnetic interaction makes the planet distant from the Sun. This is the general mechanism explaining why the orbit is not circular.

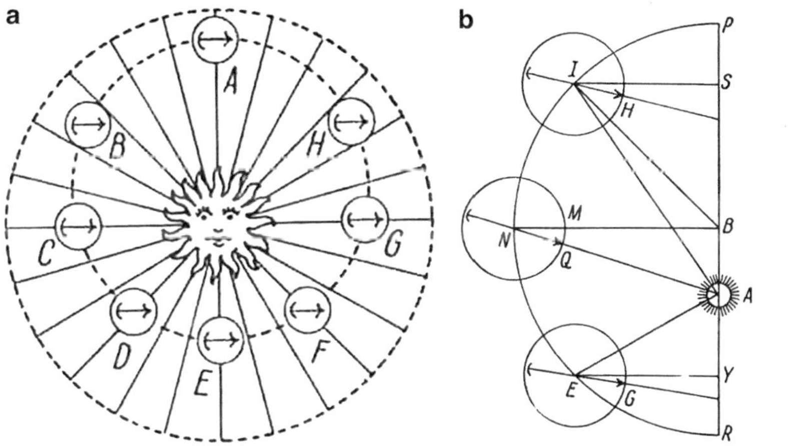

Fig. 16.3 (**a**) The mechanism explaining planetary *libration*. Particulars drawn from the *Epitome* (KGW, VII, particular of p. 337. With permission of *Bayerische Akademie der Wissenschaften*). (**b**) Specification of the same mechanism. Particulars drawn from the *Epitome* (KGW, VII, particular of p. 339. With permission of *Bayerische Akademie der Wissenschaften*)

However, things are not so easy because:

(A) The *virtus motrix* is far stronger than the attraction–repulsion between the Sun and the planetary magnetic axis. This is an important statement for the subjects we also face in the following section (see, e.g., in the *Epitome*, KGW, VII, pp. 337–338),
(B) The magnetic axis of the planet has a sort of magnetic inertia and hence tends to maintain its position.

The conclusion is that the tail of the planet's magnetic axis is directed exactly towards the Sun only when the planet is in the quadratures. This situation is clearly explained by Kepler in Fig. 16.3a. Analogous considerations are valid when the *discors* pole is oriented towards the Sun.

But what is the intensity of the libratory force? Let us follow Kepler's argument (Fig. 16.4):

Initial clarification: Kepler imagined that the planets have magnetic fibers of different kinds.[50] Those that are important in our context are supposed to be perpendicular to the apses-line RP (see Fig. 16.4). In an initial phase of his reasoning, Kepler supposed that the position of the fibers remains unmodified.

Positions: A = sun; I, E = different positions of the planet's center; EG, IH = magnetic fibers that remain perpendicular to RP during the planet's revolution; AI, AE = vector radius Sun–planet. Let EI and IO be traced parallel to RP. Let the perpendiculars to these lines be traced from F and C; they touch EI and IO in L and K, respectively.

[50] As far as the difference among the magnetic fibers is concerned, a good explanation is given by Koyré 1961a, b, c, Chapter *L'Epitome*, in particular Note 43 and connected text.

ponderum linea manet Horizonti parallela, sed iugum diuidit longitudinem brachiorum non in medio, sed propius grauiori ponderi, sic vt brachia permutatam habeant proportionem ponderum.

Vt si manubrium librae DA. sit aequale brachiis BA. AC. statera sic formabitur, pondera ista ex B. C. dependentia suspensura ad aequilibrium Horizontis. Ex D. perpendicularis in BC. ducta, quae sit DK. erit manubrium: et brachia BK. KC; et vt DF. prius ad FE. sic hic BK. ad KC. Tunc vt BK. minus brachium ad KC. maius, sic pondus H. minus ex C. suspendendum, ad pondus I. maius ex B. suspendendum.

Monendus est lector, difficilem esse experimentationem mechanicam; quia mechanicè caueri non potest pondus et crassitudo ipsorum brachiorum: debebant autem geometricè constituere meram lineam sine pondere et latitudine. Cui impedimento, quomodo ex parte occurrendum, videatur in
† ARCHIMEDE.

Teneo mensuram fortitudinis, seu incrementi librationis in quolibet situ fibrarum corporis planetae, petendam à complemento anguli fibrae cum radio Solis; quia verò difficulter patescere videtur hic angulus, eò quòd non tantum corpus continue transfertur de loco in locum, sed etiam ejus fibrae inclinantur; mensura haec incerta eoque inepta videtur ad vsum.

Imò propter hanc ipsam inclinationem fibrarum, angulus iste in arcum orbitae potest conuerti, vt ex hoc arcu prodeat idem sinus eadem scilicet mensura; quà ratione ad vsum illa fit accommodatissima.

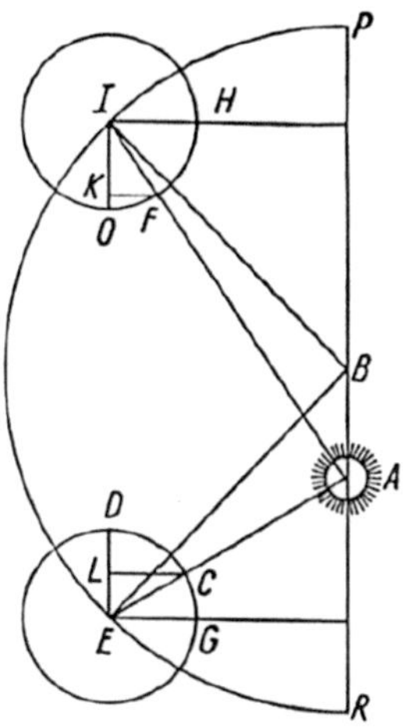

Doce et demonstra hanc conuersionem anguli dicti in Orbitam.

Memineris initio, cùm planeta est in Ap-
614 sidibus,[1] hoc est, in principio Orbitae: angulum inter radium Solis et fibram esse rectum. Rursum libro IV. fol. 593. ostensum est, fibram NQ illius figurae in ipsum Solem A. dirigi, seu cum radio Solis NA. vniri, consumpto hoc angulo, cum est peractus quadrans orbitae PN. ab Apside P. vt ita arcus orbitae ab Apside metiatur complementum huius anguli. Restat igitur hoc demonstrandum, etiam angulos intermedios fibrae cum Sole vt HIA. inter rectum et nullum, à mediis arcubus orbitae, vt PI. inter nullum et quadrantem, sic compleri, vt iuncti faciant 90.

Demonstraturque sic; fol. 596. est dictum, sicut est IS. ad NB. sic esse sinum anguli HIS. ad sinum anguli QNB. ferè. Id captus causâ sic vsurpatum fuit de IS. et NB. quamuis, vi speculationis physicae verum sit potius de sinibus angulorum IAP. NAP. Iam verò etiam sinus AIB. est ad sinum ANB. anguli, sicut sinus anguli IAP. ad sinum anguli NAP. (Vt enim BI. ad BA. sic sinus

47 Kepler VII

Fig. 16.4 Beginning of Kepler's explanation on the relation between *libratory force* and true anomaly (KGW, VII, Book V, p. 369. With permission of the *Bayerische Akademie der Wissenschaften*)

Reasoning: when the planet is in P and in R, its magnetic axis is perpendicular to the radius-vector Sun–planet and hence, there is no magnetic attraction. While the planet is moving from P to R, its radius-vector is not perpendicular to the magnetic axis and hence the Sun exerts a magnetic attraction. Its intensity is calculated like this: if IH and EG represent the global attractive force of the Sun, the ratio between this force and its part acting on the planet in I is as IH (= EG) to KF and that acting on E is as EG(= IH) to LC. The true or coequated anomaly is the angle IAP, which is congruent to BIK. Hence, IFK is complementary to the true anomaly. Because, assuming IH = IF = 1 (*sinus totus* in the language of then-astronomy), KF is the cosine of IFK, consequently the libratory force is proportional to the cosine of the true anomaly.

Our explanation is based on Kepler's account.

This result has been obtained supposing that the fibers remain perpendicular to the apses-line, but, according to Kepler, this is not the case. Kepler (KGW, VII, pp. 369–370) refined his reasoning until he reached the conclusion that, considering the deflections of the magnetic fibers, the *libratory force* is as the sinus of the eccentric anomaly, which characterizes an elliptical orbit. We do not have room to analyze Kepler's complete reasoning, hence we refer to the exhaustive synthesis given by Caspar[51]: let us consider Fig. 16.3b.

Let

$$BP = IB = NB = 1,$$

that is, the radius of the circle circumscribed to the orbit.

Let it be:

$$P\hat{A}I = u_1; P\hat{A}N = u_2.$$

From the triangles BIA and BNA, it follows:

$$\frac{IB}{BA} = \frac{\sin u_1}{\sin AIB}; \frac{NB}{BA} = \frac{\sin u_2}{\sin ANB} \quad (16.1)$$

Hence

$$\frac{\sin u_1}{\sin u_2} = \frac{\sin AIB}{\sin ANB} \quad (16.2)$$

But Kepler had previously proved that

$$\frac{\sin HIS}{\sin ANB} = \frac{\sin u_1}{\sin u_2} \quad (16.3)$$

[51] KGW, VII, pp. 594–595. With permission of the *Bayerische Akademie der Wissenschaften*.

The conclusion is that the angles HIS and AIB are equal. Therefore the angle between the radius-vector IA and the fiber IH has the property:

$$A\hat{I}H = B\hat{I}S = 90 - \beta \tag{16.4}$$

where β is the eccentric anomaly. However, because the *libration* is proportional to the cosine of the angle between the radius vector and the fibers, then—according to this reasoning—the *libration* is proportional to the sine of the eccentric anomaly, which happens for an elliptical orbit. In Kepler's eyes, this should have been a physical demonstration of the orbits' ellipticity.

Kepler went on to determine the whole amount of *libration* in every point of the orbit. By means of the reasoning, of which we expound the initial phases, Kepler was able to concretely construct the planetary path.

To calculate the whole amount of *libration* Kepler reasoned as follows (we refer to Fig. 16.5): he constructed the circle PDRZ, whose diameter is the apses-line PR. Therefore, the circle touches the planetary path in P and R. In the other points, it is external to the orbit. Kepler then traced the perpendiculars from the points P, K, G, D, N, S to the apses-line (Kepler supposes these points are infinitely near; see Fig. 16.5, KGW, p. 370, line 23). Hence KX is the sinus of the eccentric anomaly KBM, if B is the center of the circle (this hypothesis is not necessary, but useful for

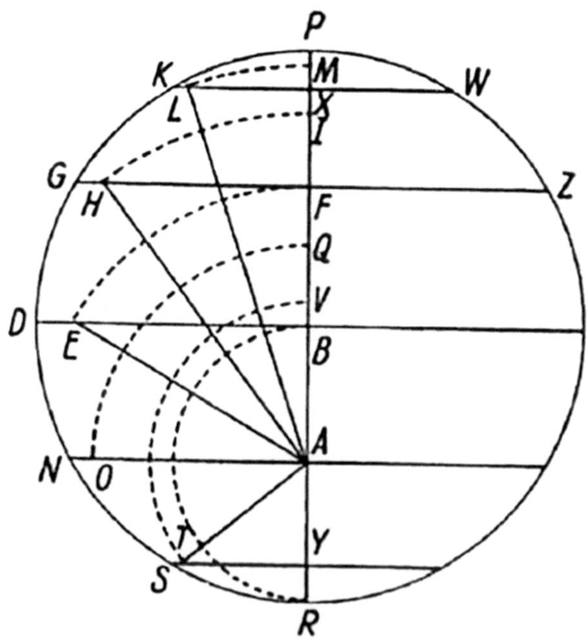

Fig. 16.5 The figure used by Kepler to calculate the whole amount of libration (KGW, VII, particular of p. 370; with permission of the *Bayerische Akademie der Wissenschaften*)

our aims and compromises nothing of our explanations), GF is the sinus of the eccentric anomaly GBP, and so on. The sinuses of the eccentric anomalies are as the librations in the single points of the orbit, therefore their ratio determines the increments of the librations from one arch to another arch. Thus, let PM be the increment of *libration* from the arch PK to the arch KG, MI the increment of *libration* from the arch KG to the arch GD, FQ the increment of *libration* from the arch GD to DN, and so on.

This means:

1. The segments PM, PI, PF, PQ, … represent the whole amount of libration, respectively, for the arches PK, PG, PD, PN, … of the eccentric circle.
2. If we pose a compass in the Sun A and from the point M we trace a circle with radius AM, when the compass touches the line KX (in the point L), L is then the point of the planetary path, whose true anomaly is PAL, corresponding to the eccentric anomaly PBK. By taking the center A and the radius AI, if the compass touches the line FG in H, then H is a point of the path, and so on. Hence, it is possible to construct the trajectory by points.

Given these considerations and by means of a theorem due to Pappus (KGW, VII, p. 371, line 24), Kepler was able (KGW, VIII, Book V, Chapters II–III, pp. 370–375):

1. To prove that the whole amount of the libration is proportional to the versed sine of the eccentric anomaly.
2. To construct the orbit.
3. To show that the constant of proportionality is the eccentricity of the ellipsis.

Continuing on this train of thought, Kepler arrived at the area law and at his famous equation (written in modern form)

$$\alpha = \beta + e \cdot \sin \beta$$

where α is the mean anomaly, which is the measure of the time necessary for a planet to reach a point P of eccentric anomaly β starting from the aphelion, and e is the eccentricity.[52]

Historiographical Accounts and Interpretations

1. Stephenson moved a series of criticism to Kepler's reasoning: we have seen that first Kepler supposed the magnetic force was as the cosine of the true anomaly and, afterwards, he proved this is equivalent to claim that it is as the sine of the eccentric anomaly. However, Stephenson claims, if this were the case, then a consequence

[52] KGW, VIII, Book V, Part II, Chapters IV–V, pp. 390–396. See also Caspar's *Nachbericht* (KGW, VII, pp. 598–600) and Stephenson (Stephenson [1987] 1994a, pp. 154–172).

would have been that the fibers were directed towards the Sun when the eccentric anomaly is 90° (Stephenson [1987] 1994a, pp. 151–154) although this does not happen because the angle ANB (see Fig. 16.3b) is not null. Kepler never claimed that its assertion, according to which the deflection is proportional to the sine of the eccentric anomaly, is in contradiction to the supposed behavior of the *librating force*. Furthermore, Stephenson claims that Kepler's step from the expression of deflection in the function of the true anomaly to the expression in the function of the eccentric anomaly is not sound. Moreover, Kepler is ambiguous and unclear as to whether he was referring to a circular orbit or to an elliptical orbit. In substance, it seems that Kepler, knowing the orbits are ellipses by observation and by means of kinematical astronomy, adapted his dynamical astronomy to this truth. Hence, dynamical astronomy should be a consequence of kinematic astronomy and not vice versa. Thus, Stephenson claims that Kepler's deduction is not correct because of the ambiguity in defining the deflection force in every point of the orbit and because:

> [...] once the deflection was separated from any specific causal mechanism, the shape of the orbit was not determined physically. Kepler's physics was not adequate to support his "first law". (Stephenson [1987] 1994a, p. 170)

Nevertheless, *if we suppose that Kepler's theory were adequate to determine the first law, then another more crucial question arises: in what sense could this deduction be called physical?*

The *libration force* in each point of the orbit had been posed by Kepler proportional to a geometric quantity (the sinus of the eccentric anomaly or cosine of the true anomaly), that is typical of kinematical astronomy, albeit the eccentric anomaly referred to an ellipsis is a novelty of Kepler. Therefore, no physical quantity comes into play in determining the *libration force*, and no physical-dynamical feature of each planet or of the Sun (i.e., mass, density). The eccentricity, the trigonometric functions of an angle, are merely geometrical, not physical-dynamical quantities. If a force, or whatever name one wishes to use, has to be a physical-dynamical quantity, it cannot be defined merely through geometrical or kinematical quantities. Some detectable physical effects of the force have to be indicated in order to get the possibility to provide a measure of the force itself and of its composing quantities, as is the case, for example, with Hooke's law (1660). In substance, it is necessary to define the concept of force and such a definition cannot be based on merely geometrical or kinematical quantities. It is not a priori necessary to define the force as Newton did, but a physical-dynamical definition is necessary[53]. However, these characteristics are missing both in the Keplerian *virtus motrix* and in the *libratory force*. Thus, the notion of force in Kepler's astronomy really should have a heuristic role in the discovery of the area law and, for the rest, a merely psychological role.

[53] We have no room to deal with the problem of the relationships between physics and mathematics inside physics, in particular with regard to the definition of the forces. We suggest to the reader Pisano 2011; Barbin and Pisano 2013; and Pisano 2014.

On the whole:

> In the process of construction of the equations, no dynamic consideration is proposed: a Ptolemaic [scholar] could have been accepted the Keplerian way to proceed without particular problems.[54]

It is true that in the construction of the equations no dynamical consideration exists, but, at the same time, Kepler's whole procedure is framed inside a general context in which his ideas on the forces and the kinematical results are continuously superimposed. Hence a Ptolemaic astronomer would not have accepted Kepler's procedures, although probably he would have accepted the results. From a scientific point of view, it is hence correct to relegate Kepler's physical astronomy to a heuristic role, but from a historical point of view, Kepler's physical ideas were important because they represented a fundamental conceptual break with the previous way of conceiving of astronomy.

Davis (1992b, 1992e, 2003; see also 1992c, 1992d, 2007) provides an interesting interpretation, which is far different from the one offered by the current of thought we have analyzed. Davis starts from a consideration concerning the concept of speed: Kepler claimed that, given an orbit, the speed is inversely proportional as the distance Sun–plane. Normally this has been interpreted as if Kepler were referring to the modulus of what today we call the vector velocity or the tangential velocity. Davis (1992b, and more specifically Davis 2003)[55] claims and tries to prove that the concept of tangential velocity is extraneous to Kepler's astronomy, and, more in general, the use of the tangent is. Actually, what is familiar to Kepler is the separation of a line into two perpendicular components. In the specific case of the velocity in a component along the vector-radius (the radial velocity) and one along the perpendicular to the vector-radius (the transverse velocity), Davis argues (Davis 1992b, pp. 116–120) that Kepler claimed only the transverse velocity is inversely proportional to the Sun–planet distance, not the modulus of the tangential velocity, which is exactly equivalent to the area law. Therefore, independently of the real demonstration given by Kepler for the area law, his idea was that transverse velocity is as the opposite of the distance. In *Epitome*, Book V, First Part, Chapter IV (KGW, VII, pp. 375–379), concerning the area law, Kepler develops a complex argumentation which is impossible to report here.[56] In this argumentation, he claims that only the component of the motion perpendicular to the radius vector has to be considered as far as area law is concerned (KGW, VII, p. 378, lines 10–14). Caspar (KGW, VII, pp. 597–598) interprets Kepler's statement in a modern manner and shows (by not-difficult mathematical argumentations) that, from Kepler's assertion, the correct

---

[54] "Nel processo di costruzione delle equazioni non interviene mai alcuna considerazione dinamica: un tolemaico avrebbe potuto accettare senza grossi problemi il modo di procedere di Keplero" (Petroni 1989, p. 192, pp. 190–196).

[55] Davis (2003) expounds and tries to prove a well-defined historiographical thesis: it is improper to speak of speed or velocity in Kepler. Concepts such as those of movements or change of movements cannot be equated, in a Keplerian perspective, with our modern concept of velocity or speed. In the paper a series of interesting considerations on the use of Euclidean geometry in Kepler is developed.

[56] For this problem, see Bussotti 2015, Section 6.1.2.1, pp. 121–125.

relation according to which the velocity is inversely proportional as the perpendicular to the tangent can be deduced. Koyré (1961a, b, c, pp. 321–323) argues that, although Kepler's relations are mathematically equivalent to Newton's, Kepler did not recognize this equivalence. He adds reasonable arguments in favor of the thesis he could not recognize. Anyway, Davis can base his interpretation on what Kepler claimed in the mentioned chapter of the *Epitome* and on her studies on the missing use of the concept of tangential velocity in Kepler.

However, Davis argues that Kepler also proved the law of orbits' ellipticity (Davis 1992e). The global force that moves the planets in Kepler is almost the inverse of the Sun–planet distance; we have to say *almost* because the *virtus motrix* is exactly the inverse of such distance and the libratory force is, in Kepler, a perturbation of the motive virtue, so that the inverse proportionality force–distance is almost linear. It is well known from Newton, and it is also provable by means of modern rational mechanics, that such a force cannot produce an orbit that is an ellipsis in which the Sun is in one of the two foci. This happens if and only if the inverse square law is valid. Therefore, a demonstration of orbits' ellipticity cannot be achieved by means of a Keplerian force.

16.5.2.3 What Force Is Responsible for the Movements of the Planetary Apses–Line?

The effects of the two analyzed forces are far the most important ones in Kepler's physical astronomy. Nevertheless, Kepler had the intention to construct a very physical system and hence he tried to explore the causes that determine the movements of the *apses-line* and the planets' latitude. Chapter XLII of the *Astronomia Nova* concerns the former problems and interrelated questions such as the correction of the mean motion (KGW, III, pp. 275–282). The mechanism of the libratory force is explained in this case, too. According to Kepler (Ibidem) the magnetic force of the planet acts on the apses-line so that, at aphelion, it induces on the apses-line a clockwise rotation (in *antecedentia*), and at the perihelion (Fig. 16.3a) counterclockwise (in *consequentia*). The pushing of the force in *antecedentia* is stronger than in *consequentia*, thus the apses-line moves forwards. This explanation looks like a hysteron–proteron: because the apses- line moves forwards, then one invents a mechanism that determines it. In this case, too, we are in a situation in which kinematics looks like the foundation, and not only an *observative guide*, for dynamic rather than vice versa. Although in a modern perspective, the movement of the apses-line is the movement of the reference system, in Kepler the force acting on the apses is the libratory force, which is also responsible for the radial movement or the planets.

16.5.2.4 What Force Determines the Movement in Latitude of the Planets?

The theory of latitude is expounded by Kepler in *Astronomia Nova* (KGW, III, pp. 389–394) and in the *Epitome* (KGW, VII, pp. 343–348). For our aims, we refer to the following page from *Astronomia Nova* (Fig. 16.6):

Theory: Kepler imagined that the planet has a latitude axis (the arrow) with a constant orientation to the fixed stars. When the movement in longitude has the same direction as the point of the arrow of the latitude axis, the planet shifts towards North, whereas it shifts towards South in the other case. *Commentaries:* Such an ingenious explanation appears *ad hoc*. It shares this problem with the explanation of the apses-line movement. Furthermore the planet has two axes:

1. The magnetic axis responsible for the movement towards and away from the Sun.
2. The latitude axis. In Kepler's theory their relations are not completely clarified.[57]

In this case, too, Kepler tried a dynamical explanation of a planetary movement (the one in latitude), but the action of the latitude-force is not quantified and, once again, no real physical variable is used to determine the effect of the force.

Moreover, when the planet moves in longitude from D to B (let us imagine that ABCD is the projection of the orbit of the Sun's equator plane), the arrow tends to make the planet move towards North because the point of the arrow drives the latitude motion. On the contrary, from B to D the tail of the arrow predominates and the planet shifts in the South direction.

Finally, Kepler thought of three kinds of forces to explain the planetary movements:

1. The *virtus motrix* or *virtus promotoria* of the Sun, which gives an account of the main movement of the planets, the one that could be called the mean movement around the Sun
2. The interaction between the magnetic force of the Sun and that of the planets for the moving towards and away of each planet from the Sun and for the movement of the apses-line
3. The latitude force (materialized by the latitude axis) for the movement in latitude of the planets

[57] Cfr.: Stephenson [1987] 1994a, pp. 130–137.

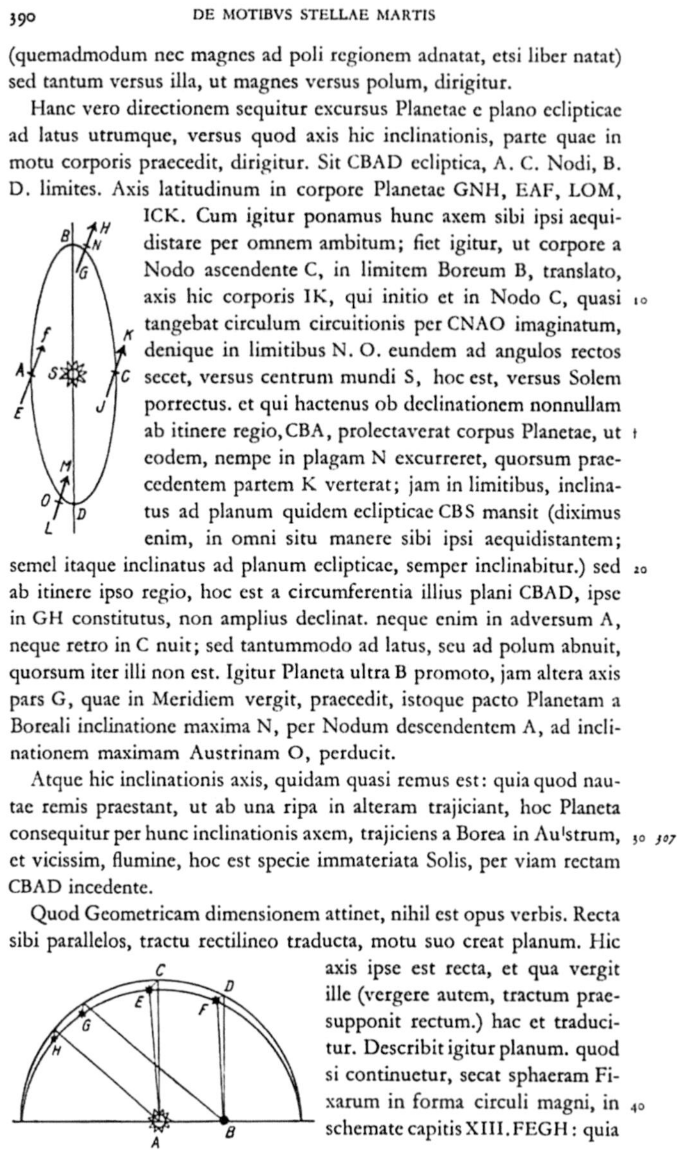

390 DE MOTIBVS STELLAE MARTIS

(quemadmodum nec magnes ad poli regionem adnatat, etsi liber natat) sed tantum versus illa, ut magnes versus polum, dirigitur.

Hanc vero directionem sequitur excursus Planetae e plano eclipticae ad latus utrumque, versus quod axis hic inclinationis, parte quae in motu corporis praecedit, dirigitur. Sit CBAD ecliptica, A. C. Nodi, B. D. limites. Axis latitudinum in corpore Planetae GNH, EAF, LOM, ICK. Cum igitur ponamus hunc axem sibi ipsi aequidistare per omnem ambitum; fiet igitur, ut corpore a Nodo ascendente C, in limitem Boreum B, translato, axis hic corporis IK, qui initio et in Nodo C, quasi tangebat circulum circuitionis per CNAO imaginatum, denique in limitibus N. O. eundem ad angulos rectos secet, versus centrum mundi S, hoc est, versus Solem porrectus. et qui hactenus ob declinationem nonnullam ab itinere regio, CBA, prolectaverat corpus Planetae, ut † eodem, nempe in plagam N excurreret, quorsum praecedentem partem K verterat; jam in limitibus, inclinatus ad planum quidem eclipticae CBS mansit (diximus enim, in omni situ manere sibi ipsi aequidistantem; semel itaque inclinatus ad planum eclipticae, semper inclinabitur.) sed ab itinere ipso regio, hoc est a circumferentia illius plani CBAD, ipse in GH constitutus, non amplius declinat. neque enim in adversum A, neque retro in C nuit; sed tantummodo ad latus, seu ad polum abnuit, quorsum iter illi non est. Igitur Planeta ultra B promoto, jam altera axis pars G, quae in Meridiem vergit, praecedit, istoque pacto Planetam a Boreali inclinatione maxima N, per Nodum descendentem A, ad inclinationem maximam Austrinam O, perducit.

Atque hic inclinationis axis, quidam quasi remus est: quia quod nautae remis praestant, ut ab una ripa in alteram trajiciant, hoc Planeta consequitur per hunc inclinationis axem, trajiciens a Borea in Au|strum, 307 et vicissim, flumine, hoc est specie immateriata Solis, per viam rectam CBAD incedente.

Quod Geometricam dimensionem attinet, nihil est opus verbis. Recta sibi parallelos, tractu rectilineo traducta, motu suo creat planum. Hic axis ipse est recta, et qua vergit ille (vergere autem, tractum praesupponit rectum.) hac et traducitur. Describit igitur planum. quod si continuetur, secat sphaeram Fixarum in forma circuli magni, in schemate capitis XIII. FEGH: quia

Fig. 16.6 The mechanism explaining the planets' latitude motion in *Astromomia Nova* (KGW, III, p. 390. With permission of *Bayerische Akademie der Wissenschaften*)

16.6 Conclusion

16.6.1 Koyré and Kepler

Koyré's studies of astronomical revolution mainly concern both *Nicolas Copernic, Des révolutions des orbes céleste* (Koyré 1934) and *La révolution astronomique: Copernic, Kepler, Borelli* (Koyré 1961c). In addition, Koyré also wrote significant and mature ideas on foundations of astronomy and mechanics in *From the Closed World to the Infinite Universe* (Koyré 1957), where he dealt with the role played by infinity in mathematics and in physics. This aspect is also important for the mathematization of Nature in the Renaissance and for the connected birth of the astronomical revolution. In *Du monde de « à–peu–près » à l'univers de là précision* (Koyré 1961a) he presented the historical and scientific features connoting the birth of modem science, also taking into account human work, specifically where artisans' works are concerned (Pisano 2008; Pisano and Gaudiello 2009a, b; Pisano and Bussotti 2014a, b). Particularly interesting, for our aims, is the inquiry on the infinite/finite size of the universe as explained in *From the Closed World to the Infinite Universe* (Koyré 1957). It mainly concerned Kepler's *De stella nova in pede Serpentarii*[58] (Kepler [1606]1859; KGW, I, pp. 147–292).

According to Koyré, and based on our arguments presented in this chapter, Kepler did not accept an infinite size of the universe (Ibidem). The reasons were religious beliefs and historical–scientific convictions (Koyré 1957, pp. 58–59). For the sake of brevity we only present, as final remarks, the latter question, thus taking into account Koyré's scientific thought (Koyré 1957, III, pp. 58–87).

From a logical point of view, a cosmological acceptance of an infinite size of the universe means that the universe should likely be a uniform physical system (Kragh 1991, 2008), as well. However, at least for an observer living in the seventeenth century, this was not the case. Let us think, for example, of the fixed stars and the solar system. Kepler thought that, because all fixed stars had the same distance from the Earth inside a cosmic *cava* sphere (Kepler [1606] 1859, pp. 688–691), a human being could not arrive at conceiving the concept of infinity in a physical perspective (Ibidem). In other words, if for any reason, some stars infinitely far really exist, then these stars should be infinitely big, otherwise their observation would be impossible, or alternatively, they should be simply invisible to the human eye.

According to Koyré (Koyré 1957, III), this is a clear case of a metaphysical conviction, which has a profound consequence on the way in which a scientist develops some aspects of his physical ideas: Kepler, because of metaphysical convictions of which Koyré speaks profusely, thought the universe was finite. Nevertheless, to prove the finiteness of the universe, he resorted to reasoning, part of which we refer to in the translation, adapted from Koyré (see below quotations related to the footnotes 60–62). Fig. 16.7 This is undoubtedly a *physical* reasoning, but it is plausible that Kepler developed it because of his metaphysical points of view. This reasoning

[58] See also *Epitome Astronomiae Copernicanae* (KGW, VII) discussed above.

JOANNIS KEPPLERI
Sac. Cæſ. Majeſt. Mathematici
DE
STELLA NOVA
IN PEDE SERPENTARII, ET
QUI SUB EJUS EXORTUM DE
NOVO INIIT,
TRIGONO IGNEO.
LIBELLUS ASTRONOMICIS, PHYSICIS, METAphyſicis, Meteorologicis & Aſtrologicis Diſputationibus,
ἐνδόξοις & παραδόξοις plenus.
ACCESSERUNT
I. DE STELLA INCOGNITA CYGNI: Narratio Aſtronomica.
II. DE JESV CHRISTI SERVATORIS VERO Anno Natalitio, conſideratio noviſſimæ ſententiæ LAVRENTII SVSLYGÆ Poloni, quatuor annos in uſitata Epocha deſiderantis.
Cum Privilegio S. C. Majeſt. ad annos XV.

PRAGAE
Typis PAULI SESSII, impenſis AUTHORIS.
ANNO M. DCVI.

Fig. 16.7 Conceptualization of the infinite and of the universe in Kepler. Image source: Google Books, public domain

does not appear to be refined or profound, whereas the profundity of Kepler is well known. This means that he constructed such an argument basing his ideas on convictions derived from different sources. These sources are religion and metaphysics.[59] In this sense, we might hypothesize that Kepler followed, in his metaphysical reasoning, the Aristotelian school.

Nevertheless, (as Koyré also argued) what would happen if the size of the universe were considered infinite? By taking into account the historical-scientific context in Kepler's epoch, if the size of the universe were infinite then it should have (according to Kepler who did not accept an infinite universe) a uniform geometrical–cosmological and well-defined structure. But, Kepler did not accept this view, which is due both to religious and rational reasons. For example, according to Kepler the universe cannot be uniform because the solar system breaks this hypothetical uniform distribution. In addition, according to Kepler (and Koyré) a metaphysical consequence arises: Kepler essentially thought that (a) the stars were placed at the same distance from the earth and (b) the infinite was unthinkable for a human being. If *ad absurdum*, the stars were infinitely far from the Earth, they should be infinitely large or simply would not be observable by human eyes; and, the total number of stars should be less than the number of human beings. On the other hand, the idea that only a finite number of stars is thinkable within a finite space is also dubious because the (Aristotelian) *vacuum* without a *body* is *nihil* and consequently it cannot exist. In any case, according to Kepler, these aspects are merely metaphysical. On the other hand, a conceptualization of a finite universe would imply a finite number of geometrical modelings. From that, it follows that the solar system should be unique.

On our side, we have focused on the aspects concerning the interplay between physics–mathematics and metaphysics because they are historically significant for our research (Pisano 2016a, b). In Kepler's words:

> [...] there is a sect of philosophers, who (to quote the judgment of Aristotle, unmerited however, about the doctrine of the Pythagoreans lately revived by Copernicus) do not start their ratiocinations with sense-perception or accommodate the causes of the things to experience: but who immediately and as if inspired (by some kind of enthusiasm) conceive and develop in their heads a certain opinion about the constitution of the world; once they have embraced it, they stick to it; and they drag in by the hair which occur and are experienced every day in order to accommodate them to their axioms. These people want this new star and all others of its kind to descend little by little from the depths of nature, which, they assert, extend to an infinite altitude, until according to the laws of optics it becomes very large and attracts the eyes of men; then it goes back to an infinite altitude and every day so much smaller as it moves higher.[60]

[59] This discussion also refers to other metaphysical aspects that, for the sake of brevity, we avoid facing. For example: the Aristotelian conviction that the series of natural numbers is potentially infinite, but it cannot be considered as an actually infinite entity (Aristotle 1999, *Physics*, Book III, Chapter VI; *Id.*, (1801), *Metaphysics*, Book IX, Chapter VI); the problematization of infinity by Plato (also Pythagoreans), where the universe could be represented by a finite arrangement of natural numbers (Aristotle 1999, *Physics*, Book III, Chapter IV).

[60] Kepler [1606] 1859, p 687 (KGW, I, Chap. XXI, pp. 251–252).

> […] We shall show them that by admitting the infinity of the fixed stars they become involved in inextricable labyrinths. Furthermore we shall, if possible, take this immensity away from them: then, indeed, the assertion will fall of itself.[61]
>
> […] To begin with, it can most certainly be learnt from astronomy that the region of the fixed stars is limited downwards; […] moreover it is not true […] that this inferior world with its sun differs in no way in its aspect from any one of the fixed stars; that is, of one region or place from another. For, be it admitted as a principle that the fixed stars extend themselves in infinitum. Nevertheless it is a fact that in their innermost bosom there will be an immense cavity, distinct and different in its proportions from the spaces that are between the fixed stars. So that if it occurred to somebody to examine only this cavity, even ignorant of the eight small bodies which fly around the centrum of this space at a very small distance from it, and did not know what they are, or how many.[62]

Finally, Kepler's universe, as discussed by Koyré, is a brilliant attempt to build something new but, at the same time, it is still connected with several *topoi* of Aristotelian conceptions:

> The conception of the infinity of the universe is, of course, a purely metaphysical doctrine that may well—as it did—form the basis of empirical science; it can never be based on empiricism. This was very well understood by Kepler who rejects it therefore—and this is very interesting and instructive—not only for metaphysical, but also for purely scientific reasons; who even, in anticipation of some present-day epistemologies, declares it scientifically meaningless. (Koyré 1957, p. 58)
>
> […]. All that is not new, nor specific to Kepler: it is the traditional teaching of Aristotelian scholasticism. Thus we have to admit that Johannes Kepler, the great and truly revolutionary thinker, was, nevertheless, bound by tradition. In his conception of being, of motion, though not of science, Kepler, in the last analysis, remains an Aristotelian. (*Ivi*, p. 87)

On this problem, our view is that, on some occasions, Koyré overestimated the influence exerted by Aristotle on Kepler.

Koyré follows an interesting historical method of inquiry with which, generally speaking, we agree; though not all of his conclusions are completely agreeable. Particularly, when he has to prove a thesis, he refers directly to the authors, he often reports long quotations and comment on them, and he is convinced that a general scientific-historical thesis, too, can be sustained only relying upon specific textual and technical evidence. We have chosen the example of Kepler not by chance. Koyré dedicated the most important and widest part of his *La révolution astronomique: Copernic, Kepler, Borelli* to Kepler. In this book (Koyré 1961c) he pointed out the novelties of Kepler's approach by presenting a broad series of textual evidence. He used the same approach in *From the Closed World to the Infinite Universe* (Koyré 1957) and *in Du monde de «à–peu–près» à l'univers de là precision* (Koyré 1961a). A part of the research presented in these books is devoted to point out the reasoning by which Kepler thought the universe is finite. Koyré aimed to prove that, for this aspect, Kepler, who was the main protagonist of the astronomic revolution, was, in a certain sense, linked to ancient science. Therefore, here one can see the same method, applied to the same author, to prove two different and broad theses related to the history of science.

---

[61] *Ivi*, p 688 (KGW, I, Chapter XXI, pp. 252–253).

[62] *Ivi*, p 689 (KGW, I, Chapter XXI, p. 253).

Without having any claim to assert, in this chapter, theses that go beyond the history of science, we have tried to clarify the concept of force in Kepler, basically relying upon Kepler's works, referring to several quotations and commenting on them. Although we are perfectly aware that a selection of passages is already an interpretation, we have tried to keep Kepler's assertions separated from our commentaries and interpretations, as the reader has seen. The literature on Kepler is nowadays incomparably broader than in Koyré's epoch, thence we have obviously, referred to literature, also.

16.6.2 Concluding Remarks

Kepler based the explanation of the planetary movements on three forces:

1. The *virtus motrix* of the sun.
2. The *libratory force* given by the attraction and repulsion between the magnetic axis of a planet and the magnetic pole, posed by Kepler on the solar surface.
3. The force, which determines the motion in latitude given by the latitude axis of each planet.

Kepler tried to construct a cosmological and physical system in order to explain the planetary paths clearly, but his conception of force is not adequate because the ideas that the solar virtue acts in inverse proportion to the distance and forces produce velocities do not lead to the determination of the correct planetary orbit. Kepler excluded gravity as a cause of planetary motion, even though his notions on this force were far more advanced than those of his contemporaries. Actually, the concept of central force is problematic in a Keplerian perspective because the *virtus motrix* acts as the Sun rotates and not because of the mass of the Sun; it is not an attractive force. This implies that Kepler did not have a physical model comparable with that deriving from universal gravitation; nevertheless he tried to propose a physical model. The conclusion is that Keplerian kinematics cannot be deduced from Keplerian dynamics.

There are some more general questions, connected to the epistemology and methodology of science and mathematics, that deserve attention to understand fully Kepler's development of the concept of force. From a philosophical point of view, Kepler thought that the structure of the universe was based on geometrical archetypes. This is evident starting from his early work *Mysterium Cosmographicum*. In the second editions of this work (1621), almost at the end of his complex intellectual itinerary, Kepler confirmed this opinion writing:

> Finally, almost each of the books on astronomy I have published from that time could be referred to a chapter of this little book, of which they contain an explanation or an integration.[63]

---

[63] "[...] denique quicquid fere librorum Astronomicorum ex illo tempore edidi, id ad unum aliquod capitum, hoc libello propositorum, referre potuit, cuius aut illustrationem aut integrationem contineret" (KGW, VIII, p. 9, lines 25–28).

Coherently with his foundational–philosophical convictions, the methods that Kepler used in astronomy were invariably connected to geometry and to the possibility to construct the points of the figures he was looking for by rule and compass (Davis 1992a, p. 98). As each astronomer does, Kepler knew trigonometry very well and he was one of the first to understand fully the importance of the concept of the logarithm (KGW, IX). However, in his conception, all these means were necessary instruments to insert inside geometry in order to make calculations possible or quicker, but the ground of astronomy was geometry. On the other hand, trigonometry can be considered a part of geometry.

The use of algebra (Katz and Parshall 2014) both from a symbolic and conceptual point of view, is almost missing in Kepler's astronomical works, although algebra achieved an important development at the end of the sixteenth century, basically thanks to François Viète (1540–1603). Furthermore calculus had not yet been developed. Obviously, this creates difficulties in considering instantaneous sizes. Therefore, the mathematical means used by Kepler were relatively poor, in comparison both to the means available in his lifetime and with those available a few decades later. Despite this, Kepler reached his results through a genial usage of these means.

In connection to the force problem in Kepler, the previous last considerations raise an epistemological and methodological question concerning the general influence of mathematical knowledge in physics. Probably if Kepler had known how to solve the differential equation deriving from the idea that planetary attractive force is inversely proportional to the distance, he would have reached the conclusion that this hypothesis was not tenable. This means that there is a strong connection between the development of physics and the state of mathematical knowledge. For, as soon as the results are transcribed into mathematical terms and mathematics provides a precise solution to the posed question—in our case the determination of the orbit in a two-bodies problem when the forces are a function of a power of the distance between the two bodies—the correct hypothesis is identified. However, if this transcription is missing, the qualitative treatment does not indicate with precision which are the consequences of the hypotheses and hence one is not able to exclude the wrong one(s) and to keep the right, or at least the possible, one(s).

The way in which Kepler dealt with the forces problem introduces us to another interesting aspect of Kepler's *Forschung*. The style of Kepler is personal: it is completely different from that of the scientists and mathematicians of his epoch because Kepler refers not only to his results, but also to his mental processes, his failures and his successes, so that his works (*Mysterium Cosmographicum* and *Astronomia Nova* are paradigmatic in this sense) are difficult to read. The treatment is not linear, the use of analogies is huge, and the whole picture is as ingenious as it is complicated. This implies that, despite Kepler's general ideas being known, the direct reading of an entire work by his was not common. In the initial part of the seventeenth century a standard scientific style did not yet exist: let us think of the style of Fermat or of Desargues. The former wrote his ideas and results in number theory using almost no symbolism, and the latter used such a personal language for projective geometry that his results were rediscovered decades later. But, other scientists and mathematicians, including Viète, Galileo, Descartes, and in part John Wallis (1616–1703), began slowly to use mathematical symbolism—this process needed a long period to assume a more or less general accepted form—in a systematic manner. This tendency prevailed for reasons that can be easily understood, so that the direct reading of the works of scientists who used a very personal way of writing and exposing their results began to diminish. Strictly connected with this phenomenon is the fact that many of the most personal of Kepler's views, such as those on the concept of force, were not generally known in the second part of the seventeenth century and they likely had scant influence on the birth of the modern theory of forces, the Newtonian one (Kuhn 1962).

Finally, we have clarified our opinion, according to which the concept of force in Kepler has an important heuristic, historical, and conceptual role as far as Kepler tried to construct a physical system, but that his attempt failed. We have also explained the limits and the merits of Kepler's conception of force. With respect to the literature, we have provided theoretical advances about the Keplerian concept of force and the mathematical–physical structures of his laws. Therefore, the research on the concept of force in Kepler is an interesting subject for many reasons:

(a) Kepler's ideas in themselves
(b) Physical consequences of such ideas
(c) Different scientific styles used by scientists to explain their results
(d) Analysis of how scientific ideas are spread
(e) Analysis of the relations between science and society

In this chapter, we have dealt with problems (a) and (b). We have only outlined, in this final remark, the others, to which we are going to dedicate further research.

Acknowledgments We want to express our warm gratitude to Daniel Di Liscia for his valuable comments, to English proofreaders and to anonymous referees for their good blind peer–reviewed job. Their comments were of great help. We also want to express our sincere thankfulness and appreciation to Peter Michael Schenkel (*Kepler-Kommission*) who on behalf of the *Bayerische Akademie der Wissenschaften* gave us permission to use images from the official and prestigious KGW edition.

References

Abattouy M (2006) The Arabic transformation of mechanics: the birth of science of weights. Foundation for Science Technology and Civilisation 615:1–25.

Abattouy M, Renn J, Weinig P (2001) Transmission as Transformation: The Translation Movements in the Medieval East and West in a Comparative Perspective. Science in Context 14:1–12.

Aiton E J (1977) Kepler and the "Mysterium Cosmographicum". Sudhoffs Archive 61/2:73–194.

Aiton E J (1978) Kepler's Path to the Construction and Rejection of his First Oval Orbit for Mars. Annals of Science 35/2:173–190.

Alvarez C, Dhombres J (2011) Une histoire de l'Imaginaire mathématique. Hermann, Paris.

Aristotle (1801) The metaphysics of Aristotle, Translated from the Greek; with Copious Notes in which the Pythagoric and Platonic dogmas respecting numbers and ideas are unfolded from ancient sources. Thomas Taylor (ed). Davis–Wilks–Taylor, Chancery–Lane, London.

Aristotle (1999) Physics. Translation by Waterfield R. The Oxford University Press, Oxford.
Baigre BB (1990) The justification of Kepler's ellipse. Studies in History and Philosophy of Science 21/4:633–664.
Barbin E, Pisano R 2013 (eds) The Dialectic Relation between Physics and Mathematics in the XIXth Century. Springer, Dordrecht.
Barker P, Goldstein BR (1994) Distances and velocity in Kepler's astronomy. Annals of Science 51:59–73.
Barker P, Goldstein BR (2001) Theological Foundations of Kepler's Astronomy. Osiris 16:88–113.
Beer A, Beer P (1975) (eds) Kepler. Four Hundred Years, Vistas in Astronomy 18. The Pergamon Press, Oxford.
Bialas V (2003) Keplers Vorarbeiten zu seiner Weltharmonik. In Pichler 2003, pp. 1–14.
Blum J, Helmchen W (1987) Von Kepler zu Newton. Von den Planetenbahnen zum Gravitationsgesetz. Praxis Mathematica 29/4:193–199.
Boner P (2006) Kepler's Living Cosmology. Bridging the Celestial and Terrestrial Realms. Centaurus 48/1:32–39.
Boner P (2008) Life in the Liquid Fields. Kepler, Tycho and Gilbert on the Nature of Heavens and Earth. History of Science 46/3:275–297.
Boner P (2009) A Tenuous Tandem: Patrizi and Kepler on the Origins of Stars. Journal for the History of Astronomy 40/4:381–391.
Boner P (2011) Kepler's Imprecise Astrology. In Mehl 2011, pp. 115–132.
Boner P (2013) Kepler's Cosmological Synthesis: Astrology, Mechanism and the Soul. Brill, Boston.
Bruhn S (2005) The musical Order of the World: Kepler, Hesse, Hindemith. Hillsdale, New York.
Buridan J (1509) Acutissimi philosophi reverendi magistri Johannis Buridani subtilissime questiones super octo phisicorum libros Aristotelis diligenter recognite & revise a magistro Johanne Dullaert de Gandavo antea nusquam impresse.
Bussotti P (2011) The circulation of Kepler's cosmological ideas in Italy during Kepler's lifetime. In Mehl 2011, pp. 197–217.
Bussotti P (2015) The Complex Itinerary of Leibniz's Planetary Theory. Springer-Birkhäuser, Basel.
Bussotti P, Pisano R (2013) On the Conceptual and civilization Frames in René Descartes' Physical Works. Advances in Historical Studies 2/3:106–125.
Bussotti P, Pisano R (2014a) Newton's Philosophiae Naturalis Principia Mathematica "Jesuit" Edition: The Tenor of a Huge Work. Accademia Nazionale Lincei-Rendiconti Matematica e Applicazioni 25/4:413–444.
Bussotti P, Pisano R (2014b) On the Jesuit Edition of Newton's Principia. Science and Advanced Researches in the Western Civilization. In Pisano 2014 3/1:33–55.
Bussotti P, Pisano R (2017) Historical and Philosophical Details on Leibniz's Planetary Movements as Physical–Structural Model. In Pisano R, Fichant M, Bussotti P and Oliveira ARE. The Dialogue between Sciences, Philosophy and Engineering. New Historical and Epistemological Insights. Homage to Gottfried W. Leibniz 1646–1716. The College Publications, London, pp. 49–92.
Caspar M ([1948] 1962) Kepler. Collier, New York.
Cohen FH (1994) The Scientific Revolution: A Historiographical Inquiry. The University of Chicago Press, Chicago.
Cohen FH (2011) How Modern Science Came into the World: Four Civilizations, One Seventeenth-Century Breakthrough. The University of Amsterdam Press, Amsterdam.
Crombie AC (1963) (ed) Scientific change: historical studies in the intellectual, social and technical conditions for scientific discovery and technical Invention, from antiquity to the present–Symposium on the history of science, University of Oxford, 9–15 July 1961. Heinemann education books, London.
Davis AEL (1981) A Mathematical Elucidation of the Bases of Kepler's Laws. The University Microfilms International, Amm Arbor–Mi.
Davis AEL (1992a) Kepler's Resolution of Individual Planetary Motion. Centaurus 35/2:97–102.

Davis AEL (1992b) Kepler's "Distance Law"–Myth, not Reality. Centaurus 35/2:103–120.
Davis AEL (1992c) Grading the Egg. Kepler's Sizing–Procedure for the Planetary Orbit. Centaurus 35/2:121–142.
Davis AEL (1992d) Kepler's Road to Damascus. Centaurus 35/2:143–164.
Davis AEL (1992e) Kepler's Physical Framework for Planetary Motion. Centaurus 35/2:165–191.
Davis AEL (2003) The Mathematics of the Area Law: Kepler's successful proof in Epitome Astronomiae Copernicanae (1621). Archive for History of Exact Sciences 57/5:355–393.
Davis AEL (2007) Some plane geometry from a cone: the focal distance of an ellipse at a glance. The Mathematical Gazette 91/521:235–245, 520–522.
De Gandt F (1995) Force and geometry in Newton's principia. The Princeton University Press, Princeton–New Jersey.
Dealmbre JB ([1817] 1965) Histoire de l'astronomie ancienne. Johnson Reprint Corporation, New York and London.
Delambre JB ([1819] 1965) Histoire de l'astronomie du Moyen Âge. Johnson Reprint Corporation, New York–London.
Delambre JB ([1821] 1969) Histoire de l'astronomie modern. Johnson Reprint Corporation, New York and London.
Dhombres J (1978) Nombre, mesure et continu : épistémologie et histoire. Cedic–Nathan, Paris.
Dhombres J (2013) Pierre–Simon de Laplace (1749–1827). Le parcours d'un savant. Hermann, Paris.
Dijksterhuis EJ ([1950] 1961) The Mechanization of the World Picture. The Oxford University Press, Oxford.
Donahue WH (1981) The dissolution of celestial spheres 1595–1650. Arno Press, New York.
Donahue WH (1993) Kepler's first thoughts on oval orbits: text, translation and commentary. Journal for the History of Astronomy 24/1–2:71–100.
Donahue WH (1994) Kepler's invention of the second planetary law. British Journal for the History of Science 27/92–1:89–102.
Donahue WH (1996) Kepler's approach to the oval of 1602, from the Mars notebook. Journal for the History of Astronomy 27/4:281–295.
Dreyer JLE ([1906] 1953) History of planetary systems from Tales to Kepler. Dover, New York.
Elena A (1983) On the different kinds of attractive forces in Kepler. Archive Internationales d'Histoire des Sciences 33:22–29.
Escobar JM (2008) Kepler's Theory of the Soul: A Study on Epistemology. Studies in History and Philosophy of Science 39:15–41.
Fabbri N (2009) Cosmologia e armonia in Kepler e Mersenne. Contrappunto a due voci sul tema dell'Harmonices Mundi. Olschki, Firenze.
Field J (1979) Kepler's rejection of solid celestial spheres. Vistas in Astronomy 23/3:207–211.
Field J (1988) Kepler's geometrical cosmology. The Chicago University Press, Chicago.
Field J (2009) Kepler's Harmony of the World. In Kremer R, Wlodarczyk J (eds). Johannes Kepler. From Tübingen to Zagan. Institute for the History of Science–The Polish Academy of Sciences–The Copernicus Centre for the Interdisciplinary Studies. Instytut Historii Nauki PAN, Warszawa, pp. 11–28.
Gaukroger S, Schuster J, Sutton J (2000) (eds) Descartes' natural philosophy. Routledge, London and New York.
Gilbert G (1600) De magnete, magneticisque corporibus, et de magno magnete tellure Physiologia nova, plurimis et argumentis et experimentis demonstrata. Petrus Shout, London.
Gillispie CC, Pisano R (2014) Lazare and Sadi Carnot. A Scientific and Filial Relationship. 2nd edition. Springer, Dordrecht.
Gingerich O (1975) Kepler's place in astronomy. In Beer and Beer 1975, pp. 261–275.
Gingerich O (2011) Kepler's Trinitarian cosmology. Science and Theology 9/1:45–51.
Goldbeck E (1896) Keplers Lehre von der Gravitation. Olms, Hildesheim–New York.
Granada MA (2009) Novelties in the Heavens between 1572 and 1604 and Kepler's Unified View of Nature. Journal for the History of Astronomy 40/4:393–402.
Granada MA (2010) "A quo moventur planetae?". Kepler et la question de l'agent du mouvement planétaire après la disparition des orbes solides. Galilaeana VII:111–141.

Grössing H (2003) Astrologie bei Johannes Kepler, Georg von Feuerbach und Johannes Regiomontanus. In Pichler 2003, pp. 63–68.
Grössing H (2005) Gedanken zu Keplers Astrologie. In Boockmann F, Di Liscia D, Kothmann H (eds). Miscellanea Kepleriana. Festschrift für Volker Bialas zum 65. Geburtstag. Rauner, Augsburg, pp. 175–182.
Guerlac H (1963) Some historical assumptions of the history of science. In Crombie 1963, pp. 797–817.
Guidi Itokazu A (2006) A Força que Move os Planetas: Da Noção de Species Immateriata na Astronomia de Johannes Kepler. CHPC 3/16/2:211–231. Retrieved: http://www.cle.unicamp.br/cadernos/pdf/AnastasiaGuidiItokazu.pdf.
Guidi Itokazu A (2007) Da Potência Motriz Solar Kepleriana como Emanação Imaterial. CHPC 3/17/2:303–324. Retrieved: http://www.cle.unicamp.br/cadernos/pdf/%288%29Anastasia%20Guidi.pdf.
Haase R (1998) Johannes Keplers Weltharmonik. Der Mensch im Geflecht von Musik, Mathematik und Astronomie. Eugen Diederichs, München.
Heinzmann G (2009) Some Coloured Remarks on the foundations of Mathematics in the 20th Century. In Rahman, Symons, Gabbay and van Bendegem 2009, pp. 41–50.
Holton G (1956) Johannes Kepler's universe: its physics and metaphysics. The American Journal of Physics 24/5:340–351.
Hoyer U (1979) Kepler's celestial mechanics. Vistas in Astronomy 23/1:69–74.
Ihmig KN (1990) Trägheit und Massebegriff bei Johannes Kepler. Philosophia Naturalis 27/2:156–205.
Jaballah HD (1999) La formation du concept de force dans la physique moderne. Vol. II. Contribution à une épistémologie historique. Alpha Edition, Tunis.
Jammer M (1957) Concept of Force: a Study in the Foundations of Dynamics. The Harvard University Press, Cambridge–Ma.
Jardine N (1984) The birth of history and philosophy of science: Kepler's A Defense of Tycho against Ursus, with essays on its provenance and significance. The Cambridge University Press, Cambridge.
Jardine N (2000) Koyré's Kepler/Kepler's Koyré. History of Science 38/4:363–376.
Juste D (2010) Musical Theory and Astrological Foundations. In Wuidar L (ed). Kepler: The Making of the New Aspects. Music and Esotericism. Brill, Leiden, pp. 177–195.
Katz VJ, Parshall KH (2014) Taming the unknown. History of algebra from antiquity to the early twentieth century. The Princeton University Press. Princeton–New Jersey.
Kepler J ([1606] 1859) De stella nova in pede Serpentarii [Prague, Ex Officina calcographica Pauli Sessii]. Opera Omnia. Vol. II, Frankofurti et Erlangae [see also: KGW, I, pp. 147–292].
Kepler J (1596) Prodromus dissertationum cosmographicarum seu mysterium cosmographicum. In KGW, I.
Kepler J (1609) Astronomia Nova ΑΙΤΙΟΛΟΓΕΤΟΣ seu phisica coelestis tradita de commentariis de motibus stellae Martis ex observationibus G. V. Tychonis Brahe. In KGW, III.
Kepler J (1618–1621) Epitome astronomiae copernicanae. In KGW, VII.
Kepler J (1621) Mysterium Cosmographycum (editio altera). In KGW, VIII.
Kepler J (1624) Chilias logarithmorum ad totidem numeros rotundus, praemissa demonstratione legitima ortus logarithmorum eorumque usus. In KGW, IX, pp. 275–352.
Kepler J (1625) Supplementum chiliadis logarithmorum, continens praecepta de eorum usu. In KGW, IX, pp. 353–426.
Kepler J (1981) Mysterium Cosmographicum (The Secret of the Universe). Abaris Book, New York.
Kepler J (1992) New Astronomy. Translation of Astronomia Nova by William H. Donahue. The Cambridge University Press, Cambridge.
Kepler J (1993) Le Secret du monde. Gallimard, Paris.
Kepler J (2010) Kurze Darstellung der kopernikanischen Astronomie. Nachwort von Eberhard Knobloch. Deutsche, Übersetzung von Eberhard Knobloch und Otto und Eva Schönberger. Königshausen & Neumann. Würzburg.

KGW Kepler J (1937–2012) Johannes Kepler, Gesammelte Werke. Van Dyck W, Caspar M. et al. (eds). Revised April 2013. 10 Vols. Deutsche Forschungsgemeinschaft und Bayerische Akademie der Wissenschaften. Beck'sche Verlagsbuchhandlung, München.
Knobloch E (1997) Die gesamte Philosophie ist eine Neuerung in alter Unkenntnis. Johannes Keplers Neuorientierung der Astronomie um 1600. Berichte zur Wissenschaftsgeschichte 20/2–3:135–146.
Knobloch E (2012) La rivoluzione scientifica. I protagonisti: Johannes Kepler. Enciclopedia Treccani. Istituto della Enciplopedia Italiana, Roma., Chapter XXVII.
Koyré A (1934) Nicolas Copernic, Des révolutions des orbes céleste. Alcand, Paris.
Koyré A (1957) From the Closed World to the Infinite Universe. The Johns Hopkins University Press, Baltimore.
Koyré A (1961a) Du monde de « à–peu–près » à l'univers de la précision. M Leclerc et Cie–Armand Colin Librairie, Paris (*Id.*, Les philosophes et la machine. Du monde de l'« à–peu–près » à l'univers de la précision. Études d'histoire de la pensée philosophique).
Koyré A (1961b) Message d'Alexandre Koyré à l'occasion du centenaire de la naissance d'Émile Meyerson. Bulletin de la Société française de philosophie 53:115–s.
Koyré A (1961c) La révolution astronomique : Copernic, Kepler, Borelli. Hermann, Paris.
Koyré A (1963) Scientific Change. In Crombie 1963, pp. 847–857.
Koyré A (1966) Études galiléennes. Hermann, Paris.
Koyré A (1971) Études d'Histoire de la pensée philosophique. Gallimard, Paris.
Koyré A (1973) Études d'histoire de la pensée scientifique. Gallimard, Paris.
Koyré A (1986) De la mystique à la science. Cours, conférences et documents 1922–1962. In Redondi P (ed). Cours, conférences et documents 1922–1962. EHESS Éditions, Paris.
Koyré A (1965) Newtonian Studies. The Harvard University Press, Cambridge–MA.
Krafft F (1975) Nicolaus Copernicus and Johannes Kepler: New Astronomy from old Astronomy. In Beer and Beer 1975, pp. 287–306.
Krafft F (1991) The New Celestial Physics of Johannes Kepler. In Unguru 1991, pp. 185–227.
Kragh H (1991) Cosmonumerology and empiricism. Astronomy Quarterly 8:109–126.
Kragh H (2008) The Moon that Wasn't. Birkhäuser Verlag, Basel.
Kuhn TS (1962) The Structure of Scientific Revolutions. The Chicago University Press, Chicago.
Lindberg DC (1986) The Genesis of Kepler's Theory of Light: Light Metaphysics from Plotinus to Kepler. Osiris 2nd ser.:25–42.
Malet A (1990) Gregoire, Descartes, Kepler and the law of refraction. Archives Internationales d'Histoire des Sciences 40:278–304.
Martens R (2000) Kepler's Philosophy and the New Astronomy. The Princeton University Press, Princeton–New Jersey.
Mehl E (2011) (ed) Kepler. La Physique céleste. Autour de l'Astronomia Nova. Les Belles Lettres, Paris.
Menschl E (2003) Die Faszination der Harmonie – Pythagoras und Kepler. In Pichler 2003, pp. 127–139.
Mittelstrass J (1972) Methodological Elements of Keplerian Astronomy. Studies in History and Philosophy of Science 3:203–232.
Nagel E (1961) The Structure of Science: Problems in the Logic of Scientific Explanation. Harcourt–Brace & World Inc., New York.
Naylor RH (1976) Galileo: the Search for the Parabolic Trajectory. Annals of Science 33:153–172.
Newton I ([1713] 1729) The Mathematical principles of natural philosophy. Translated by Motte Andrew. Motte B, London.
Newton I ([1726; 1739–1742] 1822) Philosophiae Naturalis Mathematica Principia, auctore Isaaco Newtono, Eq. Aurato, perpetuis commentariis illustrata, communi studio Pp. Thomae Le Seur et Francisci Jacquier ex Gallicana minimorum familia, matheseos professorum. Editio nova (in three volumes). Duncan, Glasgow.
Newton I (1687) Philosophiae Naturalis Principia Mathematica. Imprimatur S. Pepys, Reg. Soc. Preses. Julii 5. 1686. Londini, Jussi Societatus Regiae ac Typis Josephi Streater. Prostat apud plures Bibliopolas. Anno MDCLXXXVII.

Petroni AM (1989) I modelli, l'invenzione e la conferma. Franco Angeli, Milano.
Peuerbach G (1473) Theoricæ novæ planetarum, id est septem errantium siderum nec non octavi seu firmamenti. Nürnberg.
Pichler F (2003) (ed) Der Harmoniegedanke Gestern und Heute. Peuerbach Symposium 2002. Trauner, Linz.
Pisano R (2008) A history of chemistry à la Koyré? Introduction and setting of an epistemological problem. Khimiya 2/17:143–161.
Pisano R (2009) On method in Galileo Galilei' mechanics. In Hunger H (ed). Proceedings of 3rd Congress of the European Society for the History of Science. The Austrian Academy of Science, Vienna, pp. 174–186.
Pisano R (2011) Physics–Mathematics Relationship. Historical and Epistemological notes. In Barbin E, Kronfellner M and Tzanakis C (eds), Proceedings of the ESU 6 European Summer University History And Epistemology in Mathematics. Verlag Holzhausen GmbH–Holzhausen Publishing Ltd., Vienna, pp. 457–472.
Pisano R (2013a) History Reflections on Physics Mathematics Relationship in Electromagnetic Theory. In Barbin and Pisano 2013, pp. 31–58.
Pisano R (2014) (ed) Isaac Newton and his Scientific Heritage: New Studies in the History and Historical Epistemology of Science. Special Issue. Advances in Historical Studies 3/3.
Pisano R (2016a) What kind of Mathematics in Leonardo da Vinci and Luca Pacioli? Bulletin of the British Society for the History of Mathematics 31:104–111.
Pisano R (2016b) A Development of the Principle of Virtual Laws and its Framework in Lazare Carnot's Mechanics as Manifest Relationship between Physics and Mathematics. Transversal: International Journal for Historiography of Science 2:166-203.
Pisano R, Bussotti P (2017) (eds) Homage to Galileo Galilei 1564–2014. Reading Iuvenilia Galilean Works within History and Historical Epistemology of Science. Special Issue Philosophia Scientiae 21/1.
Pisano R, Bussotti P (2012) Galileo and Kepler: On Theoremata Circa Centrum Gravitatis Solidorum and Mysterium Cosmographicum. History Research 2/2:110–145.
Pisano R, Bussotti P (2013) Notes on the Concept of Force in Kepler. In Pisano, Capecchi and Lukešová 2013, pp. 337–344.
Pisano R, Bussotti P (2014a) Galileo in Padua: architecture, fortifications, mathematics and "practical" science. Lettera Matematica Pristem International 2/4:209–221.
Pisano R, Bussotti P (2014b). Historical and Philosophical Reflections on the Culture of Machines around the Renaissance. How Science and Technique Work? Acta Baltica Historiae et Philosophiae Scientiarum 2/2:20–42.
Pisano R, Bussotti P (2015a) Historical and Philosophical Reflections on the Culture of Machines around the Renaissance: Machines, Machineries and Perpetual Motion. Acta Baltica Historiae et Philosophiae Scientiarum 3/1:69–87.
Pisano R, Bussotti P (2015b) Introduction to Exploring Changes in How the Histories of the Exact Sciences from the 18th to through the 20th Century Have Been Written: Interpreting the Dynamics of Change in these Sciences and Interrelations Amongst Them—Past Problems, Future Cures? Advances in Historical Studies Special Issue 4/2:65–67.
Pisano R, Bussotti P (2016) A Newtonian Tale Details on Notes and Proofs in Geneva Edition of Newton's Principia. Bulletin–British Society for the History of Mathematics 32:1–19.
Pisano R, Capecchi D (2013) Conceptual and Mathematical Structures of Mechanical Science in the Western Civilization around 18th Century. Almagest 4/2:86–121.
Pisano R, Capecchi D (2015) Tartaglia's science weights. Mechanics in sixteenth century. Selection from Quesiti et invention diverse: Books VII–VIII. Dordrecht, Springer.
Pisano R, Capecchi D, Lukešová A (eds) (2013) Physics, Astronomy and Engineering. Critical Problems in the History of Science. International 32nd Congress for The SISFA–Italian Society of Historians of Physics and Astronomy. The Scientia Socialis UAB & Scientific Methodical Centre Scientia Educologica Press, Šiauliai University, Lithuania.
Pisano R, Fichant M, Bussotti P, Oliveira ARE (2017) (eds.) The Dialogue between Sciences, Philosophy and Engineering. New Historical and Epistemological Insights. Homage to Gottfried W. Leibniz 1646–1716. The College Publications, London.

Pisano R, Gaudiello I (2009a) Continuity and discontinuity. An epistemological inquiry based on the use of categories in history of science. Organon 41:245–265.
Pisano R, Gaudiello I (2009b) On categories and scientific approach in historical discourse. In Hunger H (ed) Proceedings of ESHS 3rd conference. The Austrian Academy of Science, Vienna, pp. 187–197.
Pulte H, Mandelbrote S (2011) (eds.) The Reception of Isaac Newton in Europe. Continuum–Publishing Corporation, London.
Rabin SJ (1997) Kepler's Attitude toward Pico and the Anti-astrology Polemic. Renaissance Quarterly 50/3:750–770.
Rabin SJ (2005) Was Kepler's species immateriata substantial? Journal for the History of Astronomy 36:49–56.
Radelet–de Grave P (1996) Entries: Stevin, Kepler, Leibniz, Huygens. Dictionnaire du patrimoine littéraire européen, Patrimoine Littéraire Européen. Vol. 8, Avènement de l'Équilibre européen 1616–1720. De Boeck Supérieur Bruxelles, p. 18, pp. 745–755, pp. 1020–1027.
Radelet–de Grave P (2007) Kepler (1571–1630) et Cavalieri (1598–1647) astrologues, ou le logarithme au secours de l'astrologie. In Daelemans F, Elkhadem H (eds). Mélanges offerts à Hossam Elkadem par ses amis et ses élèves. Archives et Bibliothèques de Belgique, Bruxelles, pp. 297–315k.
Radelet–de Grave P (2009) Guarini et la structure de l'Univers. Nexus Network Journal 11/3:393–414.
Rahman S, Symons J, Gabbay DM, van Bendegem JP (2009) (eds) Logic, Epistemology, and the Unity of Science. Springer, Dordrecht.
Scaliger JC (1577) Julii Caesaris Scaligeri exotericarum exercitationum liber XV de Subtilitate, ad Hieronymum Cardanum. Claudium Marnium et haeredes Ioannis Aubrii, Frankfurt.
Schuster JA (2000) Descartes opticien: the construction of the law of refraction and the manufacture of its physical rationales, 1618–1629. In Gaukroger, Schuster and Sutton 2000, pp. 258–312.
Schuster JA (2013) Descartes–Agonistes. Physico–mathematics, Method and Corpuscular–Mechanism 1618–1633. Springer, Dordrecht.
Schwaetzer H (1997) "Si nulla esset in terra anima". Johannes Keplers Seelenlehre als Grundlage seines Wissenschaftsverständnisses. Ein Beitrag zum vierten Buch der Harmonices Mundi. Studien und Materialien zur Geschichte der Philosophie 44. Hildesheim, Zürich–New York.
Shank JB (2008) The Newton Wars and the Beginning of French Enlightenment. The University of Chicago Press, Chicago.
Small R (1804) An account of the astronomical discoveries of Kepler. Mawman, London.
Stephenson B ([1987] 1994a) Kepler's Physical Astronomy. The Princeton University Press, Princeton–New Jersey.
Stephenson B (1994b) The Music of the Heavens. Kepler's Harmonic Astronomy. The Princeton University Press, Princeton–New Jersey.
Tartaglia N (1537) La noua scientia de Nicolo Tartaglia: con una gionta al terzo libro, per Stephano Da Sabio, Venetia.
Tartaglia N (1554) La nuova edizione dell'opera Quesiti et inventioni diverse de Nicolo Tartaglia brisciano, Riproduzione in facsimile dell'edizione del 1554 (1959). Masotti A (ed). Commentari dell'Ateneo di Brescia, Tipografia La Nuova cartografica, Brescia.
Treder HJ (1975) Kepler and the Theory of Gravitation. In Beer and Beer 1975, pp. 617–619.
Unguru S (1991) (ed) Physics, Cosmology and Astronomy, 1300–1700: Tension and Accommodation. The Boston Studies in the Philosophy of Science. Vol. 126. Kluwer Academic Publishers, Dordrecht–Boston–London, pp. 101–127.
Voelkel JR (1999) Johannes Kepler and the New Astronomy. The Oxford University Press, Oxford.
Voltmer U (1998) Rhythmische Astrologie. Johannes Keplers Prognose-Methode aus neuer Sicht. Urenia, Neuhausen am Rheinfall.
Westfall RS (1971) The Construction of Modern Science. Mechanism and Mechanic. Wiley & Sons Inc., New York.
Westman R (2001) Kepler's Early Physical-astrological Problematic. Journal for the History of Astronomy 32:227–236.

Chapter 17
Koyré Versus Olschki–Zilsel

Diederick Raven

Abstract In 1939 Koyré introduced the notion of the scientific revolution (SR) as a catch phrase that deals with the "[…] profonde transformation intellectuelle dont la physique modern […]" (Koyré 1939, p. 12; 1943b, p. 400) that he alleged happened at the time of Galileo. For Koyré these changes are due to "pure unadulterated thought" because, as expressed in his 1943 critique of the Olschki–Zilsel position, science "is made not by engineers or craftsmen, but by men who seldom built or made anything more real than theory" (Koyré 1943b, p. 401). We now know Galileo did quite a lot of experimentation; hence, this statement by Koyré is no longer acceptable. In this paper, I will assess the Koyré argument against the Olschki–Zilsel position. Central to my argument is that only by applying a comparative framework, such as developed in my forthcoming book *The European Roots of Science*, it is possible to throw light on this vexed issue.

Keywords Koyré • Olschki • Zilsel • Scientific revolution • Comparative epistemology • Comparative history of science

17.1 Introduction

> [T]he Creator conceived the Idea of the Universe in his mind (we speak in human fashion, so that being men we may understand), and it is the Idea of that which is prior, indeed, as has just been said, of that which is best, so that the Form of the future creation may itself be the best: it is evident that by those laws which God himself in his goodness prescribes for himself, the only thing of which he could adopt the Idea for establishing the universe is his own essence […] so that it [the universe] might become capable of accepting this Idea, he created quantity. (MC, pp. 93–5; KGW, i, 24:1ff)[1]

[1] In referring to Kepler's work, the following acronyms will be used: "MC" for *Mysterium cosmographicum* to refer to Kepler (1981) [1596], "HM" for *Harmonices mundi* to refer to Kepler (1969) [1619] and "KGW" to refer to Kepler's Gesammelte Werke (Kepler 1937).

D. Raven (✉)
Utrecht University, Utrecht, The Netherlands
e-mail: d.w.raven@uu.nl

R. Pisano et al. (eds.), *Hypotheses and Perspectives in the History and Philosophy of Science*, DOI 10.1007/978-3-319-61712-1_17

Edgar Zilsel (1892–1944) is best known for the thesis named after him which holds that modern science could emerge only when the hybridization of the conceptual resources of the scholars with the manual skills of the artisans was made possible at the time of Galileo. Zilsel started research on this theme in 1929, possibly a year earlier, but he would not publish anything of it until he arrived in America in 1939 (Raven 2003). His best-known statement of it in his essay *The Sociological Roots of Science* was only published in 1942 in the journal *The American Journal of Sociology* (Zilsel 1942, 2003 Chapter 2). From a historiographical point of view, the Zilsel thesis is an answer to a problem Olschki first formulated in his *Die Literatur der Techniek* of 1919. In this book, which is the first of his three volumes comprising *Geschichte der neusprachlichen wissenschaftlichen Literatur* (1919–1927), he analysed the emergence in the Renaissance period of a new genre of books written in the vernacular dealing with nature knowledge. Olschki believed that by studying these texts it was possible to "lay bare the cultural preconditions of the development of science". In his interpretation, this new genre:

> […] arose when the secularization of the forms and conceptions of life forced men to draw the sciences, which had removed themselves far from the world, into the sphere of practical and mental activity […]. This is why scientific literature in the vernacular starts with applied and empirical sciences, so as to find, once having arrived beyond the limits of practical necessities, the road towards purely scientific abstractions in an independent way. The end point of this development, which this history of the rise and formation of early scientific prose is devoted, is to be found in the work of Galileo and of Descartes, whose creations and discoveries are not the emanation of ancient and medieval methods of inquiry but rather the further development and triumph of an idea. (Olschki 1919, p. 6)

What Olschki tries to come to grips with is the relation of modern science to the artisanal tradition of the preceding and contemporary "artisans". To put it in dramatic terms, what has Galileo's spectacular description of the arsenal of Venice – right at the start of the First Day of the *Discorsi* – got to do with the theoretical weightiness latter on? This is what Lynn White (White 1978, p. 123) has christened the Olschki problem.

About the same time Zilsel published his *American Journal of Sociology* (*AJS*) essay, Olschki published two essays dealing with Galileo's life and work (Olschki 1942, 1943). Why Olschki felt the urge to publish these two essays needs clarification, but Koyré was clearly challenged by them and felt obliged to make some strong counterstatements. The best known of these undoubtedly are:

> […] it is thought, pure unadulterated thought, and not experience or sense-perception, […] that gives the basis for the 'new science' of Galileo Galilei. (Koyré 1943a, p. 346, 1968, p. 13)

And

> Their science [i.e. Galileo's and Descartes's] is not made by engineers or craftsmen, but by men who seldom built or made anything more real than a theory. (Koyré 1943b, p. 401, 1968, p. 17)

Both these statements need to be seen in conjunction with Koyré taking at face value Galileo's reply to his empirically minded Aristotelian opponent who challenged him if he did "make an experiment?" to which Galileo's replies "No, and I do not need it, as without any experiment I can affirm that it is so, because it cannot be otherwise" (Koyré 1943a, p. 346, 1968, p. 13). Echoes of this response by Galileo can be found in Koyré's indirect dealing with the Olschki problem when he wrote Galileo "did not learn *his* business from the people who toiled in the arsenals and the shipyards of Venice. Quite the contrary: he taught them *theirs*" (1943b, p. 401, 1968, p. 17; emphasis in the original), a position that is no longer assailable (Renn and Valleriani 2001).

The scandal of current historiography of science is that ever since the day of Olschki, Zilsel and Koyré, there has not been any noticeable progress on dealing with what happened in the trading zone between artisans and intellectuals and why it happened at all.

By far the best and most comprehensive analysis of the radical change in mechanics of the seventeenth century to date is Bertoloni Meli's outstanding *Thinking with Objects* (Bertoloni Meli 2006). But at no point does he even begin to offer what is crucial to dealing with objects via instruments: a systematic treatment of the relation between artisanal and theoretical knowledge. The problem shows up in Pamela Long's *Openness, Secrecy, Authorship* (Long 2001) as well. The issue in question is that books written by artisans and humanists (Alberti, Machiavelli) on mechanical arts like mining, metallurgy and mathematics may provide a common ground for princely rulers and social mobile artisan practitioners. But her claim that "such authorship had broad *epistemological* significance" (Long 2001, p. 176) is highly problematic. She claims that "authorship created discursive forms out of skill based practices" in order to "connect the world of empirical practice to the world of learning" (*Ibidem*) in that by "formulating their principles in treatises", these artisan authors "created potentially learned disciplines out of arts previously primarily concerned with craft production and construction" (Long 2001, p. 246). No doubt that authorship helped to raise the social status of the artisans and was beneficial to the social position of mathematical practitioners in general. But like Bertoloni Meli, the issue Long never addresses is why there was a *need* to "create disciplines of knowledge out of practices formerly primarily based on craft skill" (Long 2001, p. 176). One way to see the depth of the problem at hand is that with artisanal knowledge talk about (rational) principles is not in any way meaningful simple because skill-based practices aren't subject to (rational) principles (Raven 2013). Strongly suggesting there is more to the social emergence of modern science than the rise of the social status of the artisan and more than the artisans emulating the intellectual way of communicating, i.e. via texts.

The aim of this essay is to assess anew this classical argument about the pros and cons of the artisan–scholar thesis while taking as my starting point three assumptions. The first is that Koyré's take on the topos is flawed. The same holds true for Zilsel's position. The crucial weakness of the Zilsel argument is that it never could throw light on the question of why a craft – a fabricative, material product-making activity grounded in manual dexterity that aims at getting things build – should suddenly start to worry about (theoretical) utterances being true. Or to formulate the problem in a different way, Zilsel never sees the need to answer the question why truth questions become important when dealing with artisanal knowledge.

My third starting assumption is that Olschki's descriptive take on the topos of the joining of brain and hand comes close to the truth. The important point is that he never accepted the implication of the Zilselian argument: erode the social barriers between scholars and artisans, and experimental science naturally emerges. Unlike Zilsel, Olschki was acutely aware that what the scholars got out of this social intercourse was something "fundamentally different" than what had gone on before in the artisanal tradition. The difference is that the artisanal experience of the workshop becomes a suitable source of knowledge for drawing "the preliminary lines of the *theoretical* foundations of the mechanical arts". It is because of this that both the questions and solutions are "fully independent" from "the direct tradition of the workshops" (Olschki 1927, pp. 156–157, emphasis added). Thinking with objects and instruments did start to take hold at the time of Galileo, but nobody seems to be able to explain what is happening in the trading zone in which intellectuals, humanists and artisans interacted and even more important why it is happening at all. What needs to be elucidated is what is the relation between artisanal knowledge and theoretical knowledge and why the two suddenly become relevant to each other. We need to be able to illuminate why artisanal experimentation – which aims at producing new configurations of interlocking artefacts as a sustainable result – is transformed into scientific experimentation, the aim of which is to test an explanatory principle or to verify a theory. It is my claim that only when viewed from a comparative angle is it possible to suggest answers to these important queries.

17.2 Simon Stevin vs Galileo Galilei

If we are to progress in any way on the artisan–scholar thesis, we need to be able to illuminate what Olschki described as what is "fundamentally different". The prime observation is that the practical problems treated in the vernacular artisans' literature and the ones Galileo was struggling are indeed different: practical versus theoretical.

The prototype artisan, who most clearly fits Zilsel's argument, is the Dutchman Simon Stevin (1548/9–1680). Stevin is one of the best known of a group of people referred to by Zilsel as superior artisans. Galileo is my exemplary figure. His career provides the best guidance on what is central to the Zilsel thesis. What is special about Stevin's life and work is that he started as an artisan and went to university (Leiden) at the relative late age of 35 but acquired international fame through writing books (*De Thiende* 1585) and rose to high social prominence. He became chief engineer to Maurits, Count of Nassau (1567–1625), in 1592; he became director general of the Dutch authority for public water works ("waterstaet") and later in life became quartermaster general of the army of the States-General (Dijksterhuis 1943).

Galileo on the other hand started his career as a low-paid mathematics professor who had to make ends meet by turning his house into a lodging place for his students. Later in life, in 1610, did he raise to fame as the natural philosopher attached to the court of Grand Duke Cosimo II de' Medici (1590–1621). For Galileo, this transition from mathematical practitioner to natural philosopher always was significant; it legitimated the explanations he gave of natural phenomenon (Biagioli 1993). Mathematicians did the measuring and in case of astronomy provide "a calculus which fits the observations" (Osiander); philosophers provided the reasons, i.e. causes, of the phenomenon. Philosophers not only had a higher social and intellectual status than the mathematicians reflecting a cultural preference that providing explanations was the socially higher esteemed activity (Westman 1980).

The difference between Stevin and Galileo is what interests me here. Stevin, like Leonardo before him, always remained the artisans' engineer in today's parlance. Galileo frequented the artisans' workshops but always aspired to become a philosopher and became a secular theologian. Galileo's crime was a social one: he ventured into the domain of the intellectuals, claiming that his way of investigating God's acts provided a better understanding than theirs. In Galileo's life and work, more than in Stevin's, you see what Olschki's "radically different" actually is: the seizure of questions that belong to the domain of the theologians and answers them in part in terms of what artisanal knowledge would tell you is the case.

The opening scene of the *Discorsi e dimostrazioni matematiche intorno a due nuove scienze* (hereafter *Discorsi*) is significant here. Why is it that the dimensions of stocks, scaffolding and bracing used to launch a big vessel are different to those used for smaller ones? The answer "one cannot argue from the small to the large, because the many devices which succeed on a small scale do not work on a large scale" (Galilei 1954, p. 2) is not exactly satisfactory. This is a typical artisanal answer to a typical artisanal observation. A home-based version of it is if you need a teaspoon of salt to cook 1 Kg of potatoes, you do not need ten teaspoons to cook 10 Kg of potatoes. (Try it if you do not believe me.) Artisans maybe satisfied if they can work out how to change the dimensions of scaffolding in relation to the size of the vessel; the natural philosophers aren't satisfied with vague rules of thumbs; they want to know the reasons, the causal generative mechanism, behind it. Galileo's argument is that with increasing dimensions geometrically similar beams are not equally strong but instead become successively weaker – eventually the bigger beams break under the actions of their own weight.

The significance of the opening scene is this: an artisanal way of going about building a ship raises a question that requires a theoretical explanation. For me, that exactly is the kernel of the Zilsel thesis: artisanal knowledge becomes in need of theoretical elucidation and justification. Artisanal knowledge is appropriated by natural philosophers, and along the way it is transformed into conceptual knowledge.

What happens in this process of appropriation – usurpation if you think that knowledge domains that are not experimentally validated have merit of their own – is the artisanal tradition of learning through acquisition of skills and the university tradition of learning conceptually mediated knowledge, by grasping the underlying theoretical principles, become intertwined. The net result of this amalgamation of a bookish and intellectualist tradition that is interested in universals and truth with a manual, mundane and practical tradition that is interested in particulars is:

(a) The empirical and quantitative methods of the artisans get transformed into the verification practices of controlled variable experimentation.
(b) The empirical found rules of thumb are transformed into the theoretical idea of laws of science.
(c) Experimental practices are placed in a hierarchical subordinate position to that of the theoretical mode of understanding.

Without a doubt the increase in social status of the artisans, from the twelfth century on, is a highly significant and characteristic phenomenon of Latin Europe – it is beyond the realm of possibilities in *ru* China (Kim 1982). But the crucial step is that the maker's knowledge tradition is accepted as yielding legitimate and valid knowledge. Francis Bacon as the propagandist of this idea is the all-important figure here. The focus of attention in the Zilsel thesis needs to be shifted to reaching out the intellectuals to the artisans: reaching down instead of rising up. The artisans did not become natural philosophers – which hardly ever happened – but the other way around: the academically trained natural philosophers became artisans – which also hardly ever happened with the exceptions of Boyle, Huygens and Newton – but they always had at their disposal a socially invisible hook-like figure (or an even more invisible instrument maker) beavering away in a laboratory, somewhere hidden at the back of a grand home, who had the required artisanal skills, did all the dirty work and produced the relevant experimental data. *Experiment-based natural philosophy was created when the artisanal conception of knowledge was appropriated by and incorporated into the theoretical conception of knowledge of the scholars.*

It is in this light that I propose the following cultural reformulation of the Zilsel thesis: the (monistic) artisanal conception of knowledge fused with the (dualistic) intellectualist conception of knowledge.

Because the two learning strategies do not cohere, they all the time tend to drift apart. Confluence does not apply on a personal or group level but only on an institutional level. All that is required is that institutionally the empirical content of theoretical concepts is somewhere validated and experimental data theoretically are somewhere illuminated. Without a doubt, culturally the bias always was towards esteeming theoretical curiosity and understanding: *la più degna è la scienzia* (since theory is the most worthy) are the words of *Il libro dell'arte* (Cennini 1932, I, 17). Galileo is valued more than Brunelleschi, Newton more than Harrison, Einstein more than Edison.

17.3 Deus Omnium Opifice

From a historical comparative perspective, the significance of Copernicus is not his rejection of the Ptolemic theory and/or his alleged genius in terms of replacing it with a helio-oriented theory – helio-oriented because Copernicus uses a "mean sun" which is merely a geometrical point located outside the real sun but close to it; it was Kepler in his first important work *Mysterium Cosmographicum* (1596) who replaced this "mean sun" with the real sun, i.e. a body capable of physically influencing the other planets in the solar system. This traditional view misses a very crucial point: the Ptolemic *theory* never was. It was a set of complex mathematical rules arranged in such a way that each planet had to be dealt with separately and individually. There was no single mathematical connection between these rules. What is radical with Copernicus is that we are for the first time presented with a single *mathematical theory*.

This point is easily confirmed if we look at Copernicus's own words. In his dedication to Pope Paul III, i.e. his own preface, Copernicus mentions as his second motivation for considering his hypotheses that the astronomers:

> […] in determining the motions not only of these bodies but also of the other five planets, they do not use the same principles, assumptions, and explanations of the apparent revolutions and motions. For while some employ only homocentrics, others utilize eccentrics and epicycles, and yet they do not quite reach their goal. For although those who put their faith in homocentrics showed that some nonuniform motions could be compounded in this way, nevertheless by this means they were unable to obtain any incontrovertible result in absolute agreement with the phenomena. On the other hand, those who devised the eccentrics seem thereby in large measure to have solved the problem of the apparent motions with appropriate calculations. But meanwhile they introduced a good many ideas which apparently contradict the first principles of uniform motion. Nor could they elicit or deduce from the eccentrics the principal consideration, that is, the structure of the universe and the true symmetry of its parts. On the contrary, their experience was just like some one taking from various places hands, feet, a head, and other pieces, very well depicted, it may be, but not for the representation of a single person; since these fragments would not belong to one another at all, a monster rather than a man would be put together from them. (Copernicus 1992, p. 4)

In a word Copernicus' principal consideration for "revolutionizing" astronomy is to be found in replacing the monster created by the astronomers by *unifying* the parts under the heading of symmetry, a notion that initially not needs to be understood from a mathematical perspective but in its etymological sense of *agreement in dimensions, due proportion, arrangement* (Hallyn 1993). As Rheticus points out in his exposition of the Copernican argument:

> […] there are only six moving spheres that revolve about the sun, the centre of the universe. Their *common measure* is the great circle that carries the earth. (Rosen 1959, pp. 146–147; author's *italic*)

In the Aristotelian ideal of knowledge, "knowledge" means lying bare the causes of the phenomena. Despite all the critique this ideal was subjected to in the Middle Ages, we see that the likes of Copernicus, Clavius, Kepler and Galileo all slowly edge towards arguing for the reality of astronomical hypotheses and hence reluctantly accept a physicalization of astronomical theories. In Kepler, this is clear from the title of his 1609 work *Astronomia Nova, Aitiologetos, sue physica coelstis* which in translation renders *New Astronomy Based upon Causes or Celestial Physics*.

But the emphasis on causality and its realistic interpretation implies one's acceptance of the Aristotelian idea of a physical science, the specification of an efficient cause, specifying a concrete "physical cause" that actually is operative in the natural world. This leaves open at least two points: first, the know ability of the universe and, second, what to take as "physical" cause.

As for the first question, Copernicus's *ratio motuum machinae mundi qui propter nos ab optimo et regularissimo omnium opifice conditus esset* contains this all-important phrase *propter nos*: the world is knowable because God made the world *for us*. Normally these two words get all the attention, and rightly so, less attention is given to the following: in *De revolutionibus* Copernicus expresses the idea of God always in terms of *opifix* – artisan or manufacturer. Apart from the just quoted *optimo et regularissimo omnium opifice* (the best and most systematic Artisan), we have *opificem omnium* (maker of everything), *a divina providential opificis universorum* (the divine providence of the Creator of all things) and *divina haec Optimi opificis fabrica* (the divine handiwork of the most excellent Artisan). If we combine this with the Platonic motto ἀγεωμέτρητος οὐδείς εἰσίτω (Let no one ignorant of geometry enter here), which is on the title page, the inference has to be that Copernicus' *Deus* is an artificer, an architect and a mathematical artisan who created the world with mathematical principles in mind. In the makers' conception of knowledge, making implies knowing. With Copernicus we have that making along mathematical lines implies knowing in mathematical terms. Mathematics is the key to understanding the world created for us.

Because Copernicus is not more forthcoming on this point, if he is elaborating at all, he always exercises extreme caution; we need to look elsewhere for elucidation. Despite not having published something that resembles a philosophical essay or monograph, Kepler is very articulate on the Christian assumptions of his philosophy of science. Geometry for him is coeternal and coessential to the Creator:

> For Geometry […] coeternal with God and shining in the divine Mind, gave God the pattern. […] by which he laid out the world so that it might be best and most beautiful and finally most like the Creator. (HM, III, p. i; KGW, p. vi, 104/105:37–3)

In *Mysterium Cosmographium* (1596), he had already stated that:

> […] just like a human architect, God has approached the foundation of the world according to order and rule and so measured out everything that one might suppose that architecture did not take Nature as model but rather that God had looked upon the manner of building of the coming (i.e. yet to be created) human. (MC, pp. 53–55, KGW, p. i, 6:7–10; KGW, p. viii, 17:10–14)

Like with Copernicus, Kepler's *Deus artifex* was a geometer, mathematician in today's parlance, who created the world in concord with the norms of the quantities provided by geometry. People being created in the image of God (*Imago Dei*) have a mind that is able to understand God's creation. Hence peoples mind can recognize these geometrical principles and patterns. In a letter to his former mathematics teacher Michael Maestlin, who at the time of writing had become both a friend and mentor, Kepler would argue that *ut oculus ad colores, auris ad sonos, ita mens hominis ad quaevis sed ad quanta intelligenda condita est* (As the eye was created for colour, the ear for tone, so as the intellect of humans was created for the understanding not just any thing whatsoever but of quantities) and continues:

> [...] it grasps a matter so much the more correctly the closer it approaches pure quantities (*nudae quantitates*) as its source. But the further something diverges from them, that much more do darkness and error appear. It is the nature of our intellect to bring to the study of divine matters, which are built upon the category of quantity; if it is deprived of these concepts, then it can define only by pure negations. (KGW, p. xiii, nr 64:12–19)

Arguing that Kepler subscribed to the view that the ultimate structure of the cosmos was imprinted on the human mind is, I believe, stretching things a bit. *Ad quanta intelligenda condita*, the mind is created for understanding quantities. The creation being the material embodiment of God's ideas, the mind is attuned to grasp these ideas: *hominum mentes, Die simulachra* (mens's spirits, simulacra of God's spirit; KGW, p. ii, 16:9). Kepler, in his *sacro furore* (HM, V, p. i; KGW, vi, p. 290:3), believed he was able "[...] to think the thoughts of God over again [...]" (Caspar 1993:62; Barker and Goldstein 2001). In his *Mysterium* he would claim to have discovered God's blueprint for the universe, i.e. the archetypical assumption behind the way the solar was the way it was. Bewildering as it may seem to the modern reader, the nesting of the polyhedra needs to be understood as confirming a realist understanding of the Copernican heliocentred system of the world. Kepler argues to be able to explain why there are only six planets; the answer is there are only five polyhedra. (These polyhedra are examples of what in Kepler's terminology are called archetypical (geometrical) ideas. God created the universe, more in particular the planetary system with these five polyhedra in mind.)

What enabled Kepler to argue for a realist interpretation of Copernicus theory? For starters we have the teleological principle that "man is the goal of the world and all creation" (MC, IV:107, KGW, p. i, 30:11–12; KGW, p. viii, 52:11–12) or formulated differently "most causes for the things in the world can be derived from God's love for man" (MC, IV:107; KGW, p. i, 30:8–9, KGW, p. viii, 52:8–9). Additionally there is a metaphysical principle: "the mathematical things are the causes of the physical because God from the beginning of time carried within himself in simple and divine abstraction the mathematical things as prototypes of the materially planned quantities" (MC, XI:125, n. 2; KGW, p. viii, 62:30–33; aesthetic beauty is a matter of course included in the use of the mathematical ideas; the beauty is founded in the clarity, simplicity and elegance of the mathematical ideas used in the design of the world. The Platonic regular solids are a good example of what is meant here). Finally, there is an epistemological principle: "each philosophical speculation must take its point of departure from experiences of the senses" (MC, XII:141, n. 7; KGW, p. viii, 72:16–17).

Kepler aspired to become a theologian, ended up a brilliant mathematician, a very gifted mathematical astronomer and a creative philosophical astronomer. Yet, as he discovered much to his surprise, close scrutiny of the world leads to the contemplation of God and hence easily yielded an anagogical knowledge of nature, which raises us to what is eternal – interpretation of his natural philosophy: *Geometria una et aeterna est., in mente Dei refulgent* ("Geometry is one and eternal, a reflection out of the mind of God", KGW p. iv, 308:9–10); hence, *Deus ecce mea opera etiam in astronomia celebrator* ("Even in astronomy my work worships God", KGW, p. xiii, nr. 23, 40:6).

Without a doubt Kepler's outlook and inspiration – the mathematical harmony of the universe is the embodiment of a theological order – are Christian through and through. In his own words, "In der Schöpfung greife ich Gott gleichsam mit Händen, die Astronomie hat Verherrlichung des weisesten Schöpfers zum Gegenstand" ("In creation I can reach God who speaks with his hands, astronomy has the glorification of the wisest Creator as its subject"; Holton 1956). Without this Christian inspiration, it would not come to fruition. Needham in his *The Grand Titration* is struggling with the problem why on the one hand in China the idea of laws of nature is lacking and why on the other hand the West is so confident that the secrets of the Cosmos are intelligible to mortal human beings in a rational way. He toys with the idea that the concept of a divine legislator – which is also absent in China – may have been a necessary element and suggests that China could only come up with the idea of laws of nature if it had passed through a Western style "theological" stage. In similar vein, he quite bluntly asks:

> The Problem is whether the recognition of such statistical recognition and their mathematical expression could have been reached by any other road than that which Western science actually travelled. Was perhaps the state of mind in which an egg-lying cock could be prosecuted at law necessary in a culture, which should later have the property of producing a Kepler? (Needham 1969, p. 330)

In his *Dioptrice*, Kepler remarks he offers the:

> [...] friendly reader, a mathematical book [...] that assumes [...] a particularly intellectual alertness and *cupiditatem incredibilem cognoscendi rerum causas* (an unbelievable desire to learn the causes of things). (KGW, p. iv, 334:5–8, author's *italic*)

What Kepler is expecting from his readers is obviously an accurate description of Kepler's mind-set. But my argument is that it is not just a typical feature of Kepler's mind-set, it is a European craze to be fascinated, not to say obsessed, with learning to know the causes of things. In the spring of 1536, at the age of 22, Rheticus (1514–1574) publicly accepted the professorship of mathematics at the University of Wittenberg in 1536 with a lecture on arithmetic, the subject central to his teaching assignment. When it comes to expressing the inquisitive attitude required of students, he writes:

> But it is characteristic of the honourable mind not to love anything more ardently than truth, and, inspired by this desire, to seek a genuine science of universal nature, of religions, of the movements and effects of the heavens, of the causes of change, not only of animated bodies but also of cities and realms, of the origins of noble duties and of other such things. (Rheticus 1999, p. 91)

This quest for "the causes of change" in the natural world is characteristic for the European tradition. Echoes are to be found everywhere. As examples of such echoes, one can point to Virgil's *Felix qui potuit rerum cognoscere causas* (Blessed is he who has been able to win knowledge of the causes of things, *Georgics*, II, 490). Dramatic and equally illustrative is Raphael's identification of philosophy with *causarum cognitio* (knowing the causes), which is the official name of his fresco, located in the Stanza della Segnatura and generally known as *Scuola di Atene*.

In case there is doubt about the correctness of Raphael's point of making philosophy equivalent with *causarum cognitio*, I provide you with Hobbes' definition of philosophy, as put forward in *De Corpore*:

> [...] philosophy is such knowledge of effects or appearances, as we acquire by true ratiocination from the knowledge we have first of their causes or generation: and again, of such causes or generations as may be from knowing first their effects. (Hobbes 1994, p. 186)

The crucial point in my argument is that only the European civilization developed and prioritized this idea of knowledge being equivalent about given (physical) causes. I haven't had the space to deal with this in any more detail. But let me briefly deal with the Chinese notion "zhi", 智, to get a feel for the cultural distance between the Sinitic ideal of what knowledge is and the European one.

Zhi is rendered as knowledge, and sometimes it is rendered as wisdom, but the important point to grasp is that zhi is about moral wisdom realized in practice; it is about knowing correctly what *to do*; it is about realization. What emerges from any detailed elaboration of what the indigenous Chinese conception of "knowledge", 智, zhi, is about is that it is part and parcel of the idea of self-cultivation as a "ceaseless process of inner illumination and self-transformation". A process that entails a transforming act upon oneself and via objectless awareness (intellectual intuition) is directed at a communion with 天道 (t'ien tao, Way of Heaven). This is a nondiscursive enterprise, and to achieve a comprehensive breakthrough in the intellectual grasp, of the nature of the world and the things in it, requires an interfusion and identification of the subjectivity of a human and the objectivity of things. Sure enough, various *ru* schools had different ideas and different understandings about various aspects of what is involved in this "ceaseless process of inner illumination". The crucial idea here is that the Chinese construe life as "an unending stream reaching in all direction, into infinity".

The contrast with the European ideal type of what knowledge is of course massive. At its core is the Aristotelian statement δι αποδείξεως ειδεναι we know through demonstration yielding the notion of a *scientia demonstrativa*: knowledge of something equals understanding its necessitating causes. A different ideal would require a different meta-strategy of learning. Alternatives of course were available – the artisans' way of immersion into a practice or memorisation, the learning strategy that is dominant in the Islam (Qur'ān is of course "the speech" (*kalām*) of Allāh but its literal meaning is "the recitation"), as well as the humanist way of model emulation (*historia magistra vitae est*) which happens to be the one *ru* China opted for – but none of the scholastics ever aspired to make one an alternative to *demonstratio propter quid* (demonstration of the reason why). They all could agree with Grosseteste's remark *diligens inspector in rebus naturalibus potest dare causas omnium effectuum naturalium* (the diligent investigator of natural phenomena can give the causes of all natural effects; Grosseteste 1912, IX, p. 65).

17.4 The Christian Roots of Science

Hooykaas (1972, p. 75) hits the nail on the head when he writes "the rise of modern science is to a large extent the rise of experimental science". My reformulation of the Zilsel thesis amounts to the same thing. But there is an advantage to my formulation. I am interpreting the Zilsel thesis through the prism of a theory that is devised to handle the differences between civilisations. This comparative dimension carries over into the reformulation of the Zilsel thesis and is responsible for what to some may look like a convoluted formulation. But the advantage of this formulation is that it allows an opening up of the vexed question why it happened only in Europe. The artisanal way of learning is a way of learning available to any civilisation (Raven 2013). The intellectualist way of learning isn't; it's tied to quite specific European-nurtured assumptions such as:

- The cosmopolis is a universe.
- The cosmopolis has a definite invariant underlying rational order.
- Humankind possesses a divine attribute to discern the truth and falsehood with regard to the cosmopolis.

These assumptions have a validity and legitimacy within the indigenous Christian tradition, but not outside it! Toby Huff, my fellow traveller in this respect, has argued this quite clearly for the Islamic case (Huff 1995, 2000). The Sinitic case has already been touched upon above.

What drove this appropriating process? After all late medieval natural philosophy was "in a very important way […] *not* about nature" (Murdoch 1982, p. 174; emphasis added; see also Wallace 1988, pp. 213–214; Sorrell et al. 2010). What necessitated the intellectuals to close the books of Aristotle and open the book of nature? The crucial riddle here is as Pamela Smith (Smith 2004, p. 239) correctly formulates it, "Why did the intellectuals feel there was need to accept the makers' traditions in the first place"? (*Ibidem*) Smith talks about the artisans laying down the foundations for a new epistemology, a new *scientia*, but this formulation is misleading, not because there is no epistemological foundation to artisanal knowledge – true as this remark might be – but because the *scientia* isn't new but the conception of *naturae* is. It is this new conception of nature that transformed the contemplative discipline of natural philosophy into an active one. Latin Europe inherited from the Greek *φυσιολόγοι* the idea that there is something about "nature" that requires theoretical elucidation, i.e. there is something *erklärungsbedürftiges* about the cosmopolis. Medieval Europe nurtured this idea in the sense of a *scientia naturae*. In Latin Europe, there always was this ethos to come up with a unified understanding of nature, an ethos that has no counterpoint in *ru*-ist China. *Ru*-ist thinking is interested in the study of how men can best be helped to live together in harmony and good order, not in explaining nature.

The same holds true in Islamic thinking as expressed in the elegant phrase of Rahman (Rahman 1979, p. 32) "the basic *élan* of the Qur'ān is moral".

What does a diachronic reading of the reformulated Zilsel thesis suggest? Why do the scholastics believe that the Greek conception of nature and its associated conception of knowledge need revision? The problem is not that no agreement among the *physiologoi* was ever reached on what to take as the constituents of nature. The problem is that the Greek conception of nature as partaking in the divine and hence as animated and having an agency (teleos) of its own runs counter to indigenous Christian assumptions. The same goes for the Greek idea that the object of knowledge of the human intellect is the essence of the material thing. Can a Christian agree with the assumption that the intelligent comprehension of form is sufficient for the understanding both of what is and what happens in the actual world?

The Greek philosophical ideas about what knowledge is and the indigenous Christian tradition only meet in earnest when the Greek corpus of text became available to Latin Europe in the eleventh century. As the scholastics were to discover, the christening of Aristotle proved to be a much harder job than christening of Plato – a job already done by the early church fathers. Initially the biblical view was only superimposed on, but could not overcome, the Aristotelian conception. The symbiosis of the Aristotelian philosophy of nature and the Christian understanding of God's infinite liberty and power "rested on an unequal consideration of the attributes of God, on a subordination of the omnipotence and, even more, of the omnipresence of God to the spiritual self-sufficiency and the constant and constitutive dependence of the world on the inner life of the divinity" (Blumenberg 1987, p. 164). The indigenous Christian revolt was led by Etienne Tempier, Bishop of Paris, and culminated in the condemnations of 1270 and 1277. Between the Christian God who was able to create at a single stroke the world with the multiplicity of beings it holds and the Greek demiurge for whom effects proceed one by one and according to a necessary order, "no conciliation was possible" (Gilson 1955, p. 407).

The radical shift from a *natura naturans* (creative nature) to *natura naturata* (created nature) had to be matched by a shift from a realist reading regarding concepts to a nominalist one. With the Platonic demiurge "productive and theoretical insight converge" (Blumenberg 1983, p. 152). This is compatible with the scholastic understanding of universals as *universale ante rem* (universal having an existence prior to things) and construing the individual as the repetition of a universal. For a Christian the world is non-necessary because its origins are from nothingness. God's *potentia absoluta* (absolute power) requires a denial of universals and the assertion of the priority of reality over concepts. In other words, the concept of an absolute will is incompatible with the question of the reasons for its acts. Indigenous Christian reasons are behind the rejection of the Greek idea that the intelligibility of nature is located in nature's own intelligence in favour of the idea that the intelligibility of nature is located in something other than nature: in the *Deus artifex* (God the artificer), the divine Creator and Ruler of nature. In this process nature is robbed of its necessity and endowed with contingency, necessitating an empirical approach to knowledge. An epistemology based on universals, in which concepts possess a binding force as exemplary entities independent of things and which is the scholastic way of saying that nature is cognitively accessible to man due to reason's experience with itself, is replaced by a nominalist one – only individuals exist and concepts are mere words. Man, imputing order on nature, replaces the idea that the order of nature is adapted to the needs of reason.

Indigenous revolt? But the doctrine of *creatio ex nihilo* is "not demanded by the text of the Bible" (May 1994, p. 24). As Von Rad points out:

> One most not deprive the declaration in [Genesis] verse 1 of the character of a theological principle. If one considers vs 1–2 or 1–3 [at the beginning when God created heaven and earth, DR] as the syntactical unit, the word about chaos would stand logically and temporally before the word about creation. To be sure, the notion of a created chaos is itself a contradiction. (Von Rad 1970, p. 48)

In short in the book of Genesis, it is all clearly stated that God created order from chaos and secondly that the creation – from pre-existent materials! – is dependent upon God as well as subordinate to him. Still the notion of God as the superior artisan of the universe – *summus namque opifex universitatem* in the words of Honorius of Autun (1080–1154), *Elucidarium* (*PL*, CLXXII, Liber XII, cap ii, 1179) – is distinctly Christian. It is a Christian theological innovation although of a defensive nature, a theological reaction to Gnosticism.

The root metaphor of *liber naturae – scriptus digito Dei*, written by the finger of God (Hugh of St Victor: c. 1096–1141) *Erudit Didascalia* (PL, CLXVI, VII, 4, 814, echoing Exodus (31:18) where it is written that when God had finished communing with Moses, he gave him two tables of testimony, tables of stone, written with the finger of God) – needs to be understood in conjunction with the metaphor of God as the superior artisan; the readability of the book of nature is an expression of the idea that the universe is an intelligible entity. Key to the makers' tradition of knowledge is the idea that the only reality with which an inquirer can have any commerce is reality as he constructs it to be. *Verum esse ipsum factum* – the truth is what is made. Nature is readable because God created it (for man, *proter nos*, as Copernicus would have it). The Bible, Wisdom (11:21), states that God had ordered all things *mensura et numero et pondere* (measure, number and weight) as well (*Ivi*, 13:1) that God may be known as *artifex* (artist). Suggestions that God had used mathematics as the language for his public manuscript can be traced back to Robert Grosseteste, Bishop of Lincoln (c. 1175–1253) and Roger Bacon.

God created the universe; consequently it is knowable. Artisanal knowledge is created knowledge therefore knowable. At his point Galileo's question of "What has philosophy got to do with measuring anything?" is relevant. Measurements, weights and numbers are the route to the knowable world. Natural philosophers sought causes, not quantitative relations, hence the transformation of these numbers into the triangles, circles and other geometrical figures and symbols of the mathematicians. And this brings us directly to Galileo's social crime: arguing the book of philosophy è [un libro] *scritto in lingua matematica* (Galilei 1968, VI p. 232).

In 1930 Zilsel wrote that the search for the idea of scientific laws is what sets Europe apart from other civilizations: "For four centuries the search of scientific laws is progressing. This and only this is what Europe is, modernity is, science is" (Zilsel 1930, p. 421). This quote illustrates the huge significance the concept of scientific law had for Zilsel at quite an early point in his thinking. I have never felt that his ideas on the emergence of notion of scientific laws (Zilsel 2003, Chapter 6) sit easily with the thesis named after him. One reason is that a key idea of modern science is to be located outside the realm of the artisans: in the Judeo-Christian notion of a divine lawgiver. That is to say, "[t]he very idea of a law of nature, from the moment of its birth, was underpinned by theological considerations" (Harrison 2008, p. 14).

Exploring the ramifications of this idea is easy if one accepts the congruity of grace and nature – for the likes of Boyle, Galileo, Kepler and Newton, the study of nature is an act of worship; Kepler (KGW, XIII, p. 193) sees himself as a "priests of the highest God in regard to the book of nature" – an option that would in effect derail to a large extent Zilsel's sociological argument. Zilsel "explains" the issue of the emergence of the notion of physical law as "caused" by the rise of the absolute state. Seen from a comparative angle, this cuts no ice. If ever there was an absolute state, Imperial China must be it. *Ru* thinkers never even came close to develop a notion of scientific laws. Latin Europe developed this notion as an alternative to the teleological notion of causes it had inherited from the Greeks and exemplified in Aristotle. For a secular theologian like Descartes (Descartes 1904, p. 380), the laws of nature are the rules and ideas God had used in creating the world: *quia deus* sic *voluit, quia* sic *disposuit* (because God so willed and so ordered). Its emergence is part and parcel of the reconfiguration of nature necessitated by the Tempier condemnations. Only with Descartes do the secular theologians arrive at a fully developed notion of laws of nature and is the reconfiguration of nature completed (Henry 2004).

At the centre of my comparative dealings with the Koyré, Olschki–Zilsel dispute is two robust claims. One, as already mentioned above, is the idea that knowledge equals understanding its necessitating causes. The second claim, likely to be the more contentious one, but merely an extension of the Aristotelian idea that "we know through demonstration", is that it is only within a *societas Christiana* that the meeting of hand and brain yielded experimentally calibrated conceptual elucidation of the universe, a.k.a. experimental science. Christian is the root of science because it is the christening of the Aristotelian concept of knowledge that explains why empirical science as we come to know it originates in Latin Europe and nowhere else. This process of christening the Aristotelian knowledge construct, and in the very process transforming it beyond recognition, is clearly at work when Kepler writes (KGW, p. xvi, nr. 448:4–7) that he wants to "provide a philosophy or physics of celestial phenomena in place of the theology or metaphysics of Aristotle". For the Greeks, immanent necessity ruled the cosmos, and they matched this, to achieve a knowable cosmos, by conceiving of reason as an ordering principle inherent in reality. The most radical formulation of this idea is by Parmenides, ταὐτὸν δ' ἐστὶ νοεῖν τε καὶ οὕνεκέν ἐστι νόημα, "thinking and being are the same" (DK, fragment 8–34). What Ocham and his fellow nominalist theologians argued for was that God's will, his *potentia absoluta*, which commanded the final cause of the Creation, is impenetrable. By severing any link between the final course of reality and its material result, i.e. it is created effect, rational intelligibility of nature became impossible. Nature was ordered by having an order imposed on it instead of the rational being an expression of the immanent order of it. What Anneliese Maier (Maier 1967, p. 403) refers to as "the methodological split between theology and natural science" is made possible when the incommensurability between God and man, between *potentia absoluta* and *potentia ordinata*, is incorporated into epistemology. This requires a realisation that from a human point of view the regularity of the world, its law-like nature, is but contingency from the Creator's point of view. It is this contingency that necessitates the empiricism of science. In terms of the Aristotelian notion of knowledge, this means that final causes are out and at best only efficient causes are to be had.

17.5 Conclusion

Why could only in a *societas Christiana* the trading zone where artisans and intellectual meet each other yield this idea of modern (experimental) science? The Zilselian argument that modern science can be depicted, in the words of Koyré (Koyré 1963, p. 852), as "a promotion of arts and crafts, as an extension, as an *ancilla praxi*" is equally unsustainable as the Koyré claim that the likes of Galileo didn't perform any experiments. All I need for now is why the Zilselian argument is flawed.

Artisanal knowledge is skill based, and skills are better taught than talked about. This of course reflects that skills are performative actions and can be executed independent of discursive understanding of what one is doing. Because it is skill based, it is learned by mimicking specific behavioural routines, and this requires a desired level of manual dexterity. Because skills, behavioural routines, are central to the craft of the artisans, their knowledge is at once a form of knowledge and a form of practice. In other words, artisanal is monistic and as such is markedly different from the propositional knowledge conception of the intellectuals which is dualistic. But the dualism so characteristic of the intellectualist theoretical conception of what knowledge is – knowledge is about something that a subject has – is fully dependent upon comes into existence with, the Greek notion of ἐπιστήμη, epistêmê. (On this difference see Table 17.1.) From a compare perspective, the crucial thing is that only the European civilisation produced this dualistic knowledge conception.

Gernet is outspoken that "nothing in the Chinese traditions resembled in the least the radical opposition between the perceptible and the rational" (Gernet 1980, p. 10). The famed Sinologist Julia Ching is of the same view, but she takes it one step further by making the following radical statement:

> [w]ithout any subject/object distinction, there can be no scientific thinking. (Ching 1977, p. 244)

Prima facie Ching, who clearly knows a thing or two about *ru* thinking, seems to be saying that a Western style dualism is "responsible" for scientific thinking. A provocative suggestion here is why I think she is right.

Table 17.1 Difference between the way artisanal knowledge, on the left, and conceptual knowledge, on the right, conceptualize the relation to the cosmopolis

Ontology of events	Ontology of principles
Immanence: order realized	Transcendence: order instantiated
Humans are "organism-persons" (Ingold) relating to a cosmopolis, taken up a view by dwelling *in* it	Human are composites of mind and body apprehending nature by grasping a view *of* it
The world is an environment constituted through the unfolding relations to a being	The world is an external nature "waiting to be given meaningful shape and content by the mind of man" (Shalins 1976, p. 210)
Active interaction is fundamental to the production of knowledge	Observation – detached contemplation – is the causal bridge between the passive mind of the self and the external world, where the facts somehow manifest themselves

The *ru* are not into truth but are into devising a system of practical morality; in a world practice that concentrates on acting righteously, truth doesn't come into play. It doesn't come into play because of the cardinal Chinese idea of the consanguinity between man and nature, and the monism it entails implies there is no higher level of being from which man or nature can grasp, grasp in a necessarily abstract way that is. (Epistemological worries have no room to life in a monistic model of knowledge.) Comprehending nature and man, Chinese style implies showing the manifestations of the reasonableness of life in action.

Ching grasped something of profound importance in understanding the difference between the European and Sinitic civilisations that is relevant for this essay. The point she makes is that because a dualistic knowledge conception was not available in *ru* China the idea of a theoretically, i.e. conceptually, unified understanding of the cosmopolis in the Sinitic civilisation never was on the table. For Westerners, of course, it is. Blumenberg (Blumenberg 1983, pp. 232–233) is absolutely correct when he writes "the 'theoretical attitude' may be a constant in European history since the awaking of the Ionians' interest in nature". This constant is missing in *ru* China and explains the many times observed absence of theoretical explanations in China. In China, the marrying together of practical and theoretical skills into one kind of understanding could never happen.

It is a truism to say that Greek philosophy is close to being co-existent with the idea of *λόγος* (logos), which is best understood as the rational pattern of the world process. In German, this is referred to as *Weltvernunft* and expresses the crucial idea that there is a rational structure to the cosmopolis and the idea of philosophy is to come up with a theory that articulates this immanent structure as best as possible.

This idea of *λόγος* that the articulation of the rational is an expression of the immanent order of cosmopolis sits very uncomfortable with the Semitic idea of the world being created by God. Therefore, the Greek conception of nature had to be replaced by the one that is compatible with Christian assumptions. These Christian assumptions where found in the idea of God as a *Deus Opifex*, superior artisan, who used mathematical principles to construct the cosmos. Because God created the physical universe, the cosmos is an intelligible entity cognitively accessible to those created *Imago Dei*.

In his Cecco dialogue of 1605, Galilei (Galilei 1976, p. 38) casually asks a highly significant question: “What has philosophy got to do with measuring anything?” Koyré’s notion of “Géométrisation de l’espace” (Koyré 1939, pp. 12–15) is all important to answer it. This famous notion is meant as a substitution of Ptolemy’s concrete and finite space for the abstract infinite one of Euclid. This formulation however hides the radical point Koyré wants to make: “La nature ne répond qu’aux questions posées en langage mathématique, parce que la nature est le règne de la mesure et de l’ordre” (*Ivi*, p. 156). (The nature replies to questions posed in mathematical language if and only if nature is the domain of measure and order.) Geometrization implies measurability. The gist of the Galileo quote is a methodological one; it is about a theory of measuring; the readability of nature is due to it being measurable. A measurable nature is quantifiable and hence subject to mathematical treatment. Measuring nature is another word for experimentation. Or, as Lord Kelvin puts it, “to measure is to know”, (Thomson 1889, pp. 73–74) or as the Dutch have it, “meten is weten”.

References

Barker P, Goldstein BR (2001) Theological Foundations of Kepler’s Astronomy. Osiris 16:88–113.
Bertoloni Meli D (2006) Thinking with Objects. Johns Hopkins University Press, Baltimore.
Biagioli (1993) Galileo Courtier. University of Chicago Press, Chicago.
Blumenberg H (1983) The Legitimacy of the Modern Age. MIT Press, Cambridge.
Blumenberg H (1987) The Genesis of the Copernican World. MIT Press, Cambridge.
Caspar M (1993) Kepler. Dover, New York.
Cennini (1932) Il Libro dell’arte. Vol I. The Yale University Press, New Haven.
Ching J (1977) Confucianism and Christianity. Kodansha International, Tokyo.
Copernicus N (1992) On the Revolutions. Translated by Rosen E. The Johns Hopkins University Press, Baltimore.
Descartes R (1904) Meditationes de Prima Philosophia. Œuvres de Descartes, Vol. VII. Cerf, Paris.
Dijksterhuis EJ (1943) Simon Stevin. Martinus Nijhoff, ‘s-Gravenhage.
Galilei G (1954) Dialogues Concerning Two New Sciences. Translated by Crew H, de Salvio A). Dover, New York.
Galilei G (1968) Le opere di Gallileo Galilei. Barbèra, Firenze.
Galilei G (1976) Against the Philosophers. Translated by Drake S. Zeitlin & Ver Brugge, Los Angeles.
Gilson E (1955) History of Christian Philosophy in the Middle Ages. Sheed and Ward, London.
Gernet J (1980) Christian and Chinese Visions of the World in the Seventeenth Century. Chin Sci 4:1–17.
Grosseteste R (1912) Die Philosophischen Werke des Robert Grosseteste. Aschendorfsche Verlagsbuchhandlung, Münster.
Hallyn F (1993) The Poetic Structure of the World. Zone Books, New York.
Harrison P (2008) The Development of the Concept of Laws of Nature. In Watts F (ed). Creation: Law and Probability. Ashgate, Aldershot, pp. 13–36.
Henry J (2004) Metaphysics and the Origins of Modern Science. Early Science and Medicine 9/2:73–114.
Hobbes T (1994) Human Nature and De Copore Politico. The Oxford University Press, Oxford.
Holton G (1956) Johannes Kepler’s Universe. American Journal of Physics xxiv:340–351.
Huff T (1995) Islam, Science and Fundamentalism. Journal of Arabic, Islamic, and Middle Eastern Studies 2/2:1–27.

Huff T (2000) Science and Metaphysics in the Three Religions of the Book. Intellectual Discourse 8/2:173–198.
Hooykaas R (1972) Religion and the Rise of Modern Science. The Scottish Academic Press, Edinburgh.
Kim YS (1982) Natural Knowledge in a Traditional Culture. Minerva 20/1:83–104.
Kepler J (1937) Johannes Kepler Gesammelte Werke. Beck, München.
Kepler J (1969) Harmonices Mundi Libri V. [reprint] Forni, Bologna.
Kepler J (1981) Mysterium Cosmographicum. Translated by Aiton EJ. Abaris Books, New York.
Koyré A (1939) Études galiléennes. Hermann, Paris.
Koyré A (1943a) Galileo and the Scientific Revolution of the Seventeenth Century. Philosophical Review 52/4:333–348.
Koyré A (1943b) Galileo and Plato. Journal of the History of Ideas 4:400–428.
Koyré A (1963) Commentary. In Crombie AC (ed). Scientific Change. Heinemann, London, pp. 847–857.
Koyré A (1968) Metaphysics and Measurement. Gordon & Breach, Yverdon.
Long PA (2001) Openness, Secrecy, Authorship. The Johns Hopkins University Press, Baltimore.
Maier A (1967) Ausgehendes Mittelalter. Vol II. Edizioni di Storia e Letteratura, Roma.
May G (1994) Creatio ex Nihilo. Clark, Edinburgh.
Murdoch JE (1982) The Analytic Character of Late Medieval Learning. In Roberts LD (ed). Approaches to Nature in the Middle Ages. Center for Medieval and Early Renaissance Studies, Binghamton, NY pp. 171–213.
Needham J (1969) The Grand Titration. George Allen & Unwin, London.
Olschki L (1919) Die literatur der Techniek. Winter, Heidelberg.
Olschki L (1927) Galilei und seine Zeit. Niemeyer, Halle.
Olschki L (1942) The Scientific Personality of Galileo. Bulletin of the History of Medicine XII:248–73.
Olschki L (1943) Galileo's Philosophy of Science. Philosophical Review 12:349–65.
Rahman F (1979) Islam. University of Chicago Press, Chicago.
Raven D (2003) Edgar Zilsel in America. In Hardcastle GL, Richarsson AW (eds). Logical Empiricism in North America. The University of Minnesota Press, Mineapolis, pp. 129–48.
Raven D (2013) Artisanal Knowledge. Acta Baltica Historiae et Philosophiae Scientiarum 1/1:5–34.
Renn J, Valleriani M (2001) Galileo and the Challenge of the Arsenal. Nuncius 16(2):481–503.
Rheticus (1999) Preface to Arithmetic. In Kusukawa S (ed). Orations on Philosophy and Education. The Cambridge University Press, Cambridge, pp. 90–97.
Rosen E (1959) Three Copernican Treatises. Dover, New York.
Shalins M (1976) Culture and Practical Reason. University of Chicago Press, Chicago.
Smith PH (2004) The Body of the Artisan. The Chicago University Press, Chicago.
Sorrell T, Rogers GAJ, Krayer J (2010) (eds) Scientia in Early Modern Philosophy. Springer, Dordrecht.
Thomson W (1889) Popular Lectures and Addresses. Macmillan, London.
Von Rad G (1970) Genesis: A Commentary. Translated by Marks JH. SCM Press, London.
Wallace WA (1988) Traditional Natural Philosophy. In Schmitt CB, Skinner Q, Kessler E, Kraye J (eds). The Cambridge History of Renaissance Philosophy. Part II. The Cambridge University Press, Cambridge, pp. 199–235.
Westman RS (1980) The Astronomer's Role in the Sixteenth Century: A Preliminary Study. His Sci 18:105–47.
White L (1978) Medieval Religion and Technology. The University of California Press, Berkeley.
Zilsel E (1930) Soziologische Bemerkungen zur Philosophie der Gegenwart. Der Kampf 23:410–24
Zilsel E (1942) The Sociological Roots of Science. America Journal of Sociology 47:544–62.
Zilsel E (2003) The Social Origins of Modern Science. The Boston Studies in the Philosophy and History of Science. Vol. 200. Kluwer Academic Publishers, Dordrecht.

Chapter 18
Alexandre Koyré: History and Actuality

Marlon Salomon

Abstract When it comes to highlighting Koyré's contribution to the methodology of the history of sciences, analysts have frequently insisted on presenting him as an author who innovatively strove to study antiquated scientific theories in the setting of their own time. That new attitude toward the past should not, however, obscure the role that the actuality [*actualité*] or modernity of science performed in the elaboration of a new conception of history. My hypothesis is that the elaboration of this new conception of history did not stem from any methodical distinction between past and present, but, quite the contrary, from a new way of articulating the actual and the no-longer actual. The theoretical and methodological novelty of his conception of history was intrinsically connected to the actuality of science. Thus, Koyré's perspective was inserted in the debates then promoted by historians and philosophers about the intricate links between present and past. In 1946 he stated that "the reality of time" could only be revealed through transformations and, at the same time, that there was only past from the stance of a present. Therefore, an epistemological transformation in the present called for a new history of sciences. As early as 1935, Koyré had declared that the history he was writing was inextricably connected to the epistemological transformations then in course. A revolution in the history of the scientific thought, such as the one being seen in the interwar period, not only marked a deviation in the course of science but also made it possible to think about its path from a new perspective. From Einstein's finite (though unlimited) world, it was possible to think from a different perspective the infinite universe of the moderns and the ancient-medieval closed cosmos.

Keywords Historiography of sciences • The actual and the non-actual in history • Methodology of history of sciences • History and contemporary time

M. Salomon (✉)
Faculty of History, Federal University of Goiás, Av. Esperança, s/n, Campus Samambaia, 74690-900 GO, Brazil
e-mail: marlonsalomon@ufg.br

R. Pisano et al. (eds.), *Hypotheses and Perspectives in the History and Philosophy of Science*, DOI 10.1007/978-3-319-61712-1_18

18.1 Introduction[1]

When it comes to highlighting Koyré's contribution to the methodology of the history of sciences, analysts have frequently insisted on presenting him as an author who innovatively strove to study antiquated scientific theories in the setting of their own time or, as he put it himself in 1951, "to put the studied works back in their intellectual and spiritual environment, to interpret them in relation to the mental habits, preferences and dislikes of their authors" (Koyré 1982c [1951], p. 14).

As we know, however, that approach was not restricted to the history of sciences because from early on he sought to elaborate it in relation to other areas of the history of philosophy. In 1933 he wrote that:

> What is most difficult – and most necessary – when studying thinking other than our own, is – as one great historian has so admirably shown – less a question of finding out what the thinker in question knew or did not know, than of forgetting what we ourselves know or believe we know. Let us admit that it is sometimes necessary not only to forget truths that have become an integral part of our thinking, but to adopt certain modes, certain categories of reasoning or at least certain metaphysical principles which, for people in a past era, were just as valid and formed bases just as solid for their reasoning and research as the principles of mathematical physics do for us. (Koyré 1971 [1933], p. 77)

That is why, to Pietro Redondi, the originality of the author of *Études Galiléennes* lies less in his philosophical elaboration and definition of the scientific revolution concept than in his "historical methodology, which makes it possible to imbue his general interpretations with content" (Redondi 1986, p. XXIII). It was undoubtedly in this new way of approaching the past of science, considering it in its own conceptual structure, internal coherence, and systematic aspect, that postwar historians found a new key to interpreting the old theories, those theories that had fallen into disuse and been set aside in the present. It was necessary, from then on, to avoid "modernizing" the science of the past, that is to say, to avoid translating into our clear precise terms of today those imprecise, fuzzy, and obscure old notions and in that way avoid anachronism, and avoid seeking in the past those who, supposedly ahead of their time and their contemporaries, heralded the way forward that future science was to take, for, in so doing, we falsify history.

That new attitude toward the past should not, however, obscure the role that the actuality [*actualité*][2] or modernity of science performed in the elaboration of a new conception of history. His criticism of anachronism, of the figure of the precursor,

[1] This paper is part of a broader study of Alexandre Koyré's conception of history, and it enjoys the support from the National Council for Scientific and Technological Development Council (CNPq is the Portuguese acronym) in the form of a productivity grant. Translated by Martin Charles Nicholl

[2] The French word "actualité" refers to the quality or state of that which is current, the circumstances of the present (actual) moment. Many authors have translated it as "actuality" although the usual meaning attributed to the word in English is not identical with its meaning in French. For reasons that the reader will readily perceive in the course of the text, we too have decided to adopt this option. Similarly, we have opted to translate the French "actuel" (the antonym of "inactuel") with the word "actual."

and of the modernization of the past – which presupposes a critical distancing in relation to the time which the historian finds himself in – does not make Koyré the spiritual inheritor of the nineteenth-century historical objectivism or of the promoters of history as "pure science." My hypothesis is that the elaboration of this new conception of history did not stem from any withdrawal or methodical distinction between past and present, but, quite the contrary – however paradoxical it may sound – from a drawing closer of the past and present and from a new way of articulating the actual [*actuel*] and the non-actual [*inactuel*]. The theoretical and methodological novelty of his conception of history was inseparably connected to the actuality of the science and in at least two aspects.

Before passing on to the analysis, however, I would like to register the fact that the mode of articulating the relation between past and present found itself profoundly altered in the interwar period in the wake of those transformations that were leaving their mark on historical thinking at the time. The postulated radical separation between past and present, which was intended to be the means of establishing history as a rigorous, reliable science in the nineteenth century, was being questioned on all sides. "If we intend to acquire knowledge of the past," wrote Fustel de Coulanges, in 1875, "we need to begin by removing from the spirit everything that has to do with the present" (Coulanges 2003 [1875], p. 307). It was precisely that requirement, widely shared in the nineteenth century by those promoting history as a pure and objective science, that was being questioned, and it meant the collapse of the ideal of history as the science of the past. The possibility of reforming historical thinking involved a new alliance of the world of the living and the world of the dead. "It is in the light of what is living that it [history] should interrogate what is dead" wrote Lucien Febvre (2009 [1949], p. 373). Present and past were so enmeshed in one another that it was no longer possible to achieve an absolute separation of the two. As Marc Bloch (quoting Michelet) insisted, it was not just that "the very idea that the past as such could be the object of science is absurd," but it is equally true that whoever "limits themselves to the present, to the actual alone, will not even understand the actual itself" (Bloch 2006 [1942], pp. 875–876). So it was not only by means of the present that the historian got to know and interpreted the past, but rather, historical knowledge became the means to "organizing the past." Again it is Febvre who adds "organizing the past in the light of the present" (Febvre 2009 [1949], p. 373). That, in his view, was the veritable "social function" of history.

However, we must not reduce that new way of articulating the actual and the non-actual – whose genealogy undoubtedly dates back to the Bergsonian speculation on time – to a mere historiographic institution or historical school of that period. For example, that entanglement immediately raised the problem of historical objectivity and, accordingly, of the epistemological legitimacy of historical knowledge. That was to be the very theme of the thesis Raymond Aron presented at the end of the 1930s – and that Koyré would later comment on in 1946 – regarding the limits of historical objectivity. It was therefore a much broader problem linked to the epistemic transformation of historical thinking (Gattinara 1998). The historiography of sciences was not indifferent to that problem; we are reminded, for example, of the works of Gaston Bachelard, Helène Metzger, Robert Lenoble, and Alexandre Koyré.

Nevertheless, inside those works, more than in any other historiographic domain, the problem presented itself in an entirely singular and, at the same time, radical way. Singular insofar as the study of sciences' past could not be indifferent to the actuality, especially to the radical epistemological transformations in course with the advent of the theory of relativity and quantum theory, because the definition of what the science whose history needed to be written once *was* found itself undergoing a profound alteration. The present transformed the object that the historian needed to address, and that transformation was not without effect on its status in the past. And radical, insofar as, within the study, the historian could not economize on truth.

18.2 The Thickness of Time

In 1946, reviewing of a set of texts that addressed historical knowledge theoretically, Koyré coined a definition of what made history as an experience of time possible: "The reality of time only reveals itself to us through change." It is only with transformations that the thickness of time becomes accessible to thought. History as a science and as a reality "only exists where there is change. History as such does not exist, nor does historical science exist in societies which, if not entirely immutable, are sufficiently stable for changes to remain imperceptible." Time, as something susceptible to perception and as the problem object to be construed, constitutes itself through change. Thus, it follows that "in an aristocratic society with a constant economic and social structure, it is the political events (wars, dynastic changes etc.) that appear forming the contents of its history"; that times of economic transformation recognize in the past the importance of economic factors; that an era of giddy technological transformations considers the study of the historical aspect of the technical phenomenon important; and that in an era of intense social conflicts, importance is placed on the past of class struggles (Koyré 2011 [1946], p. 60).[3]

That argument, however, was already implicit in the notes accompanying a study published in 1935, which was later to constitute the first book of the *Études Galiléennes*. In the note, Koyré situated his work on the birth of the new science in the seventeenth century in relation to the modernity of science or the epistemological transformations that were in course in those years. "In the light of the scientific revolution of the last ten years, it is preferable to reserve for it the epithet 'modern' and to designate pre-quantum physics as 'classic'" (Koyré 1992 [1939], p. 14, note 3). It is not a case of seeing it as a simple question of semantics: the scientific revolution of the early twentieth century redefines what modernity or scientific actuality is.

If we think on that note in the light of the 1946 definition, we will see that Koyré makes implicit in the text that which makes the constitution of a new kind of history of sciences possible in the interwar period. A period marked by profound epistemological transformations makes it possible to recognize new layers in the time of

[3] Thus, it is the speed or slowness of the changes that makes it possible to talk about the "acceleration" or "deceleration" of time.

sciences, and those layers, in turn, allow its history to be thought on from a new point of view. Thus, it is not the discovery of new archives, documents, or empirical evidence that makes it possible to rewrite history; quite the contrary, it is the transformations in the present that trigger the rewriting process and that allow us to understand not only the forces mobilizing historians to revisit to the archives but also the very "discovery" of documents that had, until then, gone unnoticed.

The importance of actuality, however, is not limited to that aspect. Koyré, along with many of his contemporaries, characterizes his own times as revolutionary. He refers to his own contemporary setting in that way. He refers to his times in the light of the concept of revolution, a revolution in the history ofscientific thought and a radical transformation in the structure of the science of the real and of the very understanding as to what constitutes the real that those sciences are dedicated to.

18.3 The History Book

Thus, it is not science alone that becomes young again in that revolution. A new possibility for thinking the past is constituted with such rejuvenation. That is because the expression "revolution in the history of scientific thought" plays with a dual dimension that is often not sufficiently underscored. The first dimension is the one that immediately stands out: a revolution in the history of scientific thought marks a discontinuity in the trajectory of a given knowledge domain. But it also marks, to the same extent, a transformation in the way that history is being rewritten. That does not mean to say that the novelty and youthfulness of the new conceptions demand a space for themselves in the already written history of sciences or call for the opening of a complementary chapter of the history that has already been written. It is not a case of patching up the existing written history of sciences and including its most recently written chapter, believing that the (The) History book history book of sciences has already been written and that the events of the beginning of the twentieth century require that it be reedited so that its most recent chapter may be included. Such a conception presupposes that time only relates to the book in an exterior and cumulative manner; it presupposes not only the idea of a definitive book that cannot be rewritten but that the past has an autonomous existence with no relation to time or to the present. It is the presupposition of a chronological time.

To Koyré, historical reason conjugates itself with the problems and the trading in ideas of its times.[4] "The historian projects the interests and scale of values of his

[4]That was an idea shared at the time among those that had broken with nineteenth-century scientism. In that regard, it is interesting to note that Koyré, in spite of coming from the field of philosophy, situated himself in the movement of historiographic renovation then in course. "It is obligatory for those that dedicate themselves to historical studies [...] to achieve a total dissociation from things of the present." The truth of history itself would be distorted if the past were approached "in the light of today's preoccupations" (Coulanges 2003 [1875], p. 302). Here is what Lucien Febvre wrote quite severely about the history book in 1942. "Let us not act as if the historians conclusions were not necessarily contaminated by the contingency. Of all the stupid formu-

time and it is with the ideas of his time – as well as his own ideas – that he carries out his reconstruction" (Koyré 1982 [1961], p. 371). Thus, Koyré makes it clear where his perspective is situated. It not only lies in the times of a science formed by new ideas, but those times are precisely what makes it possible. Thus, it constitutes itself as a breach with the book of a time that is no longer its own.

In that sense, the past itself becomes altered under the impact of transformations in the present because the "discovery made 'in the present' provokes a reconstruction and a reinterpretation of the past; it provokes the discovery in the past of things which up until this moment had gone unnoticed by us" (Koyré 2011 [1946], p. 60). While the chronology is inalterable and it is impossible to eliminate the facts when the past is the object, nevertheless, "the history is not inalterable" (Koyré 1982b [1930], p. 16). Therefore, what the (The) History book history book calls for is something else. The past is not an autonomous realm in relation to the present. It is not only historical science that is in movement but the very object that the historian addresses. The past that has gone or that has been transformed into a book is never definitively immobilized. It is never identical to itself because we do not find ourselves definitively fixed in relation to it. "That is precisely why history renews itself, and nothing changes faster than the immutable past" (Koyré 1982 [1961], p. 371).

A revolution also marks a transformation of that history, of that book. It is a rehash of the order of the books. It indicates what contemporaries experienced with the advent of a new time/period – which explains why it often involves the destruction of certain symbols and bibliographic monuments. Under the impact of that revolution on the present, the very history of sciences is transformed, and the past is plowed up. Thus, history is not a science of the past, of a cutout clear-cut time strip, of an object disconnected from the present, and of a strip of a past time definitively separated from the present time. To Koyré there only exists a past "of" a present "in" a present. "It is on the basis of the present that our past takes form for us."[5] If revolutions reconfigure the present and redefine what contemporary actuality is, then that new present demands a new past for itself.

---

las, that of 'the book that can never be rewritten' runs the risk of being the stupidest. Or rather: that book will not be written over again, not because it has attained the last word in perfection, but because it is a child of its own time." He concludes by saying: "History, is the child of its time. I do not say so to detract from it, philosophy is also the child of its time and so is physics. The physics of Langevin is not that of Galileo, which, in turn is not that of Aristotle" (Febvre 1970 [1942], pp. 11–12).

[5] And it is because there are, in the present, different (not necessarily actual) perspectives that different pasts are constituted and because the "entities analyzed [...] are viewed from the standpoint of different 'presents'" that the frequently divergent perspectives of the past are constituted (Koyré 2011 [1946], p. 52).

18.4 The Actual and the Virtual

Would that not be an archeological condition for the constitution of a history of the scientific revolutions and of scientific thought itself, just as Koyré stated in his writings? But would it not also be an archeological condition that could explain the very structuring and formation of a historiographic field of the sciences, starting in the 1930s?[6] Is it not that which would make it possible to understand why the history of scientific thought comes into being in the form of a history of its revolutions – a fact that is often overlooked nowadays?[7] That perhaps is the horizon in which we can set Koyré's work. His endeavor to write a different history of modern science – that from Copernicus to Newton – was not indifferent to the epistemological reconfiguration of the interwar period. Indeed, to be more precise, we should not speak of modern science alone, because Koyré wrote a history of modern physics, astronomy, and cosmology – not by chance the very fields of knowledge involved in profound transformations at the time. It is not merely a question of declaring that Pierre Duhem and the positivist tradition were mistaken in their interpretation of the origins of modern science but, rather, to state that the past or history, insofar as they do not exist as such, exists only in a given time and for a certain someone. It is the youthful beginning of this new era in which Koyré and his contemporaries were living that demands a different past for itself, different from that of Duhem's time – that is, what would make it possible to partly understand the flagrant interest and development of a historiography of the sciences beginning in the 1930s. The present in which the perspective of those historians had been developed and dissolved and a new plane of thinking came into being.[8] It is the history of that plane which needs to be written and of that modernity which as yet has no past and demands for itself the book of its own history. This is the new plane that mobilizes the past, redefines its mileposts, dislodges certain theories, and unearths other that had been forgotten. For history:

> […] modifies itself insofar as we change. Bacon was modern when empiricist thinking was in vogue. However, in an era like ours when science is increasingly mathematics, he is no

---

[6] We could think about whether renovation of the schools of historical thinking found their conditions of possibility in that. Lucien Febvre arrived at that conclusion by other ways. In texts of that time, the 1930s and the early 1940s, he insisted on that point: the times of a new science not only make it possible for a new idea of the history of sciences to exist, but they demand it. There lies the crux of the problem: in the 1930s, the historiographic renovation that needed to be carried out (and not just in the history of the historians) was a demand of the current actuality.

[7] And regarding archeological condition, I would like, in the wake of Michel Foucault, to articulate two issues: first, an analysis of what kind of transformation would make the constitution of new objects of knowledge and of a domain or field in which they can be formulated possible and, second, seeing that Koyré states that the thickness of time only reveals itself when there are transformations, an analysis and identification of the different time strata and layers that make up the past of a science.

[8] Therein lies the reason for Koyré's critical stance and even opposition to Aldo Mieli and the spiritual inheritors of Pierre Duhem-type history who wrote and promoted a history of science that was absolutely indifferent to the events that shocked and transformed the modernity of science in the period between the wars.

> longer modern. Nowadays, it is Descartes that figures as the first modern philosopher. Thus in each historical period and at each moment of evolution, history itself is to be rewritten and research into our ancestry is to be conducted in a different way. (Koyré 1982b [1930], p. 16)[9]

Accordingly, history cannot be explained by a simple appreciation of the past. It is modernity that demands it; it is the latest happenings that require it. History is the mark of what is new. If the past does not exist as such, it is because history is always a virtuality existing in relation to the actual. The transformation of the actual mobilizes that virtuality so that the book of history is not the sign of any immobilization of the past by means of the written word, but rather a space for making the virtual actual.[10]

18.5 A New Past for a New Science

We could say that Koyré's work philosophized on the history both *of* this new era and *in* the new era. His books on the history of scientific thought constitute that past and provide the history with its own book, the book it was calling for but did not yet have. Obviously, Koyré did not write a history of quantum physics or of Einstein's cosmology, but he did invent a new past for the era of a new science.[11]

From that stems his 1933 study on Nicholas Copernicus and the redefinition of the milestone marking the modern world which had to be shifted from the Fall of Constantinople or the Discovery of the Americas to the publication of *De Revolutionibus Orbium Coelestium*, in 1543, considering that the mark made by Copernicus was to prove far more profound than either of those two events. "It marked the end of a period that encompassed the Middle Ages and Antiquity. Starting with Copernicus and with Copernicus alone, man no longer finds himself at the center of the world. The universe no longer spins at his behest" (Koyré 1934,

[9]That was why, in the 1930s, Bachelard and Koyré were to write histories of sciences marked by trajectories that were the complete reverse of those described by the empiricist tradition. In its trajectory, science did not head toward the empirical or concrete, but, instead, toward the mathematical and abstract. Ever since his interpretation of Galileo in 1935, Koyré had defended that thesis. Just a few years later, in 1938, Bachelard was to invert the Comte's law of three stages in his preliminary address "Discours préliminaire" to *La Formation De L'Esprit Scientifique*. In Koyré's case, however, it was a conclusion drawn from a historical study because there was no finality in that trajectory that would allow for him to refer, as Bachelard did, to a "law."

[10]Hence, the conception of the history book then formed as being always a book to come, contrary to the nineteenth-century conception. That recognition marked a whole generation of historians in the between-wars period. In 1945, in his analysis of the new conceptions of history, science historian Robert Lenoble transformed it in the question he posed: "There is a question that bothers me, all the more so because I have never seen it clearly posed. How can one explain the constant writing of new history books?" (Lenoble 1945, p. 195). In a future paper on *the history book*, that moment could be defined as the "temporalization of the book."

[11]Perhaps that is why there was such repeated insistence on delineating a space of resonance between his work and that of Gaston Bachelard, whose efforts were dedicated, not by chance, to writing the philosophy that the new scientific spirit called for.

p. 1).[12] In that same scenario, we can situate the 1935 study on Galileo and his polemic with the supporters of continuism. Modern science comes into being with Galileo because he alone was able to solve the philosophical problem of the role of mathematics in science and nature. Thus, the milestone marking the foundation of modern science is displaced from Bacon, and Galileo becomes the threshold for the birth of the mathematical form of scientific thought.

The introduction of the concept of revolution into the history of sciences marked an important divergence from one of the principal reference figures for that history at the beginning of the twentieth century, Pierre Duhem. In the view of the author of *Études Galiléennes,* the pathway that led Galileo to mathematicize physics was opened up by an event that radically transformed the foundations of our science and our concept of the world at the beginning of the seventeenth century. It was in fact a rupture, a discontinuity in the trajectory of science up to that point. Thus, Galilean mechanics did not prolong the science of *impetus* that had been produced at the heart of medieval Christianity, nor can it be held to have been the culminating point of a knowledge that had undergone continuous linear development ever since the beginning of the seventeenth century. The latter was precisely the thesis that Pierre Duhem had defended almost two decades earlier. In Duhem's view, the science that apparently came into being with Galileo and which was to be further developed by Torricelli, Descartes, Beekman, etc. could not be considered a "creation." Galileo and his contemporaries merely needed to further develop a mechanics whose fundamental principles and propositions had been established in the Christian Middle Ages in the period beginning with William of Ockham right through to the mid-sixteenth century. That is precisely the reason behind the title of the third volume of *Études sur Léonard de Vinci* by Pierre Duhem, which was *Les précurseurs parisiens de Galilée* (Duhem 1913), and it illustrated, in a masterly way, the thesis he was putting forward. Duhem contended that modern science was no more than the updating and development of ideas already present in the medieval germ that gave rise to it. As can be seen, the Koyréan concept of revolution marks a historiographic and philosophical stance taken in the light of Duhem's interpretation: a science founded on qualitative concepts and ideas could never lead to a form of knowledge supported by a quantitative ideal and founded on mathematics (Clavelin 1987). Only a revolution could explain such a transformation. To Koyré, Duhem modernized the non-actual forms of knowledge and did not recognize the actuality of the non-actual in the past. He sought to show that such non-actual was in fact the germ of a potential actuality. The past, in reality, already contained the traces of what would come to be. History was merely its happening. To some extent, the course of history was already mapped, just as Galileo's predecessors had avowed.

Such considerations show us that there is, importantly, a relation between Koyré's work and his actuality. Furthermore, the concept of revolution finds itself precisely at the crossroads of that two-way street. After World War II, the concept coined by Koyré in the 1930s acquired considerable importance, especially in the United States where it helped in the development and consolidation of the professionalization of

[12] The introduction was published a year before, with the title "Copernic" (Koyré 1933).

the historiography of the sciences. That operative aspect of the concept of revolution in the historical analysis of the sciences – with which Koyré's name became definitively associated – was not its only aspect at the time it was formulated. Alongside its historiographic importance, a veritably philosophical dimension of the concept right from the moment it came into being and one that is given little attention by those that dedicate themselves to studying it and who have only interested themselves in its later effects on historiography should be added. A concept is not endowed with a history alone; it belongs, in the same way, to a topography. Thus, it is necessary to relate it to a problem that was particularly important in Europe between the wars and from which it is inseparable. We should not forget that the time in question was thought of by its contemporaries as one of profound crises.

18.6 The Topography of a Concept

In the period between the wars, Europe witnessed the emergence of a problem which in innumerable domains was referred to as a crisis: a crisis of culture, of civilization (Paul Valéry), of history (Henri Berr), of the spirit (Lucien Febvre), of conscience (Paul Hazard), of sciences (Edmund Husserl), of progress (Georges Friedmann), and, at the outside, of reason. In regard to what interests us here, namely, science, the current conception was that there was a crisis in its very foundations. The new radical discoveries in physics had destroyed what had hitherto been seen as its solid and definitive fundamentals. The bases of what were methodologically and philosophically considered to be science were collapsing before everyone's eyes. Crisis was, therefore, the name attributed to the mode of *problematization* of that experience in which a whole generation found itself implicated. Simultaneously with the realization of the impossibility of elaborating a single global unified theory of reality, the ideal vision of a deterministic science imbued with predictability began to crumble. From the philosophical or simply epistemological point of view, the crisis of science consisted not of a simple methodological issue but, more profoundly, of a blow struck at reason itself, hence the pessimisms and the theme of irrationalism. That is why the discourses of the period are permeated by a profound disquiet and sense of unease and portray the terrain in which they configure their intellectual topography as a quicksand of uncertainty.

However, we should not reduce it or consider it to be merely an effect of the great scientific discoveries of that period. Ever since the end of the nineteenth century, the identification of science with progress, the understanding that science progressed "as a cumulative and mechanical process," had been called into question. The possibility of a *faillite de la science* had indeed been announced (Rasmussen 1996, p. 94). However, the *faillite de la science,* as Léon Brunschvicg was to say later, was more in the nature of a collapse of a kind of philosophy of science (Gattinara 1997, p. 24) and had little to do with what he considered to be the crisis in science identified with the crisis of mechanism and determinism, both of which, in the same way, had been called into question ever since the end of the nineteenth century. That shows us how, on the one hand, the radical discoveries in physics at the beginning of the twentieth century took place in theoretical and philosophical terrain in which certain scientific concepts were the object of criticism, and, on the other hand, it shows us how the idea of crisis should be endowed with a plural dimension and not seen as "a historical category with a clearly defined outline" (Gattinara 1997, p. 24).

There were multiple reactions to that crisis, differing according to the respective domains – for it must be reiterated that it was not a crisis in the sciences alone – and differing according to the country under consideration. So much so that it would be more accurate to refer to crises in the plural as Enrico Castelli Gattinara proposed. To that Italian philosopher, the originality of the French "reaction" to the crisis was typified by what he called "the history question," that is, the transformation of the problem of the crisis in the sciences and of the philosophical and epistemological problem of rationalism that it gave rise to into a question in the domain of history. If the epistemology and philosophical study of scientific thought were then to become historical, it was because the complex of problems in rationalism had become a question of history.[13] With the advent of the crisis, reason lost its a priori foundations. The absolute, definitive, static, and architectonic image of reason collapsed completely. That meant it ran the risk of becoming an absolutely incoherent and meaningless figure. If history had become unavoidable for the French philosophy of sciences, it was because it founded reason itself in history. A new conception of reason was thereby constituted, at once open, polemic, and dynamic, whose footprints could be tracked by studying a science in movement. Thus, the affirmation of the historicity of reason implied temporalizing its very foundations.[14] It meant that from there on, every a priori would become historical. By analyzing reason in the context of its historicity and attributing to it the capacity of mobility, it was possible to think of it in terms of an open movement whereby it would become possible to recognize the transformation of its fundamental categories. It was then, in history, that reason found its meaning and its own coherence.

[13] While a new way of articulating epistemology and history had been fostered ever since the work of Léon Brunschvicg, Émile Meyerson, and Abel Rey, it was only with Koyré and Bachelard that reason came to be inscribed in the opening of history, marking, as Gattinara put it, a "point of no return" in the epistemology and philosophy of the sciences in France (Gattinara 1998, pp. 55–57). What is interesting in the work of Castelli Gattinara is to show how those "two generations" actually faced sets of problems common to both of them so that it is possible actually to think of them on the same plane. That mode of understanding marks an explicit withdrawal from the historiography of the 1960s and 1970s which analyzed those "two generations" in terms of the opposition between the continuism and discontinuism and between the immobilism and the dynamism of reason. In that respect, it is in alignment with Gerard Jorland's interpretation when he showed that Koyré's theory of the history of thought was a synthesis of those of Brunschvicg (creating activity) and Meyerson (*cheminement*) (Jorland 1981, pp. 90–102).

[14] That conception moves away from the Pierre Duhem's history of reason whereby reason evolved teleologically from its primitive forms toward its more complex forms. The idea of movement in the former conception is not the same as that of Duhem for whom reason had an origin that could be located in certain germs that held, in potential, what it would come to be. That form of evolution is not only guaranteed but it is also the work of a certain *Sagesse*.

Thus, in Gattinara's view, while the French philosophic tradition of the period criticized Cartesianism and positivism, it never let itself get carried away with a radical criticism of reason (Gattinara 1998, p. 15). On the contrary, by becoming an open, dynamic, and provisional rationalism through history, it could actually defend itself from that type of criticism and steer clear of the antiscientific vogue that was sweeping Europe at the time. We can see that the French reaction was quite different from another contemporary reaction, the one elaborated by logical empiricism, which sought to found reason on logic, and we can also see, why, from the outset a dynamic understanding of reason and the historical study that it implied philosophically was destined to clash with the logical analysis of the language. Nevertheless, beyond the geographic and linguistic differences and the profoundly irreconcilable philosophical conceptions, they shared a common epistemic ground.

Koyré's concept of scientific revolution belongs to that ground and is graved on that topography of thought. The theme of the crisis is transformed in the problem of the revolution. It is not a question of the pessimistic identification of the shattering of certainties or regret for the loss of those references that were the ballast for a secure way of thinking. It was in fact a question of thinking through a transformation and the new possibilities it opens up to thinking – that is, what partly explains the power and the audience of that concept in the postwar period. It was necessary to analyze that which some viewed with pessimism others with optimism and still others with skepticism, as being a radical event, singular and unprecedented. If it is true that the sciences are immersed in movement and that they are nomadic and not monarchic, then they are founded on change. In addition, we must not forget that change is what makes it possible to rewrite history.

18.7 The Historicity of Reason: Archives of an Invention

In this context, we have just situated the 1933 text on Copernicus and the 1935 text on Galileo. There is yet another text, however, that is even more explicit in regard to this problem – the 1937 text on Descartes – so that these three texts seem to echo that topography and even to exhibit a certain problematic unity.[15] Furthermore, they are important insofar as it was with them that Koyré gave form to his way of understanding the history of scientific thought. And the status that Koyré attributed to Descartes right at the outset of the text is a direct reference to that same problem. "Philosophic *actuality* [*actualité*] comes from as far back as philosophy itself. And it is possible that there is no philosophical thinking today more *actual* [*actuel*] than that of Descartes" (Koyré 1962 [1938], p. 163). So then, where would that actuality of the Cartesian text come from? Its important novelty lies in the fact that it was constituted in the seventeenth century as a reaction to a profound crisis of the time which could best be summed up in two words: uncertainty and disarray. However,

[15] With those texts, reason is inscribed in what it would come to be; his analysis clearly shows how, in Koyré, the history of science is articulated with the history of reason.

"it concerned a crisis in the culture, not a personal crisis of Descartes" (Koyré 1962 [1938], p. 189). And Koyré does not dissimulate the relation of that crisis to the crisis in the interwar period, so that it was the problem of Descartes' time "and of ours" (Koyré 1962 [1938], p. 175).

We can readily see why, to Koyré, the problem of those years in which he was writing his first epistemological studies was the same as that of Descartes' time. The Cartesian injunction had become more actual than ever because "the world [had] once more become uncertain" (Koyré 1962 [1938], p. 229). A new era of uncertainty and disarray was being experienced. That did not mean that it was a question of wishing to recuperate Cartesianism, but rather of reflecting on how the abandonment of the principles of scientific and philosophical knowledge led to a crisis. The crisis that history identified, however, did not correspond to the disappearance of reason, to the decadence of western Europe, or to the end of history. It was necessary, first of all, to think of it in terms of a *revolution*. Thus, we can see, as the first aspect, the importance of actuality in Koyré; it makes it possible to revisit the past in the light of a new perspective, but the new perspective that is constructed, on the basis of that return to the past, in turn, makes it possible to analyze the present differently.

18.8 The Actual and a Preliminary Attitude to the Past

If, however, the radical transformations that sciences were undergoing at that time make it possible to understand or even to reconstitute the horizon in which the perspective of the author of *La philosophie de Jacob Boehme* was situated, how can we understand the fact that his writing of history is not founded on the norms of that new knowledge and neither does it take up, once more, the methodic program of the nineteenth century? How can we understand why the actuality of science does not become the court of reason that should be the judge of its past, in the same way and at the same time as happened with Gaston Bachelard's historical epistemology? The answer is that to Koyré the actuality allows for the existence of at least two perspectives of science's past whose differences and exact profiles needed to be clearly delineated, on pain of transforming the history of scientific thought into a mere philosophical reflection with no object. Thus, what is really at stake is the object and the very history of science as a unique field of knowledge.

That distinction is addressed specifically at a conference in 1954. The history of scientific or technical thought is declared to be "a graveyard of errors, or even a collection of *monstra* rightly relegated to the junk room and only good for the demolition site, a *graveyard of forgotten theories* or a chapter in the *Geschichte der menschlichen Dummheit*" (Koyré 1991 [1955], p. 205). Here is the first perspective of the past of science that stems immediately from the effect of the actuality. Koyré makes it clear that, rather than corresponding "to" the history of scientific or technical thought, this is in fact a very precise disposition in regard to its past. Although it is indeed a recurrent attitude, that is the spontaneous conduct of the technologist or the scientist, the one who, departing from the horizon of his current work, casts his eye on the past:

> It is quite normal for one who from the standpoint of the present or even of the future to which his work is heading, casts a glance backwards at the past – a past that has long become *superseded* [*dépassé*] – to find that the old theories look like incomprehensible monstrosities, ridiculous and deformed. Indeed, in his search back through the current of time, he only encounters them at the moment of their death; old, shriveled and sclerotic. In short, he sees them as something like Rodin's statue *The Old Courtesan*. (Koyré 1991 [1955], p. 205)

It so happens that the scientist's perspective is founded in the laboratory and in its latest language. On the basis of the perspective inscribed in that sphere, all that can be found in the past are errors and misshapen bodies, not just misshapen but sclerotic and lifeless, and the corpses of theories only awaiting burial in the mausoleum of forgotten theories. There is no room for them in the working spaces of the science of current times. They cannot even be treated as theories because they are unworthy to figure in the current halls of reason. The only space allotted to them by the current rationalism is in the showcase of curiosities. To the current scientific rationalism, in the present-day halls of science, the past of science is something petrified, a corpse reminiscent of the decrepit body of the formerly beautiful woman sculpted by Rodin, an image which Koyré specifically evokes in his analysis. The old theories are no longer legitimate from the point of view of logic nor worthy of interest from the rational point of view. A museum of human stupidity would doubtlessly be the most suitable place for lodging those negative figures of reason.

Accordingly, Koyré refers to that attitude as "normal." It is in fact a view of a non-actual form of thought based on the rules of actual thought. It is something evident. How could one do otherwise than to analyze the history of science on the basis of that evidence, according to which the conclusions scientists arrived at in the past were either wrong or inadequate? Furthermore, it must be borne in mind that new scientific theories come into being in opposition to and struggling against the then current forms of knowledge and their associated conceptions of the world (Koyré 1991 [1955], p. 204). That means that they come into being destroying existing forms and burying them. Koyré's historical-philosophical efforts are inseparable from a critique and a denaturalization of that evidence.

The date of that conference also merits attention. It was only 3 years earlier, in 1951, that Gaston Bachelard had given his celebrated conference *L'actualité de l'histoire des sciences* at the *Palais de la Découverte* in Paris, and it was to mark the methodological formation of the new generations of French epistemologists. Right at the beginning of the conference, he declared that "given its revolutionary discoveries, contemporary science could consider itself to be a *liquidation of the past*" (Bachelard 1972 [1951], p. 137) – note that 3 years later, Koyré was to refer to a "demolition site." What is more important, however, is that Bachelard was declaring that the science historian needed to "learn his trade" in the actuality of the science whose history he intended to analyze. "The drama of the great scientific discoveries" would be more readily understood after having "watched the fifth act" (Bachelard 1972 [1951], p. 142). Thus, the only thing in the past of interest to the epistemologist was that which persisted and was present in current reality. That was because, to Bachelard, "science evolves in the direction of a manifest progress." From that, it follows that the history of science should "necessarily be a determination of the successive values of the progress of scientific thought" (Bachelard 1972 [1951], p. 139). To determine those values, the historian should, contrary to all recommendations, pass judgment. And it was to be in the knowledge available in the present that he should learn his skills as a magistrate and the jurisprudence on which he should found his sentences. "The historian of science, in order to be able to judge the past, must understand the present."[16]

Bachelard not only based his historical epistemology on the present of science, but from the philosophical point of view, he also condemned any study of outdated theories. In another text published in the same year, he argued that on the one hand, the history of a theory like the phlogiston theory that was based on a fundamental error is of not the slightest interest from the rational point of view. "An epistemologist could only take an interest in it insofar as it contained motives for the psychoanalysis of the objective knowledge" (Bachelard 1951, p. 36). On the other hand, dedicating attention to the coherence of long-vanished systems of thinking such as those of Ptolemy "would be merely [doing] the work of a historian" (Bachelard 1951, p. 36). Such historical work would be mere erudition with no epistemological interest and would be of no rational importance, given that it would be studying the prehistory of science and dealing with "a vanished spirit."

Moreover, while Koyré makes no allusion to that 1951 text at any time, we cannot fail to note the implications of his pronouncements for the history of scientific thought. As we know, the Bachelardian perspective was based on the modernity of the *Palais de la Découverte*, in the sphere of "living science." It was there that he raised his epistemology because it was there that a clear limit was traced between science and non-science, a distinction which, for philosophy, determined the separation of primitive values and rational ones. Although that is not my objective, I cannot fail to point out a conception in that argument that underscored the long-standing opposition between history and philosophy – obviously it was not here that a philosophical history would find its terms – an opposition which for inversely opposing reasons, on the history side, found its promoters among the enthusiasts of the *Annales* School. At the same time, it should be remembered that, ever since the 1930s, Koyré had been dedicating himself to the study of theories that Bachelard considered *perimées* (out of date) and to the study of a time period that Bachelard had inscribed in the prehistory of his epistemological calendar. Koyré, however, right from the outset of his studies of the past of science had kept aloof from such psychisms and insisted on the philosophical worth of the historical study of science (Redondi 1983). That is why it is interesting to note that, in 1954, Koyré established his distinction between the two attitudes to the past of science and, at the same time, formulated another methodological role for the present to play in his historical study. That is precisely

[16] Without doubt, "the ephemeral nature of the modernity of science" implied that this theory of history was founded on quicksand and he recognized the implication (Bachelard 1972 [1951], 142).

where the problem arises: how could the present be theoretically important for the history of sciences if, from the point of view of scientific opinion, it led to a sclerotic vision of the past and, from the point of view of philosophy, to a comprehension that failed to recognize the existence of rational values in the past?

18.9 A Second Attitude to Address the Past

Koyré made a distinction between that common, "normal" attitude in relation to the past of science and that of the historian of scientific and technical thought.

> It is only the historian who will actually encounter her [*The Old Courtesan*] in the prime of her glorious youth, in all her splendor and beauty. Only the historian, by recreating and having recourse to the evolution of science, learns the theories of the past as they were at the moment they came into being, and lives, with them, the creative *élan* of the thought. (Koyré 1991 [1955], p. 205)

The point of view of his analysis does not coincide with that of the scientist, and so he is enabled to apprehend the old theories, not at the moment of their death agony but at the moment of their birth. That means the historian of scientific thought has to analyze the past from a point of view other than that of the actuality of the non-actual. It means that the historian must remove his perspective from the science laboratory, re-situate it in a different space, and construct it on the basis of another language.

That does not mean to say that Koyré disregards the importance of the present in the history of sciences, but the norms governing current knowledge should not instruct the historian to the point of orientating his analysis of the past of science. Neither does it mean to say that he must take up the Fustelian program whereby "it must be required of whoever dedicates himself to the study of history […] a renunciation of the present, to totally forget the issues of the current time."[17] In theoretical terms, the present has a different function. A history of sciences is only possible because "we ourselves have lived through two or three profound crises in our ways of thinking" (Koyré 1982c [1951], p. 13). Contemporary scientific revolutions cause a fracture in the evolution of science, and that fracture introduces a displacement leading to the formation of a plane of thinking based on which it becomes possible to distinguish past ways of thinking with their own structures and governed by their own norms. Our own way of thinking is affected and qualified by contemporary revolutions.

> We no longer live in the world of Newton's, or even Maxwell's ideas so we are now capable of examining them from the inside and from the outside; of analyzing their structures, perceiving the causes of their inadequacies. We are better equipped to understand the sense of mediaeval speculation on the composition of the continuum and the "latitude of forms", the evolution of the structure of thinking in mathematics and physics that took place in the last century, in their efforts to create new forms of reasoning." (Koyré 1982c [1951], p. 13)

[17] At that same conference, he stated: "In my view, for the historian, the scenario of actual current life constitutes more of a danger than a help" (Coulanges 2003 [1875], p. 307).

It is precisely that exteriority, thought from the outside, which is the key to Koyré in the present. It makes it possible to differentiate planes or structures of thinking in the past, the exact outlines that enable the delimitation of the configurations of various different ontologies (Jorland 1981, pp. 27–70). At the same time, the new forms of reasoning make it possible to differentiate certain "modes of reasoning" and "metaphysical principles" that the historian needs to adopt when "studying thinking other than our own." It is what makes it possible to realize that "the solid bases for reasoning and research have not always been the same," as he wrote in 1933. Hence, the importance of transformations which make a new history possible insofar as the time thickness of that history and its layers is exposed by the fracture that the transformations imply. Einsteinian ideas provoke a displacement of the present in relation to Newton's synthesis or even a double displacement if we include Aristotelian physics and one that delimits its own temporal thickness and constitutes the exteriority that makes it possible to analyze it in another perspective. The constitution of that point of view marks a fundamental difference in the way that the ancient theories had been analyzed up until that time.

To Koyré, it is precisely that exteriority that enables an understanding of the bad name that the Aristotelian conceptions were given in the seventeenth and eighteenth centuries. At a time when Newtonian physics prevailed with its notions of an infinite universe in which space corresponded to that of Euclidean geometry, the Aristotelian concept of the cosmos seemed patently absurd and even incomprehensible. "It was hard to understand it because, involuntarily, thinkers replaced it with the image of a great round sphere floating through an infinite vacuum – an image that irrevocably disfigured the former one – something that no longer happens today" (Koyré 1991 [1949], p. 26). In our day we can not only understand the structure of the Aristotelian Cosmos in a radically different way, but we can also understand the reasons behind the negative attitudes to it that marked the men of the seventeenth and eighteenth centuries. That is because in the Einsteinian Universe (just as in the Aristotelian Cosmos), the world is coextensive to physical space, and there is nothing outside it, neither full nor empty [*ni plein ni vide*] (Koyré 1991 [1949], p. 25). On the other hand, given that historians of that time "considered the Newtonian conceptions to be not only valid but also evident and even natural, so the very idea of a finite Cosmos seemed to be ridiculous and absurd" (Koyré 1991 [1955], p. 206). *Today*, in the light of the Einsteinian conception of a finite (but unlimited) cosmos, we know that there was nothing monstrous or absurd in the Aristotelian concept and our view of it has become radically transformed.

After Einstein, it has become possible to analyze the past of science from a standpoint that transforms our understanding of the past.[18] The present transforms the past. We have instruments at our disposal that our predecessors never had. The historian can now situate himself both inside and outside of ancient and medieval physics but, at the same time, inside and outside of Newtonian physics as well. It is

[18] In the same way that Galileo's physics had avenged Plato on Aristotle, so, in Koyré's view, Einstein's which reduces physics to the real and geometrical had avenged Descartes on Newton (Koyré 1962 [1938], p. 228).

precisely that perspective that was impossible before Einstein when Aristotelian physics was analyzed in the light of a mathematical physics and Newtonian physics in the light of its own principles. That was what made their truths appear evident and even natural. That is without a doubt a highly important postulate of Koyré's that is patently present in his work. From the standpoint of the finite-unlimited Einsteinian world, we are able to think the closed and finite (ancient and medieval) cosmos and the open and infinite (modern) world from another standpoint. In a world marked by the imprecision of *quanta*, it becomes possible to analyze the world of "more or less" (ancient and medieval) and the universe of precision (modern) in a new perspective. That is the actual that makes Koyré's history of scientific thought possible. The current scene of scientific life, as can be seen, is more than just a helping hand.

Unlike the scientist who always judges the past on the basis of current categories of science, the historian seeks not only to "put himself in the situation of a contemporary of Galileo's" but also to analyze it on the basis of its own categories and its own language. My hypothesis is that Koyré distinguishes between the notions of *present* and *actuality* – and furthermore, that is the singularity of his position in the debates on the entanglement of past and present that mobilized historians in the interwar period and that make it possible to distinguish it from that of Lucien Febvre's presentist conception. Virtually, it is possible to think the non-actual as actual. The present makes a new history of sciences possible and at the same time provides the historian with intellectual tools that his predecessors did not have which enable him to understand the past. However, from that point of departure, he seeks to understand the theories of the past in the light of the current actualities of their time. They are only non-actual in our time, but that does not mean the historian cannot understand them in the setting of their own actuality. Thus, actuality does not necessarily coincide with "present." That proviso, which enables a distinction to be made between the historian's perspective and that of the scientist or even from that of the judgmental Bachelardian perspective, would prove to be of fundamental importance for the constitution of the history of sciences as we will see below.

What does situating itself at one and the same time in the interior and on the outside of a plane of thought mean if not situating itself *at the limit*? Koyré's analysis is a historically founded philosophical reflection about the limits of thought. Hence, the importance of the subject of revolutions in his work – revolutions as being transformations of the limits of thought – but also of the images of monsters evoked in his epistemological reflection, because it is well to remember that monsters are always figures wandering around the limits. Thus, the present not only makes Koyré's writing of history possible but also performs an important function in it. That is why it is so difficult, from the point of view of an analysis of the historicity of his works, to assign any of the traditional labels to him: he is neither an internalist nor an externalist. Koyré is a thinker of the limits.

18.10 An Analysis Domain

We should not forget that the study of the history of sciences in the interwar period enjoyed very little intellectual prestige – and the way a person like Gaston Bachelard referred to it can leave no doubt as to the existence of that low prestige. In 1934, Koyré insisted on that state of affairs: "Indeed why waste your time studying the mistaken doctrines of the past? That is a very common attitude among the promoters of scientific knowledge and even among the philosophers" (Koyré 1934–1935, p. 522). Koyré's distinction of the two attitudes to the past that actuality made possible is extremely important insofar as it seems to contain or presuppose the existence of another, which I also need to dwell on. Without doubt, it is the one which in the historiographic analyses has ensured an outstanding role for Koyré in the constitution of a historical and philosophical field of the study of science. One of its merits is the fact that it perceived that unless that distinction were made, it would be impossible, first, to clearly delineate the difference of that perspective of science's past from that of the scientist himself or, to be more precise, to have produced the fracture that made it possible to clearly delineate a domain of analysis of science that did not coincide with or was even opposed to the opinions and assessments that the scientists might have of the past of their activity, based on their current work. Second, that the distinction made it possible to distinguish *the object of science* from *the object of the history of science* as Georges Canguilhem was later to express it in his work. In Koyré's view, the object of science and the object of the history of science are no longer identical with one another (Canguilhem 1983 [1966]),[19] for as he wrote in 1929, "scientific thought is concentrated on things, not on its own self" and not on the historicity of its activity and discourse. The methodological corollary stemming from that is that "scientists epistemology is not usually worth much," because the historical and philosophical interpretations of their own theories that scientists present in no way constitute a guarantee of the historical or philosophical meaning of those interpretations. In Koyré's view, Einstein was "an outstanding example of the philosophical incompetence of the scientist" (Koyré 1929, p. 156).[20]

That differentiation is the condition for the elaboration of a domain for the analysis of science that is autonomous in relation to the current scientific work. It makes it possible for history and philosophy to withdraw from the way a scientist judges the past – and even the present – of scientific activity; it also clearly defines the tasks of the historian of thought. It is equally true that it was decisive in bringing about the constitution of a specific field for the history of sciences. From the moment, a

[19] Pietro Redondi also situates the establishment of that distinction in Koyré (Redondi 1986, p. X).

[20] Just a few years later, he was to state exactly the same thing but his time regarding the works of Eddington and of Heisenberg. "It would be almost impertinent to praise Sir A. Eddington's work. One can only regret that its *philosophical* part [...] does not match the standard of the scientific part" (Koyré 1935–1936, p. 457). In his book on the transformations in the fundaments of science, "Heisenberg adds some historical remarks on the roles played by Copernicus, Cristopher Columbus, Galileo, Newton. Our respect for Heisenberg forbids us to comment on them" (Koyré 1935–1936, p. 458).

new object of knowledge is demonstrated and delimited, and a new discipline and a new genre of knowledge can be constituted. From the stance of an archeology of knowledge, it would not be incorrect to situate the birth of the history of science in the interwar period or to attribute a relevant role in its constitution to the author of the *Études Galiléennes*. It was indeed a new object of knowledge quite distinct from what the early twentieth-century historians of science had inherited from the erudite history of the nineteenth century (e.g., we can think of George Sarton). But it was also a program whose subtle distinction actually marked a huge difference from the French rationalism of the beginning of the twentieth century. We have in mind, for example, the epistemological reflection of two *maîtres à penser* of that time, Léon Brunschvicg and Émile Meyerson, who considered that history played a fundamental role in philosophical thinking on the sciences. History may have been fundamental, but it was not the objective or finality of philosophical work (Gattinara 1998). It was just the means, a locus where material could be garnered and examples found that would enable Brunschvicg to demonstrate the dynamism of reason and Meyerson, on the contrary, to demonstrate its identity, that is, its constitutive and constant aspects. It is a kind of history very similar to that which would be found much later in Bachelard and in the French historical-epistemological tradition (Bachelard 2003 [1938], pp. 10–11). That was what made the Koyrean program so radical for, in it, the history is transformed into the very purpose of the philosophy. It is in that sphere that we must place the statement made by Thomas Kuhn in 1968 to the effect "that the establishment of the then prosperous contemporary tradition in the History of Science" could be credited "above all" to Alexandre Koyré. Here is what Kuhn declared that the author of *La Révolution Astronomique* had bequeathed to the history of science:

> […] I and my colleagues learned to recognize the structure and coherence of other systems of [scientific] ideas different from our own." (Kuhn 2011 [1968], p. 35)

That meant to recognize clearly the profile and the limits of the *object* of the history of sciences which is not to be confused with the object of science. It is exteriority that makes the forms of thought that are not our own stand out in all their singularity and be recognized as different rather than deformed or seen as a clinical object.

18.11 The Problem of the Forms: Final Remarks

This paper on the figures of actuality in Koyré's work makes it possible, in the end, to consider his writing of history as the history of forms of thought. The images he repeatedly evokes to explain it are all linked to forms: the monsters, Rodin's sculpture, and the disfigured bodies. Besides those images, he also refers to phantoms: "absolute time just like absolute space, realities that Newton accepted unhesitatingly […] became, to Einstein, phantoms with no consistency whatever and no meaning" (Koyré 1991 [1955], pp. 213–214). Gaston Bachelard also evoked such images in his analyses. But the phantoms and monsters that Koyré refers to have a distinct status from those that Bachelard exhibits in his chamber of horrors – note that Koyré spoke of a "collection of monsters" in his reference to the perspectives of scientists founded on their own actuality – and they require no psychoanalysis of the objective knowledge; they are, instead, susceptible to historical analysis.[21] It is because Einstein transforms nature into a metaphysical principle – "nature is the measure of things as they really are", not God or Man – based on which he thinks on time and space, that those Newtonian notions appear to us in our actuality to be so many ghosts with no consistency or meaning. Einstein no longer thinks on the same plane of thought as Newton did. It is with the advent of this new plane of thought that the Newtonian truths become inconsistent phantoms.

A monster is a body without proper form or, rather, misshapen and deformed. A phantom is a form without a body, a mere apparition. But those deformed and spectral figures never are that way, intrinsically, in themselves. The Newtonian conception of absolute space only becomes a specter when viewed in the light of the plane of thought opened up by Einstein. All those figures and images that Koyré evokes are temporal figures and images. It is not the eternity of truth that matters, but its consistency. The consistency does not stem from the firm imposition of a form to a content. Outside the plane of thought in which it was composed, it loses consistency and meaning. Outside of that scheme, it becomes deformed, loses its body, and ages, and what it has to say no longer makes any sense. Outside of that plane, it not only errs but is errant insofar as it no longer lies within the bounds of "truth." So the problem is not one of *stating* the truth or not, but one of *being* [at the time] within the truth or not[22]: "We should not mock", writes Koyré, the Aristotelian argument against the mobility of the Earth. "From the standpoint of Aristotelian physics it is totally just; so much so that on the basis of that physics, it is irrefutable" (Koyré 1982 [1955], p. 187). It is only outside of that sphere that it becomes a figure with no limits, a limit figure. That is why it needs to be addressed on the basis of other limits. One of the difficulties and characteristics of the history of sciences is that it is not simply work that handles the dead, but in fact has to deal with monsters and ghosts.

What we see are apparitions of forms that are always inchoate. It is impossible to actually *see* them because a phantom never has a clearly defined appearance; it is unclear, taciturn, and more like an obscure and somber thought. To Koyré, however, those phantoms are not mere primitive, archetypical, prelogical, or psychic images. Neither is it the sleep of reason that produces monsters. Quite the contrary, they are the products of reason's regular daily work. It is not at night that they are engendered in dreams, but during the working day, by a reason at once active, creative, and dynamic. Such figures do not stem from the unconscious, even though it could be said that they are symptoms of a transformation. It is transformation that is

[21] François Delaporte showed very clearly how, by means of those images, Michel Foucault, in a dialogue with Gaston Bachelard, sought to distinguish his archeology of epistemology from that of the author of *La formation de L'Esprit Scientifique* (see Delaporte 2011). It was Delaporte who called attention to the importance of teratological images in epistemological reflections.

[22] Delaporte showed the importance that the notion, *être dans le vrai* [being within the truth], coined by Koyré, acquired in the philosophy of Sciences in France (Delaporte 2011, p. 69).

always pushing them outside the limit into the shadow zones of knowledge, turning them into obscure uncertain shapes devoid of clarity. Similarly, a phantom is a figure at the limits or rather a disfigure at the limits, a being that wanders at the frontiers between no longer alive and not yet dead. That means to say that reason does not affix itself to a definitive image nor petrify itself in static architecture. Instead, it *de*forms, *re*forms, and *trans*forms, and it *meta*-morphosizes. Hence Koyré's transformation of the image of the past, based on which it would be necessary to capture the past forms of thought that are obscure to us today. If it is only the historian of thought that is capable of capturing the ancient theories at the moment of their "creation," the past of that history takes on the aspect of an artist or sculptor's studio, a place where the thought *takes form*. That is because it is not a question of reconstituting the theories alone, but rather of reconstituting the time itself. The temporal strata in which the theories took shape need to be reconstituted. That brings Rodin to mind when, referring to a sculpture, he said that the surface was the extremity of a volume, the point of emergence of a volume. Koyré's stance in relation to the theories of the past is the same as the sculptor's: the historian needs to place himself at the level of the surfacing, the emergence of a shape, and the birth of a thought. That is why, in his view, the sciences historian is a "thinker-creator" (Koyré 1991 [1955], p. 205). He recreates, in the present, forms that are not offered by any specific time at all. Therein lies the extreme difference of such images from those of the cemetery or the chamber of horrors. All those figures enable us to realize how Koyré's work is in fact a study on the forms and limits of thought, hence the centrality of the concept of revolution in it.

Thicknesses and layers: history has something in it of geology, paleontology, and archeology. That is why, in 1946 Koyré wrote that if we "wished to make a forced comparison between history and the natural sciences" (Koyré 2011 [1946], p. 59), then it should not be made with physics but rather with those sciences that address natural history, that handle fragments of worlds and forms of life that have long since disappeared, and that have been embedded in deposits and layers. In that way he formulated a time model for the history of sciences made up of layers, strata, and planes. Time, in that history model, was no longer structured on a chronological scale based on the sequence of scientific discoveries. It is no longer a question of setting science against time or deducing the laws that give it meaning by identifying their precursors or individuals that were ahead of their time. Instead, it means articulating the scientific ideas with their temporal strata, and it was in their actuality that the thickness of those strata found their limits.

References

Bachelard G (2003 [1938]) A formação do espírito científico. Contribuição para uma psicanálise do conhecimento. Transl. Estela dos Santos Abreu. Contraponto, Rio de Janeiro.

Bachelard G (1951) L'activité rationaliste de la physique contemporaine. PUF, Paris.

Bachelard G (1972 [1951]) L'actualité de l'histoire des sciences. L'engagement rationaliste. PUF, Paris, pp. 137–152.

Bloch M. 2006 [1942]. Apologie pour l'histoire ou métier d'historien. In Becker A, Bloch E (eds). L'histoire, la guerre, la résistance. Gallimard [Quarto], Paris, pp. 843–985.

Canguilhem G (1983 [1966]) L'objet de l'histoire des sciences. In Canguilhem G (ed). Études d'histoire et de philosophie des sciences. Vrin, Paris, pp. 9–23.

Clavelin M (1987) Le débat Koyré–Duhem, hier et aujourd'hui. History and Technology 4:13–35.

Coulanges F (2003 [1875]) A história, ciência pura. In Hartog F (ed). O século XIX e a história. O caso Fustel de Coulanges. Translated by Roberto Cortes de Lacerda. EdUFRJ, Rio de Janeiro, pp. 304–309.

Delaporte F (2011) Foucault, Canguilhem e os monstros. In Salomon M (ed). História, verdade e tempo. Translated by Marlon Salomon e Raquel Campos. Argos, Chapecó, pp. 51–74.

Duhem P (1913) Études sur Léonard de Vinci. Les précurseurs parisiens de Galilée. Tome III, Hermann, Paris.

Febvre L (1970 [1942]) O problema da descrença no século XVI. A religião de Rabelais. Trans. de Rui Nunes. Início, Lisboa.

Febvre L (2009 [1949]) Vers une autre histoire. In Febvre L (ed). Vivre l'histoire. Édition établie par Brigitte Mazon. Edition Robert Laffont [Collection Bouquins], Paris, pp. 357–374.

Gattinara EC (1998) Les inquiétudes de la raison. Vrin, Paris.

Gattinara EC (1997) L'idée de la synthèse : Henri Berr et les crises du savoir dans la première moitié du XX. Revue de Synthèse 117:21–37.

Jorland G (1981) La science dans la philosophie. Les recherches épistémologiques d'Alexandre Koyré, Gallimard, Paris.

Koyré A (1933) Copernic. Revue philosophique 116/7–8:101–118.

Koyré A (1991 [1955]) Da influência das concepções filosóficas sobre a evolução das teorias científicas. Estudos de História do Pensamento Filosófico. Translated by Maria de Lourdes Menezes. Forense Universitária, Rio de Janeiro, pp. 201–214.

Koyré A (1992 [1939]) Estudos Galilaicos. Transl. Nuno Ferreira da Fonseca. Dom Quixote, Lisboa.

Koyré A (1934) Introduction. In Copernic N. Des Révolutions des Orbes Célestes. Traduction, avec introduction et notes par A. Koyré [Textes et Traductions]. Alcan, Paris, pp. I-VIII.

Koyré A (1962 [1938]) Introduction à la lecture de Platon suivi d'Entretiens sur Descartes. Gallimard, Paris.

Koyré A (2011 [1946]) Filosofia da história. In Salomon M (ed). Alexandre Koyré, historiador do pensamento. Translated by Fábio Ferreira de Almeida. Ricochete, Goiânia, pp. 49–61.

Koyré A (1982a [1955]) Galileu e a Revolução Científica do século XVII. Estudos de História do Pensamento Científico. Forense Universitária, Rio de Janeiro, pp. 181–196.

Koyré A (1982b [1930]) O pensamento moderno. Estudos de história do pensamento científico. Translated by Márcio Ramalho. Forense Universitária–Brasília, EdUNB, Rio de Janeiro, pp. 15–21.

Koyré A (1982c [1951]) Orientação e Projetos de Pesquisa. Estudos de História do Pensamento Científico. Forense Universitária, Rio de Janeiro, pp. 10–14.

Koyré A (1991 [1949]) O vácuo e o espaço infinito no século XIV. Estudos de História do Pensamento Filosófico. Translated by Maria de Lourdes Menezes. Forense Universitária, Rio de Janeiro, pp. 23–69.

Koyré A (1971 [1933]) Paracelso. Mystiques, spirituels, alchimistes du XVIe siècle allemand. 2ª ed. Gallimard, Paris, pp. 75–129.

Koyré A (1982 [1961]) Perspectivas da história das ciências. Estudos de história do pensamento científico. Translated by Márcio Ramalho. Forense Universitária–Brasília, EdUNB, Rio de Janeiro, pp. 370–379.

Koyré A (1929) Review of André Metz. Une nouvelle philosophie des sciences: le causalisme de Émile Meyerson. Revue philosophique de la France et de l'étranger 108/7–8:154–156.

Koyré A (1935–1936) Review of Arthur Eddington: New pathway in science. Recherches Philosophiques 5:455–457.
Koyré A (1934–1935) Review of Signification de l'histoire de la pensée scientifique. Recherches Philosophiques 4:522.
Koyré A (1935–1936) Review of Werner Heisenberg. Wandlungen in den Grundlagen der Naturwissenschaft. Recherches Philosophiques 5:457–458.
Kuhn T (2011 [1968]) As relações entre a História e a Filosofia da Ciência. A tensão essencial. Translated by Marcelo Amaral Penna–Forte. EdUNESP, São Paulo, pp. 27–44.
Lenoble R (1945) Les nouvelles conceptions de l'histoire. Revue Philosophique 135/7–9:193–208.
Rasmussen A (1996) Critique du progrès, « crise de la science » : débats et représentations du tournant du siècle. Mil neuf cent 14:89–113.
Redondi P (1986) Préface. In Redondi P (ed). Koyré A. De la mystique à la science. Cours, conférences et documents 1922–1962. Édition EHESS, Paris, pp. IX–XVII.
Redondi P (1983) Science moderne et histoire des mentalités. La rencontre de Lucien Febvre, Robert Lenoble et Alexandre Koyré. Revue de Synthèse 111–112:309–332.

Chapter 19
The Pitfalls and Possibilities of Following Koyré: The Younger Tom Kuhn, "Critical Historian," on Tradition Dynamics and Big History

John A. Schuster

Abstract Late in his career, Thomas S. Kuhn practiced more as a philosopher of science than as a historian of science. However, his earlier work—leading up to *The Structure of Scientific Revolutions* and during the majority of his tenure in the Princeton history of science group—focused on "mapping" the shape of the history of the physical sciences and on modeling the dynamics, or "motor," of scientific traditions. This paper examines the younger Kuhn's excursions in map and motor design. It views Kuhn as a "critical historian," that is, a historian who constructs explanatory categories in order to apply them to large-scale narratives, evaluation of which can suggest modification of those guiding categories.

The younger Kuhn's map and motor design was largely shaped by the work of his historiographical idol, Alexandre Koyré. Kuhn's creative articulation of Koyré's position explains his innovations concerning Scientific Revolutions (plural), his loosening of Koyré's central category of "metaphysics," and his invention of the crucial conception of "normal science." Additionally, Kuhn's devotion to Koyré explains some historiographical pitfalls and blind spots that bedeviled his historical work: for example, his ignoring early modern natural philosophizing as an institution and culture in its own right and his failure to capitalize on his correct insight into the nature of scientific discovery as the nonrevolutionary yet tradition-modifying core process in the sciences. The paper is concerned with Kuhn's work as a critical historian and his legacy for younger historians, not with philosophical debates about his texts.

Keywords Thomas S. Kuhn • Alexandre Koyré • The Scientific Revolution • Discovery • Historiography of science • Kuhnian normal science • Kuhnian revolutionary science • Metaphysics of science • Experimental sciences • Sociology of scientific knowledge • Internalist/externalist debate

---

J.A. Schuster (✉)
Unit for History and Philosophy of Science & Sydney Centre for the Foundations of Science, University of Sydney, Sydney, NSW, Australia
e-mail: drjaschuster@gmail.com

R. Pisano et al. (eds.), *Hypotheses and Perspectives in the History and Philosophy of Science*, DOI 10.1007/978-3-319-61712-1_19

19.1 Introduction: Tom Kuhn, Critical Historian (Circa 1958–1977)

Late in his career, Thomas S. Kuhn practiced more as a philosopher of science than as a historian of science. However, his earlier work—leading up to *The Structure of Scientific Revolutions*[1] and during the majority of his tenure (1964–1979) in the Princeton history of science group—focused on what I shall call "mapping" the shape of the history of the physical sciences and on modeling the dynamics, or "motor," of scientific traditions.[2] This paper examines the younger Kuhn's excursions in map and motor design. It views Kuhn as a "critical historian," that is, a historian who deliberately constructs explanatory categories in order to apply them to large-scale narratives, evaluation of which can suggest modification of those guiding categories.[3]

I argue that much about the younger Kuhn's map and motor design was shaped by the work of his historiographical idol, Alexandre Koyré. Kuhn's creative articulation of Koyré's position explains his innovations concerning Scientific Revolutions (plural), his loosening of Koyré's central category of "metaphysics," and his invention of the crucial conception of "normal science." Additionally, Kuhn's devotion to Koyré explains some historiographical pitfalls and blind spots that bedeviled his historical work: for example, his ignoring early modern natural philosophizing as an institution and culture in its own right and his failure to capitalize on his correct insight into the nature of scientific discovery as **the** nonrevolutionary yet tradition-modifying core process in the sciences. There is a complex dialectic to be unraveled in the younger Kuhn's creative yet self-consciously devoted relation to Koyré. Similarly, there is a complex dialectic of relation between post-Kuhnian historians and sociologists of science and the younger Kuhn, whose historiographical thinking unintentionally pointed toward innovations that were to be Kuhnian in spirit but mostly rejected by Kuhn himself. Note that I say post-Kuhnian historians and

---

[1] Hereafter cited as *SSR*, from the 2nd edition of Kuhn (1970). The first edition was published in 1962.

[2] One could take the terminus point for Kuhn's activity as a critical historian as 1977, with the publication of Kuhn (1977a) and particularly the essay, Kuhn (1977c), which first appeared in 1976. In 1979, Kuhn moved to MIT., but he had been on leave and away from Princeton quite a bit in the previous two years. I restrict the use of the term "young" for Kuhn up through 1962. "Younger" covers that period and on through 1977.

[3] Any field of history, including history of science, consists of two interacting levels: One is where we craft together our public, published products—narratives imprinted with explanations of what was happening and why. The other level is where one designs the categories that are being both applied and revised in our narrative/explanations. Some areas of history require more concentrated attention to that second level than others—history of science is one of them. Hence, a critical historian is one who explicitly attends to the formation of categories and evaluates the goodness of the narrative/explanations in which they are deployed, with an eye to modification and improvement. The younger Kuhn certainly was a critical historian, especially concerned with the legacy of Koyré.

sociologists, not philosophers. We are concerned here with Tom Kuhn as a practicing and explicitly theorizing historian.

Before we begin, several caveats need to be set out. First of all, this essay is not an exercise in textual hermeneutics concerning Kuhn's published references to Koyré. I envision something broader, deeper, and more implicit in what Kuhn was doing (and, as we shall see, sometimes signaled in Kuhn's teaching history graduate students about Koyré and the challenges of big, highly theorized history of science). Second, I avoid any hasty identifications of the younger Kuhn with particular post-Kuhnian developments. For example, Kuhn is not treated here as a "forerunner" of a "post-Kuhnian" sociology of scientific knowledge (hereafter SSK) or of historical epistemology. We shall see that Kuhn's own concern with the micro-dynamics of traditions does not map very well onto later SSK and where it does this best, as in the nature of discovery and continuous change in scientific research traditions, Kuhn, for reasons we shall explore, did not pursue those insights. Finally, this study is in part intended to help shape the outlook of young scholars in the field of history of science about historiography and about Scientific Revolution studies. In this sense, the paper continues and radically updates the kind of questioning of Kuhn's intentions, accomplishments, and relations to Koyré that I and other apprentice historians of science, then under his direct tutelage at Princeton, used to pursue in the 1970s.[4]

19.2 The Young Tom Kuhn on "A Role for History"

The young Kuhn, critical historian, appears in the first chapter of *SSR*, "A Role for History." Kuhn tells us that the stimulus for his project was his observation that his studies of the development of science simply failed to match up with accounts given by abstract methodologies of science or by science textbooks. He did not mean that brute historical facts speak for themselves. Rather, he meant (1) that scientists and philosophers have no monopoly on purveying frameworks of interpretation to historians and (2) that historians are their own best purveyors, building up revisable models of process that are applied to narration and open to revision in the light of

[4] I was a graduate student in the Princeton Program in History and Philosophy of Science, Department of History, from September 1969 to August 1973. From September 1973 to July 1974, I was an "instructor" in the HPS Program and the Department of History. Michael S. Mahoney was the chief supervisor of my doctoral dissertation, *Descartes and the Scientific Revolution: An Interpretation*, and Kuhn was co-supervisor. In August 1974, I began my first regular academic appointment in the Division of History and Philosophy of Science, Department of Philosophy, University of Leeds. Its senior member, Jerome R. Ravetz, was a keen student of the work of Kuhn, and his seminal, *Scientific Knowledge and Its Social Problems* (1971), had enlarged and improved the Kuhnian model for "normal science" in ways paralleling the contemporary initial development of post-Kuhnian interpretative sociology of scientific knowledge, in the hands of scholars such as Mulkay (1979), Collins (1975), Barnes (1974, 1982), and Shapin (1982, 1992). My first comment on these Kuhn/post-Kuhn developments was Schuster (1979).

further inquiry and debate. Kuhn intended to be one of these historian-purveyors—to suggest a model for the dynamics of process of the sciences and to map in narration large regions of the history of the sciences.

Kuhn also states that he intends to correct the idealizations and abstractions of philosophy of science. But, rather than conclude that he was mainly or entirely a philosopher at that moment, perhaps we should work on the assumption that these corrections were intended to come from providing precisely what I have called a mapping and motor. If that is so, it is also plausible that the young Kuhn underestimated the difficulties involved in proposing to professional philosophers that his historiographical conceptualizations could serve in their own field as a substitute for their work. For purposes of this study, therefore, I assume that as far as Kuhn was concerned at this stage of his career, what were at stake were properly historical technique and problematics, that is, the professional concerns of intelligent and critically aware historians. If philosophers were to be corrected and even if Kuhn at some deep level viewed himself as a philosopher, it is still the case that he, perhaps naively, thought that his winning position in philosophy would arise from this kind of critical historical work. Certainly, having moved to Princeton two years after the 1962 publication of *SSR*, he was for the next few years mainly involved in teaching apprentice historians of science who were officially located in the great Princeton Department of History and therefore also studying the economic and social history of their preferred historical period.

As one of those apprentices, exposed to Kuhn's teaching and informal historiographical lessons in graduate courses (called seminars at Princeton), I can report that many of us thought we were being shown deep problems, and solutions, about the shape of the history of Western sciences and how their dynamics should be studied and set out in narrative *cum* explanations. Additionally, most of us saw affinities between Kuhn's historical theorizing and the grand, layered structures of explanation (cultural/ideological, social, and economic) being offered by Princeton's two leading experts on European early modern history, Lawrence Stone and Theodore K. Rabb. Both of these formidable scholars were interested in long time periods, critical awareness of interpretive frameworks, as well as the historiography of the Scientific Revolution itself.[5] Kuhn, who held an endowed chair in that same Department of History, presumably felt, as we did, the gravitational pull of deep critical thought about history.[6]

[5] Consider synthetic works of these authors: Stone (1972) and Rabb (1975).

[6] I am not suggesting that Kuhn himself was accomplished in social, political, or economic history. He was not, and he often pointed that out to the graduate students, telling us explicitly that whatever history of science problem we worked on, we would require gigantic loadings of knowledge of "context, context, context," knowledge he was in no position to impart.

19.3 Alexandre Koyré as Historian of Science: Kuhn's Point of View

To the young Thomas Kuhn, Alexandre Koyré was both a professional and historiographical model. Kuhn had much to say, in print and in the seminar room, about his hero. Emulation of Koyré accounts to a large degree for the goals of Kuhn's work in critical history of science, as well as for many of its strengths and weaknesses. Here we examine Koyré himself as a critical historian of science, rather than as a philosopher, and we do that from Kuhn's point of view, rather than as an exercise in modern Koyré exegesis.[7]

Koyré belonged to the second generation of a group of Continental neo-Kantian historians of philosophy, such as Ernst Cassirer and Léon Brunschvicg, who had turned their attention to the conceptual development of science. Neo-Kantianism in the historiography of philosophy had stressed the sympathetic understanding of the *sui generis* inner coherence and rationality of earlier systems of thought, the historical transformation and "progress" of certain concepts across such systems, and the ways in which categorical systems shape experience. Koyré held that the development of modern science depended upon a revolution in ideas, involving the establishment of a new metaphysics or set of deep conceptual presuppositions, which in turn shaped theorizing and experience in the emerging fields of modern science, especially classical mechanics and Copernican astronomy. By the mid-1950s, Koyré had become, in Anglo-American circles, the leading exponent of such "internalist" historiography of science, posed in opposition to Marxist or materialist "externalist" approaches.[8]

Galileo's revolutionary constitution of classical mechanics was Koyré's exemplary case of the emergence of modern science.[9] According to Koyré, Galileo had no need to apply some putatively universal and efficacious method, because research procedures and strategies always follow from within one's particular metaphysics. Belief in a universal, transferable, and workable scientific method is, according to Koyré, a myth.[10] Galileo succeeded in founding the first version of classical

---

[7] Kuhn gave early expression to his appreciation of Koyré's contribution to the maturation of the history of science profession in Kuhn (1977b, e). These essays are cited as reprinted in Kuhn (1977a). They were each originally produced in 1968. When Kuhn addressed each year's crop of new history of science graduate students, he would make a point of bringing in his well-worn, pre-World War II copy of Koyré's *Études galiléennes* [the English translation only appeared in 1978]. He would intone, "Nobody is leaving here until they have read all of this." Presumably he would know from one's work whether the exercise had been done.

[8] The best discussion of the "internalist vs. externalist" debate is Shapin (1992); see also Schuster (2000).

[9] For the emerging Anglophone profession of history of science following World War II, Koyré's treatment of Galileo (Koyré 1939, 1978), then only available in its original prewar French edition, became the exemplar of how to practice the history of science.

[10] Koyré (1956). As we shall see, this principle was followed and deepened by Kuhn, especially through his insistence that a large segment of any living scientific research tradition was passed onto apprentices and applied by them, in the form of tacit, craftsperson-like "knowledge."

mechanics because he worked within the correct sort of metaphysical framework, a kind of nonmystical "Platonism," a conviction that the basic furniture of the world consists in mathematical objects, moved according to simple mathematical laws. If Galileo experimented (a proposition Koyré often appeared to doubt, with the exception of thought experiments), the experiments were shaped by cognition and action themselves constrained by and constructed within this metaphysics.[11] For Koyré, the sort of Platonic metaphysics he attributed to Galileo was the only viable framework for scientific advance. Other frameworks might have virtues, but not scientific ones. Thus, Aristotelian natural philosophy, itself coherent as a categorical framework, could never structure experience and reasoning so as to produce modern mathematical physics. It was too closely enmeshed with the categories of natural language and everyday life (Koyré 1939, 1978).

Koyré's subordination of practice to metaphysics meant that he explained all his key facets of the Scientific Revolution—Galileo's mechanics, Copernican Revolution and Newtonian synthesis—on the same basis. There was no account of the dynamics of everyday practice within a scientific tradition nor of how traditions of scientific practice interact nor, especially for the internalist Koyré, an account of how traditions relate to contexts.[12] That, at least, is how the matter must appear to today's readers of Koyré. But, the situation was different when, following World War II, Koyré's influence was at its height. To understand this seemingly paradoxical point, one must turn to Koyré's detailed practice as a historian, rather than to his grand internalist historiographical pronouncements.

As a working historian of science, Koyré showed by example that scrupulous, critical explication of primary sources resides at the center of any serious understanding of the history of science. Koyré insisted that at a textual level, the conceptual structure, and faults, of a scientific thinker should be laid bare, setting aside anything that thinker might have said on a meta-level about "method" or program. One must avoid what Kuhn, following Koyré, always termed "preface history" and the simplistic "island hopping" among unproblematically linked sequences of correct scientific ideas, which, for Koyré and Kuhn alike, constituted the always to be avoided "Whig" history.[13] Close textual analysis, studying the primary texts "with all senses open", as Kuhn often said, not only meant eliciting the conceptual structures in play but also, tellingly, examining the ambiguities and errors of the author. These, Koyré stressed, showed as much or more about the inner texture of the actor's

[11] There is more to say about Koyré on Galileo's experimentation, which we defer to Sect. 19.5.

[12] Whether and how Koyré's historiography handles the contexts of his great revolutionary figures is a more complex issue than it may appear at first sight. See Barnes 1974, Chapter 4 for a brilliant and suggestive early post-Kuhnian discussion. For example, it has been noted, by Barnes and others, that Koyré's notion of the "metaphysical framework" embraces the intellectual and philosophical "contexts" of science. This has prompted the question of whether Koyré was an internalist or externalist. That question however is misplaced. It is just a matter of the location of what we may in post-Kuhnian terminology call the cognitive/social frontier (Schuster 2000, p. 335).

[13] Koyré and Kuhn had their own problem with a different species of Whiggism, which was built into their respective approaches and little noted until the emergence of post-Kuhnian discussions in the 1970s and 1980s, as we shall see later, Sect. 19.10 and Note 44.

thought than those smooth skeins of conceptualization with which we might still fully agree. As Kuhn recognized, Koyré was importing into history of science techniques of textual study and explication previously pioneered in the history of philosophy.[14] This was what had been missing in the amateurish historical writing of superannuated scientists and other enthusiastic hero-worshipping scholars. Such hermeneutical sophistication, combined with Koyré's internalist pronouncements, was judged to be the key to the training of the new generation of professional historians of science.

However, to return to our main point, a slippage occurred just here regarding the understanding of Koyré's work. As we have seen, Koyré lacked a model of scientific process that would bridge the gap between conceptual revolutions, between, say, Galileo and Newton. Nevertheless, it certainly "seemed" that Koyré had provided one. Koyré could supply a sequence of close, textual analyses of secondary figures, making it seem that the actual dynamics of scientific development had been uncovered. This kind of picture emerged in Koyré's most important sustained historical works, such as *The Astronomical Revolution* (1973) or *A Documentary History of the Problem of Fall* (1955). They presented accomplished, sympathetic readings of the theoretical structures of this, then that, scientist. All this occurred in the reader's mind under the sign of Koyré thundering historiographical pronouncements. It looked as though Koyré not only had adduced the blueprint for historiography, but also had described the dynamics of modern science, while providing a map of the great conceptual rupture points.

Writing about the debate at the historiographical level between internalists and externalists in the 1950s and 1960s, Steven Shapin remarked, with characteristic acuity, that on the internalist, mainly Koyréan side, we did not have a serious attempt to understand the living continuity and dynamics of scientific traditions. "A style of research and writing," he wrote, "does not amount to a theory of scientific change" (Shapin 1992, p. 346). That insight is as penetrating today as it was in 1992 when Shapin first said it. But to repeat, at the height of Koyré's acclaim, his dazzling style of textual explication was taken for a model of how sciences proceed. This is where the young Tom Kuhn, critical historian, entered. Kuhn, alone among his contemporaries, understood that there actually was no Koyréan theory of science dynamics and that one was needed. Moreover, as we are about to see, Kuhn's brilliance in this regard resided not so much in providing a more detailed general model of how revolutions unfold, than in offering a model of the dynamics of everyday, continuous, and in the end revolution-triggering scientific research: the model, that is, of "normal science" in any given scientific tradition. Kuhn saw that Koyréan textual analysis was one necessary thing and that an explicit model of what I am calling "the motor" of scientific dynamics was quite another.

[14] In his overview of how history of science had evolved, Kuhn indicated that reading Koyré and other early philosophically acute historians such as Emile Meyerson and Léon Brunschvicg crystallized his ability to sympathize with outmoded structures of thought, such as Aristotle's. They showed him that past philosophical and scientific systems have their own *sui generis* rationality, coherence, and cogency which the historian must penetrate (Kuhn 1977b, p. 11; 1977e, p. 108).

19.4 Kuhn's Core Premises as Historian: Post-Koyréan and Neo-Koyréan

Three premises stand out in the historical theorizing of the young Kuhn. They are properly historiographical rather than epistemological or philosophical. The first premise is that there is no such thing as Science (capital S). You cannot say "Science began with the Greeks" or "Modern science started in the 17th century." Kuhn is interested in the *histories of the sciences*. He views Science (capital S) as an invention of poor historical thought, preface history if you will.[15] The second premise is that there is no universal, efficacious scientific method. In this, he followed and articulated Koyré's position: Kuhn reinforced Koyré's view by explicitly and consistently holding that the sciences are many and that each science, in any given normal period, has its own disciplinary research culture. This in itself spoke strongly against the myth of a universal method. Kuhn, however, famously reinforced this claim by insisting that a large portion of any such normal research culture consists in tacit knowledge of how to apply the paradigm to certain species of problem and of the criteria for selecting problems and evaluating their solutions.[16] The third premise is that even though the sciences are many, there is a common pattern of development and change displayed in the history of each science. This is the pattern of normal science—problem- or puzzle-solving or "mopping up"—under the aegis of a ruling paradigm, until such time as serious anomalies are recognized in the application of that paradigm, crisis ensues, alternatives are proposed, and eventually an incommensurable new paradigm is installed to govern normal research until the next crisis and rupture.[17]

Leaving aside the difficulties—historical and philosophical—of Kuhn's model of revolution, we should note that Kuhn's model of the common, recurring pattern of

---

[15] Anyone who has taught entry-level history of science knows import of this premise as well as students' proclivity to slip, even when doing history, into talk about Science, capital S. However, if properly introduced, it transforms their reading of historical and philosophical literatures that miss this point. See Schuster 1995a, Chapter 15, p. 155; 2013b, pp. 284–285.

[16] Kuhn's views on the tacit component of paradigms are usually linked to those of Polanyi (1958), whom he cites in this connection early in *SSR* (44 n. 1). Recently doubt has been cast upon the reliability of Kuhn's recollections about the timing and import of his reading of Polanyi (Jacobs 2009). The earliest and most impressive post-Kuhnian articulation of the theme of scientists' activity as "craftsmen's work" was in Ravetz (Ravetz 1971), who interestingly mainly cites Polanyi in this connection rather than Kuhn (see Ravetz's index entries on Kuhn and Polanyi). It is also important to note that while the Kuhn/Koyré anti-method position is extremely important for the historiography of science, it did not go far enough. That is, they left the issue of method as one of ironic denial. They did not ask why scientists regularly profess to believe in a general scientific method, what political and rhetorical roles such belief plays, and how "method talk" functions as a misleading species of discourse. See Schuster 1986, 2013a, pp. 70–77, pp. 265–273.

[17] For an explication of the phases in Kuhn's model of revolution, its onset, process, and resolution, see further Schuster 1995a, Chapter 16, pp. 161–165. This exposition is aimed to help entry level students of history or sociology of science, not particularly to facilitate participation in the philosophical debate about Kuhn on revolution. Note also that Kuhn's conception of the "first paradigm," founding a new disciplinary tradition, poses its own problems, and we shall deal with this in Sect. 19.10.

change in any given mature scientific tradition accomplishes four key articulations in relation to Koyré. Firstly, it provides a blow-by-blow account of what presumably goes on inside a Koyréan revolution. Koyré had not dissected revolutions. Kuhn now did so. Secondly, and this particularly interests us here, it provides a post-Koyréan model of the dynamics of ordinary, day-to-day, garden-variety scientific practice that eventually conduces to an internally generated crisis and finally overthrow of the very basis of that ordinary practice. Koyré had only wanted to deal with revolutionary genius. As discussed above, if we avoid mistaking textual analysis for a theory of scientific change, we see that Koyré had nothing to say about what goes between moments of grand rupture. In a way completely unknown to Koyréan historiography, Kuhn's model of normal science points us to the continuity and dynamics of any given scientific tradition, between (now explicated) Koyréan ruptural moments. Thirdly, Kuhn achieved some serious mapping of the macro history of science: just as Kuhn was providing, de novo versus Koyré, a model of the dynamics of normal research, so too he was offering a detailed if preliminary map of the history of the sciences, especially the physical sciences, where Koyré had offered instances of revolution. Once you have decided that the history of science is the history of multiple scientific traditions/disciplines and once you have also provided a generic pattern of change applying to each tradition over time, you have implied a map of the history of the sciences. The younger Kuhn firmly grasped these points and, as we discuss below in Sects.19.7, 19.8, 19.9, 19.10 and 19.11, pursued them in detail.

Finally, Kuhn introduced the possibility of metaphysical pluralism. No longer is there only one metaphysical stance—Platonism of the Koyréan genre—that can stand behind genuine "science." *SSR* teaches that every single paradigm in the history of the various sciences has possessed a particular metaphysics, which need not have been of Platonic type. Thus, Aristotelianism recovers its status as a possible metaphysical core for arguably genuine science.[18] However, the necessary presence of a metaphysical dimension in every scientific tradition, during each and every of its normal science periods, remains a Koyréan element. Similarly, as we shall see below, Kuhn's paradigms retain a destiny of fulfillment, due to the presumed cognitive pressure of their respective metaphysical dimensions—a Koyréan legacy in Kuhn's thought.

19.5 The Historical Problem of Experiment and Experimental Hardware

Koyré's idealist and metaphysically driven vision of knowledge meant he minimized the role of experiment and experimental hardware. This stance comported with his anti-Marxist agenda in historiography of science. Marxist historians of early modern science stressed the causative role of technics, craftsmen, and

[18] This is apparent in the status accorded to Aristotelian-backed Ptolemaic astronomy as a competitive paradigm. For a textbook treatment of how Kuhn released the constraints on what might be construed as the metaphysics of a given paradigm, see Schuster 1995a, Chapter 11, Chapter15, p. 157 and Schuster 2013b, pp. 193–208, pp. 290–291.

instrumental intervention against the background of rising commercial capitalism, state formation, and military competition. Koyré's contrasting account denied such considerations. Conceding experiment a subsidiary, uncreative role in confirming the results of predictions deduced from mathematicized theory, he also granted importance to Galileo's thought experiments, as a way of detecting contradictions in Aristotelian and Scholastic physics.[19]

Kuhn had been trained as a professional physicist. He also sympathized with Robert Merton's attempt to demonstrate—with a non-Marxist emphasis—the importance of technics and men of practice in the rise of experimental, Baconian science in seventeenth-century England (Merton 1970). Kuhn broadened the question, asking how and why new experimental sciences emerged in the eighteenth century. For Kuhn to entertain the importance of experiment and experimental sciences was a large step away from Koyré's program. To that end, Kuhn offered several fruitful initiatives about the role and nature of experiment. Because we shall later focus on the rise of new experimental sciences, as part of our study of Kuhn's larger mapping and motor modeling project, we briefly mention these issues here and recur to some of them later.

[1] Toward a Politics of Testing: In his brilliant paper, "The Function of Measurement in Modern Physical Science," originally published in *Isis* in 1961, Kuhn took the notion of confirmatory experiment and historicized it. Deploying his emerging concept of normal science, he focused on scientists' judgments about the significance of small discrepancies of "fit" between prediction and test results. Training into the ruling theory transfers to apprentice practitioners a sense of the standard of accuracy expected in their particular field. Research problems exist where there are challenges to bring "fit" within acceptable limits and to extend the adequately accurate predictions to new domains of test data. In extreme cases, the value placed on increasing precision might justify a change of ruling theory. In Kuhn's account, experiment undertaken at the research front takes on a new cast: not as providing simple confirmation or refutation of ruling theory, but rather as an activity where the goodness and meaning of results can be expertly debated. This was a far cry from Koyré, who expected a good theory, like Galileo's, to be confirmed (or not even submitted to test).[20] It did not take much imagination to see that in the search for improved "fit," a research community could renegotiate the ruling theory and hence its predictions and/or that the understanding of experimental hardware

[19] "Galilean epistemology [...] is both *a priorist* and experimentalist at one and the same time (one could even say that it is the latter because it is the former) [....] [Galileo's] experiments...are designed on a theoretical basis and of which the function is to confirm or refute the application to reality of laws deduced from principles which themselves have a quite different basis" (Koyré 1978, p. 106).

[20] Kuhn 1977h, Schuster 1979, pp. 305–306. Koyré of course knew that "strict agreement" between experimental results and mathematically mediated theoretical predictions is "strictly impossible" (Koyré , 1978, p. 107). Rather, Galileo, who according to Koyré knew this as well, "[...] was not at all looking to found his theory [of motion] on facts gained in the realm of experience: he knew perfectly well that this is impossible. [...]. Experiment can confirm that [a theoretical assumption] is a good assumption. It can do this within its limited means; or rather, within the limits of our means" (ibidem).

could be consensually modified to alter the resulting data toward better "fit." Within a decade and a half the micro-sociological dimensions of such activities became a topic of early SSK research.[21]

[2] Theory Loading of Instruments: Kuhn, perhaps taking on board his reading of Popper (1959) or Hanson (1958), was adamant that experimental hardware and procedures were "loaded" by the theoretical commitments of the ruling paradigm. This was a good start, but, of course, it has been shown by later historical and sociological research to be too simple: instruments are internally complex and may bear, especially in black-boxed fashion, the loading of older theories which are no longer in play. Theory-loaded instrumental practices from other realms of science may be used, relatively unproblematically, as tools of exploratory experimental work in another field.

[3] Thought Experiment as Creative: Kuhn, like Koyré, emphasized cases of thought experiment, but he worked out a more sophisticated account, by considering some of the psychological and cognitive issues involved. Kuhn argued that thought experiments do not merely reveal inconsistencies in previously held conceptual schemes. They must be understood to concern the relation of concepts to nature. They can "alter one's knowledge of the world." Accordingly, in these cases, the results are not a catastrophic "Gestalt shifts" but rather conceptual "reforms" or "modifications."[22] The processes of conceptual change developed in this paper do not fit the more extreme formulations in *SSR* about ruptural conceptual change. Kuhn virtually admits that he missed the significance of these ideas, conceding that this paper had "little influence" on *SSR* (Kuhn 1977a, Preface, p. xx).[23]

As with Kuhn's potential politicization of testing [1], this articulation opened the possibility of seeing normal, work-a-day research as having feedback consequences for small but significant modification of the ruling paradigm. However, this potentially fruitful gambit was swallowed up by his official *SSR* model of problem-solving normal science within a more or less frozen paradigmatic frame.[24]

---

[21] In effect, Kuhn started from Koyré's skepticism about the possibility of strict agreement of data with prediction and opened a realm of social and historical inquiry into exactly what the level of expectation about theory/data "gap" was in a given discipline at a given moment. Post-Kuhnians then focused on the continual possibility of renegotiating that level in the course of consequential work (i.e., creative normal science, possibly leading to paradigm-modifying "discoveries"). Thus, in SSK, the two strands of Kuhnian insight—which Kuhn had relegated to the margin of his thought—about discovery [which we discuss in the next section] and about reasonable agreement in experiment were woven into a broader vision of studying the micro-politics of testing and negotiation of the significance of claimed "results." The works of Collins (Collins 1975) and Pinch (Pinch 1985) are canonical in this regard. See Schuster 1995b, Chapter 6, for a textbook treatment of the politics of testing, devised in the wake of Collins' and Pinch's work.

[22] Kuhn 1977i, originally published in 1964, especially, p. 251.

[23] Again, these ideas of significant conceptual change in the course of normal science accord with Kuhn's ideas about experimental testing (Kuhn 1977h) and his pre-*SSR* view of significant discovery within normal research (Kuhn 1977g) to be discussed below. Taken together, these three themes foreshadow more the subsequent development of SSK than the contents of *SSR*.

[24] There was one other area where the younger Kuhn innovated about experiment, beyond Koyré: following Popper, Kuhn saw that test results were often offered as "proof" of the falsity of one paradigm and truth of its competitor. But Kuhn had looked sufficiently closely at experiment as an

It is worth noting, finally, that although Kuhn vastly extended the empire of experiment within a basically Koyréan historiography, he failed to cash out the more radical implications of his approach for developing a micro-politics of experiment and testing in science. This was accomplished in the brilliant, and often misunderstood, efflorescence of early, or what I term "classical," post-Kuhnian sociology of scientific knowledge. Two things need to be noted about these developments. Firstly, once the early SSK writings were assimilated, they tended to occlude the significance of Kuhn's early moves about experiment. Secondly, as we shall see in the next section, Kuhn's official statement of his motor model, his account of normal science in *SSR*, more or less required that the potential growth points of a politics or sociology of experiment were going to be marginalized.

19.6 Modeling the Motor of Traditions: Normal Science—Importance and Limits

19.6.1 Creative and Political Normal Science or Trivial and Dogmatic Normal Science

Here we are going to examine Kuhn's motor of scientific research, his model of normal science, its strengths versus Koyré, and its own possibilities and implications, many of which Kuhn missed because of his Koyréan bearings. An emphasis on living scientific traditions and their dynamics, the continuous process leading to revolutions, had been missing in Koyré, but it was exactly what Kuhn knew about as a practicing physicist and via close study of the history of physics.[25] Kuhn's original intention was somehow to model, in terms of what he dubbed "paradigms," the structure and dynamics of given scientific fields or traditions—how the members went on with work at the coalface and with what resources, to what standards, toward what aims, and under which constraints. The traditional rather abstract and timeless philosopher's questions about the structure of theories, methods, progress, etc. would be deferred in the light of this enterprise, focused on the how's and why's of the ever-moving, working coalfaces of the individual, mature, scientific traditions or fields.

Note, however, that what I have just said might seem a bit overblown—despite Kuhn's clear improvement over Koyré—when we recall that in *SSR* normal science

expert enterprise and at cases of theory debate, to conclude that "crucial" tests did not in themselves determine the outcome of theory contestations, the Popperian position. Kuhn held instead that "crucial tests" were offered as evidence, among other evidence and argument, by one side in its contest for support against its opponents. Crucial tests were a phenomenon within a wider cognitive and social process of heightened theory debates, not the determinative end points of them.

[25] Kuhn (1977a, Preface pp. 17–18) tells us that his paper on "Function for Measurement" (Kuhn 1977h) was largely finished by spring of 1958. Kuhn says it comes very close to describing what became "normal science," adding, "Though I had recognized for some years that periods governed by one or another traditional mode of practice must necessarily intervene between revolutions, the special nature of that tradition-bound practice had in large part previously escaped me."

is rather uncreative. The "mopping up" or "puzzle-solving" that characterizes normal science is straightjacketed by the ruling paradigm, until and unless some anomaly or anomalies happen to trigger a crisis and the remainder of the flight path of revolution.[26] Clearly there is a tension between the letter of *SSR* on normal science and its "almost" obvious wider possibilities as a sociopolitical model of some degree of creativity and change within a ruling paradigm. This reflects on the nature of the Kuhn/post-Kuhnian dialectic. We can appreciate what Kuhn did and did not achieve concerning normal science by looking at an early post-Kuhnian insight about Kuhn which emphasized the—at the time—paradoxical point that the central achievement of Kuhn was to conceptualize normal science, rather than to produce his highly contentious and disputed account of revolution.

The best and earliest post-Kuhnian exponent of this point of view was Barry Barnes, then a rising star in the Edinburgh Science Studies Unit. Barnes took seriously Kuhn's style of critical history and attempted to integrate it with SSK. Barnes grasped that Kuhn was concerned with understanding traditions of scientific practice, so he stressed instances of Kuhn's historical work rather than his grand modeling in *SSR*. Therefore, at the beginning of his seminal (especially for historians!) Barnes (1982), he started from the rules of exegesis implicit in Kuhn's own historical practice as displayed in his numerous specialist studies: (1) the historian must place a scientist in his subculture or tradition; (2) the historian must treat utterances charitably as internally coherent, where coherence is judged in terms of the actor's relevant contexts and culture; and (3) the only causes asserted of utterances and actions must arguably have been present and active in the contexts in question. Barnes then fleshed out the nature of a tradition of specialized scientific practice in a post-Kuhnian way, viewing *normal practice within a tradition* as a process of social negotiation of conceptual change, possibly small-scale conceptual change.[27]

[26] For an account of what Kuhnian normal science involves, designed for a first or second year introductory course on HPS, see Schuster 1995a, Chapter 15, pp. 157–158 and Schuster 2013b, pp. 291–292.

[27] "The continuation of a form of culture implies mechanisms of socialization and knowledge transmission, procedures for displaying the range of accepted meanings and representations, methods of ratifying acceptable innovations and giving them the stamp of legitimacy. All of these must be kept operative by members of the culture themselves [...]. When there is a continuing form of culture there must be sources of cognitive authority and control. Kuhn was initially almost alone among historians in giving serious attention to these features of science. The result of this attention […] is to display just how profound and pervasive is the significance of the sub-culture in science, and the communal activity of the organized groups of practitioners who sustain it. The culture is far more than the setting for scientific research; it is the research itself [...]. Science is not a set of universal standards... Scientific standards themselves are part of a specific form of culture" (Barnes 1982, pp. 9–10; see also Barnes 1972). The English sociologist M. D. King (1971) had incisively made related points around the same time: addressing specifically Kuhn's paper on the "*Historical Structure of Scientific Discovery*" (Kuhn 1977g) which we are about to discuss, King observed that the paper issued a revolutionary challenge to positivist philosophy of science and to orthodox Mertonian sociology of science. It threatened the latter by hinting that the object of study in the sociology of science is not the cluster of transtheoretical Mertonian social norms of science, but rather the institutional, social, and political processes by which explanatory frames are produced, maintained, and altered by significant (but not catastrophic) discovery.

Barnes helped shape the emergent SSK consensus that Kuhn's stark differentiation between normal and revolutionary phases in the history of a tradition of scientific practice was too strong.[28] This view considers "normal research" paradigms as constantly subject to partial renegotiation and modification. If a problem can be solved only by advocating a shift in some aspect of the paradigm, however so slight, then one can say that the problem solution involves feedback alterations to the paradigm. Such alterations are carried over into the next rounds of problem-solving where further alterations may be suggested. Of course, such bids to alter the paradigm slightly must be accepted by the relevant community. We may term a noticeable alteration of the paradigm, which has been negotiated into place, as a "discovery." Moreover, given this new suggestion that normal science lives by producing significant discoveries, then "revolutionary science" need not be so wild as Kuhn thought, and indeed it need not exist at all. Perhaps a revolution is just a case of relatively large modification of a paradigm in which the innovators legitimate their actions with a "rhetoric" of revolutionary overthrow of the bad old theory.

With these sorts of insights, one was in new territory. The focus shifted onto the micro-politics and organizational dynamics of mature scientific fields, a step that further discomfited anti-Kuhn scholars not used to thinking this way about how science works. But clearly, if one wished to be a post-Kuhnian historian, there was now no alternative. Critically minded historians of science of my generation grasped this eagerly and passed it on, at least to their first generation of students. By this point, the debate, at least among those interested in being historians rather than philosophers, had suddenly and gratifyingly slipped into a post-Kuhnian key. The strict theory of *SSR* was well left behind, and indeed so was Kuhn, who for many years showed little sympathy for these new "sociologists," even though they were arguably legitimate heirs of Kuhnianism. The capping irony is that the younger Kuhn, when practicing as a critical historian, had hit upon elements of the later post-Kuhnian view of normal science, only to overwrite them with the grand pronouncements of *SSR*.

19.6.2 The Young Kuhn on Creative Normal Science and the Process of Significant Discovery

In late 1961, on the eve of publication of *SSR*, Kuhn finished his dazzling paper on the "Historical structure of scientific discovery." The core ideas in it predate his famous paper on energy conservation of 1957 (Kuhn 1977a, Preface, pp. xvi–xvii)—so the paper reflects very early theoretical ideas prior to *SSR*.[29]

---

[28] This view appeared early in the SSK tradition: Jerry Ravetz (1971) articulated Kuhn at length in this way; I later suggested such an SSK articulation of Kuhn (Schuster 1979). And, of course, this is what Barnes (1982) did so brilliantly. Barnes also devoted a chapter to exposition of this emerging post-Kuhnian conception of discovery, which per force entailed erosion of the stark normal versus revolutionary science dichotomy.

[29] The paper in question (Kuhn 1977g) is the first in the "meta-history" part of *The Essential Tension*.

What interested Kuhn was the idea that "significant discoveries" are not simple "events" in which a new fact or law is slotted cumulatively into a growing edifice of scientific knowledge. Significant discoveries arise from complex historical processes, and they ecologically alter the structure of knowledge through which they are produced, rather than simply adding to it. First, a striking difficulty must stand out against a well-developed framework of theory, theory-guided technique, and expectation. "Anomalies," Kuhn was already writing:

> [...] do not emerge from the normal course of scientific research until both instruments and concepts have developed sufficiently to make their emergence likely and to make the anomaly which results recognizable as a violation of expectation. (Kuhn 1977g, pp. 173–174)

There follows a struggle—or negotiation as we would say—both theoretical and experimental, to render the anomaly lawlike and to fit it into the accepted categories of explanation. This work issues in what we might term a "feedback" effect upon the web of techniques and concepts against which the anomaly arose (Kuhn 1977g, p. 175ff). This is the process and the product known colloquially as "discovery." The upshot can be part of an "upheaval" of established "theory and practice," such as the discovery of oxygen; or it can be subtle, like the effect of the discovery of Uranus upon the expectation that similar patterns of anomaly would henceforth best be handled by postulating additional planets.

Kuhn here saw significant discovery as a theory-bound and theory-altering process involving subtle or not so subtle ecological renegotiations of the preceding frame of concept and practice. He came very close to what was later taken as the dynamic, and creative, nature of normal science, characteristic of post-Kuhnian thinking. What Kuhn's vision lacked was the later sense of the competitive character of the normal scientific goings-on, and, *a fortiori*, the insight that what one needs to study are the organizational dynamics and micro-politics of the expert field. Kuhn never realized the potential of the notion of significant discovery—and it was precisely his work on revolutionary change, growing from normal science as puzzle-solving and "mopping up," that occluded and marginalized it.[30] Significant discovery cuts across the black and white categorization of normal and revolution-

[30] We see this happening in *SSR*, especially Chapter 6 'Anomaly and the Emergence of Scientific Discoveries." Here an elaborated version of Kuhn's (1977g) account of discovery is offered [mainly] as a moment in the account of anomaly and the emergence of new paradigms. He offers the oxygen/phlogiston case, as in Kuhn (1977g), but now of course it subserves a model of revolution; he offers the case of the Leyden jar, which subserves an account of emergence of a "first paradigm" in electrical science [cf. Sects. 19.10 and 19.11], and he offers the case of Roentgen and x-rays, an example where paradigm change was not immediately in the offing. There is a vestige of the original sense of discovery at Kuhn (1970, pp. 52–53): "[The process of discovery] then continues with a more or less extended exploration of the area of anomaly. And it closes only when the paradigm theory has been adjusted so that the anomalous has become the expected." Readers concentrating on the stock Kuhnian theory of revolution may well read right through that passage, as I did many times, before I reread it in the light of Kuhn (1977g) and the emergence of the SSK approach to discovery. It is then clear that Kuhn is slipping back toward a view of potentially creative normal science. But the opportunity is lost as the chapter and book flow on with the model of revolution (the next chapter is titled "Crisis and the Emergence of Scientific Theories" and the one after that "The Response to Crisis").

ary science. Kuhn never seriously thought through the idea that scientific actors are skilled interpreters, negotiators, and indeed hermeneuts, always involved in the competitive making and breaking of paradigm altering "significant discoveries."[31] In short, Kuhn had his embryonic conception of significant discovery before he had worked out most of *SSR*. But *SSR*—very much influenced by and reacting to Koyré's historiography of ruptures played out under determining metaphysical umbrellas—marginalized significant discovery in favor of the interplay of excessively routine normal science and excessively ruptural revolutionary science.[32]

19.7 Kuhn's Mapping: The Scientific Revolution, Classical and Baconian Sciences

We have surveyed Kuhn's relation to Koyré on the issue of the "motor" of practice in the sciences and reflected on what happened to Kuhn's mechanics of normal science in the hands of early SSK, observing the irony that Kuhn, had he been less Koyréan, might have come closer to the post-Kuhnian position.[33] We now turn to the mapping side Kuhn's early critical history project. To the extent that Koyré left a map at all, it was rather unexplicated, concentrated on the Scientific Revolution, and consisted in locating three temporally splayed sub-revolutions: the Copernican in astronomy, the Galilean in classical mechanics, and the final Newtonian synthesis and transformation. These played out against the background of a "nonscientific" Scholastic Aristotelianism. It was a map of high points, with little attention paid to the interrelations, temporal and cognitive, among the three towering moments, offering no continuous narrative of change and transformation.

In contrast, Kuhn stepped in with a detailed and challenging map of the period and beyond. Kuhn's map resembled in detail none of the other contemporary sketches of "the Scientific Revolution," in part because it was a map meant to deploy and further articulate his model of tradition dynamics. Thus, it amounted to a brilliant correction of Koyré. However, it also harbored serious problems: it was hindered by lack of mobilization of a notion of significant discovery and creative normal science, by failure to understand what natural philosophy was, and by ambiguities and hesitations about the origins of new disciplines and the nature of initial

[31] Kuhn lacked the interest in micro-sociology or to be precise Schutzian phenomenological sociology, to go further with this. Many of the early SSK scholars drew upon Schutz in particular (Schutz 1970; Schutz and Luckmann 1973). Kuhn always seemed devoted to Parsonian/Mertonian structural/functional, large-scale normative and consensual views of groups and institutions.

[32] Moreover, late in his career, after years of rejecting the work of SSK, Kuhn wrote a piece that again seems to return to that earlier point of accepting what had been his own proto-SSK position. He reflects on the intervening development, but then leaves his position up in the air, not cashing out the conclusions, such as others had already done by the early 1980s! (Kuhn 2000).

[33] This is not to say that SSK devotees were broadly committed to developing general historical models of motors and applying them to large mappings and narratives in the history of scientific practice. Barnes and Shapin were, but many were not.

paradigm-founding achievements. As several of Kuhn's essays show, he saw the Scientific Revolution as having advanced on two loosely connected fronts (Kuhn 1977b, f, pp. 136–137; 1977h, pp. 213–221).

1. Radical change in the preexisting "classical sciences," such as the mathematically oriented ones of geometrical astronomy, optics, mathematics, and the study of motion, as well as in the "biomedical" ones of medical theory/physiology and anatomy
2. The initial development of some experimentally oriented "Baconian sciences," the fields that, in *SSR*, mature to their first paradigms at various stages in the eighteenth century: electricity and magnetism, optics (from Newton), heat theory, and chemistry

We shall examine two disciplinary case studies in this mapping: in Sect. 19.8, the Copernican Revolution from (1) and, in Sect. 19.10, case of the birth of electrical science from (2).

19.8 Mapping Test Case I: The Copernican Revolution

Let's consider a potential mapping in the area of Kuhn's classical mathematical sciences—the case of the Copernican Revolution. I am not talking about the pre-*SSR* story in Kuhn (1957). We are dealing strictly with the mapping inherent in *SSR*, a revolutionary change: the Copernican system defeated and displaced the Ptolemaic system. If we take *SSR* seriously, we must envision a showdown between two set-piece, finished and "incommensurable" paradigms in competition, backed by their respective teams.[34] This conforms to popular and even professional readings of Kuhn, but it is not a picture of Copernicanism supported by the slightest reflection about the debates among the key players. Nor is it a view that comports with the findings of SSK about the concept of significant, negotiated discoveries within a continuous tradition of practice. The following four paragraphs sketch the way I approach a correction of Kuhn in undergraduate teaching, so that Kuhn can be better understood, even as students can move beyond Kuhn in terms of SSK understandings and historiographical sophistication.[35]

One begins by realizing that there never was a single, agreed Copernican system—not one to be fleshed out and not one to be worked toward. What we have is a

[34] Here I take "incommensurability" to mean "*there is no single, agreed, and overriding criterion of goodness accepted by both sides*," rather than "*there are no criteria at all available to the two sides*," thereby defusing philosophical anxieties and allowing us to get on with historiographical discussion. On how to interpret "incommensurability" in Kuhnian theory without falling into complete irrationality, see Schuster 1995a, Chapter 16, pp. 164–165; 2013b, pp. 304–305. On the "competing teams" conceit and a figural representation thereof, see Schuster 1995a, Chapter 25, pp. 236–237, and figure 4, p. 240; Schuster 2013b, pp. 482–483.

[35] Schuster 1995a, Chapter 25, p. 237–238, and figures 5–7, pp. 240–241; Schuster 2013b, pp. 484–486, and figures 26.5–26.7.

ramifying series of variants—conflicting and in part contradictory variants. Copernicus' Copernicanism is not the Copernicanism of any other Copernican and so on—down to Newton. Even within the group that strict *SSR* theory would account a Copernican paradigm "team," what we see are competing bids for individual hegemony. Each player fashioned a significantly different claim at the coalface of astronomical debate, depending upon his judgment of the state of play; his personal cognitive, technical and material resources, skills and investments, as well as his judgment of how best to defeat, co-opt or marginalize competing versions. Kepler and Galileo did not simply accept some straitjacket offered by Copernicus in 1543.[36] Galileo did not even practice mathematical astronomy; and his version of Copernicanism is not a repeat of Copernicus' theory. Nor did Kepler march shoulder to shoulder with Galileo as a straitjacketed follower of Copernicus. His version of Copernicus did not parallel Galileo's version — in fact their versions were in competition. These men were creative players and negotiators, battling to get the best advantage for themselves, and so they had consequentially different versions of "Copernicanism."

For a while—1590 to 1630—the most successful move proved to have been played by Tycho Brahe. He was not a pure Ptolemaist, and he was not a Copernican at the level of self-accounting. Rather, he was a clever professional negotiator, saying in effect that he wanted to preserve as much as possible of the basic Aristotelian system while accommodating, indeed co-opting, the best parts of the Copernican bid —especially Copernicus' claimed finding of several "harmonies of cosmic structure" that are used to support the physical truth of the system. The Tychonic system is in a way both geocentric and heliocentric and houses Copernicus' harmonies of cosmic structure with equal mathematical ease. Tycho did not thereby produce a timid or unimaginative compromise, but rather a brilliantly constructed gambit. One can view him as the most radical Ptolemaist of the day or the most conservative Copernican. He himself, of course, preferred the idea that his was a significantly different, third, and true theory. But that does not mean Tycho had a third, straightjacketed paradigm, perhaps missed in *SSR*! Tycho's bid was within the same field of discourse as Copernicus', and we can further tie his claims back to that evolving field. After all, Kepler would go out on a limb for elliptical planetary orbits, because he preferred to stay within the improved error limits established by Tycho's data. Kepler and Tycho were maneuvering in a common, evolving field.[37] And the same point applies to the "revolutionary" Copernicus: just as Tycho was in part a Copernican astronomer, so Copernicus was almost entirely a Ptolemaic astronomer. Almost everything that Copernicus did was in that tradition, except for the uncovering of the cosmic harmonies. Copernicus can be seen as a radical version of a Ptolemaic astronomer.[38]

So, there were no opposing paradigms with backing teams—only a historical process of negotiation, revision, and alteration in one tradition and field of competition.

---

[36] Schuster 1995a, Chapter 25, p. 237, and figure 5, p. 240; Schuster 2013b, p. 484 (figure 26.5).

[37] Schuster 1995a, Chapter 25, pp. 237–238, figure 6, p. 240; Schuster 2013b, p. 485 (figure 26.6).

[38] Schuster 1995a, Chapter 25, p. 238, figure 7, p. 241; Schuster 2013b, p. 486 (figures 26.7).

The only important thing to understand is the process of bidding, counter-bidding, and negotiating in attempts to establish the longevity of one's own claims. There was no emergent essence "Copernicanism," no final goal given in advance—the entire history displays a dynamic process within one complexly evolving tradition.

Thus, one can achieve a revision of the Koyréan Kuhn by going more deeply into the issue of the nature of a tradition and its dynamics and by employing the "lost" Kuhnian concept of "significant discovery." The remaining issue is whether so far we have conceptualized the tradition of astronomical practice in historically acceptable terms, thus paving the way for an improved mapping. And the answer is "not yet." There is still more to be said, which again bears directly on Kuhn's model, his mapping, and his deeply seated Koyréan orientations: this is the problem of the historical category of natural philosophy.

19.9 Forming the Category of Natural Philosophy, Modeling Its Dynamics, and Re-mapping the Copernican Revolution

Kuhn, like Koyré before him, failed to recognize that in order to understand the Scientific Revolution, one must take serious historical cognizance of the then competing varieties of natural philosophizing. He endemically ignored the role in the history of the sciences of the great families of natural philosophies—of mechanism, neo-Platonism, and Aristotelianism, within which families' particular variants flourished and competed. Post-Kuhnians have come to realize that conflict among instances of such systems defined the rhythm and moments in the Scientific Revolution.[39] But, according to Kuhn, such philosophies of nature variously provided only the "metaphysical" elements of the sciences—"new intellectual ingredients," as he calls them, of the now being revolutionized classical mathematical sciences and of the about to emerge Baconian experimental sciences (Kuhn 1977c, p. 53). Kuhn occluded the historical problem of natural philosophy and its varieties within his own neo-Koyréan conception of multiple and various metaphysical umbrellas for the various existing and nascent sciences. He thereby missed a major element in any mapping of the history of the sciences in the period, including the Copernican Revolution.

---

[39] I first emphasized this point in Schuster (1990) and Schuster and Watchirs (1990). Much earlier Robert Lenoble (Lenoble 1943) put the conflict of varieties of natural philosophy onto the map of the Scientific Revolution. Kuhn owned a first edition of this work—as I learned when I borrowed his copy, since it was not in Princeton's Firestone Library. Lenoble was followed in the 1960s by P. M Rattansi (Rattansi 1964). Later Easlea (Easlea 1980) and Ravetz (Ravetz 1975) tried to popularize this view. My early work followed from these initiatives and from discussions with my then Cambridge colleague Andrew Cunningham, circa 1978–1979. Later attempts to delineate the category of natural philosophy include Andrew Cunningham (Cunningham 1988, 1991), Cunningham and Williams (Cunningham and Williams 1993), Peter Dear (Dear 2001), and Peter Harrison (Harrison 2000, 2002).

In recent work, I have argued that one must go beyond even the conception of jostling instances of the general types of natural philosophizing. One must look at natural philosophizing as an institution and cultural enterprise (Schuster 2013a, pp. 10–13, pp. 37–70, pp. 77–88, 2013c, pp. 21–28). That is, one must appreciate that natural philosophy was not just the Scholastic Aristotelianism of the universities, so that the eclipse of Aristotelianism did not mark the death of natural philosophy per se. It was an entire subculture and field of contestation. When one "natural philosophized," one tried systematically to explain the nature of matter, the cosmological structuring of that matter, the principles of causation, and the techniques for acquiring or justifying such natural knowledge. All natural philosophers and natural philosophies constituted one subculture in dynamic process over time. Early modern natural philosophers learned the rules of natural philosophizing at university while studying hegemonic neo-Scholastic Aristotelianism. Even alternative systems—neo-Platonic, chemical, mechanistic, and later Newtonian—followed the rules of this game. The Scientific Revolution, in its most turbulent phase, in the early and mid-seventeenth century, was a set of transformations, a civil war, inside the seething, contested culture of natural philosophizing. That culture then continued to evolve under internal contestation, and external drivers, and eventually fragmented into more modern-looking, science-like disciplines and domains, plural, over a period of 150 years from 1650 (Schuster and Taylor 1997).

Along with the post-Kuhnian study of early modern natural philosophy has come attention to those disciplines then thought to be subordinate to it, such as the traditional "mixed mathematical sciences" of hydrostatics, statics, geometrical optics, geometrical astronomy, and harmonics. These of course are precisely Kuhn's classical mathematical sciences. If we are to understand Kuhn's claimed revolutions in these domains, including the Copernican Revolution, we need to grasp the relations between the dynamics of these fields and that of the supervening field of natural philosophy. Strict Aristotelians insisted that the mixed mathematical sciences were of instrumental value only: they could not touch upon questions of matter and cause, natural philosophical categories. But, this "rule" was increasingly bent and questioned, and nowhere more importantly than in the debate over realist Copernicanism.

In my model of the game of natural philosophizing, I use the term "articulation" to denote moves made by a natural philosopher to bring at least the higher theoretical levels of a subordinate science, such as geometrical optics or astronomy, into coordination with the matter, cause, and cosmic structure elements of his own natural philosophy (Schuster 2013a, pp. 42–43, pp. 51–55, 2013c, pp. 21–23). In the case of geometrical astronomy, virtually all natural philosophical contenders tried to articulate a preferred version of astronomy—Ptolemaic, Tychonic, or Copernican—to their preferred natural philosophy.[40] Such an articulation amounted

[40] It is important to note that such articulation existed in the relation even of Aristotelianism to Ptolemaic astronomy, regardless of what the strictest Aristotelians might have said. The fine details and elaborate geometrical tools of Ptolemaic astronomy fell outside any plausible realistic interpretation, offered merely appearance-saving geometrical models, and could not provide natural philosophical explanations in terms of matter and cause. However, the fundamental concepts of

to specifying the cosmological element of the natural philosophy in question. But of course, only part of the selected version of astronomical theory was "articulated upon," since no natural philosophy ever presumed to explain the full scope of astronomical practice and the elaborate geometrical detail of planetary models.

The macro history of astronomy in the period depends upon the concatenation of the ways it was articulated upon, and thus practiced, under competing natural philosophical claims. This allows further refinement of our previous correction of Kuhn on the Copernican revolution. Not only was astronomy, as we have just seen, a single agonistic tradition of practice—rather than a site of paradigm versus paradigm combat—but we can now affirm that astronomy existed in complex articulation relations with that other encompassing discipline and field of contention, natural philosophy. Hence, the Copernican debate was not about astronomy alone nor about astronomy on the one hand and a separate domain of "world views" on the other. It was precisely a battle about articulations of varieties of natural philosophy onto Copernicanism, or not. It was about the direct challenge in the field of natural philosophizing posed by realist Copernicanism—because realist Copernicanism only existed in articulations of non-Aristotelian natural philosophies onto Copernicanism. This is why the Copernican debate was a "hot spot" in the desperate natural philosophical struggle of the early seventeenth century, and in turn it helped fuel that crisis.[41] The supporters of *realist* Copernicanism needed to adduce a framework of non-Aristotelian natural philosophy, a new theory of matter and cause, adequate to explaining the heliocentric cosmos. The entire late sixteenth- and early seventeenth-century debate over realist Copernicanism (culminating in the emergence inside Kepler's and Descartes' respective philosophies of nature of a discourse of "celestial physics") was a phenomenon of competition at a now inflamed site within the natural philosophical field—no realist Copernicanism, no inflammation (Schuster 2013a; Schuster and Brody 2013). But why be a realist Copernican, unless you intend a quite radical overhaul of Aristotelian natural philosophy (and its rules) as such?

This sketch of an alternative mapping of Kuhn's terrain of the Copernican Revolution (as we must read the latter from *SSR*) begins to illustrate the price paid by Kuhn for not recognizing the role of systematic natural philosophies per se in shaping the contents and directions of the sciences, both existing and nascent.[42] Our case study of Kuhn on the Copernican Revolution also shows, on the one hand, the complex dialectic involved in his modifications of Koyré, which were executed while he was still conditioned by allegiance to Koyré, and, on the other hand, the vast distance separating our own best practice historiography, mapping, and notions

---

Ptolemaic astronomy were shaped by Aristotelian natural philosophy: the finite Earth-centered cosmos, the distinction between the celestial and the terrestrial realms, and the primacy of uniform circular motion.

[41] "Hot spot" like "articulation" is a term of art in my model of the dynamics natural philosophy. Schuster 2013a, Chapter 2, section 2.5.4.

[42] For example, Kuhn speaks of atomism as a "new intellectual ingredient" providing the metaphysics of the cluster of classical sciences.

of tradition dynamics, from those of the younger Kuhn, post-Koyréan historical theorist. Finally, note that all these matters—emergent from our approach to the younger Kuhn as a historical theorist—are far from those usually canvassed in philosophical explications and criticisms of Kuhn.

19.10 Mapping Test Case II: Kuhn on the Rise of New Experimental Sciences

We can now turn to the second branch of Kuhn's map, dealing with the emergence of new experimental sciences in the wake of the seventeenth century Scientific Revolution. Here again, Kuhn filled a gap left by Koyré's historiography in a manner partially shaped by his own Koyréan leanings. Additionally, Kuhn's effort left problems and ambiguities, some of which he himself could have detected. Most importantly, he left two different accounts of how the new sciences arose.

19.10.1 Kuhn's Version One: Baconian Sciences Born in Ruptural Emergence of a First Paradigm

In *SSR*, Kuhn attacked the problem of the rise of new Baconian experimental sciences by invoking what may be termed a "two-place" historiography. Kuhn identifies the genesis of an experimental science with the emergence of a first "paradigm," against a field of "pre-paradigmatic" enterprises in the relevant domain of research. He seeks points of rupture, where such sciences are born, placing such events in the context of the preexistence of what is "not science" (Schuster and Watchirs 1990, pp. 3–7). At the beginning of *SSR*, Kuhn deploys the case of electrostatics to illustrate this theory of paradigm formation (Kuhn 1970, pp. 13–15). The pre-paradigm stage of electrical research was dominated by competing "schools," with differing "metaphysical" commitments (e.g., Cartesian, Newtonian) and differing preferred problems and starting points for surveying the burgeoning field of known electrical phenomena (e.g., stressing attraction/repulsion phenomena first and foremost or the phenomena of conduction, flow, and sparking). For Kuhn, the founder of the first real electrical paradigm was Benjamin Franklin, initially an adherent of the pre-paradigm school concerned mainly with conduction effects. This school "tended to speak of electricity as a 'fluid' that could run through conductors, rather than as an effluvium that emanated from non-conductors." They could deal with simple conduction effects, but not very well with the known attraction/repulsion effects. The work of Franklin and his successors, however, issued in a successful first paradigm:

> […] that could account with something like equal facility for very nearly all these effects and therefore (provided) a subsequent generation of electricians with a common paradigm for its research. (Kuhn 1970, p. 15)

The "exemplar" of the Franklinists, their exemplary problem solution, out of which their full paradigm was articulated, was their "successful" explanation of the Leyden jar, an early capacitor. According to Kuhn, the basis of Franklin's exemplary, paradigm-forming problem solution was the possibility of conceptualizing electrostatic induction and hence the possibility of distinguishing between, on the one hand, the flow and conduction of electrical fluid and, on the other hand, its exertion (upon itself) of repulsion at a distance.

For Kuhn, these insights constituted the essence of the subsequent Franklinian paradigm, as he makes quite clear by conflating the establishment of Franklin's paradigm with the widespread comprehension and acceptance of the critical clarification of inductive effects (and of conduction versus distance effects): by packing these understandings of what we might term "the condenser as such" into the paradigm from the point of its "assimilation," Kuhn endows Franklin's paradigm with a destiny of unproblematically maturing, through articulation of its own pre-given conceptual and technical resources.[43] That is, once the rupture to this first genuinely scientific paradigm has occurred over against the field of pre-scientific/pre-paradigmatic "schools," a kind of epistemological predestination seems to hover around Franklin's paradigm. Assuming, as Kuhn does, that Franklin was quite clear about distinguishing induction from conduction, and, in general, flow effects of the electrical fluid from its distance effects, then Kuhnian historiography can gaze almost Whiggishly at the cumulative articulation of Franklin's paradigm. Kuhn sees this maturation and articulation of Franklin's paradigm as consisting essentially in the progressive mathematization of electrostatics later in the works of Coulomb, Cavendish, and Poisson. In neo-Koyréan fashion, Kuhn assumes that development from the ruptural moment of finding the "first paradigm" will consist in virtually preprogrammed moves, the central process being the all but inevitable drive to mathematicize the field (and so achieve in the experimental science in question the realization of the Koyréan goal of mathematization).[44]

19.10.2 Kuhn's Version Two: Baconian Sciences Born from Continuous Process (of "Scientificity")

Interestingly, in his later essay on "Mathematical Versus Experimental Traditions in the Development of Physical Science" (Kuhn 1977c), Kuhn provided from the same sources a model based more on continuity and process. This theoretical ambivalence and confusion can be traced back to Kuhn's grappling with his Koyréan

[43] Kuhn 1970, p. 21, pp. 28–29; 1977c, pp. 47–8; 1963, pp. 356–357. The latter paper, known as perhaps the most radical of Kuhn's historiographical efforts, was pointedly *not* included in Kuhn 1977a.

[44] I say "almost Whiggishly" because what we have here is a specifically Koyréan/Kuhnian variant of Whiggism. Kuhn and Koyré each rightly claimed that they did not practice the form of (inductivist and often "method-based") Whiggish narrative which they often criticized in others. Nevertheless, where Koyré had invoked a kind of destiny toward mature, comprehensive mathematization under his preferred ruling "Platonic" metaphysics, Kuhn played a Koyréan variation, with each first paradigm in an experimental field harboring a destiny toward mature and comprehensive mathematization.

heritage and with the demand to be more historically accurate. The upshot is a model quite at odds with the vision of normal science in *SSR*.[45]

Kuhn's continuity story of the Baconian sciences starts from the idea that in the Renaissance and seventeenth century, there existed in addition to the classical mathematical and biomedical sciences a heterogeneous collection of partially overlapping areas of what we should now term physical and chemical inquiry into the nature of matter, light, heat, magnetism, chemical reactions, and the like. Facts and experiments about such matters were not pursued within any of the existing, mature classical sciences, but were partially and variously the subject matter of alchemy, natural magic, and the crafts (Kuhn 1977c, pp. 32–33, p. 36, p. 46). From about 1650, these domains began to be reformed and recrystallized under the aegis of the triumphant rhetoric of Baconian method and experiment (Kuhn 1977c, pp. 41–46). The characteristic products of these emergent Baconian sciences were natural or experimental "histories," in which corpuscular-mechanism-guided experiment only loosely if at all and, correspondingly, corpuscular-mechanism was largely unaffected in detail by experimental outcomes.

According to Kuhn, the Baconian sciences, largely subsisting independently of the cluster of classical sciences, underwent a three-stage process of maturation over the next century and a half. Initially, from the late seventeenth century to well into the eighteenth century, the pattern of Baconian compilation of "histories" dominated. The Baconian sciences "remain underdeveloped," if we take the criterion of development to be "the possession of a body of consistent theory capable of producing refined predictions" (Kuhn 1977c, p. 47). But, toward the middle of the eighteenth century, more systematic forms of experiment developed in these fields, focused upon revealing sets of phenomena to which corpuscular and cognate explanatory concepts were increasingly applied. The theories in these domains remained mainly qualitative, but they could be matched up to specific experiments with degrees of conformity not seen earlier in the century (ibid.). This second stage therefore corresponds to that identified with the acquisition of first paradigms in the model of *SSR*. Finally, in the last third of the eighteenth century, some portions of these relatively mature fields became more quantitative and were subjected to sophisticated mathematical articulation (Kuhn 1977c, pp. 47–48).

In sum, Kuhn's second story of the Baconian sciences is a *continuity*-oriented narrative of the "pre- and post-paradigm" states of these sciences. His account veers away from the official categories and two-place historiography of *SSR*. Hence, Kuhn is assuming that these sciences really existed from their initial seventeenth-century points of crystallization and that, in general, there were less mature sciences in the pre-paradigm stage and more mature sciences in the post-paradigm stage. Kuhn's continuity, therefore, can only be a continuity of some kind of scientificity.

Evidence of Kuhn's difficulties with the pre-paradigm/paradigm distinction appears in his description (mentioned earlier) of the competing pre-paradigm schools of electricians, each characterized by its own metaphysical commitments and preferred domain of problems and phenomena (Kuhn 1970, pp. 14–15). This, of

[45] Material in the next five paragraphs derives from Schuster and Watchirs (1990).

course, makes the pre-paradigm schools seem very paradigm- or science-like, and it is a description holding for the pre-paradigm stages of other fields as well.[46] Replying to critics in his 1970 "Postscript" to the second edition of *SSR*, Kuhn responded in the spirit of his earlier ambivalent assertion that pre-paradigm researchers "were scientists, [but produced] something less than science" (Kuhn 1970, p. 13), arguing that the transition to maturity of a science "need not be associated with the first acquisition of a paradigm," for the members of the schools of the pre-paradigm period of a field "share the sorts of elements which I have collectively labeled a paradigm" (Kuhn 1970, p. 179). What changes, Kuhn now suggested, is not "the presence of a paradigm but rather its nature. Only after the change is normal puzzle-solving research possible" (ibidem).[47]

This certainly seems a distinction without much of a difference, when we recall Kuhn's views on the effectiveness and maturity of the fluid school of electricians even before Franklin.[48] Kuhn's difficulty in grounding his two-place historiography in the relevant pre-eighteenth-century facts led him in the direction of these vague intimations of "continuity." But there was a deeper cause of his problem. It had to do with the issue of proper historiographical categories. That is, the underlying cause of Kuhn's problem resides in his failure, again, to deploy a category of natural philosophy. We saw how the absence of this category—shaped by his Koyréan proclivities—vitiated much of his understanding of the Scientific Revolution. Now we shall see how that absence permitted (and demanded) his use of the term "Baconian sciences" to articulate the continuity that looms through his failure to enforce his own pre-paradigm/paradigm distinction.

19.11 Articulating the Category of Natural Philosophy: Rethinking the Motor and Map of the Emergent Experimental Sciences

Our argument in this section has three steps. First, recall that rather than taking seriously the historical reality of natural philosophizing as an institution and activity, Kuhn spoke of "intellectual ingredients," neo-Koyréan metaphysical groundings for

[46] Similarly, in discussing pre-paradigm schools in physical optics, Kuhn (1970, pp.12–13), even observes that they each had different respective "paradigmatic observations."

[47] He continues, "Many of the attributes of a developed science which I have associated with the acquisition of a paradigm I would therefore now discuss as consequences of the acquisition of the sort of paradigm that identifies challenging puzzles, supplies clues to their solution, and guarantees that the truly clever practitioner will succeed" (ibidem).

[48] In effect, Kuhn still needed a rupture, a transition to maturity, but he was left with two weak indicators of its supposed occurrence: relatively greater consensus—the end of interschool debate—apparently dependent upon relatively greater puzzle-defining and puzzle-solving power (which also instills confidence). But, given the paradigm-like virtues of the pre-paradigm schools, it is still not possible to see, in Kuhn's terms, why and how such points of superiority are made out and enforced upon the debating parties.

various scientific pursuits. In his continuity story of the maturation of the Baconian experimental domains between the late sixteenth and mid-eighteenth century, Kuhn granted varied roles to such "ingredients": Hermeticism promoted interest in electricity and magnetism; Paracelsianism and Helmontianism promoted the status of "chemistry"; and Baconianism, with its rhetoric of utility and of experiment sanitized of magic and alchemy, thereby effectively crystallized the experimental "sciences" and rendered them fit for the injection of the ingredient of corpuscular-mechanism at the hands of Boyle, Hooke, and others. Corpuscular-mechanism, as an ingredient, had equivocal effects upon the Baconian sciences: it completed the purge of magic, but its explanatory resources were only beneficial where they were "adapted" to the needs of "particular areas of experimental research" (Kuhn 1977c, p. 47, pp. 53–54). So, Kuhn shifted his "intellectual ingredients" around to help explain moments in the trajectory of maturation of *already existing* Baconian sciences.

The second step is to conceptualize how theoretical concepts are embodied within experimental hardware and expressed in their outputs.[49] The evolution of best practice in history and sociology of science since Kuhn indicates the way forward: we follow Gaston Bachelard's (Bachelard 1938, 1949) monumental insight that the phenomena or objects of inquiry of experimental sciences are manufactured in and by experimental hardware, because such hardware are themselves developed, used, and understood as materializations of conceptual structures—what he called *phénoménotechniques*. Bachelard insisted that embodied concepts can only be mathematical and noncontingent, but post-Kuhnian sociology of scientific knowledge scholars have stressed the fluidity and negotiability of meanings embodied in hardware (Collins 1975; Pinch 1985; Shapin 1982; Barnes 1982).This allows us to speak not only of couplings of *mathematical theory* and hardware, but of *discourse-hardware couples*, which are contingent products, outcomes of social processes of closure, and subject to possible renegotiation of the elements on either side of the couple.[50] This "softening" of Bachelard in recognition of continuous processes of construction and negotiation brings out the *historical* character of experimental practices and allows them to be related to our model of natural philosophical processes.

Third, one can ask the following: If early experimental scientists, or better natural philosophers, coupled discourse to hardware, where did the discourse come from?[51] The answer obviously is that in early experimentalism, the discourse at stake in materializations was of natural philosophical provenance. This is trivially obvious: couplings of hardware and discourse *without* use of natural philosophical

[49] Material on Bachelard in this paragraph evolved from findings first put forward in Schuster and Watchirs (1990, pp. 7–11, pp. 21–25). A fully articulated, pedagogical model of how thus to deal with the issue of the theory loading of experimental hardware is presented in my textbooks: Schuster 1995a, Chapter 14, pp. 145–146 and Schuster 2013b, pp. 269–77.

[50] All this goes well beyond the insights of Popper, Hanson, and then Kuhn about "theory loading" of experience and experiment. Arguably, it was Bachelard, not Kuhn or the others, who specifically stimulated the efflorescence of SSK inquiry into experiment.

[51] Material in the remainder of this section was first presented in Schuster and Watchirs 1990, pp. 14–29.

discourse were called crafts and practical arts. From the early seventeenth century, the natural philosophical field was increasingly characterized by an imperative to articulate competing natural philosophical utterances onto hardware, in order that the reports of the outputs of hardware could serve to support other natural philosophical claims.[52]

All this permits us to sketch how experimental sciences crystallized out of the dynamics of the field of experimental natural philosophy. The imperative to articulate competing natural philosophical utterances onto hardware exerted subtle pressure toward creating domains of relative specialization, that is, sets of couples increasingly articulated in terms of more localized and specific swathes of originally natural philosophical discourse: increasingly, clusters of hardware-discourse couples emerged within the natural philosophical field—constituting such experimental domains as electricity, magnetism, and heat. (Kuhn read this process toward domain formation as eruption of "first paradigms" in the experimental sciences.) As this happened, natural philosophers increasingly abandoned the aim of systematic natural philosophical discourse to the requirements of engaging the flow of practice in these narrower spaces. More local explanatory "dialects" emerged out of the originally wider natural philosophical field of discourse.[53] This model avoids Kuhn's penchant for ruptures, but also rationalizes his unconsummated intimation of continuity, by explaining the formation of experimental sciences out of the dynamics of the already existing natural philosophical culture. Discipline formation was an unintended process within the natural philosophical field and helped to fragment and eventually evaporate it.

Kuhn played himself into difficulties by failing to theorize natural philosophy as a historical category. He neglected to consider whether the dynamics of natural philosophy had anything to do with the Baconian sciences and their long gestation process to "full maturity." Kuhn had defined his pre-paradigm schools primarily in terms of metaphysical predilections—in my terms essentially natural philosophical commitments. Kuhn had trouble factoring such schools into his strict *SSR* model: on the one hand, he had to depict some schools as too "paradigm-like," while, on the other hand, he had to introduce into the schools a suitable degree of "Baconian" fact gathering, so that they would remain pre-paradigmatic. The model of domain and dialect formation cuts across Kuhn's tortured variety of "schools." When Kuhn sought science-making breakthroughs in one or another school, he was looking for the wrong thing in the wrong place. His alternative strategy of tracing the long-term maturation of the "Baconian sciences" also missed the mark, misconceiving the object continuously in play—experimental natural philosophy—and its peculiar dynamics. Slightly improving upon Koyré, by multiplying the possible metaphysical bases of various "scientific" traditions, meant continuing Koyré's avoidance of the presumably nonscientific and not important realm of natural philosophy.

---

[52] This aspect of the model being offered was developed in Schuster (2002, 2013a, c) and Schuster and Taylor (1997, pp. 519–24).

[53] For case of electro-statics up to, through and beyond Franklin, Schuster and Watchirs (1990, pp. 30–36).

19.12 Conclusion

We have analyzed the younger Kuhn as a "critical historian," who desired to understand the motor or dynamics of change in the sciences and to produce an initial map of how those changes had played out over time. As a critical historian, Kuhn was trying both to articulate sympathetically and revise radically what he took as Koyré's riding orders for the history of science profession. The younger Kuhn was the way he was as a historian largely because of his complex grappling with what he saw as the historiographical heritage of Koyré. It would have been easy simply to list the similarities and differences between Koyré and the younger Kuhn. But, the real issues come out when one looks—as we have—in greater depth at exactly what was going on in Kuhn's mapping and motor exercises. That allowed us to identify key areas—concerning discovery, experiment, and natural philosophy—where Koyré and Kuhn each underplayed or missed opportunities that were later picked up during early post-Kuhnian developments in sociology of knowledge and Scientific Revolution studies.

It is worth underscoring the "history-centric" nature of our inquiry and our findings. All the issues we have dealt with—Kuhn's relations to Koyré and the relations of post-Kuhnian thought to Kuhn himself—reside squarely in the realm of historiographical theorizing. We have not examined issues about Kuhn doing philosophy. Nor have we had to concern ourselves with philosophical evaluation of Kuhn's work. All the matters we have considered arose when a critical historian of science, the young Tom Kuhn, engaged in category and model formation and modification, endeavoring, as we said at the beginning, to be his own purveyor of key historiographical categories.

Of course, such critical history did not concern the older Kuhn. From the time of his being philosophically ambushed by Popper and the Lakatosians at the Bedford College Conference in 1965 (Lakatos and Musgrave 1970), Kuhn began to retreat from creative thought about historical process, slowly sinking into a maelstrom of unending and unwinnable debate with philosophers about rationality and method. Kuhn's occluded and aborted trajectory in critical history was taken up by younger historians and sociologists of science, who often knew their philosophy of science, but who wished to advance post-Kuhnian critical history of science/sociology of science without first having to try to legitimate themselves in philosophical circles, as the older Kuhn increasingly did.

References

Bachelard G (1949) Le rationalisme appliqué. Presses Universitaires de France, Paris.
Bachelard G (1938) La Formation de l'Esprit Scientifique. Vrin, Paris.
Barnes B (1972) Sociological Explanation and Natural Science: A Kuhnian Reappraisal. Archives Européennes de sociologie 13:373–391.
Barnes B (1974) Scientific Knowledge and Sociological Theory. Routledge & Kegan, London.

Barnes B (1982) T. S. Kuhn and Social Science. MacMillan, London.
Collins HM (1975) The Seven Sexes: A Study in the Sociology of a Phenomenon. Sociology 9:205–224.
Cunningham A (1988) Getting the Game Right: Some Plain Words on the Identity and Invention of Science. Studies in History and Philosophy of Science 19:365–389.
Cunningham A (1991) How the Principia got its Name. Or, Taking Natural Philosophy Seriously. History of Science 24:377–392.
Cunningham A, Williams P (1993) De–Centring the 'Big Picture': The Origins of Modern Science and the Modern Origins of Science. British Journal for the History of Science 26:407–432.
Dear P (2001) Religion, Science and Natural Philosophy: Thoughts on Cunningham's Thesis. Studies in History and Philosophy of Science 32:377–386.
Easlea B (1980) Witch–Hunting, Magic and the New Philosophy: An Introduction to the Debates of the Scientific Revolution 1450–1750. Harvester Press, Sussex.
Hanson NR (1958) Patterns of Discovery. The Cambridge University Press, Cambridge.
Harrison P (2000) The Influence of Cartesian Cosmology in England. In Gaukroger S, Schuster J, Sutton J (eds). Descartes' Natural Philosophy. Routledge, London, pp. 168–192.
Harrison P (2002) Voluntarism and Early Modern Science. History of Science 40:63–89.
Jacobs S (2009) Thomas Kuhn's Memory. Intellectual History Review 19/1:83–101.
King MD (1971) Reasons, Tradition and the Progressiveness of Science. History and Theory 10:3–31.
Koyré A (1939) Études Galiléennes. Hermann, Paris. English. [Koyré A (1978) Galileo Studies. The Harvester–Hassocks, Sussex].
Koyré A (1956) The Origins of Modern Science. Diogenes 16:1–22.
Koyré A (1955) A Documentary History of the Problem of Fall from Kepler to Newton. Transactions of the American Philosophical Society 45/4:329–395.
Koyré A (1973). The Astronomical Revolution. Methuen, London. [From: original French Edition 1961]
Kuhn TS (1957) The Copernican Revolution. Vintage, New York.
Kuhn TS (1963) The Function of Dogma in Scientific Research. In Crombie AC (ed). Scientific Change. Heineman, London, pp. 347–369.
Kuhn TS (1970) The Structure of Scientific Revolutions, 2nd edition. University of Chicago Press, Chicago.
Kuhn TS (1977a) The Essential Tension: Selected Studies in Scientific Tradition and Change. University of Chicago Press, Chicago.
Kuhn TS (1977b) The Relations between the History and the Philosophy of Science. In Kuhn 1977a, pp. 3–20.
Kuhn TS (1977c) Mathematical versus Experimental Traditions in the Development of Physical Science. In Kuhn 1977a, pp. 31–65.
Kuhn TS (1977d) Energy Conservation as an Example of Simultaneous Discovery. In Kuhn 1977a, pp. 66–104.
Kuhn TS (1977e) The History of Science. In Kuhn 1977a, pp. 105–126.
Kuhn TS (1977f) The Relations between History and the History of Science. In Kuhn 1977a, pp. 127–161.
Kuhn TS (1977g) The Historical Structure of Scientific Discovery. In Kuhn 1977a, pp. 165–177.
Kuhn TS (1977h) The Function of Measurement in Modern Physical Science. In Kuhn 1977a, pp. 178–224.
Kuhn TS (1977i) A Function for Thought Experiments. In Kuhn 1977a, pp. 240–265.
Kuhn TS (2000) The Trouble with the Historical Philosophy of Science. In Conant J, Haugeland J (eds). The Road Since Structure. Thomas S. Kuhn. Philosophical Essays, 1970–1993, with an Autobiographical Interview. The University of Chicago Press, Chicago, pp. 105–120.
Lakatos I, Musgrave A (1970) (eds) Criticism and the Growth of Knowledge. Cambridge University Press, Cambridge.
Lenoble R (1943) Mersenne ou la naissance du mécanisme. Vrin, Paris.

Merton RK (1970) Science, Technology and Society in Seventeenth Century England. Harper & Row, New York. [Original publication in: Merton RK 1938, Science, technology, and society in seventeenth-century England. Osiris 4:360–632].
Mulkay M (1979) Science and the Sociology of Knowledge. Allen & Unwin, London.
Pinch T (1985) Towards an Analysis of Scientific Observation: the Externality and Evidential Significance of Observational Reports in Physics. Social Studies of Science 15:3–36.
Polanyi M (1958) Personal Knowledge: Toward a Post-Critical Philosophy. University of Chicago Press, Chicago.
Popper KR (1959) The Logic of Scientific Discovery. Basic Books, London.
Rabb TK (1975) The Struggle for Stability in Early Modern Europe. Oxford University Press, New York.
Rattansi PM (1964) The Helmontian-Galenist Controversy in Seventeenth Century England. Ambix 12:1–23.
Ravetz JR (1971) Scientific Knowledge and its Social Problems. Oxford University Press, Oxford.
Ravetz JR (1975) Entry: Science, History of. Encyclopedia Britannica. 15th edition. Vol. 16. pp. 366–372.
Schuster JA (1979) Kuhn and Lakatos Revisited. British Journal for the History of Science 12:301–317.
Schuster JA (1986) Cartesian Method as Mythic Speech: A Diachronic and Structural Analysis. In Schuster JA, Yeo RR (eds). The Politics and Rhetoric of Scientific Method. Reidel, Dordrecht, pp. 33–95.
Schuster JA (1990) The Scientific Revolution. In Olby RC, Cantor GN, Christie JRR, Hodge MJS (eds). The Companion to the History of Modern Science. Routledge, London, pp. 217–242.
Schuster JA (1995a) The Scientific Revolution: An Introduction to the History and Philosophy of Science. Open Learning Australia. Located at http://descartes-agonistes.com/
Schuster JA (1995b) An Introduction to the History and Social Studies of Science. Open Learning Australia. Located at http://descartes-agonistes.com/
Schuster JA (2000) Internalist and Externalist Historiographies of the Scientific Revolution. In Applebaum W (ed). Encyclopedia of the Scientific Revolution. Garland Publishing, New York.
Schuster JA (2002) L'Aristotelismo e le sue Alternative. In Garber B (ed). La Rivoluzione Scientifica. Instituto della Enciclopedia Italiana, Roma, pp. 337–357.
Schuster JA (2013a) Descartes–Agonistes: Physico–Mathematics, Method and Corpuscular–Mechanism, 1619–1633. Springer, Dordrecht.
Schuster JA (2013b) 科学革命: 科学史与科学哲学导论. (上海科学技术出版社, 上海) [The Scientific Revolution: Introduction to the History & Philosophy of Science; Translated by An Weifu]. Shanghai Scientific and Technological Education Publishing, Shanghai.
Schuster JA (2013c) What was the relation of Baroque Culture to the Trajectory of Early Modern Natural Philosophy. In Gal O, Chen-Morris R (eds). Science in the Age of Baroque. Archives internationales d'histoire des idées 208:13–45.
Schuster JA, Watchirs G (1990) Natural Philosophy, Experiment and Discourse: Beyond the Kuhn/Bachelard Problematic. In Le Grande HE (ed). Experimental Inquiries: Historical, Philosophical and Social Studies of Experimentation in Science. Kluwer, Dordrecht, pp. 1–47.
Schuster JA, Taylor ABH (1997) Blind Trust: The Gentlemanly Origins of Experimental Science. Social Studies of Science 27:503–536.
Schuster JA, Brody J (2013) Descartes and Sunspots: Matters of Fact and Systematising Strategies in the Principia Philosophiae, Annals of Science 70/1:1–45.
Schutz A (1970) Reflections on the Problem of Relevance. Yale University Press, New Haven.
Schutz A, Luckmann T (1973) The Structures of the Life-World. Heinemann, London.
Shapin S (1982) History of Science and its Sociological Reconstructions. History of Science 20:157–211.
Shapin S (1992) Discipline and Bounding: The History and Sociology of Science As Seen Through the Externalism-Internalism Debate. History of Science 30:333–369.
Stone L (1972) The Causes of the English Revolution 1529–1642. Harper and Row, New York.

Chapter 20
Alexandre Koyré and the Traditional Interpretation of the Anthropological Consequences of the Copernican Revolution

Jean-François Stoffel

Abstract Koyré's work being situated, chronologically speaking, between the end of the definitive structuring phase (1925) for the traditional interpretation of the Copernican Revolution and the appearance of its first and most accomplished lines of questioning (1969), we felt it appropriate to examine to which extent his work endorses this interpretation or, on the contrary, to which extent it makes way for its reassessment. An analysis of his interpretation of the Copernican Revolution in itself raises a number of issues which, although sparse and often obscure, attest to a certain distancing with respect to the traditional interpretation without, however, leading one to consider the need for an entire reassessment of it. Thus, even though Koyré, being aware of the axiological reversal that occurred between geocentrism and heliocentrism, is able to appreciate all the ambivalence of the geocentric position of our abode (the best and the worst), he however fails to perceive the full importance of the planetary centrality that geocentrism already bestows upon the Sun, just as he overlooks the consequences resulting from this axiological inversion, namely, the availability of various interpretation frameworks for the anthropological consequences arising from the transition from geocentrism to heliocentrism. While his analysis of the Copernican Revolution therefore provides us with ample incentive to question the pertinence of the traditional interpretation, his thesis on "the spiritual revolution of the 17th century" seems, however, inclined to uphold the merits of this interpretation. Nevertheless, given that the first consequence which he draws from it, namely, the divorce between the world of science and the world of life, seems to us to be philosophically plausible albeit of little historical foundation and that the second, the retreat of the Divine, seems, on the contrary, historically sound although neither philosophically nor theologically credible, we thought it more reasonable to conclude that the Koyrean work does not serve to confirm the traditional interpretation in itself but only its suitability with respect to the thinkers

J.-F. Stoffel (✉)
Haute École Louvain en Hainaut, 136 rue Trieu Kaisin, 6061 Montignies-sur-Sambre, Belgium
e-mail: jfstoffel@skynet.be

R. Pisano et al. (eds.), *Hypotheses and Perspectives in the History and Philosophy of Science*, DOI 10.1007/978-3-319-61712-1_20

of the twentieth century, amongst which Koyré is undoubtedly an emblematic figure.

Keywords Anthropocentrism • Copernican revolution • Destruction of the cosmos • Divorce between science and life • Geocentrism • Heliocentrism • Retreat of the divine • Sun • Topography

20.1 Introduction

According to traditional interpretation, the anthropological consequences of the Copernican Revolution are manifest: Man was dethroned and banished from his privileged position at the heart of the Cosmos, to find himself flung into an unknown corner of the universe, on an earth henceforth considered as merely another planet devoid of distinguishing features. This traditional interpretation, having first appeared in the seventeenth century, and for which the main idea has just been iterated, became increasingly structured until adopting its final form between 1835 and 1925, largely under the influence of positivism, Darwinism and Freudianism (Stoffel 2012).

Alexandre Koyré made his appearance on the scene of scientific publications in 1912 (Koyré 1912); therefore, all that remained was for this interpretation to be endorsed a few years later by Sigmund Freud (1856–1939)—firstly in 1917[1] and again in 1925 (Freud 1992 [1925], 134–135)—for it to become definitively established, along with the introduction of the famous lineage "Copernicus–Darwin–Freud", and to take on the physiognomy that is still recognised throughout the world today. Born in 1892, so towards the end of the structuring phase for this traditional interpretation, Koyré, who died in 1964, remained unaware of his most accomplished lines of questioning. Having decided to limit ourselves to those works dealing explicitly with this interpretation—and therefore having had to put aside such important works as those of Arthur O. Lovejoy (1873–1962) (Lovejoy 1936), which Koyré knew, or those of Hans Blumenberg (1920–1996) (Blumenberg 1975) and Fernand Hallyn (1945–2009) (Hallyn 1987), which he didn't know—we were able to identify similar lines of questioning in the 1969 article by Claude Savary (Savary 1969), which sparked a controversy with Raymond Petit (Montpetit 1970 and Savary 1970); secondly, from 1990, with the forceful and persistent denunciation by Rémi Brague (Brague 1990, 1994, 1997, 2001, 2008), particularly with respect to the falsity of that part of the interpretation pertaining specifically to geocentrism; thirdly, if we may be so bold as to add, with our own work, which from 1998 (Stoffel 1998, 2001, 2002, 2005, 2012) strove to bring to light the inadequacies of this interpretation, both with respect to its appreciation of geocentrism as well as heliocentrism; from 2001 with the popular articles of Dennis R. Danielson (Danielson 2001,

[1] Cf. Freud 1996 [1917], 46–47, and, the same year but more briefly, Freud 2000 1917, 295.

2009) that drew attention to this historiographical cliché; and, lastly, in 2014 with the article by Daniel Špelda (Špelda 2014), which pursues the criticism of the geocentric part of this interpretation. Koyré's work, therefore, positions itself broadly speaking within this period during which the traditional interpretation, henceforth definitively determined, established itself as a certainty, and although it was no longer as ideologically charged as was characteristic of the nineteenth century, neither had it been explicitly questioned.

Taking into account the particular chronological position of the Koyrean work, the primary objective of this chapter consists in examining the ties between his work and this interpretation. More precisely, we intend to question to which extent the Koyrean work supports this interpretive framework or, to put it differently, to which extent it initiates and prepares its reassessment.

Our secondary objective aims to "test" our interpretation of the Copernican Revolution by checking it against the Koyrean way of thinking. We will also endeavour not so much to *explain* this reasoning—since the work of Gérard Jorland (Jorland 1981) remains essential reading in this respect—as to *comment* on certain of his observations in light of our own research. It will then be up to our readers to assess the pertinence of our commentary and, consequently, the interpretation that governs them.

In order to carry out our "test", Koyré seemed an obvious choice.

Firstly, his intellectual itinerary, which led him to study the history of philosophical, religious and scientific thought, must have rendered him particularly capable with respect to analysing the consequences of that which was not only a change of cosmological system but also—to use a term that he was particularly fond of,[2] at least until he became familiar with the American History of Ideas towards the end of the 1940s—of *Weltanschauung* and consequently a mutation during which these different forms of thought became particularly intertwined. It was also essential to choose a historian not only capable, as he was, of mastering so many different forms of thought but also concerned with taking their constant interaction into consideration, as he continually invited us to do.

Secondly, in this historic analysis of the *Weltanschauung* shift, his concern for compassion, his wariness with respect to translations, his prudence towards the idea of precursors, his commitment to adopting the way of thinking of the authors he studied, his willingness to also pay an equal amount of attention to mistakes and failures, and his cautiousness when it came to not lapsing into "retroprojection" (Koyré 1957, 292)—all these methodological characteristics make him, for once and for always, a remarkable "master of reading" (Belaval 1964, 675), according to the delightful expression by Yvon Belaval (1908–1988), which must have helped in guarding him against many an error of perspective.

Thirdly, even if Koyré, as a historian of scientific thought, had barely studied the medieval era—his article on *Le vide et l'espace infini au XIV*^{e} *siècle* (1949; Koyré 1986e [1949]) constitutes a noteworthy exception—his first works on Saint Anselm

---

[2] Without an exhaustive account, between 1928 (Koyré 1928) and 1948 (Koyré 1986f [1948]), we found this term employed in 12 different publications.

of Canterbury (1033–1109; Koyré 1984 [1923] and 1927), on Jacob Boehme (1575–1624; Koyré 1979 [1929]) and on the "mystics, spiritualists, and alchemists of 16th century Germany" (Koyré 1971a [1933], 1971b [1928]), as well as the fact that he remained a highly active literary critic of medieval publications (Stoffel 2000) after his reorientation towards the history of scientific thought (1933), must nonetheless have also equipped him with the ability to appreciate the medieval–Aristotelian facet of the traditional interpretation in addition to his renascent and modern side.

Fourthly, having himself lived through, as he explicitly stated:

> […] two or three profound crises in our manner of thinking [is undoubtedly why he aptly felt] more capable than [his] predecessors of understanding the crises and polemics of yesteryear. (Koyré 1989 [1951], p. 248)

such as those that draw our attention today.

Lastly, the undeniable success of his characterisation of the seventeenth-century revolution, namely, the destruction of the Cosmos and the geometrisation of space, and of his determination of the resulting consequences, in this case the retreat of the Divine and the divorce between the world of science and the world of life, cannot leave one indifferent. For all these reasons, we felt compelled to choose Koyré and to dare to confront his way of thinking. The fiftieth anniversary of his death provides us with this opportunity and is, we esteem, a way of paying tribute to him by entering into a public dialogue with him, albeit critical at times.

Before thoroughly examining the Koyrean analysis of the Copernican Revolution in itself, in the second part of this article, we first need to outline his understanding of the broader revolution, of which it is a part, in order to put it into perspective. Koyré most frequently describes this more fundamental revolution, which was brought about by the Copernican Revolution, despite itself, as "the spiritual revolution of the 17th century". In reference to the original characterisation that he bestows upon it, we will call it the "ontological revolution".

20.2 The Ontological Revolution

20.2.1 Importance and Violence

20.2.1.1 Importance

The traditional interpretation attaches particular importance to the Copernican gesture (understood to be the decentring of the Earth and the centring of the Sun), as it embodies the first humiliation of humankind which was soon to be renewed, in their respective spheres, by Darwinism and Freudianism. Copernicus (1473–1543), and through his modern science, thus found themselves attributed with the full responsibility for a violence directed not only against human pride but, even more broadly, against humankind itself.

Without explicitly subscribing to this traditional interpretation that associates the loss of terrestrial centrality to a "vexation" considered "narcissistic" in nature (Freud 1996 [1917], pp. 46–47), Koyré also attaches a particular importance to the Copernican Revolution per se and to the entire process that follows, including the ontological revolution (suffice it to say, by means of this deliberately imprecise circumlocution, that we acknowledge the instability of his own terminology). He even goes one step further in emphasising the impact of it: both chronological, since, according to him, the year 1543 "not only marks the end of the Middle Ages and Classical Antiquity", but that of "a period that embraces both the Middle Ages and Classical Antiquity" (Koyré 1933b, p. 101, 1961, p. 15), and conceptual, since he considers this revolution to be (and herein lies, as we already mentioned, his greatest originality) (Jorland 2006, p. 159) an expression "of a much more profound and serious process", namely, a change in *ontology* (Koyré's own term)—that obliged humankind to transform and replace "the very structure of its thought" (Koyré 1988 [1962], pp. 10–11).

20.2.1.2 Violence

What seems even more important to us to emphasise for our present purposes is that Koyré echoes, in his own way, not only this importance but also this violence as evidenced by his choice of vocabulary, which constantly and continually surfaces in his writing, starting from his prefaced article *Copernic* (1933) up until his *La révolution astronomique* (1961). The following is proof of this. Copernicus, he wrote, "*wrested* the Earth from its foundations" (Koyré 1933b, p. 102, 1965 [1950], p. 8), "*dislodged* [it] from the centre of the Universe" (Koyré 1985e [1951], p. 94) and "*flung*" it (Koyré 1933b, p. 102, 1965 [1950], p. 8) or, if one prefers, "*hurled* it into the heavens" (Koyré 1961, p. 16), which sufficiently demonstrates—with all due respect to Arthur Koestler (1905–1983)—the "incredible" (Koyré 1933b, p. 102) and "incomparable boldness" that he dared to display. In line with these statements, which recur after a 30-year interval, Koyré, in 1950, still does not know if he should write that the mutations and transformations—or the "crisis"—(Koyré 1957, p. 1) of the scientific revolution of the seventeenth century were merely "accomplished", or, worse, "suffered" by the human spirit (Koyré 1943c, p. 404, 1965 [1950], p. 5). We note, however, that 7 years later, in his most prominent work *Du monde clos à l'univers infini*, he chose to invert these two verbs and in doing so favoured the state of submission that was required (Koyré 1957, p. vii, 1988 [1962], p. 9). Equally significantly, our historian, it seems, finds himself obliged to leave his readers the choice, as to the description of the "disastrous effects" of this revolution, between an interpretation which is nostalgic and one that is not, since "disastrous effects" there certainly were (Koyré 1957, p. 281)! And how, indeed, could Koyré otherwise qualify the "total separation between the world of values and the world of facts" (Koyré 1988 [1962], p. 12) that he describes to us and which leads notably to "the fatal blow" delivered to the geocentric and anthropocentric world (Koyré 1933b, p. 102, 1961, p. 17)? As regards the new Cartesian science, he wrote as early as

1937 that it "did not merely settle for *banishing* man, and Earth, from the centre of the Cosmos: this Cosmos was *broken, destroyed, annihilated*", in such a way that after having offered man "a desperate picture" of the universe, this Cartesian science not only leads to a "decisive" victory but also to a "tragic" one, since it left "no room for either man or God" (Koyré 1987 [1938], p. 172, p. 210). He also added, a year later in his *Études galiléennes*, that this victory—which had "cost more dearly" than any other (Koyré 1986d [1939], p. 291)—had, in concert with Galilean science, caused outer space to lose the cosmic value it once held (or could have held). Lastly, by way of one final example in support of our argument, we would like to draw attention to the fact that, according to our author, Nicholas of Cusa (1401–1464) could well be attributed with "the *merit*, or the *crime*, of having asserted the infinity of the universe" (Koyré 1957, p. 6).[3]

20.2.1.3 Nostalgia

In correlation to the expression of this pervasive violence—demonstrated, as we have just noted, by the virulence of the chosen vocabulary—Koyré, throughout his work, seems to indulge in discrete expressions of nostalgia for the old world view, accompanied by fleeting hopes that the current situation may not be permanent.

Does he not specify that the term "Cosmos", understood in the sense of totality, which had "completely lost all meaning during the Classical Period of physics seems to have gained new meaning again since Mr. Einstein" (Koyré 1986d [1939], p. 18)? In the same vein, does he not note that contemporary cosmology, still with respect to Einstein (1879–1955), seemed to have "turned back to a finitist conception" of the world (Koyré 1957, p. 288) and that with Eddington (1882–1944), it seemed to pursue this "disinfinitisation of the universe" (Koyré 1952a, p. 135)? Does he not qualify as "happy" those eras where man, in search of the essence of his being, could start by questioning the world and his rightful place within the heart of it, before—the Cosmos having become "uncertain" and having gone "to pieces"—having to turn to himself in order to find the answer to this question (Koyré 1987 [1938], pp. 177–178)? Does he not mention the fact that one cannot write good poetry from either good philosophy or good science, while one can do so very well with magic (Koyré 1947a, p. 94)? Philosophy and science, in order to be good, first had to disenchant, dehumanise and even render the world inhuman by geometrising it, while poetry, to be successful, instead needed an enchanted world, such as that "of primitive thought, of pre-logical thought, of infantile thought". With this in mind, does he not conclude that one would need to "de-geometrise geometry", and

[3]This latter juxtaposition of contradictory adjectives could certainly be the view of a historian who knew to which point an event can produce interpretations that are not only different but even opposing—and, incidentally, we will herein be the first to highlight the necessity of systematically endorsing, with regard to the Copernican gesture, the different forms of interpretation available in as much as their appraisals might have differed—but the numerous extracts we have just reiterated do not suggest that these examples fit this sole motivation and that they in any way reveal the inner convictions of Koyré himself.

therefore completely subvert it, in order to be able to create good poetry (Koyré 1943a, p. 511)? Does he not relate, in 1934–1935, Alfred North Whitehead's notion (1861–1947) according to which

> […] the error that vitiated, or at least that impeded the progress of scientific and philosophical thought […] was its excessive mechanisation [or] the reduction of the physical to the geometric […] that is to say the negation, express or implied, of change [a point of view that contemporary physics has now fortunately completely abandoned, which thus allows one, according to this illustrious British philosopher] to conceive of the reality of qualities? (Koyré 1935, p. 398)

Fifteen years later, still in the same vein, does he not take a certain pleasure in showing us how even Henri Poincaré (1854–1912) himself was succumbing to "the invincible might of the—absurd—trend of extreme geometrisation" (Koyré 1950b, p. 322)?

Apart from those passages in which our historian is the spokesperson for the intoxication experienced by Giordano Bruno (1548–1600) before the spectacle celebrating the collapse of the closed world of the Ancients (Koyré 1957, p. 43), the only Koyrean text that we found, which differs from those we just mentioned, in that it brings to light the positive aspects of the disenchantment of the world—such as the reintegration of man into the world, the legitimisation of the pursuit for worldly happiness or even the invitation extended to man for achieving his destiny while still in this earthly realm—is a report published in 1947 from the book *La pensée européenne au XVIII*[e] *siècle* by Paul Hazard (1878–1944) (Koyré 1947b). It should, however, be noted that, from our point of view, such statements in no way disprove the theory of the humiliation brought upon man, since—whether this humiliation is experienced as destructive (as is the case in the majority of Koyrean texts) or, on the contrary, as liberating (as in the rare example shown here in the body of his work)—the fact remains that, either way, there is still humiliation. It is, therefore, evident that Koyré chose to favour the destructive interpretation of this humiliation.

After having brought this common theme running through the Koyrean works and the traditional interpretation to light, namely, that of a veritable assault wrought upon man, we will now go into more depth concerning the causes that, according to Koyré, are responsible for this state of affairs—but not without first reiterating the two principal characteristics of the revolution from which these causes arise.

20.2.2 Characteristics

20.2.2.1 Introduction

As early as his *Études galiléennes* (Koyré 1939), Koyré provides the educated public with two characteristics, which were to become canonical, of the seventeenth-century scientific revolution: the "geometrisation of space" and the "dissolution of the Cosmos"—that is to say, on the one hand, "the substitution of the abstract space of Euclidean geometry by the concrete space of pre-Galilean physics" and, on the other hand, the "disappearance, within scientific reasoning, of any consideration deriving from the Cosmos" (Koyré 1986d [1939], p. 15).

Systematically addressing the various instances[4] of this characterisation and analysing the variants that they reveal, in order to provide us with a better understanding of the gradual development of this Koyrean characterisation, would certainly prove interesting and even necessary since many articles by our author include (more or less revised) extracts of previous publications. This is, however, not our objective: suffice it to note here, firstly, the early appearance of this characterisation within Koyré's work on the history of scientific thinking and, secondly, its rapid achievement of a state of completion, as of 1943, that seemed to shelter it from any subsequent significant modification.

20.2.2.2 Precocity

In support of this first statement, we would like to mention that the latter of the two characteristics which was outlined in 1939, namely, the "dissolution of the Cosmos", obviously dates back to the lessons that Koyré consecrated to Galileo (1564–1642) during the academic years 1933–1934 and 1934–1935 (Koyré 1986b, p. 42, pp. 43–44), in so far as it seems to constitute the *proprium* of the Florentine astronomer: While Copernicus was developing his astronomy based on a particular concept of the Cosmos and while Kepler (1571–1630) was *still* including cosmological considerations in his scientific reasoning, Galileo himself voluntarily put an end to all considerations of this sort by a revolutionary gesture, which is therefore as much of a "rupture" as a "deviation" (with respect to Kepler, his contemporary) (Jorland 1981, p. 248). By this distancing of the Cosmos, understood as a "closed unit of hierarchical order", in favour of the universe, conceived as an "open ensemble linked by the unity of its laws" (Koyré 1986d [1939], p. 165)—although distanced, the term "Cosmos" remained "but with an entirely new meaning" (Koyré 1943c, p. 404)—"the most profound revolution achieved or suffered by the human mind since the invention of the Cosmos by the Greek" (Koyré 1986d [1939], p. 12, 1943c, p. 404) occurred, of which Galileo, contrary to Kepler (Koyré 1986b, p. 42, p. 44; 1986d [1939], p. 15), constitutes, together with Giordano Bruno (Koyré 1986d [1939], pp. 171–181), one of the most emblematic figures.

The first of these two characteristics, namely, the "geometrisation of space", made its first, clearly recognisable appearance in two conferences delivered in 1935 and 1936: "Aristotle's physical space", "a set of qualitatively distinct locations", affirmed Koyré, "was replaced by the homogenous space, devoid of geometry" (Archimedean space) (Koyré 1986b, pp. 38–39). The fact that the statement of this process of the geometrisation of space appeared in the context of conferences dedicated to Galileo should not surprise us, since while Kepler was still sharing a qualitative view of the universe—which, incidentally, prevented him from formulating

[4]Cf. Koyré 1986d [1939], p. 15; 1943c, pp. 403–404; 1965 [1950], pp. 6–7; 1986a [1955], p. 258; 1957, p. viii. Secondarily: cf. Koyré 1986e [1949], p. 39; 1951a, p. 782; 1951b, p. 788.

the law of universal gravitation (Koyré 1968 [1951], p. 13)—the Florentine astronomer himself no longer belonged, in any way, to the Renaissance: He experienced no joy at the variety of things and felt a deep aversion to magic, which allowed him to unscrupulously divest the Aristotelian space of its qualitative characteristics and to associate the physical space with that of Euclidean geometry (Koyré 1985c [1950], pp. 58–60).

20.2.2.3 Conclusion

In support of our second statement, this time—namely, this characterisation's rapid attainment of a quasi-definitive state of completion—we would like to emphasise that it was as early as the article *Galilée et Platon* (1943), so it was less than 5 years after *Galileo Studies* that Koyré performed the following modifications: He replaced, with respect to the statement concerning these two characteristics, the expression "destruction of the Cosmos" with that of "dissolution of the Cosmos" (Koyré 1943c, p. 403); he reversed their order, almost without notable exception (Koyré 1951a, p. 782), by henceforth mentioning "the destruction of the Cosmos" before "the geometrisation of space"[5]; and, finally, he chiefly ascribed this intellectual revolution to an ontological change.

This double characterisation, so well known that it hardly needs emphasising—and which is easy to remember and easy to teach—was then to be regularly reaffirmed, up until his renowned work *Du monde clos à l'univers infini*, nearly 15 years later.

20.2.3 Consequences

This intellectual revolution, for which the two distinctive traits were just reiterated, does not simply set the mind "on a path that leads to principle of inertia" (Koyré 1986d [1939], p. 74) by rendering it "plausible and almost self-evident" (Koyré 1951a, p. 782); it also leads, according to Koyré, to far more important anthropological consequences: the separation of the world of science and the world of life, on the one hand, and the retreat of the Divine, on the other hand.

---

[5] Since these characteristics are closely linked, is this inversion really important? On the one hand, it is perhaps only a raising of the initial ambiguity, as the explanation of these characteristics does not follow the order in which they are mentioned (Koyré 1986d [1939], p. 15). On the other hand, Koyré seems to attribute a certain priority to the geometrisation of space both in his conference of 1936, "Aristotle's physical space replaces the abstract space of geometry [...] and the cosmos of medieval physics disappears" (Koyré 1986b, p. 39), and in his *Galileo Studies*, "the complete geometrisation of space, which means the infinity of the Universe, and the destruction of the Cosmos" (Koyré 1986d [1939], p. 211).

20.2.3.1 The Separation of the World of Science and the World of Life

When he was still a historian of philosophical and religious thought of the Middle Ages and the Renaissance, so before he turned towards the history of scientific thinking, Koyré became a contemporary of this world in which there was no separation between man, the world and God. Through Valentin Weigel (1533–1588) and Paracelsus (1493–1541) in particular, he echoed this "state of mind" that "was not opposed to the world but lived with it" and that even "*felt* a kinship with it" (Koyré 1971a [1933], p. 84)—since it was thenceforth "possible to establish a precise relationship between the components of the human organism and [those] of the world" (Koyré 1971a [1933], p. 87) or, better yet, to conceive that man "contains within himself everything that the world contains", being himself "divine, astral, material" (Koyré 1971b [1928], p. 157). Through Jacob Boehme, who was already familiar with the Copernican doctrine but who nevertheless confined himself to "the vulgar astronomical doctrine" (Koyré 1979 [1929], p. 72), he experienced this world where tangible qualities were still real (Koyré 1979 [1929], p. 92), and in which—a characteristic that Koyré seemed to have forgotten when he later described the reunified universe as replacing the radical splitting into two of the medieval–Aristotelian Cosmos—there was, indeed, no strict separation between the heavenly world and the terrestrial world, as it is evident that everything in the latter originates from the former (Koyré 1979 [1929], p. 91, pp. 99–100, pp. 128–129). And when our historian finally found himself, with Niccolò Machiavelli (1469–1527), "in a whole other world", this did not prevent him from recognising, through Andrea Cesalpino (1519–1603), "to what extent the image of the medieval and ancient world had been solidified, 'realised' in human consciousness" (Koyré 1985d [1930], pp. 21–22).

After having become a historian of scientific thinking, it was the process of the dissolution of this world, though impregnated within human consciousness, that Koyré would focus on describing as a necessary consequence of the scientific revolution of the seventeenth century. Indeed, he explains, the two characteristic traits of this revolution, namely, the destruction of the Cosmos and the geometrisation of space, would almost be equivalent to "the mathematisation (geometrisation) of nature and therefore the mathematisation (geometrisation) of science" (Koyré 1965 [1950], p. 7). The order of events is also important here: It is because of Galileo—who did not hesitate to privilege the rational and ideal approach to mathematics to the detriment of empirically given reality (Koyré 1985f [1956], p. 83) and who proclaimed that "the book of nature is written in a geometrical language" (Galilei 1980 [1623], p. 141), thereby bringing about the mathematisation of nature—that modern science, in order to decipher this book, was obliged in turn to mathematise science by excluding from its universe "everything that cannot be precisely measured" (Koyré 1952b, p. 85). Consequently—since "there are no qualities either in the universe of numbers or in that of geometrical figures"—these qualities were unable to maintain, in this new "realm of mathematical ontology" (Koyré 1965 [1950], p. 8), their former position in the world of Aristotelianism and common sense. They were thus to be banished from the, now purely qualitative, world of science—or at least from those sciences referred to as "exact" (Koyré 1953, p. 223)—as were to be banished "all considerations based on value, perfection, harmony, meaning, and aim" (Koyré 1965 [1950], p. 7). Similarly, there is no "more or less" in mathematics. So from then on, "the everyday world of approximation" was to be replaced "by the (Archimedean) universe of precision, of exact measures, of strict determination" (Koyré 1965 [1950], p. 5)—which, together with "the elaboration of the notion and techniques of precise measurement", would enable "the transition of the world of approximation to the 'universe of precision'" (Koyré 1989 [1951], p. 248). Lastly, "there is no change and no becoming in numbers and in figures". Additionally, classical science would have to substitute "a world of being for a world of becoming and change" (Koyré 1965 [1950], p. 8).

The substitution of the world of quality by that of quantity, the transition from the world of approximation to the universe of precision and lastly the promotion of being at the expense of changing are thus the "deepest meaning and aim [...] of the whole scientific revolution of the seventeenth century" (Koyré 1965 [1950], pp. 4–5). This certainly resulted in the reunification of the natural world—since the Cosmos, which was qualitatively and ontologically differentiated between the sublunary world of becoming and the celestial world of being, was substituted by a homogenous and unified universe, both physically and intellectually, by the identity of its mathematical laws—but at the cost of introducing a further separation, altogether more radical than the previous one, between the world of science and the world of life: The reunification of the Heavens and the Earth was brought about "by substituting for our world of quality and sense perception, the world in which we live, and love, and die, another world – the world of quantity, of reified geometry, a world in which, though there is place for everything, there is no place for man" (Koyré 1965 [1950], p. 23). "This is", to quote the famous phrase with which Koyré concluded his article *Sens et portée de la synthèse newtonienne* (1950), "the tragedy of the modern mind which 'solved the riddle of the universe', but only to replace it by another riddle: the riddle of itself" (Koyré 1965 [1950], p. 24).

This theme—that of the divorce between the world of life and the world of science—constitutes, undoubtedly even more intensely than that of the retreat of the Divine, one of Koyré's leitmotivs, as evidenced by its frequent recurrence.[6] The authors that inspired him, or that he met, are varied and would be worth studying in depth. Koyré himself mentions a few of them: Whitehead, for the impossibility of modern science to substitute the world of quality it forces us to leave behind, by something other than a world of quantity within which we cannot live (Koyré 1985c [1950], p. 52), and Émile Meyerson (1859–1933) for the tendency, at work in scientific thought, to reduce the real to the geometric in such a way that the real world—our world—in order to be explained, must be declared unreal in favour of a world constructed out of quantity, which was thenceforth considered to be the only real one (Koyré 1950b, p. 321; 1985c [1950], p. 52). As for the commentators, with respect to the substitution of the universe of precision for the world of approximation of everyday life and for the distinction between the world of being of science and the world of becoming of life, Gérard Jorland points us in the direction of Henri

[6] Cf. especially Koyré 1943c; 1943a; 1986c [1948]; 1965 [1950]; 1985c [1950]; 1952b; 1953.

Bergson (1859–1941) (Jorland 1981, pp. 61–62). Concerning the disassociation between the world of science and the world of life, Godofredo Iommi Amunàtegui detects "a conceptual intimacy born of a shared horizon" (2002, p. 74) between the article *Sens et portée de la synthèse newtonienne* and *Die Krisis der Europäischen Wissenschaften und die Transzendentale Phänomenologie* (The Crisis of European Sciences and Transcendental Phenomenology) by Edmund Husserl (1859–1938), while Soren Gosvig Olesen detects "and extension of Husserl's work" (1994, p. 22).

For our part, we would like to mention that it was in 1926—following the failure of, firstly, Pierre Duhem's endeavours (1861–1916) and, secondly, those of Bergson—that Koyré acknowledged, in the wake of Whitehead and of Meyerson (Meyerson 1931, p. 149), the absolute impossibility (he would retract this statement 25 years later) (Koyré 1985c [1950], pp. 52–53) of returning to qualitative physics (Koyré 1926, p. 465, 1933a). The necessity of renouncing the qualitative world of sensory perception, caused by the mathematisation of nature (Koyré 1943b, p. 347), was thus accompanied by the abandoning of all hope of turning back. Additionally, the "dehumanised and even inhumane world" (Koyré 1943a, p. 511)—which leads to the paradoxical (Koyré 1986b, p. 140) reduction to mathematics of a world, that of everyday life, which "is not mathematical or even mathematisable" (Koyré 1986c [1948], p. 342)—imposed itself as a seemingly inescapable fact. Following the dehumanisation of the Cosmos—which happened initially in Greece in order to make way for the development of cosmological scientific theories (Koyré 1985e [1951], pp. 87–88)—would it not have been necessary, sooner or later, to do that which had been successfully done for the Heavens, for Earth and, more precisely, for our everyday world? Be that as it may, the situation in which we find ourselves—even if not insoluble, at least unresolved (Koyré 1985c [1950], pp. 52–53)—is an "extremely difficult" and "extremely unpleasant" one (Koyré 1950a, p. 61).

20.2.3.2 The Retreat of the Divine

As a historian of medieval thinking, Koyré rightly recognised that the world of Saint Augustine (354–430) or that of Saint Bonaventure (1217–1274) was not yet what it has become for us, as for them it was but "a vestige […] of a supersensible and divine reality", for which "existence and presence are so evident, are such an indubitable fact that it is hardly necessary to seek further proof" (1984 [1923], p. 7). Additionally, in a similar context, characterised by the obvious fact that the world and the Heavens were claiming the glory of the Eternal (Ps, 19, 2), in 1953 he stated that "the difficulty — if there was indeed a difficulty — was to find a place for the world alongside God", as "the world is in grave danger of completely losing its own reality, of being resolved and absorbed by God" (Koyré 1923, p. 452). Moreover, if it has become difficult for the present-day historian to recognise this way of thinking, it is precisely because nowadays, conversely, it is faith that "is constrained to seeking a place for its God beside, or outside of the world of nature" (Koyré 1923, p. 452). From a world looking to preserve its place beside God, to a God trying to maintain his position alongside the world, the entire Koyrean history of the retreat of the Divine—inspired by *The Metaphysical Foundations of Modern Physical Science* (1925) by Edwin Arthur Burtt (1892–1989), as Pietro Redondi (Koyré 1986b, pp. xx–xxiii) rightly remarked—is already present, implicitly, in these few lines.

This history was to find its fullest meaning, nearly 25 years later, in his article *Sens et portée de la synthèse newtonienne* (1950) and above all in his book *Du monde clos à l'univers infini* (1957) since the article of 1950 concluded by noting the separation of the world of science and the world of life, which compelled humankind to find itself faced with "the riddle of itself" (Koyré 1965 [1950], p. 24), whereas the conclusion of the 1957 book puts a greater emphasis on the departure of God himself who, however, also abandons man to himself. Between these two extreme dates—1923 for its first appearance and 1957 for its final appearance—this history of the retreat of the Divine pervades all of Koyré's work. As Koyré underlined in his lectures in 1935–1936 and chiefly in 1936–1937, the loss of the central position of man in the world and the confirmation of the infinity of the universe, made the idea of a God committed to paying special attention to him, seem quite unlikely to the libertine (Koyré 1986b, 50 and 51). Those routes, now recognised as impassable in this new context, that traditionally followed the path of nature to arrive at the acknowledgement of the existence of God now oblige the apologetic to find a new itinerary by inviting man to discover "The one who is" no longer to be found in nature but in himself, in his soul or in his history[7]—which at least offers the advantage, noted Koyré wisely in an undated and unpublished text devoted to *La pensée mystique*—of rediscovering "a direct contact with God" (Koyré 1986b, p. 211), since the disappearance of the ladder that, by means of nature, led to God is identical to that which initially created the separation between Him and us. Moreover, it was not only the pathways that led to God that were thus undermined, he remarked during the same period, but also those that were to teach man who he was, since thenceforth man—no longer being able to continue questioning the Cosmos to this end, as he had done since Plato (427–347 BCE) (Brague 2002)—was obliged to turn towards himself and to query the very person asking the question (Koyré 1987 [1938], p. 178).

Ten years later, this question regarding God's rightful place and purpose in the world designed by the New Cosmology became even more explicitly interlocked with the problem of infinity and so became at the very least paradoxical. On the one hand, an infinite God needs an infinite universe, asserted Giordano Bruno, in order to be able to express "his infinite wealth and his infinite creative power", whereby "it is only an infinite universe […] that allows us to form an idea of the infinity of God" (Koyré 1986b, p. 147, 1957, p. 52). We feel it is worth mentioning that, whereas it was the *dignity of man* that seemed to demand a finite and hierarchical world (viz., geocentric) in ancient times, nowadays it is, however, the *dignity of God* that seems to require an infinite world. On the other hand—and as an extension of the Cartesian doctrine that, by denying all spiritual matters (Koyré 1957, p. 138, 1986b, p. 171), had already deprived God of any standing in the world—this infinite

[7] Cf. Koyré 1986b, p. 51, p. 54, pp. 202–204; 1987 [1938], p. 210.

world given to God to allow him to witness his own infinity would not hesitate to turn against Him: After having been stripped of his status, He would also be deprived of all action, however small, since “the very concept of creation hardly seemed applicable to an infinite Universe” (Koyré 1986b, p. 144). Koyré mentioned this in 1947–1948 before having elaborated upon this *sentiment* of incompatibility (if we may be allowed to add this crucial detail, which we will come back to in the conclusion) in the memorable and renowned pages of *Du monde clos à l’univers infini*. In short, the infinitisation of God thus leads, in a “decidedly remote yet certain” way, to the infinitisation of space, itself beholden to this same God, who was thenceforth deprived of both a place to stay and of actions to perform and who had finally “packed his bags” as he was obliged to relinquish his “ontological attributes” to “the infinite Universe of the New Cosmology” (Koyré 1957, p. 276).

20.3 The Copernican Revolution

As we have just witnessed, the Koyrean interpretation of the spiritual revolution of the seventeenth century is clear and well circumscribed: Characterised by the destruction of the Cosmos and the geometrisation of space, this revolution of an ontological nature leads not only to the divorce between the world of science and the world of life but also to the retreat of the Divine. We now propose to examine, in the work of our author, that which reveals his interpretation of the Copernican Revolution per se. Since, on the one hand, Koyrean thinking in this respect is considerably less evident and maintains a certain distance from the traditional interpretation, and since, on the other hand, this subject has not yet been covered by the commentators, we are obliged to limit our investigation to the only anthropological part of the Copernican Revolution’s traditional interpretation. Indeed, this interpretation, which is less limited than is commonly believed, is not restricted to such considerations. It also includes epistemological assertions (e.g. geocentrism, by supporting anthropocentrism, anthropofinalism[8] and anthropomorphism, has continually impeded scientific progress[9]) and, moreover, ideological claims (e.g. the abandonment of geocentrism, rendering anthropofinalism inconceivable, unavoidably led to the undermining, by its very foundations, of the entire theological edifice[10]) and, lastly, historical considerations (e.g. embarrassed by the humiliating nature of his cosmological system, Copernicus hesitated for a long time and died immediately after its publication, thus skilfully withdrawing himself from the affair) (Fontenelle Bernard de 2013 [1686], p. 159). Within its strictly anthropological part, which is certainly the most well known, this interpretation aims to determine

[8] By this term we mean the elevation of man to ultimate end and final cause of the physical world, which is established to be as his service.

[9] Here we are referring to Laplace, Comte, Flammarion, Poincaré, Haeckel, Büchner, and even Dallemagne. Cf. Stoffel 2012.

[10] This concerns the rationale proposed by Auguste Comte. Cf. Stoffel 2012.

the consequences of the Copernican Revolution for humankind by adopting a threefold approach: positional (man's position in the universe), dimensional (the size of Earth relative to the other celestial bodies and ultimately relative to the universe itself) and, lastly, kinetic (the state of rest or motion of the planet that hosts the human species). Given the magnitude of this interpretation, we are obliged to restrict ourselves even further: Even within its strictly anthropological part, we can only analyse certain aspects related to the positional perspective.

20.3.1 The Geocentric Topography

20.3.1.1 Does It Privilege the Central Position of the Earth?

If, with heliocentrism, the loss of the central position of the Earth is described, by traditional interpretation, as a humiliation inflicted upon man, it is evidently because this interpretation considers that terrestrial centrality enjoyed, in geocentrism, a favourable connotation according to those who inhabit it. And how indeed could one imagine it to be otherwise? Since the centre of a space (and even more so a sphere, as is the closed world of the Ancients) is not only the most appropriate place to be seen and heard but also to protect our most precious possessions, how could one conceive that the occupation of such a place does not constitute evidence of a rare privilege granted to the human race? And yet, if centrality is indeed ascribed a certain value with respect to the topography[11] specific to heliocentrism, it is ascribed an entirely different one with respect to that which governs geocentrism. Regarding the latter, even if the cosmic centrality attributed to Earth is certainly singular to the extent that only our dwelling can avail itself of this feature within the entire Cosmos, it is by no means, however, experienced as rewarding: In this topography, the position of the Earth, while being cosmologically recognised as central, is nonetheless symbolically considered as low and therefore demeaning. As a result of the ignorance of this fundamental difference between the topography of geocentrism (abasement of cosmic centrality despite its uniqueness) and that of heliocentrism (elevation of this same centrality), the error of the traditional interpretation was therefore to

[11] By "topography", we mean the all of the scientific, philosophic, religious and symbolic characteristics whereby a cosmological system does not only order and scientifically arrange the celestial bodies but also determines a distinct world view (that is to say a *Weltanschauung*). By defining this term—which is less restrictive than that of "axiology" used by Koyré, before he spoke of "geometric–hierarchical sensibility" (Koyré 1958, p. 63)—we thus remain faithful to our author's thinking in that he himself made the observation that "cosmological concepts, even those that we consider to be scientific, were only rarely — and even almost never — independent of those which are not scientific, namely, philosophical, magical and religious notions" (Koyré 1985e [1951], p. 87). However, we do not agree with his assertion that "a truly pure cosmology, free of any extra-scientific pretence, appeared for the first time in the work of Laplace" (Koyré 1951c, p. 31), since, as we feel we have already established (Stoffel 2012), the author of *Exposition du système du monde* draws, from the cosmology of his time, conclusions of an ideological nature which were to be continued by Comte.

have *assumed* that cosmic centrality had *always* been privileged and to have consequently interpreted the loss of this centrality, following the onset of heliocentrism, as a humiliation inflicted upon mankind.

With respect to this question, what is the conception of our author? Initially, Koyré seems to remain predictably within the bounds of the traditional, interpretive outline. Thus, in a report published in 1935–1936, our reviewer wrote: "In the Copernican cosmos man ceases to occupy the central position; it is for this very reason that earth relinquishes its own to the sun" (Koyré 1936, p. 460). By this we understand that while man occupied the central position, the Earth was the centre of the world; now that this is no longer the case, the Earth must surrender its honorary place to the one henceforth possessing this privilege, namely, the Sun. From geocentrism to heliocentrism, there was therefore no change with respect to the qualitative appreciation of cosmic centrality—being positive in either system—but simply a change with respect to the celestial body possessing the privilege of this position.

During a second phase, which we place at the beginning of the 1940s, Koyré became aware of the reversal of appreciation that had occurred between these two cosmological systems and in line with the work of Lovejoy (1936) who had briefly (Lovejoy 1936, pp. 101–102) remarked, countering that of Burtt (1925) (Burtt 1925, p. 6), that terrestrial centrality was, in geocentricism, more inclined to humiliate man than to uplift him, since it was associated with being lowermost and therefore baseness. Bolstered by this realisation of the assimilation of the centre to the bottom, Koyré emphasised for the first time, in a discourse delivered in 1944, that the geocentric position of the Earth was likely to cause man to feel both a sense of humility and of pride at the same time:

> We are […] at the very centre, in a position where, to observe the remainder, we always look towards the heavens *above* and we are *below*; the earth occupies the worst position in this cosmos. But being in the centre, now that is something, with the sky, the moon and the stars all revolving around us. *They revolve for us* because we, mankind, are the most important being of all creation. (Koyré 1986b, pp. 69–70)

Despite its brevity and discretion—it remained unpublished until 1986—this text demonstrates, with great accuracy, the complexity of the situation of the Earth in geocentrism: By its position, it is, as such, unique and yet contemptible (this being the purely positional perspective, consistently privileged by the commentators[12]); but by its resulting immobility, it is uplifting, because, conversely, it is the stars that revolve around us and thus for us (this being the kinetic perspective, too often neglected,[13] which offers the advantage of giving rise to an anthropofinalist interpretation).

---

[12] This is true, however, as much for the commentators faithful to the traditional interpretation as for those who strive to refute it.

[13] Without outlining all the consequences himself, Koyré never fails to mention, aside from the centrality of the Earth, the fact that the astral bodies, in geocentrism, revolve around it. This characteristic is significant, because it demonstrates that, contrary to all expectations, it is not the Earth that is obliged to move itself in order to search for what it needs but that, conversely, the stars must

A few years later, dedicating his first seminar of the academic year 1949–1950 to the *Problème de l'infini au XVIIe siècle*, Koyré brought to light, more forcefully yet still discretely—placing some words between brackets—all the ambivalence, already evident in his text of 1944, associated with the geocentric position of our dwelling: "Copernicus", he wrote, "deprives the Earth of its unique position (*the best and the worst*) in the centre of the world and identifies it with the planets" (Koyré 1986b, p. 146). In *Du monde clos à l'univers infini*, he develops his standpoint: the "worst" because the central position of the Earth is the lowest one possible (with the exception of hell, which is even lower) (Koyré 1957, p. 281, note 6) and this baseness renders our dwelling contemptible and assimilates it to the cloaca of the world[14] and the "best" because its centrality sets it apart from all the other heavenly bodies in such a way that, ultimately, the position of our Earth is distinguishable by its uniqueness rather than by its centrality (Koyré 1957, p. 43).

Contrary to the traditional interpretation that unduly extends, on top of geocentrism, the interpretation framework pertaining to heliocentrism (assuming a common appreciation of cosmic centrality in both topographies), Koyré is thus informed, from the 1940s, of this difference whereby cosmic centrality, in geocentrism, should rather be associated with baseness. But whereas Lovejoy, in 1936, only mentioned four authors—Aristotle, Cicero (106–43 BCE), Montaigne (1533–1592) and John Wilkins (1614–1672)—thus not including any of those representing medieval thought, Koyré in turn, as we just noted, was satisfied with merely confirming this identity, without attempting to either explain it or argue it further. Lastly, he does not seem to have fully realised that it was likely to ruin this interpretation which, in retrospect, we can qualify, today, as traditional.

20.3.1.2 Does It Feature a Centrality That Is Entirely Uplifting?

If, in geocentric topography, the cosmic centrality of the Earth is more humiliating than uplifting, must one conclude that, in this vision of the world, there is no centrality which is entirely uplifting? Not at all.[15] The Aristotelian distinction between geometric centre and ontological centre allows one to envisage, over and above the purely geometric centre of Earth, another centrality, a far more rewarding one, which is ontological in nature. Though sometimes linked to the celestial sphere of fixed stars, this ontological centrality was more commonly associated with the Sun. Given that the Sun occupies (at least according to Babylonian order) a medial position in the order of succession of the heavenly bodies (three before it and three after it), it can even avail itself, in addition to its ontological centrality, of also possessing

move in order to satisfy, as much as possible, its needs. They are therefore obliged to serve Earth and, consequently, mankind.

[14] Cf. Koyré 1957, p. 19; 1961, p. 62, p. 114, note 24; 1958, p. 52.

[15] Generally, we take the liberty of referring to our own article *La révolution copernicienne responsable du « désenchantement du monde »?* (Stoffel 2002) for all technical details and textual proof.

geometric centrality—although only planetary rather than cosmic[16]—even if, from a post-Copernican perspective such as ours, it is merely numerical.[17] If there is, therefore, a centrality that enjoys an entirely favourable connotation in geocentrism, it is indeed the planetary centrality of the Sun, which is both ontological and geometric—and by no means the cosmic centrality of the Earth (as the traditional interpretation mistakenly believes) which, being entirely geometrical, is symbolically interpreted, as we have seen, in a fundamentally ambivalent way.

What is the position of our historian on this matter? Although he was evidently familiar with the Aristotelian distinction between geometric centrality and ontological centrality—which, even though a significant fact, is however only mentioned, unless we are mistaken, once in all his work (Koyré 1933b, p. 117)—Koyré does not seem to have entirely realised the full significance, in medieval mentality, of the planetary centrality bestowed upon the Sun, which results from this distinction. Amongst other examples that we could provide, of particular importance is, from this point of view, the way Koyré deals with one of the reasons furnished by Copernicus himself for justifying the position—a purely symbolic one, as our historian never ceases to remind us[18]—of the Sun at the centre of the world. In this extremely renowned passage from *De Revolutionibus orbium cœlestium* (bk. 1, chap. 10), the illustrious astronomer, inviting the complicity of his readers, justifies this centring of the Sun by asking: "for in this most beautiful temple, who would place this lamp in another or better position than that from which it can light up the whole thing at the same time?" (Copernicus 1992 [1543], p. 22). From his teachings dating back to 1929–1930 (Koyré 1986b, p. 29) up until *La révolution astronomique* (Koyré 1961), Koyré would not change his mind with respect to either the importance of this passage or its interpretation.

With respect to its significance, Koyré, confronting the commentators that contested the reality of this Copernican reasoning—such as Lynn Thorndike (1882–1965) who, in his *History of Magic and Experimental Science* (1941), perceived in it a grandiloquent and artificial exaltation of the Sun—would always maintain[19] that Copernicus "was perfectly sincere in his adoration of the sun and in his conviction that the centre of the world was the ideal position for this *lampada pulcherrima*",

[16] We coined this expression in order to distinguish cosmic centrality (viz. the geometric centrality that a celestial body enjoys in relation to the entire Cosmos) from planetary centrality (viz. the numerical centrality that a celestial body enjoys in relation to the order of succession of most or all of the consecutive bodies comprising our solar system).

[17] Since heliocentrism reunified ontological centrality and geometric centrality into a single centrality, namely, cosmic centrality, which is both ontological and *perfectly* geometrical, it is tempting to underestimate the importance that the Ancients attributed to planetary centrality by reproaching it for being simply numerical (the order of succession of celestial bodies) and not entirely geometrical (as cosmic centrality, however, is). Thus, as written in the texts, it is only *after* a reunification of this kind that one can experience, *retrospectively*, such a feeling of inadequacy as regards the ancient planetary centrality.

[18] Cf., for example, Koyré 1933b, 117 or 1961, p. 155 and p. 404, note 22.

[19] Cf. also Koyré 1985e [1951], p. 95; 1961, p. 69, p. 114, note 25.

before concluding that there was "no reason to question the sincerity of Copernicus" (Koyré 1947a, p. 97).

As for his interpretation, this reason, presented as "the only one Copernicus has" (Koyré 1933b, p. 117) or at least as "probably the most profound", even if it is "not scientific at all" (Koyré 1985e [1951], p. 95), shows that "it is in order to brighten the world that the sun is placed in its centre", that is to say in the "*visibly* most favourable position for this purpose" (Koyré 1933b, p. 117). This also explains why the illustrious astronomer, without lingering on the system of Tycho Brahe (1546–1601), effectuated his "great astronomical reform": "the Sun being the source of light and the light being the most beautiful and best in the world, to him it seemed to comply to the reasoning that governs the world and that created it, that this light should be placed in the centre of the Universe it is responsible for illuminating" (Koyré 1985e [1951], p. 95). To put it another way (and in order to furnish one last justification for the perpetuation of this Koyrean interpretation), "it is to give light and therefore life and movement to the Universe that the Sun […] is placed at its centre […], which is *evidently* the most propitious [position] to this end" (Koyré 1961, p. 63).

This passage from *De Revolutionibus* was the subject of many a long debate that focussed primarily on the *sincerity* of the proposed argument. Given that planetary centrality already corresponds to the Sun in geocentrism, the Copernican argument should also, and above all, have been examined from the point of view of its relevance to a geocentrist. Indeed, bolstered by the assertion of Rheticus, whereby we can no longer consider the praises addressed to the Sun by the Ancients as simple poetry since heliocentrism had hence established that the Sun actually functions, in the centre of the world, as the "governor" of the planets (Rheticus 1982 [1540], p. 109), and by the assertion of Kepler, who explained that "the eminent dignity of the Sun having been fully demonstrated", "its central position could be considered to result almost *automatically* from that" (Koyré 1961, p. 288)—Koyré is naturally led to maintain, as an obvious fact, that the Sun "could not be placed elsewhere other than in the centre of the world" (Koyré 1961, p. 289) and that by finally realising this fact, heliocentrism had reinforced the heliolatry of the Renaissance by providing it with a fully apparent cosmological (even if not astronomical) foundation (Koyré 1961, p. 129).

In conclusion, the evidence, which was conceived and confirmed—"visibly", "automatically" and "evidently"—by Koyré, arises from a historical misunderstanding of centrality, both ontological and also geometrical, which *already* belonged to the Sun in geocentrism. Consequently, the argument put forward by Copernicus lacked, for a geocentrist, not only pertinence, but even led to an absurd result. It lost a great deal of its relevance since, according to geocentric topography, the Sun occupies a position that is *already* entirely adequate for illuminating everything, and this just as equitably as if it were "really", due to heliocentrism, positioned in the centre of the world. It even leads to the absurd consequence—glimpsed at, in another context, by our historian (Koyré 1961, p. 109, note 3)—of depriving the Sun—supposedly to honour it—of a central position (planetary centrality), only to relegate it—alongside hell—to the very bottom of the world (cosmic centrality associated with baseness). It is therefore only for the partisan of the new heliocentric topography, who is no longer able to appreciate the planetary centrality of the Sun in geocentrism, that the Copernican argument becomes significant. As a result thereof, for our historian, the announced obvious fact (that of the symbolic advantage consisting in placing the Sun at the centre of the world) disappeared to make way for this entirely unexpected question: How can heliocentrism (with its cosmic centrality) serve, better than geocentrism (with its planetary centrality), the solar cult, since, as witnessed in the texts, the *same* epithets and the *same* arguments traditionally used to demonstrate its importance could be applied, indifferently, to either of these two cosmological systems? The answer that we put forward—the cosmic centering of the Sun effected by Copernican heliocentrism actually strengthened, at least temporarily (Stoffel 2005), the traditional heliolatry to the extent that not only did the ancient analogies persist, but they even saw an increase in their validity—is also likely to clarify the anthropological consequences of the Copernican Revolution: Far from being solely concerned with the decentring of the Earth (which yet, alone, held the attention of the traditional interpretation), the contemporaries of these events also, and perhaps above all, paid attention to what was happening to the Sun, some to rejoice it and others to rue it!

Misunderstanding (or disregarding) the significance of planetary centrality in the geocentric topography, Koyré thus mistook for an obvious symbolic advantage that which was only one for those evaluating the centring of the Sun in light of the new heliocentric topography. We thus recognise, yet again, the same error here: a difficulty in discarding the heliocentric topography, which remained so familiar to us (despite the subsequent infinitisation of the universe), in order to understand geocentrism according to its very own reading matter and to understand the Copernican Revolution from the geocentrists' perspective. In dealing with cosmic centrality, we noticed that this difficulty could lead to ascribing value where it should not be ascribed, while dealing recently with planetary centrality, we realised that it could just as easily lead us to neglect that which must not be either!

20.3.2 *The Introduction of Heliocentrism*

20.3.2.1 A Change in Topography

Even if Koyré had always thought, with the traditional interpretation, that heliocentric topography elevated cosmic centrality, it is, as we have seen, from the beginning of the 1940s that he realised, contrary to this very interpretation this time, that this was not the case for geocentric topography. Consequently, he came to grasp the fact that, between these two subjects, a veritable axiological reversal had occurred, since the position that had been abased in geocentric topography had instead been ascribed a certain value in heliocentric topography. In *Du monde clos à l'univers infini*, our historian expressly draws the attention of his readers to this reversal:

> [...] it is on account of its supreme perfection and value — source of light and of life — that the place it occupies in the world is assigned to the sun; the central place which, *following the Pythagorean tradition and thus reversing completely the Aristotelian and mediaeval scale*, Copernicus believes to be the best and the most important one. (Koyré 1957, p. 30)

Expressed quite clearly the following year in *Les sciences exactes de 1450 à 1600*, this "radical transformation of geometric-hierarchical sensibility which, in opposition to Aristotelianism or Christianity, no longer sees the central position as the basest or the most unworthy, but, with the Pythagoreans, as the most beautiful and honourable" (Koyré 1958, p. 63) is further indicated in various places (Koyré 1961, pp. 288–289, p. 436 note 17) in *La révolution astronomique*, including the following:

> One does not always notice, or at least not often enough, that by placing the Sun in the centre of the world by virtue of its dignity, Copernicus inverses the Pythagorean concept and completely overturns the positional hierarchy of the Medieval and Ancient Cosmos, in which the central position is in no way the most honourable, but instead the most unworthy. It is, indeed, the basest and therefore suited to Earth's imperfection; the place of perfection is to be found *above* in the celestial sphere, upon which lie 'the heavens' (Paradise) whereas below Earth (under its surface) lies, quite aptly, hell. (Koyré 1961, p. 114 note 24)

20.3.2.2 Multiplicity of the Existing Topographies

Even if he therefore became explicitly aware of the topographical change, Koyré (and the vast majority of the posterior secondary literature) does not seem to have, however, educed this consequence that results nevertheless directly from it: Since there were, during the era of the Copernican Revolution, both geocentrists, naturally inclined to interpret this revolution according to a topography that is associated to the cosmological system that is most advantageous to them, and heliocentrists, also naturally leaning towards an evaluation of this same revolution according to the new topography associated to their new world system, at least[20] two different sets of reading matter are available for interpreting the same events. As a result, a historian of the Copernican Revolution must not only, in order to understand the reasoning of the speaker he is echoing, endorse his very own topography, but he must also, to obtain a comprehensive view of the various reactions present, systematically envisage, for each event, the interpretation that may result by applying *both* of these topographical perspectives.

Let us look at an example. According to geocentric topography, the decentring of the Earth effectuated by Copernicus should be interpreted as an aggrandisement and not as a humiliation—despite the contradictory assertion of the traditional interpretation—since our dwelling is removed from the ignominious position in which it found itself in order to be reunited with the other celestial bodies. According to heliocentric topography, however, this very same decentring should—pursuant to the traditional interpretation this time—be perceived as demeaning, since the Earth

[20] The geo-heliocentric and acentric topographies should not be overlooked.

is obliged to abandon its cosmic centrality at a point in time when it finally finds itself valued. Of course—and it is Koyré himself who taught this to us, following on from Meyerson—the route pursued by thought is not a straight line, as it is never purely logical, so that its veritable course is fundamentally unpredictable.[21] Additionally, since "reason does not cherish history" (Jorland 1981, p. 8), we should conscientiously ensure that our logical deductions indeed correspond to historical reality. Did Koyré do this? Let us verify this by reiterating the simplest case yet also the most important one (since it directly opposes the traditional interpretation), namely, that of the geocentrist examining, equipped with his topography, the Copernican decentring of the Earth.

20.3.2.3 The Decentring of the Earth Acccording to Geocentric Topography

According to Koyré, the rejection of the medieval–Aristotelian topography of the Cosmos began with Nicholas of Cusa who, by proclaiming *Terra est stella nobilis* (Nicholas of Cusa 2010 [1440], p. 215), already placed "on the same ontological level the reality of the Earth and that of the Heavens" (Koyré 1985c [1950], p. 54). It continued, he added in 1958, with Leonardo da Vinci (1452–1519) who, by likening the Moon to the Earth and the Earth to the Moon, proved "the nobility of our world" (Koyré 1958, p. 55). Along the same lines, the Copernican Revolution, far from humiliating man as the traditional interpretation asserted, "raises", he aptly wrote in 1951, "the Earth to the level of the stars" (Koyré 1989 [1951], p. 246). Naturally more explicit in *Du monde clos à l'univers infini*, Koyré again notes the posture adopted by Nicholas of Cusa, namely, "his denial—together with its central position—of the uniquely *low* and *despicable* position assigned to the earth by traditional cosmology" (Koyré 1957, p. 19), without seeming to realise that the attribution, by the Cusan, of a light belonging to the Earth issues from a common motivation.[22] In conclusion, our author wrote:

> The displacement of the earth from the centrum of the world was not felt to be a demotion. Quite the contrary: it is with satisfaction that Nicholas of Cusa asserts its *promotion* to the rank of the noble stars; and, as for Giordano Bruno, it is with a burning enthusiasm—that of a prisoner who sees the walls of his jail crumble—that he announces the bursting of the spheres that separated us from the wide open spaces and inexhaustible treasures of the ever-changing, eternal and infinite universe. (Koyré 1957, p. 43)

We could undoubtedly reproach Koyré for calling upon Nicholas of Cusa, who considerably predated the Copernican Revolution, to attest to how it was *actually* experienced by the geocentrists, just as one might be taken aback that he mentioned, in

[21] Cf. Koyré 1933a, p. 650; 1985g [1963], p. 399; 1961, p. 11.

[22] The opposition established by Koyré between "the depth of his metaphysical intuition" (the rejection of the baseness of the Earth) and "certain scientific concepts lagging behind the times" (its own light) does not do justice to the common motivation, expressed quite explicitly by Nicholas of Cusa, underlying both assertions.

the same sentence and by way of confirmation, the name of Giordano Bruno, since the reasons for enjoying either one diverge as do the topographies that guide them. For the Cusan, it is the desire to escape the inherent baseness of geocentric topography; for the Nolan, it is the joy of adopting an acentric topography characterised by the absence of any limits. One presents the disadvantage of preceding, chronologically, the actual decentring of the Earth effectuated by Copernicus while being contemporary, intellectually, to the way of thinking required to appreciate it; yet the other is contemporary, chronologically, to the events in question while presenting the disadvantage of being, intellectually, already far ahead since he challenges all hierarchy of any kind. Thus, what is their relevance with respect to establishing, from the geocentric perspective, how the Copernican gesture was really experienced? That being said, the statement we have just analysed—repeated differently in *La révolution astronomique* (Koyré 1961, p. 62)—provides a welcome invitation to distance ourselves from the traditional interpretation.

Koyré also specified that this positive consequence of the Copernican Revolution—namely, the promotion of our planet to the rank of a noble star—would be short-lived and would not prevent "the nihilism and despair" from setting in at the end of this evolution. Indeed, as he explicitly expressed, but only in notes, starting from his *Études galiléennes* (Koyré 1986d [1939], p. 225, note 2), the ontological levelling between the Earth and the other stars occurred, firstly, essentially by assimilating the Earth to the stars, before continuing, and, secondly, by assimilating the stars to the Earth, which then meant the secularisation of the celestial world and not, as was initially the case, the promotion of our dwelling.

Would a history of the anthropological consequences of the Copernican Revolution that is not only conscious of the coexistence of various interpretational frameworks for the same events but that also systematically focusses on these different points of view without unconsciously privileging—as strong as the significance of the traditional interpretation remains—the heliocentric topography, and this without forgetting to note the observed difference between the foreseen logical reactions and the actual detected behaviours, be worth writing?

20.4 Conclusion

20.4.1 Required Additional Research

In order to better contextualise and thus appreciate the Koyrean interpretation, not only of the Copernican Revolution but also of the ontological revolution, it would certainly be appropriate to follow up this first investigation with a study of the works of those predecessors and contemporaries that our author claims to have been inspired by. Without forgetting either Husserl or Meyerson, whose influence we have already noted, and in order to adhere to the publications that are most frequently mentioned by our historian,[23] a study of this kind would include, according to chronological order of appearance:

1. "The admirable" (Koyré 1961, p. 376 note 2) *History of the Planetary Systems from Thales to Kepler* (1906) by John Louis Emil Dreyer (1852–1926), which is "one of the best histories of astronomy ever written" (Koyré 1933b, 112, 1961, p. 41).
2. *The Metaphysical Foundations of Modern Physical Science: A Historical and Critical Essay* (1925) by Burtt, to which Koyré refers not only in his *Études galiléennes* (Koyré 1986d [1939], p. 214, note 2), in support of his thesis of the Platonism of Galileo, but also, and more significantly for our purposes, at the end of the revealing conclusion of his article *Sens et portée de la synthèse newtonienne* (Koyré 1965 [1950], p. 24, note 1) and at the beginning of his most prominent work *Du monde clos à l'univers infini* (Koyré 1957, p. 277, note 1), to the point that Redondi was prompted to underline the importance of this thinker for our author (Koyré 1986b, pp. xx–xxiii).
3. *Science and the Modern World* (1925) by Whitehead was also mentioned in the same strategic places as the work of Burtt,[24] which does not preclude Koyré having known him since at least 1943.[25]
4. *The Great Chain of Being: A Study of the History of an Idea* (1936) by Lovejoy, whose work, as we have repeatedly indicated, was of great importance to Koyré and which he qualifies as a "classic" (Koyré 1957, p. 277 note 1) as well as a "non-nostalgic treatment" (Koyré 1957, p. 281 note 4) of the consequences of the seventeenth-century spiritual revolution.
5. *The Breaking of the Circle: Studies in the Effect of the «New Science» upon Seventeenth Century Poetry* (1950) by Marjorie H. Nicolson (1894–1981) (Koyré 1957, p. 281, note 4), for which "the rupture of the circle" is presented, in Koyrean terms, as "the bursting of the sphere" (Koyré 1965 [1950], p. 7 note 1, 1986a [1955], p. 262).
6. *The Copernican Revolution: Planetary Astronomy in the Development of Western Thought* (1957) by Thomas S. Kuhn (1922–1996), merely mentioned in *La révolution astronomique* (Koyré 1961, p. 79, note 4).
7. *The Sleepwalkers: A History of Man's Changing Vision of the Universe* (1959) by Koestler—which Michel-Pierre Lerner recently viewed with a critical eye (Lerner 2012)—about which Koyré noted the "long chapter, interesting although quite malicious, about Copernicus" (Koyré 1961, p. 79 note 4), as well as the "200 very interesting pages" (Koyré 1961, p. 377 note 2) on Kepler, without

[23] Koyré also mentions J. H. Randall, *The Making of the Modern Mind* (1926); B. Willey, *The Seventeenth Century Background* (1934); E. M. W. Tillyard, *The Elizabethan World Picture* (1943); V. Harris, *All Coherence Gone* (1949) and S. L. Bethell, *The Cultural Revolution of the 17th Century* (1951). Cf. Koyré 1957, p. 277, note 1, p. 281, note 4).

[24] Cf. Koyré 1965 [1950], p. 24, note 1, 1957, p. 277, note 1.

[25] While reflecting on, by reviewing it, the introduction to *Galileo Studies* in order to create his article *Galileo and Plato*, Koyré mentions, this time, the work of Whitehead (cf. Koyré 1943c, 400 note 1).

remarking, maliciously, that this author "reproaches Copernicus—who he calls *the timid cannon*—for his lack of courage and intellectual audacity!" (Koyré 1961, p. 74 note 4).

20.4.2 His Interpretation of the Copernican Revolution

From our current research, it is however already possible to formulate a certain number of conclusive propositions, which are sure to subsequently either be confirmed, refuted or, more likely, nuanced.

While, as we indicated in the introduction, numerous elements, amongst which his intellectual journey and his historical methodology, seemed to have predisposed Koyré to providing an accurate history of the Copernican Revolution as a stage within the great seventeenth-century spiritual revolution and that, in addition to his general theme aimed at retracing the history of the passage "from the closed world to the infinite universe", a certain number of his historical interests, and even philosophical ones, seemed to have (or should have) led him in this direction, one has to acknowledge that, from our undoubtedly limited perspective, the Koyrean work proves to be, ultimately, quite disappointing.

Admittedly, it does include a range of interesting statements that are likely to nuance, or even contest, the traditional interpretation. Concerning the geocentric world view, we would like to mention his demonstration of the symbolic ambivalence of the position of the Earth (the worst by its baseness and the best by its singularity and its immobility); his insistence on the sufficient magnitude, relative to the Earth, of the medieval–Aristotelian Cosmos in order to make it understood that it is not on a "human scale"[26]; or even his statement that geocentrism is not intrinsically anthropocentric, since it is only in the Christian tradition that these two concepts come together[27] and, inversely, that heliocentrism is not necessarily more anti-anthropocentric, as evidenced by the Keplerian conviction according to which "the Universe is created for man" who constitutes "the axiological centre" and "profoundly determines its architecture" (Koyré 1961, p. 122). On the other hand, it is regrettable that his inattention towards the planetary centrality inherent to the geocentric topography which prevents him from realising that the extra-scientific reasons advanced by Copernicus in order to justify the centering of the Sun—reasons that, like a good Platonist who is careful about diminishing the historical importance of everything related to the empirical, he keeps insisting on—does not actually demonstrate this evidence that he likes to accord them. As regards the consequences resulting from the Copernican Revolution, we again refer to the allegation that the decentring of the Earth was not experienced as a degradation but as

[26] Cf. Koyré 1957, p. 32, pp. 34–35; 1958, p. 66; 1961, p. 109, note 1.

[27] Cf. Koyré 1961, p. 75, note 8. In speaking of the "geocentric world of the Greeks and the anthropocentric world of the Middle Ages", Koyré was already committed, at least in the French translation of *Du monde clos à l'univers infini*, to making this distinction (Koyré 1988 [1962], pp. 9–10).

a promotion, however short-lived, since the "equalisation from above" was soon to be followed by an "equalisation from below" (Koyré 1986d [1939], p. 224, note 2) or even the pertinent indication of the complete reversal of values between the medieval–Aristotelian and Copernican topographies. However, it is regrettable that these affirmations, which clearly attest to the distancing, worth emphasising, with respect to a certain number of theses dealing with the traditional interpretation, often remained sparse, rarely identified or uttered with little explanation or textual justification and, lastly, that they did not produce either a systematic treatment of the question or an explicit realisation of their incompatibility with the usual interpretation. From this point of view, it is intriguing to note that it is Michel Foucault (1926–1984), in a review of *De Revolutionibus orbium cœlestium*, who had most emphatically expressed the existing opposition between this book and the traditional interpretation (Foucault 1994 [1961], p. 170). Thus, even if the Koyrean work incontestably furnishes a fair amount of useful information for tracing, from the anthropological consequences of the Copernican Revolution, an interpretation other than the traditional one, this other interpretation is not really offered. Moreover, both the consequences that Koyré derives from the spiritual revolution, namely, the divorce between the world of science and the world of life on the one hand and the retreat of the Divine on the other hand, tend rather, by their tone, to reinforce the general pertinence of the traditional interpretation, of which our author effectively interprets the tragic scope by remarking that man not only lost his place in the world but also the "same world that framed his existence and the object of his knowledge" (Koyré 1988 [1962], p. 11).

It would be insulting towards Koyré to support, or even simply imagine, that he did not perceive these weaknesses. Also, without denying the inadequacy of his historical analysis of the anthropological consequences of the Copernican Revolution, one should examine the reasons for this inadequacy. It stems, we think, from his relative indifference towards this problem. As with Lovejoy, who equally neglects the anthropological consequences of the "Copernican moment" and turns towards those more decisive ones, which issue from the theory of the plurality of inhabited worlds (a theme, incidentally, that our historian completely abandons), Koyré seems to do the same but to the benefit of the two themes, which are also intimately linked, that were always at the heart of his concerns: the first question being that of infinity and the second question, for him an existential one,[28] being that of the existence of God. Indeed, taking one's time to study the anthropological consequence of the Copernican Revolution is trying to determine how the supporters of the different topographies in existence experienced, for example, the decentring of the Earth or the centring of the Sun. Yet, such considerations, while still justified in the Copernican era, will soon no longer be, nearly 100 years later, once the geometrisation of space and the spatial infinity of the universe are acquired. Hastening to the stage of the infinitisation of the universe, primarily concerned with the potential

[28] Jorland reports that, according to the testimony of his wife, Koyré, while he was studying the evidence for the existence of God in mystical theology, "[…] tried to convince himself, but in vain […]" (Jorland 1994, p. 114).

impact of this infinitisation on God, Koyré, as also evidenced by the few pages he devotes to this, simply did not have enough interest in this transitional phase from "the closed world to the infinite universe", which he knew was purely transitory.

20.4.3 His Interpretation of the Ontological Revolution

Since Koyré seemed to have somewhat neglected the consequences of the Copernican Revolution to turn resolutely towards those of the ontological revolution, what does one then make of his interpretation of these? Although this issue is beyond the explicitly announced theme in the title of this article, we will undoubtedly be permitted, given that we mentioned it in the first part, to express our feelings on the subject, which will of course be amply justified and nuanced.

As regards the first consequence, namely, the divorce between the world of science and the world of life and between the world of facts and the world of values, this seems to us to be philosophically plausible but having little historical foundation: Who are these men who really experienced the sentiment of this divorce? How and where did they express this? From which era are they? It seems to us that Koyré hardly, or even never, specifies this and that it is rather Rémi Brague (Brague 2002) who historically described, at least partially, that which our historian seemed to have been content with merely stating. Does this absence of historical proof not lead us to think that, ultimately, it is the turmoil of *his* era that Koyré was marvellously echoing?

As regards the second consequence, namely, the retreat of the Divine, this seems, contrary to the preceding one, to be historically founded but without being either philosophically or theologically so. By this decidedly rapid assessment, we suggest that Koyré offers, with his own very undeniable talent and consistent with his role as a historian—it is indeed a historian who narrates the retained arguments even if these are flawed or completely erroneous (this is a point that Koyré himself, contrary to the positive practice of history, insisted on frequently)—a line of reasoning, presented as necessarily leading to the retreat of the Divine and that was actually regarded as such, but of which one sometimes wonders if he notices the philosophical and religious falsehood.

Allow us to clarify our idea. As a historian, Koyré had the great merit of taking some issues, which could have raised a smile, seriously, for example, the following: "how many angels could fit on the tip of a needle?" or even:

> […] does human intellect reside in the Moon, or elsewhere? [because he had understood the fundamental issue, which is] to discover if the spirit, whether a being or a spiritual act […] occupies, or not, a *place* in space [and he thenceforth realised that this issue was not] ridiculous at all. (Koyré 1985a [1944], p. 25)

In his work, this issue became more specifically, as we have seen, that of the place and the action of God in the world. However if, as a historian, Koyré was entirely correct in affording great importance to this theme, one may well wonder if, as a philosopher (and perhaps also as a research agnostic), he did not commit an error in ultimately allocating too much importance to it! Since even if it is historically proven (as the Koyrean work sufficiently demonstrates), the reasoning that progressively leads the Moderns to think that their universe could no longer reserve either a place or an action for God seems, to us, to be erroneous. To be convinced of this, one only needs—as Koyré could have done (or as he possibly did, we have no idea)—to read Saint Thomas Aquinas (1225–1274) or at least his commentator Antonin-Dalmace Sertillanges (1863–1948)[29]: One thus realises that this sort of reasoning falls under a theology of the Creation, which is certainly naive, and that was already known as such during the era when Koyré wrote the touching conclusion of his *Du monde clos à l'univers infini*!

To put it briefly, no, the metaphysical Creation does not identify itself with an act that occurred once and for all at the beginning of time, because we completely dissociate metaphysical Creation and the natural beginning of the world; no, consequently, the eternity of the world is not likely to render vain and superfluous any notion of Creation. No, unless one remains in an anthropomorphic vision of divinity, the "presence" of God does not require a spatially finite material universe, no more than his "disappearance" results from the spatial infinity of this same universe, because our concept of God disqualifies him, by his transcendence, from both spatial and temporal categories. No, the progressive realisation of the autonomy of the world (in reaction to a God historically conceived as a great watchmaker) does not result in making him unnecessary or imminently inexistent, because this autonomy of the created world, far from disturbing us, rather confirms his very own order.

It would therefore seem wise to question the theology of the creation of Koyré—notably from his conference *Théologie et science* of 1947 (Koyré 1986b, p. 196)—in order to verify whether his theological thinking was indeed on par with his historical thinking.

Our initial question (to what extent did the Koyrean work subscribe to the traditional interpretation or, on the contrary, did it prepare to be challenged?) ultimately receives a rather nuanced answer. On the one hand, this work offers us, in an admittedly sparse and discrete manner, a certain number of elements that could be used against this interpretation. On the other hand, by its general tone and its most explicit theories (the divorce established between the world of science and the world of life and the retreat of the Divine), it justifies, in its own way, this interpretation: not insofar as this interpretation would adequately describe what the contemporaries of the Copernican Revolution and the ontological revolution felt, or the consequences one must really derive from these revolutions but insofar as it sufficiently expresses the lasting sentiment, of these revolutions and their place in the universe, still experienced by the twentieth-century man and, particularly, by the existentialists.[30] By a process of "retroprojection", which Koyré himself cautioned us against and which we thus apply to counter the one who taught it to us, his work seems to deal less with the turmoil of the men of yesteryear than the restless wandering of those of his

[29] Cf. Thomas of Aquinas 1948; Sertillanges 1945.

[30] Cf. Koyré 1998 [c. 1946–1950], p. 536; Jorland 1981, pp. 132–133.

era. It seems therefore—and this was our second aim, namely, to assess our interpretation in the light of that of Koyré—that his interpretation of the events, far from having to be rejected, should instead be remembered in so far as that it, first and foremost, reveals the *Weltanschauung* of the twentieth-century man. As regards the work of Koyré, who shaped us in more ways than one, it really is the least we can do!

Acknowledgements I wish to thank Vincent Ligot, Director of the Paramedical Department of Montignies, and the Haute École Louvain-en-Hainaut (HELHa) for their support in the publication of this chapter.

References

Anselme de Cantorbéry (1927) Fides quaerens intellectum id est Proslogion, Liber Gaunilonis pro insipiente atque Liber apologeticus contra Gaunilonem. Librairie philosophique J. Vrin, Paris.

Belaval Y (1964) Les recherches philosophiques d'Alexandre Koyré. Critique 20/207–208:675–704.

Blumenberg H (1975) Die Genesis der kopernikanischen Welt. Frankfurt-am-Main: Suhrkamp.

Brague R (1990) Le géocentrisme comme humiliation de l'homme. In Brague R and Courtine JF (ed). Herméneutique et ontologie: Mélanges en hommage à Pierre Aubenque. Presses universitaires de France, Paris, pp. 203–223.

Brague R (1994) Geozentrismus als Demütigung des Menschen. Internationale Zeitschrift für Philosophie 1/1:1–25.

Brague R (1997) Geocentrism as a humiliation for man. Medieval Encounters 3(3):187–210.

Brague R (2001) Geocentrizmas kaip žmogaus pažeminimas. Logos [Lithuania] 25:41–56; 26:34–43.

Brague R (2002) La Sagesse du monde: histoire de l'expérience humaine de l'Univers. Librairie Arthème Fayard, Paris.

Brague R (2008) Le géocentrisme comme humiliation de l'homme. In Brague R (ed). Au moyen du Moyen âge: philosophies médiévales en chrétienté, judaïsme et islam. Flammarion, Paris, pp. 362–396.

Burtt EA (1925) The metaphysical foundations of modern physical science: a historical and critical essay. Kegan Paul, Trench, Trubner, London.

Copernicus N (1992) Complete works: On the revolutions. Translated by E Rosen. The Johns Hopkins University Press, Baltimore, London.

Danielson DR (2001) The great Copernican cliché. American Journal of Physics 69/10:1029–1035.

Danielson DR (2009) That Copernicanism demoted humans from the center of the Cosmos. In Numbers RL (ed). Galileo goes to jail and other myths about science and religion. The Harvard University Press, Cambridge–MA Mass, London, pp. 50–58.

Fontenelle Bernard de (2013) Œuvres complètes. Vol. 1. Entretiens sur la pluralité des mondes. Cazanave C, Poulouin C (eds). Honoré Champion, Paris.

Foucault M (1994) Compte rendu d'A. Koyré : La révolution astronomique: Copernic, Kepler, Borelli (1961). In Foucault M, Dits et écrits (1954–1988). Vol. 1: 1954–1969. Defert D, Ewald F, Lagrange J (eds). Éditions Gallimard, Paris, pp. 170–171.

Freud S (1992) Les résistances contre la psychanalyse. Translated by Altounian J, Cotet P, Hanus M, Strauss M. In Freud S, Œuvres complètes: Psychanalyse. Vol. 17: 1923–1925. Bourguignon A, Cotet P (eds). Presses universitaires de France, Paris, pp. 125–135.

Freud S (1996) Une difficulté de la psychanalyse. Translated by Altounian J, Bourguignon A, Cotet P, Rauzy A. In Freud S, Œuvres complètes: Psychanalyse. Vol. 15: 1916–1920. Bourguignon A, Cotet P (eds). Presses universitaires de France, Paris, pp. 41–51.

Freud S (2000) Leçons d'introduction à la psychanalyse. Translated by Bourguignon A, JG Delarbre JG, Hartmann D, Robert F. In Freud S, Œuvres complètes: Psychanalyse. Vol. 14: 1915–1917. Bourguignon A, Cotet P (eds). Presses universitaires de France, Paris, pp. 1–480.
Galilei G (1980) L'essayeur. Translated by Chr. Chauviré. Les Belles Lettres, Paris.
Hallyn F (1987) La structure poétique du monde: Copernic, Kepler. Éditions du Seuil, Paris.
Iommi Amunátegui G (2002) E. Husserl et A. Koyré: un point de rencontre. Revue des questions scientifiques 173/1:73–78.
Jorland G (1981) La science dans la philosophie: les recherches épistémologiques d'Alexandre Koyré. Éditions Gallimard, Paris.
Jorland G (1994) Koyré phénoménologue ? In Vinti C (ed). Alexandre Koyré: L'avventura intellettuale. Edizioni Scientifiche Italiane, Napoli, pp. 105–126.
Jorland G (2006) La notion de révolution scientifique: le modèle de Koyré. In Bitbol M, Gayon J (ed). L'épistémologie française, 1830–1970. Presses universitaires de France, Paris, pp. 157–171.
Koyré A (1912) Sur les nombres de M. Russell. Revue de métaphysique et de morale 20/5:722–724.
Koyré A (1923) La pensée judaïque et la philosophie moderne. Menorah 2/27–28:452–453.
Koyré A (1926) Compte rendu de L. Rougier: La scolastique et le thomisme (1925). Revue philosophique de la France et de l'étranger 101/5–6:462–469.
Koyré A (1928) La jeunesse d'Ivan Kirêevskij. Le monde slave 5/2:213–238.
Koyré A (1929) La philosophie de Jacob Boehme. Librairie philosophique J. Vrin, Paris.
Koyré A (1933a) Compte rendu d'É. Meyerson: Du cheminement de la pensée (1931). Journal de psychologie normale et pathologique 30/5–6:647–655.
Koyré A (1933b) Copernic. Revue philosophique de la France et de l'étranger 116/7–8:101–118.
Koyré A (1935) Compte rendu d'A. N. Whitehead: Nature and life (1934). Recherches philosophiques 4:398.
Koyré A (1936) Compte rendu d'E. Brachvogel: Nicolaus Koppernicus und Aristarch von Samos (1935). Recherches philosophiques 5:459–460.
Koyré A (1939), Études galiléennes. I : À l'aube de la science classique. Hermann & Cie éditeurs, Paris.
Koyré A (1943a) Compte rendu d'A. Viatte: Victor Hugo et les illuminés de son temps (1942). Renaissance 1/3:508–512.
Koyré A (1943b) Galileo and the Scientific Revolution of the Seventeenth Century. The Philosophical Review 52/4:333–348.
Koyré A (1943c) Galileo and Plato. Journal of the History of Ideas 4/4:400–428.
Koyré A (1947a) Histoire de la magie et de la science expérimentale. Revue philosophique de la France et de l'étranger 137/1–3:90–100.
Koyré A (1947b) La philosophie au XVIIIe siècle. Europe 25/19:114–121.
Koyré A (1950a) L'apport scientifique de la Renaissance. Revue de synthèse 67:30–65.
Koyré A (1950b) Le mythe et l'espace. Revue philosophique de la France et de l'étranger 140/7–9:320–322.
Koyré A (1951a) Compte rendu d'A. Maïer : Die Vörläufer Galileis im 14. Jahrhundert: Studien zur Naturphilosophie der Spätscholastic (1949). Archives internationales d'histoire des sciences 4/16:769–783.
Koyré A (1951b) Compte rendu de D. W. Singer: Giordano Bruno, his life and thought, with annotated translation of his work On the infinite universe and worlds (1950). Archives internationales d'histoire des sciences 4/16:787–790.
Koyré A (1951c) Les étapes de la cosmologie scientifique. Revue de synthèse 70:11–32].
Koyré A (1952a) Compte rendu d'Ed. Whittaker: From Euclid to Eddington: a study of the conceptions of the external world (1949). Archives internationales d'histoire des sciences 5/18–19:133–136.
Koyré A (1952b) Un « experimentum » au XVIIe siècle: la détermination de « g ». XXIIe Congrès international de philosophie des sciences [Paris, 1949]. Vol. 8. Histoire des sciences. Hermann, Paris, pp. 83–92.

Koyré A (1953) An Experiment in Measurement. Proceedings of the American Philosophical Society 97/2:222–237.
Koyré A (1957) From the closed world to the infinite universe. The Johns Hopkins University Press, Baltimore.
Koyré A (1958) Les sciences exactes de 1450 à 1600. In Taton R (ed). Histoire générale des sciences. Vol. 2. La science moderne (de 1450 à 1800). Presses universitaires de France, Paris, pp. 11–105
Koyré A (1961) La révolution astronomique: Copernic, Kepler, Borelli. Herman, Paris.
Koyré A (1965) The significance of the Newtonian synthesis. In Koyré A, Newtonian studies, 3–24. Chapman & Hall, London.
Koyré A (1968) La gravitation universelle de Kepler à Newton. In Koyré A (1968) Études newtoniennes. Éditions Gallimard, Paris, pp. 11–24.
Koyré A (1971a) Paracelse. In Koyré A (1971a) Mystiques, spirituels, alchimistes du XVIe siècle allemand. Éditions Gallimard, Paris, pp. 75–129.
Koyré A (1971b) Un mystique protestant: Maître Valentin Weigel. In Koyré A (1971b) Mystiques, spirituels, alchimistes du XVIe siècle allemand. Éditions Gallimard, Paris, pp. 131–184.
Koyré A (1979) La philosophie de Jacob Boehme. Librairie philosophique J. Vrin, Paris.
Koyré A (1984) L'idée de Dieu dans la philosophie de St. Anselme. Librairie philosophique J. Vrin, Paris.
Koyré A (1985a) Aristotélisme et platonisme dans la philosophie du moyen âge. In Koyré A, (1985a) Études d'histoire de la pensée scientifique. Éditions Gallimard, Paris, pp. 24–49.
Koyré A (1985b) Attitude esthétique et pensée scientifique. In Koyré A (1985b) Études d'histoire de la pensée scientifique. Éditions Gallimard, Paris, pp. 275–288.
Koyré A (1985c) L'apport scientifique de la Renaissance. In Koyré A (1985c) Études d'histoire de la pensée scientifique. Éditions Gallimard, Paris, pp. 50–60.
Koyré A (1985d) La pensée moderne. In Koyré A (1985d) Études d'histoire de la pensée scientifique. Éditions Gallimard, Paris, pp. 16–23.
Koyré A (1985e) Les étapes de la cosmologie scientifique. In Koyré A (1985e) Études d'histoire de la pensée scientifique. Éditions Gallimard, Paris, pp. 87–98.
Koyré A (1985f) Les origines de la science moderne: une interprétation nouvelle. In Koyré A (1985f) Études d'histoire de la pensée scientifique. Éditions Gallimard, Paris, pp. 61–86.
Koyré A (1985g) Perspectives sur l'histoire des sciences. In: Koyré A (1985g) Études d'histoire de la pensée scientifique. Éditions Gallimard, Paris, pp. 390–399.
Koyré A (1986a) De l'influence des conceptions philosophiques sur l'évolution des théories scientifiques. In Koyré A (1986a) Études d'histoire de la pensée philosophique. Éditions Gallimard, Paris, pp. 253–269.
Koyré A (1986b) De la mystique à la science: Cours, conférences et documents (1922–1962). Edited by Redondi P. Éditions de l'École des hautes études en sciences sociales, Paris.
Koyré A (1986c) Du monde de l'« à-peu-près » à l'univers de la précision. In Koyré A (1986c) Études d'histoire de la pensée philosophique. Éditions Gallimard, Paris, pp. 341–362.
Koyré A (1986d) Études galiléennes. Hermann Éditeurs des Sciences et des Arts, Paris.
Koyré A (1986e) Le vide et l'espace infini au XIVe siècle. In Koyré A (1986e) Études d'histoire de la pensée philosophique. Éditions Gallimard, Paris, pp. 37–92.
Koyré A (1986f) Les philosophes et la machine. In Koyré A (1986f) Études d'histoire de la pensée philosophique. Éditions Gallimard, Paris, pp. 305–339.
Koyré A (1987) Entretiens sur Descartes. In Koyré A (1987) Introduction à la lecture de Platon, suivi de Entretiens sur Descartes. Éditions Gallimard, Paris, pp. 161–229.
Koyré A (1988) Du monde clos à l'univers infini. Translated by Tarr R. Éditions Gallimard. Paris.
Koyré A (1989) Pedagogical and research orientations. Translated by Virmani A. In Redondi P (ed). The history of sciences: The French debate. Orient Longman, New Delhi, pp. 245–249.
Koyré A (1998) Present trends of French philosophical thought. Journal of the History of Ideas 59/3:521–548.

Lerner MP (2012) Le « Copernic » de Koestler dans « Les Somnambules » ou de l'art de (mal) traiter les sources. Giornale critico della filosofia italiana 91/2:514–529.

Lovejoy AO (1936) The great chain of being: a study of the history of an idea. The Harvard University Press, Cambridge–MA.

Meyerson É (1931) Du cheminement de la pensée. Librairie Félix Alcan, Paris.

Montpetit R (1970) Freud, Copernic et la méprise. Dialogue : Revue canadienne de philosophie 9/1:88–92.

Nicolas de Cuse (2010) De la docte ignorance. Translated by Lagarrigue JC. Les éditions du Cerf, Paris.

Olesen SG (1994) L'héritage husserlien chez Koyré et Bachelard. Danish Yearbook of Philosophy 29:7–43.

Rheticus GJ (1982) Narratio prima. Translated by Hugonnard–Roche H, Verdet JP, Lerner MP, Segonds A. Ossolineum, Wroclaw–Warszawa–Kraków.

Savary C (1969) La révolution copernicienne: Freud et le géocentrisme médiéval. Dialogue : Revue Canadienne de Philosophie 8/3:417–432.

Savary C (1970) Narcisse et son médecin. Dialogue: Revue canadienne de philosophie 9(3):397–400.

Sertillanges AD (1945) L'idée de création et ses retentissements en philosophie. Aubier Éditions Montaigne, Paris.

Špelda D (2014) Kloaka světa: Geocentrismus, antropocentrismus a mimozemšťané. Pro-Fil 15(1):62–81.

Stoffel JF (1998) La révolution copernicienne et la place de l'Homme dans l'Univers: étude programmatique. Revue philosophique de Louvain 96/1:7–50.

Stoffel JF (2000) Bibliographie d'Alexandre Koyré. Leo S. Olschki, Firenze.

Stoffel JF (2001) Géocentrisme, héliocentrisme, anthropocentrisme: quelles interactions ? Scientiarum historia 27/2:77–92.

Stoffel JF (2002) La révolution copernicienne responsable du « désenchantement du monde » ? L'exemple des analogies solaires. Revue belge de philologie et d'histoire 80/4:1189–1224.

Stoffel JF (2005) Cosmologie versus idolâtrie: l'exemple de la désacralisation du Soleil. In Dekoninck R, Watthée–Delmotte M (eds). L'idole dans l'imaginaire occidental. L'Harmattan, Paris, pp. 195–216.

Stoffel JF (2012) Origine et constitution d'un mythe historiographique: l'interprétation traditionnelle de la révolution copernicienne. Sa phase de structuration (1835–1925). Philosophica: Revista del Instituto de filosofía de la Universidad católica de Valparaíso 41–42/1–2:95–132.

Thomas d'Aquin (1948) Somme théologique: la Création (1a, questions 44–49). Translated by Sertillanges AD. Éditions de la Revue des jeunes, Paris–Tournai–Roma.

Chapter 21
Koyré as a Historian of Religion and the New French Phenomenology

Anna Yampolskaya

Abstract The aim of this article is to explore the influence that Koyré's early work on history of religion had on the development of French phenomenology, with focus on Emmanuel Levinas and Michel Henry. Although Koyré's affiliation to the phenomenological movement is debatable, his thought owes much to Husserl's phenomenological method. In his books on St. Anselm and Descartes, Koyré focuses on the idea of God and the idea of the infinite. I trace the influence of Koyré's analysis of the infinite in its relation to the finite on the development of the idea of the infinite in Levinas. I also show that Levinassian approach to the idea of God as "the idea of the Infinite in me" goes back to Koyré's interpretation of the ontological proof of St. Anselm. Next, I explore the influence of Koyré's book on Böhme on the philosophy of Michel Henry. Koyré's reading of Böhme makes Böhme essentially a precursor of German idealism describing the Absolute that wishes to manifest itself and distinguishing between the manifestation and what is made manifest in this manifestation. Henry applies this approach to phenomena in general, which leads him to a criticism of intentionality loss of a cosmological dimension. I would like to argue in favour of a more balanced phenomenology that wants to be not only prescriptive but also descriptive and sensitive to a certain scientific dimension.

Keywords History of phenomenological movement • Ontological argument • Idea of the infinite • Cosmological dimension of human experience • Anselm • Böhme • Descartes • Husserl • Michel Henry • Levinas

A. Yampolskaya (✉)
National Research University Higher School of Economics, Staraya Basmannaya ul., 21/4c1, Moscow 105066, Russian Federation
e-mail: ayampolskaya@hse.ru

R. Pisano et al. (eds.), *Hypotheses and Perspectives in the History and Philosophy of Science*, DOI 10.1007/978-3-319-61712-1_21

21.1 Introduction

Alexandre Koyré, although a former student of Husserl, never described himself as a genuine phenomenologist; neither did he formally apply intentional analysis or phenomenological reduction in his studies of medieval or modern thought (Jorland 1994). In this paper I would like to put aside the question whether Koyré's method was truly phenomenological or not as that would require a clear definition of what phenomenology is; I would rather concentrate on the impact his work made on his contemporaries as well as on the next generation of thinkers.

It is well known that Koyré's life and work are closely linked to the history of the phenomenological movement. Indeed, young Koyré was a student in Gottingen in 1909–1911[1] where he attended Edmund Husserl's and Adolf Reinach's courses in philosophy and started lifelong friendships with Edith Stein, Jean Hering and Hedwig Conrad-Martius. Despite being close to Husserl, Koyré failed to produce a thesis in logic and foundation of mathematics that Husserl would have approved (Zambelli 1999). Eventually Koyré decides to interrupt his studies in Gottingen, and in 1912 he moves to Paris. He enrols into the *École pratique des hautes études* (*EPHE*) and reads history of religion under the supervision of François Picavet. Koyré subsequently receives his first, second and third degrees from *EPHE* for his research on the history of religious and mystical thought. Throughout his entire career, he would be returning to Section V of *EPHE*. Nevertheless, in the 1920s and 1930s, Koyré did his best to create and maintain the reputation of a leading French expert on phenomenology[2] and even of an "agent of influence" of phenomenology on French soil. In 1953, in his famous letter to Spiegelberg, Koyré declined to call himself a phenomenologist while admitting his phenomenological lineage:

> Now your question, how far I am still a phenomenologist – I don't know myself. I have been deeply influenced by Husserl [...] But, probably, he would tell that all this is very far from the meaning of phenomenology as philosophy. And that I have never understood it. *Now* I assume that he knew better than anyone else what 'phenomenology' really meant. (Jorland 1981, p. 28, *italics* added)

However, in the 1920s, he did exactly the opposite. Indeed, Koyré regularly published German versions of his research papers in important phenomenological volumes, such as *Jahrbuch für Philosophie und phänomenologische Forschung* (Koyré 1922b) and Husserl's *Festschrift* (1929b), as well as the German translation of his book on Descartes. Koyré also maintained his personal contact with Husserl: he visited Husserl several times,[3] and alongside other members of the Gottingen circle, he attended Husserl's 70th anniversary event in April 1929. On 1 March 1929,[4]

[1]For more details regarding Koyré's early biography, including new archive materials, see Drozdova 2012.

[2]Jan Patočka, who attended Koyré's courses in *EPHE*, also mentions him among his phenomenological contacts (Patočka 1976, p. vi).

[3]Summer 1921, Summer 1922 with Jean Hering, October 1928, Summer 1929 (Schuhmann 1977).

[4]According to Kojève (Koyré 1929a), in the "Husserl Chronik", the date of the *soutenance* is indicated as 23–25 February (Schuhmann 1977, p. 342).

Husserl was seated among the jury in Koyré's public defence in Sorbonne as an "honorary guest" (Kojève 1929) and witnessed "his former student's triumph" (Patočka 1976, p. vii). Koyré also made a lot of effort to introduce phenomenology to the French philosophical scene. He gave a talk at the very first French conference on phenomenology, held in 1932 in Juvisy; he also revised Levinas' and Pfeiffer's translation of *Cartesian Meditations* and was praised by Husserl as the book's "true translator".[5] Also, he smoothed out Husserl's election as a corresponding member of the *Académie des Sciences Morales et Politiques* (Schuhmann 1997: 392). In the 1930s Koyré used his position as a managing editor of the leading French periodical *Recherches philosophiques* to advertise the new German thought, especially phenomenology. However, the phenomenology he promoted was not necessarily Husserl's transcendental phenomenology but rather a kind of "fusion" between Husserlian, Heideggerian[6] and Hegelian phenomenology; Koyré's efforts to hegelianize the phenomenological method were applauded by Kojève (see Kojeve 1934)). While Spiegelberg accuses Kojève, "a Russian Marxist", of this "misinterpretation" or even "misinformation" (Spiegelberg 1982, p. 441), it was indeed Alexandre Koyré who was responsible not only for the anthropological but also for neo-Hegelian twist in early French phenomenology (cf. also Baugh 2014, pp. 25–28).

Husserl was well aware of some of these nonorthodox interpretations of his doctrine, and he even found it appropriate to forewarn his current and future students against them. He wrote to one of his prospective adepts, E. Parl Welch:

> The fact that someone was my academic student or became a philosopher under the influence of my writing does not therefore mean that he has penetrated to a real understanding of the inner meaning of *my*, the original phenomenology and its method and does research into the new horizons of problems which I have opened up, to which the future belongs (of which I have become completely certain). It is true to almost all the students of the Gottingen and the first Freiburg period, even of such famous men as Max Scheler and Heidegger, in whose philosophies I see merely ingenious relapses into the old philosophical naivités. I have to refer in this context even to my close friend Jean Hering…you would go astray if you rush at any of the literary accounts of my phenomenology (not even at the latest by Levinas […] who puts my phenomenology on the same plane with that of Heidegger and thus deprives it of its proper meaning. (Spiegelberg 1981, p. 181)

Indeed, Alsatian philosopher Jean Hering was another member of the Gottingen circle who after the First World War had much contributed to the development of the French phenomenology. He fiercely defended Husserl in his polemics with the

[5] Husserl wrote to Koyré on 22.VI.1931: "Immer wieder hörte ich die Luzidität Ihrer Übersetzung rühmen (auch in den Zuschriften), man meinte sogar, in Ihrer französischen Sprache und der ihr eigenen Durchsichtigkeit, kämen meine Gedanken zu einem wirksameren Ausdrucke als in meiner deutschen Sprache. Natürlich habe ich nachdrücklich darauf hingewiesen, dass Sie der eigentliche Übersetzer seien" (Husserl 1994, p. 359). This was a dubious compliment as Husserl did not like the translation at all: on 19 August 1932, he wrote to Roman Ingarden (1893–1970) that "die französische Übersetzung voll Hemmnisse des Verständnisses" (Husserl 1994, p. 288).

[6] Koyré published a review of *Sein und Zeit*; he was also the author of the preface to the very first French translation of Heidegger, which was *Was ist Metaphysik?* translated by Henri Corbin. The first issue of *Recherches Philosophiques* contained the French translation of *Vom Wesen des Grundes*. For more details regarding Koyré's role in the French reception of Heidegger's philosophy *de l'entre-deux-guerres*, see (Janicaud 2001, pp. 33–43) and (Geroulanos 2010, pp. 54–57).

Russian emigrant philosopher Leon Shestov; Hering's book on phenomenology of religious life (Hering 1926) was among first French publications on the subject (which were not very numerous at that point). In the bibliography, Hering includes Koyré's books on St. Anselme and Descartes under the rubric "the most important works that can serve as an initiation to the phenomenological philosophy" (Hering 1926, p. 145). Hering also claims that Koyré's approach provides a good example of how an application of the "Husserlian eidetic" method can lead to a "rehabilitation" of the ontological argument (Hering 1926, p. 136). It is worth noting that Hering includes a special section on Koyré in his post-war review of French phenomenology. There he adds some important details to his description of Koyré's phenomenological and historical method: instead of merely "studying the influences of the surroundings", Koyré "put[s] us in immediate relation not only to the era of those philosophers but also with a certain field of philosophical problems itself"; the problems "treated by the great thinkers of the past" are "often proved identical" to the problems rediscovered by phenomenologists (Hering 1950, p. 71).

21.2 The Idea of the Infinite in Me: Koyré and Levinas

Hering's opinion is of a particular importance to us, as in the late 1920s he was an informal mentor to the young Lithuanian student Emmanuel Levinas. It is the affinity between Koyré's and Levinas' interpretations of the ontological argument that I would now like to explore in detail. Levinas personally knew Koyré quite well[7] and mentioned his name with great affection and respect. They had a lot in common – both were originally from the Russian Empire, both were native Russian speakers, both were Jewish, both played an important role in spreading Heidegger's popularity in France and both changed their enthusiastic attitude to Heidegger's philosophy once his notorious involvement with the Nazism became known.[8] In the 1930s Levinas published several book reviews in the *Recherches Philosophiques*, and his most important paper of that period, *De l'evasion*, also appeared there. Levinas never referenced any of Koyré's works, apart from his post-war review of Heidegger's

[7]Levinas' biographer, Marie-Anne Lescourret, writes that Levinas and Koyré used to drink coffee together (Lescourret 2005, p. 108), and there is an uncut copy of Levinas' *En découvrant l'existence avec Husserl et Heidegger* with Levinas' inscription "*à Monsieur Koyré et à son ironie*" in *Koyré's personal library* in Paris (I would like to thank Daria Drozdova who informed to me on this matter).

[8]According to Levinas, it was Koyré who brought from Germany the unsettling news about Heidegger: "I learned very early, perhaps even before 1933 and certainly after Hitler's huge success at the time of his election to the Reichstag, of Heidegger's sympathy toward National Socialism. It was the late [*le regretté*] Alexandre Koyré who mentioned it to me for the first time on his return from a trip to Germany" (Levinas 1989). In the original version, this paper had a subtitle "Alexandre Koyré avait averti les Français", which was omitted in the English translation. In his unpublished letter to G. Spiegelberg from 10 August 1956, Koyré, describing how highly successful 1953 Cerisy's event on Heidegger was, also adds "Jean Wahl and, of course, myself, did not go there" (Fonds Koyré at the Centre Alexandre Koyré, Paris).

On the Essence of Truth;[9] however, there is a noticeable similarity between the interpretations of the ontological argument in Koyré and in Levinas. I would suggest that there are elements of the phenomenological method in Koyré's early works on Descartes and Anselm and that Levinas has mainly inherited his approach.

For both Levinas and Koyré, the infinite was a matter of greatest philosophical importance; both used the word "infinite" in the titles of their major works. Of course, there is a huge difference between the cosmological and mathematical infinity studied by Koyré and the ethical and theological infinity of Levinas' own philosophy; nevertheless, they have a lot in common in how they treat this concept.

In his early article on the paradoxes of Zeno (Koyré 1922b), Koyré claims that the scientific breakthrough of Descartes is that he not only established the legitimacy of actual infinity but, moreover, made it a theoretical foundation of finitude.[10] It was the genius of Descartes the mathematician (who is "superior to Cantor") to treat the infinite as a primary concept and to relegate the finite to a secondary position of what can be defined in terms of the infinite. Furthermore, Koyré states that Descartes was the first to understand that *the finite as such* cannot be properly grasped outside of its relation to the infinite and that this is indeed a metaphysical and not a mathematical claim. Indeed, no finite entity can grasp itself without recourse to the infinite. Descartes is driven by the desire to understand how to think the unthinkable, which cannot be imagined or conceived mentally: the ideas of the infinite and the continuous be it in the realm of philosophy, metaphysics or theology. One encounters here "the eternal problem of *me on*"; its difficulty is the difficulty of the "constitution of being itself" (Koyré 1980, p. 31). Indeed, the existence problems of the infinite and the continuous *in intellectu* or *extra intellectu* (Koyré 1980, p. 31) are of no importance to Koyré, because the only thing that matters is *how* we think the infinite and the continuous. Similar assertions can be found in Koyré's pre-war writings on Descartes. In his Cairo lectures, Koyré says that although not much remains of Descartes' metaphysics or his proofs of the existence of God, his greatest discovery – the intellectual primacy of the infinite – is still valid today.[11] It is this redirection of the focus of attention from an infinite object to the manner in which we encounter it that constitutes the principal novelty of Koyré's analysis of the ontological proofs of Anselm and Descartes. In these early works, Koyré presents Descartes primarily as a theologian and even as a "mystical

[9] At least in his published works, while there are testimonies that he used Koyré's studies on the philosophy of science in his teaching (Lescourret 2005, p. 108).

[10] "Supérieur à Cantor par la puissance et la profondeur de ses vues, il a pu établir non seulement la légitimité essentielle de l'infini actuel, et montrer l'impossibilité de le remplacer par la notion de l'indéfini, mais, en plus, il en a fait le fondement et le principe de la théorie du fini" (Koyré 1980, p. 26).

[11] "Il ne reste plus grande chose de la métaphysique de Descartes, et ses preuves de l'existence de Dieu sont allées rejoindre les preuves d'Aristote et de saint Thomas. Et pourtant, la grande découverte cartésienne, la découverte de la primauté intellectuelle de l'infini, reste vraie. Il reste vrai que la pensée enveloppe et implique l'infini, il reste vrai que la pensée finie – toute pensée finie – ne peut se saisir, ni se comprendre qu'à partir d'une idée infinie (Koyré 1962, p. 227).

apologist"[12]; he tries to show how Descartes' purely theological thesis, that is, the infinity of God and His incomprehensibility, prompts the destruction of the old scientific worldview and the establishment of a new one.[13] Koyré's approach to Descartes is strikingly phenomenological: instead of studying *what is* the infinite in itself, whether it is real or subjective, one examines *how* we think about it. In more technical terms that Koyré himself preferred not to use, one takes the infinite *not as an object which is intended but as an intentional object, that is, as the object as it is intended* (Husserl 1970, p. 113). Then one can legitimately concentrate on the intentional relation to this object, because this relation makes sense even when the object itself does not exist.

Koyré's interpretation of the ontological argument is centred on what can be called its "anthropological aspect": he puts aside the theological problem of the existence of God, while focusing on the human being as *capax Dei* (Koyré 1971, p. 31), the expression itself belongs to St. Augustin's *De Trinitate* XIV, 8, as a being endowed with the idea of God.[14] The "Third Meditation" is important for Koyré not because it is a valid proof of the existence of God. The logical structure of Descartes' proof was borrowed from St. Anselm; this structure has been made explicit already by St. Bonaventura. The argument has two parts: the discovery of the possibility of the infinite being and the transition from the possibility of this being to its existence. *Deus cogitatur – ergo Deus est*: this is Koyré's summary of the logic of Descartes (Koyré 1922a, p. 150). Like Levinas, Koyré does not pay much attention to the second part of the syllogism (the transition from a possible existence to the actual existence); he is entirely focused on the first part: the discovery of the *idea of God* as the most perfect, or infinite, being. According to Koyré, this idea constitutes the very core of Descartes' philosophical system, where both the world and the cognition are to be understood *sub specie deitatis*.[15] Indeed, *cogito* itself is not an independent reality. In the radical doubt, *cogito* is given to us; psychologically, we experience this givenness as an absolute certainty. However, the philosophical status of this givenness remains unclear: it has to be confirmed by another act of givenness, that is, the givenness of *ego sum, ego existo*, where God and I are accessed in the very same act of immediate intuition. It is in this act that we reach the unity between being and thought,[16] and that is why the act of self-reflection could become the fundament of truth.

---

[12] Méditations [...] œuvre hardie d'apologétique mystique" (Koyré 1922a, p. 1).

[13] L'infinité de Dieu et l'impossibilité de connaître ses raisons [...] – voilà tout ce qui est. nécessaire pour le fondement métaphysique de la physique des causes efficientes (Koyré 1922a, p. 1).

[14] Ferdinand Alquié is even more radical: "L'homme n'a pas d'idée de Dieu, il est. l'idée de Dieu" (Alquié 1950, p. 236).

[15] "[...] l'idée de Dieu est. au centre du système, parce que la pensée cartésienne part de Dieu et revient à Dieu, envisage le monde et la connaissance *sub specie deitatis*, en fonction et par rapport à Dieu" (Koyré 1922a, p. 61).

[16] "Nous ne pouvons point les mettre en doute — non seulement parce qu'il nous est. impossible de les penser sans éprouver chaque fois que nous les pensons le même sentiment de certitude absolue —; ce n'est. pas seulement un fait psychologique qui se renouvelle chaque fois — nous voyons bien qu'il ne peut pas en être autrement, parce qu'en nous pensant nous-mêmes nous saisis-

Also much later, in the 1950s, Koyré writes:

> The idea of the infinite plays an important part in the philosophy of Descartes, so important that all Cartesian philosophy may be considered as being wholly based upon this idea. Indeed, it is only as absolutely infinite being that God can be conceived; it is only as such that He can be proved to exist; it is only by the possessing this idea that man's true nature – *of a finite being endowed with the idea of God* – can be defined. (Koyré 1957, p. 106, *italics* added)

So, the originality of the Cartesian proof consists in "the idea of God as I find in myself, or I myself in so far as I possess this idea, or, if one wants to be even more precise, the fact that I possess the idea of God".[17] The attention is relocated from the object of thought to the very act of thinking. The crucial moment here is the manner in which consciousness intends and represents God or perhaps the gap between the representation and the represented in this act of consciousness.

Koyré emphasises that God as infinite being cannot be an object of intuition: "the infinite distance separating everything that is finite from the infinite"[18] is an "abyss" (*l'abîme,* Koyré 1922a, p. 129) that cannot be "filled" by the intuition, and a distinct perception of divine essence cannot be attained by the finite human spirit. The idea of the infinite God:

> [...] is certainly a *clear* and *positive* one—we do not reach infinity by negating finitude; on the contrary, it is by negating the infinite that we conceive finiteness, and yet it is not *distinct*.[19] It so far surpasses the level of our finite understanding that we can neither comprehend nor even analyse it completely. (Koyré 1957, p. 106, author's *italics*)

---

sons directement notre propre être, parce que dans ce cas privilégié notre pensée et notre être ne font plus qu'un, non pas que notre pensée soit absolument identique à notre être" (Koyré 1922a, b, p. 58).

[17] "[...] l'idée de Dieu en tant que réalisée en moi, ou moi-même en tant que je possède cette idée ou, si l'on veut être plus précis encore, le fait que je possède une idée de Dieu – voilà la base de démonstrations de la IIIe et IVe Méditations" (Koyré 1922a, b, p. 149).

[18] "Personne n'avait pas aussi bien compris la différence, la distance infinie qui sépare tout fini de l'infini" (Koyré 1922a, b, p. 126). Cf. in *Totality and infinity*: "[...] the distance that separates *ideatum* and idea constitutes the content of *ideatum* itself... it is it infinitely removed from its idea [éloigné de son idée]" (Levinas 1969, p. 49). For Levinas, as well as for Koyré, the most important point in the *Third Meditation* is the inadequacy between the idea of the infinite and the finite being that has this idea, that is, "[...] une [...] disproportion entre cette idée et nous–mêmes" (Koyré 1922a, b, p. 157). Of course, Levinas is concerned here not with the mathematical infinity but with the infinitely transcendent, that is, the absolutely other, but the logical structure of his argument is borrowed from Koyré.

[19] Jeangène Vilmer spotted that Koyré is literarily wrong here, as Descartes repeats several times that the idea of God is *maxime clara et distincta* (Jeangène Vilmer 2009, p. 505). But how could Koyré make such a blatant mistake? Perhaps one can find an explanation in his earlier version of the same thesis: "Dieu, dans l'infinie et absolue clarté de son essence, reste quand même inaccessible à notre entendement, qui ne peut en ce monde (in *statu viae*) en avoir une connaissance intuitive et complète. L'idée de Dieu, tout en étant la plus claire de nos idées, reste indistincte malgré, ou peut–être, à cause de sa trop grande clarté. Elle est. trop claire, trop lumineuse; *elle nous aveugle*" (Koyré 1922a, p. 23, italics added). The idea of God is too bright for us and so its light makes us blind; in the same manner as Marion's "saturated phenomena" "blind" the gaze and thus cause the "bedazzlement", *aveuglement* (Marion 2002, p. 206).

To think of God and to speak about God require an indirect representation of God. That is why Koyré claims that Anselm's proof is superior to that of Descartes in this respect. Indeed, the argument of *Proslogion* does not prove the existence of God; it rather proves only the "impossibility to comprehend" God's non-existence. "Not just the metaphysical impossibility, but first and foremost the logical impossibility", emphasises Koyré (Koyré 1923, p. 201). We are not capable to comprehend the non-existence of God because this is a thought impossible to think and a thought that contradicts inherent laws of thinking itself.[20]

In other words, Anselm's argument is not an ontological proof "in the proper sense of this term" (Koyré 1971, p. 69); it is a purely logical one.[21] This is a logical exercise pointing at intrinsic problems of thinking. What happens when the fool says in his heart *non est Deum*? According to Koyré, the fool

> [...] says something he does not even understand himself, something that has no meaning whatsoever. In fact, his statement is not just false – it is contradictory to the extent that it cannot even be thought – as long as to think means anything different from a purely verbal thought. (Koyré 1923, p. 201)

Employing phenomenological terminology, one could say that the fool stays at the level of pure indication and never reaches the level of signification, a level where meaning is achieved. Indeed, a genuine thought attains the being itself. This definition of thinking reminds one of Husserl rather than the Archbishop of Canterbury, but this is a very important point for Koyré, who inscribes St. Anselm in the general perspective of Neoplatonic mysticism. It is impossible to think the finite unless one makes the infinite one's starting point, and it is impossible to think the non-existence of God; however, the concepts of God, the infinite and the continuous, are not properly conceivable either. Above all, they are not to be conceivable rationally (Koyré 1929a, p. 305). These concepts are contradictory (Koyré 1929a, p. 303), as they cannot be properly delimited and determined (Koyré 1923, p. 132). Levinas, too, tries to escape the domain of the thinkable *otherwise than knowledge*; his philosophical task can be described as *a spiritual intrigue quite different from the gnosis*. Any philosophy of knowledge is inferior to first philosophy, the philosophy of affection by the transcendence that has an ethical character. A closer look at this ethics reveals a deep relation to a reading of Descartes' *Third Meditation* that strongly reminds that of Koyré. It is the Cartesian idea of the infinite that Levinas uses as a structure of relation to the Other.[22]

[20] Il suit de ses considérations que *l'ens majus cogitari nequit* ne peut pas être envisagé comme n'existant pas; et cela non seulement parce que c'est. une impossibilité *quoad rem*, mais aussi et surtout parce que c'est. une pensée impossible à penser, une pensée qui contredit les lois immanentes de la pensée elle-même (Koyré 1923, p. 202).

[21] Gilson reiterates this observation in his polemics with Karl Bart (Gilson 1934); later, this point has been developed by J. L. Marion (Marion 1991, p. 221 sq.).

[22] "This relation of the same with the other, where the transcendence of the relation does not cut the bonds a relation implies, yet where these bonds do not unite the same and the other into a Whole, is in fact fixed in the situation described by Descartes in which the 'I think' maintains with the Infinite it can nowise contain and from which it is separated a relation called 'idea of infinity'" (Levinas 1969, p. 48).

What are the key features of this relation to the Other? Firstly, this relation cannot be described in terms of intentionality, because it is not a relation to an object.[23] I am not able to comprehend the infinite,[24] and this very inability constitutes "the condition – or non-condition – of thought" (Levinas 1998, p. 65). The idea of the infinite cannot be attained by any intuition whatsoever; it is not controlled by the noesis–noema structure of consciousness (Levinas 1969, p. 264). Infinity belongs to the realm of the meontological (Kearney and Levinas 1986, p. 25) – that is why we are not able to represent the infinite through the intentional structures of consciousness. Secondly, this is a relation to something different that precedes me – not merely in ontological sense but above all in the order of constitution of my own subjectivity. Levinas writes:

> [...] a separated being fixed in its identity, the same, the I, nonetheless contains in itself what it can neither contain nor receive solely by virtue of its own identity. Subjectivity realizes these impossible exigencies—the astonishing feat of containing more than it is possible to contain. (Levinas 1969, p. 27)

I cannot bypass the Infinite if I try to get access to my own self – this is indeed very similar to Koyré's description of the argument of Descartes.[25] Therefore, in Levinas one also finds the "anthropological" reading of the idea of the infinite, used by him to define the "humanity of humans".[26] The idea of the infinite is remarkable not because it gives us a knowledge of God – such a knowledge cannot belong to philosophy, since God cannot become a theme of our discourse; in thematising God we reduce Him to a "conceptual idol", as Jean-Luc Marion would put it. The idea of the infinite is remarkable in the first place because it awakes the I to responsibility for the Other that constitutes me as a subject.

Thirdly, it is not the idea of the infinite as such that matters for Levinas, but "the idea of the infinite in me" (Levinas 1998, p. xiv). Here again one notices the same displacement of the meaning of the ontological argument – from God as such to the subject that finds herself or himself endowed with the idea of God.[27] Levinas, as Koyré before him, uses here a kind of phenomenological *epoche*: the thesis of God's

---

[23] To think the infinite, the transcendent and the Stranger is hence not to think an object. But to think what does not have the lineaments of an object is in reality to do more or better than think "[...] *the difference between objectivity and transcendence will serve as a general guideline for all the analyses of this work*" (Levinas 1969, p. 49, author's *italics*).

[24] Cf. "[...] le fini ne peut 'comprendre', ne peut embrasser et contenir l'infini" (Koyré 1922a, p. 137).

[25] Cf. "Nous ne pouvons nous voir sans voir Dieu, nous ne pouvons nous voir autrement que dans la lumière divine, et notre existence nous apparaît désormais donnée dans l'évidence absolue de l'intuition immédiate, justifiée et garantie par la clarté de la lumière divine qui, se manifestant comme telle, porte en elle–même sa justification et sa garantie" (Koyré 1922a, p. 59).

[26] "[...] by the idea of the infinite in us or by the humanity of man understood as the theology or the intelligibility of transcendent" (Levinas 1996, p. 149).

[27] Cf. "La base réelle de ces preuves, et leur sens profond, est. très simple – c'est. Descartes lui-même que le dit –: *la conscience de soi implique la conscience de Dieu*. Le 'je pense' implique 'je pense Dieu'. J'en eu donc une idée. Et c'est. une idée innée, une idée sans laquelle nous sommes impensables" (Koyré 1962, p. 224–225).

existence is suspended, and thus the structure of subjectivity comes into light as an instance capable of thinking God. But for Levinas this thinking is "a thought that at each instant thinks more than it thinks" (Levinas 1969, p. 62).

This expression of Anselm (*majus quam cogitari possit*), repeated by Levinas a number of times (Cf. Kienzler 1990), marks a point of divergence. For Levinas the ontological proof of the existence of God, while proving nothing, reveals to us the secret of our own subjectivity. It shows that we, the humans, are capable of the "breakup of consciousness" (Levinas 1998, p. 63). Koyré would never admit such a breakup. Undoubtedly, there are objects that are not properly representable; in such cases the corresponding acts may lack adequacy, the correlation between an act and its content may fail, and logical paradoxes may appear, but the structure of consciousness as such remains intact. There are thoughts too complicated, too difficult to think – such as the idea of the infinite and the idea of the continuous – but there are no thoughts too "big" for consciousness to contain them, there are no thoughts that would make consciousness explode.

What Levinas describes under the term of "the–idea–of–the–infinite–in–me" is the affection of the finite by the infinite; it is desire. This desire is at the same time an intrigue of three players: a kind of love triangle between the I, God and the Other (Levinas 1998, p. 68). In other words, the idea of the infinite comes in "two copies" given to us as two modes that cannot be reduced to one another: as the idea of God and the face of the Other. "The way in which the other presents itself, *exceeding the idea of the other in me*, we here name face" (Levinas 1969, p. 50). The alterity of the Other is given to me in the same manner as the idea of the Creator; it is given to me by the same structure as the idea of the infinite. It affects me simultaneously from the outside and from the inside: from the outside, because I am affected by an independent reality, and from the inside, because I am affected by my own thinking (and my own speech). This notion of thought as affection is completely foreign to Koyré. Koyré remains an essentially pre-Heideggerian philosopher; for him the subject means a free and autonomous being.

21.3 Self-Manifestation of the Absolute: Koyré and Michel Henry

In the second part of the paper, I would like to investigate the influence of Koyré on the work of Michel Henry, one of the most important French phenomenologists of the generation that followed Levinas. Henry was an attentive reader of Koyré's book on Böhme,[28] and it seems that his reading (or, perhaps, his misreading) of the great German mystical thinker played an important role in the development of Henry's own views on the history of philosophy.

[28] As Grégori Jean writes in his commentary on the transcription of one of Henry's unpublished manuscripts, Koyré's Böhme was an "[…] ouvrage, qui si l'on en croit les notes préparatoires à *L'essence de la manifestation,* avait beaucoup impressionné Henry" (Henry 2013, p. 49).

Let us recall some of the crucial points of Henry's philosophy that will be important for us. According to Henry, rigorous application of the principle of reduction requires bracketing not only the objective world and the empirical I but also the phenomena inasmuch as *they show themselves*. Phenomenology is not the study of phenomena but rather the study of their manner of showing themselves and their modes of appearance. To each type of phenomena belongs a specific mode of phenomenalisation. In particular, there are two main types of phenomenality: the mode in which we perceive the objects of the world, the earthly things, and the mode in which we experience ourselves. Although the second mode is original and authentic, it has been completely obliterated from the history of philosophy. For Heidegger the black sheep of the history of being is Descartes, a French philosopher; but Henry gives this role to a figure from *outre-Rhin*, Jakob Böhme, the *Teutonicus philosophus*. It should be noted that in the works of Henry, one finds no trace of his reading of Böhme. Whenever Henry mentions his name, he means Böhme interpreted by Koyré; mostly he refers to the chapter devoted to the idea of God in Böhme (the same chapter that Koyré has republished in German in Husserl's *Festschrift*).

Koyré's book on Böhme is not a stand-alone piece of research but rather a part of a more ambitious project, that is, a history of the German mysticism: in the 1920s Koyré subsequently teaches different courses on this subject in the *EPHE*. Koyré's works on Schwenckfeld, Franck, Weigel, Paracelsus and Böhme represent those thinkers in the greater context of development of the German thought that has reached its highest point in the German idealism and the Romanticism. Describing their theological and mystical concepts, Koyré builds a certain perspective: all these authors represent the major trend of the German thought, where the magical worldview, based on the idea of the *imagination,* is gradually transformed into the philosophical, and then scientific, concept of the universe. The key notion for Koyré is the *principle of expression*, since it is central for the medieval mystics as well as for later thinkers like Schelling or Fichte. Expression as manifestation, expression as incarnation and expression as objectivation are the principal themes in Koyré's most important work of that period, that is, his dissertation on Böhme.

Koyré's book was highly praised by his contemporaries; to the best of my knowledge, the only negative reaction was that of Berdiaev (Berdiaev 1929), who complained that Koyré "fails to understand, that it is impossible ultimately to understand Böhme, it is impossible to convey him in the language of clear concepts".[29] I tend to agree with Berdiaev that Böhme was too foreign for Koyré, who tried to "tame" Böhme by imposing on him a certain conceptual scheme that does not exhaust the depth of Böhme's heritage; sometimes it may seem that Koyré is so much annoyed by Böhme that eventually his famous ability to read medieval texts "emphatically" fails him.[30] However, it is the very power and clarity of this conceptual scheme that

[29] However, Koyré did his best not to over-clarify Böhme's thought: "Nous ne voudrions pas [...] laisser notre lecteur sous l'impression d'une trop grande clarté" (Koyré 1929a, p. 392).

[30] Jorland calls Koyré's method "empathic" (Jorland 1994). Berdiaev also praises Koyré's ability to represent empathically Böhme's thought, although that thought remained largely foreign to him: "The author has made everything possible and even impossible to penetrate the world of visions

made Koyré's book so influential. Indeed, Henry's critique of Böhme is directed not so much towards Böhme but rather towards this scheme.

In a nutshell, Koyré's vision of Böhme's work could be described as follows: Böhme is "a great metaphysical genius" (Koyré 1929a, p. 394) who is "fighting himself" in order to resolve "the central problem of metaphysics, that is, the problem of the Absolute" (Koyré 1929a, p. 307). How can one reach the Absolute,[31] transcendent to the world and to the thought? Böhme (as Sebastian Frank before him and all the German idealism after him) implicitly presupposes the difference between the expressed and its manifestation.[32] The metaphysical question "How do we think the Absolute that is not properly thinkable?" is replaced by the question "How does the Absolute manifest itself?" The Absolute desires to manifest itself and, in order to do so, creates the world. As Michel Henry puts it, the apparently theological problem (why God created the world) here is dealt with in a purely phenomenological way[33]: one can say that the Absolute manifests itself in the world and through the world and that the world was created in order to become its "medium of phenomenalisation".

The structure of the manifestation is very complex and the whole development of Böhme's thought could then be described as an elucidation of this structure. Koyré follows Oetinger's breakthrough motto: the Absolute is *ens manifestavitum sui*, or, as Koyré puts it, *mysterium manifestans seipsum* (Koyré 1929a, p. 329).[34] The heart of this mystery is the will to self-manifest. The Absolute in itself is the source of all manifestation; apart from this the Absolute has no other essence. In a sense, the Absolute has no being outside its self-manifestation. In other words, one could apply to the Absolute the motto of Herbart, (mis)quoted in *Cartesian Meditations* and *Being and Time*: "so much semblance, so much being".[35]

---

and thinking of the great Christian theosophist, the world which is so alien to him. He has revealed a substantial ability to empathise with the thinking of others (*sposobnost' vzhivanija v chuzhuju mysl'*)" (Berdiaev 1929, p. 116).

[31] Koyré translates "Ungrund" by "Absolu". This serves his main purpose: to inscribe Böhme in the tradition of Paracelsian Augustinism, or even in the Neoplatonic tradition in the broadest sense.

[32] La solution de Böhme – suivi en cela par l'idéalisme allemand – consistera à poser comme principe d'explication le rapport d'expression dans toutes ses formes. L'avantage de cette méthode réside dans la position implicite de la différence entre ce qui est. exprimé et sa manifestation (Koyré 1929a, p. 306, n. 2). See also (Henry 1973, p. 159).

[33] "[…] à l'immense question en apparence théologique: pourquoi Dieu a–t–il crée le monde? – l'extraordinaire réponse avancée appartient à la phénoménologie: Dieu a crée le monde pour se manifester" (Henry 2000, p. 66).

[34] In his 1947 review, Koyré slightly changes the wording to emphasise the importance of the revelation *ad intra*: *mysterium revelans seipsum seipso* (Koyré 1947, p. 425). One would not fail to notice the word *revelatio* in this text. Indeed, in the book written in 1929, Koyré mostly translates *Offenbarung* as *manifestation* and avoids the word *révélation*. This translation strategy, an example of which we have seen in the translation of *Ungrund* by *Absolu*, systematically represents Böhme as a philosopher rather than a religious thinker.

[35] "The participation in apodicticity appears in the *formal law* (which is itself apodictic): so much illusion, so much being (*Soviel Schein, soviel Sein*) – which is only covered up and falsified thereby and which therefore can be asked about, sought, and … found" (Husserl 1960, p. 103). "Yet so

But the very idea of self-manifestation implies a kind of "internal duality". According to the "principle of expression", by distinguishing expression and the expressed, the act of manifestation produces a certain splitting between the manifestation and the manifested.[36] Manifestation manifests something else, that is, the manifested. As Michel Henry puts it, "already with Böhme there was the thought that a manifestation [of the world as opposed to the manifestation of the life] could not but manifest something else" (Henry 1973, p. 107). In Heidegger's terms, this kind of manifestation is not a phenomenon in the "positive and primordial" sense, that is "as that which shows itself", but an appearance (*Erscheinung*): "appearance, as the appearance 'of something', does not mean showing itself; it means rather the announcing-itself by something which does not show itself, but which announces itself through something which does show itself" (Heidegger 1962, pp. 51–52). Such phenomenological structure can be described as self-reference or self-indication. Thus, God revealed *is* a phenomenon, albeit not a phenomenon "in the phenomenological sense": it does not "show itself from itself"; it rather announces itself to itself through something else, that is, through its own manifestation. According to Michel Henry, this scheme describes the phenomenological structure of the manifestation of self-reflection as opposed to the self-manifestation of one's inner life.

So, in order to self-manifest, the Absolute has to split itself into two instances: the manifested self and the manifesting self. This duplication occurs as the Absolute becomes a sort of a mirror, which is, at the same time, an eye. The eye that looks and sees, the "eye subject" and the "eye object", says Koyré.[37] Interpreting the well-known passage from "six theosophical points", Koyré imposes on the reader the language of the subject–object relation. It is no surprise that, thus interpreted, Böhme is no longer a mystic or a theosophist but looks rather like an average representative of the German idealism. As G. Jorland puts it, "Hegel *genuit* Marx, but Böhme *genuit* Hegel" (Jorland 1981, p. 196). Characteristically, Henry also holds Böhme responsible for the primacy of the idea of *Scheidichkeit*, the reflective

much semblance, so much 'Being'" (*Wieviel Schein jedoch, soviel "Sein"*) (Heidegger 1962, p. 51). Michel Henry proclaimed this motto as "[…] the third principle of phenomenology […]". For more details on the history of the maxim, see (Benoist 2010).

[36] "C'est. une idée complexe et même paradoxale que cette idée de la manifestation. Elle apparaît, à première vue, comme presque contradictoire. En effet, elle implique une sorte de la dualité interne. Ce qui se manifeste, se manifeste nécessairement par quelque chose qu'il n'est. pas, et qui n'est. pas lui; inversement, une manifestation (un phénomène) ne saurait manifester qu'autre chose que soi" (Koyré 1929a, p. 243).

[37] "L'un Absolu est donc un œil qui veut regarder et qui désire voir, car, en effet, qu'est.-ce qu'un œil qui ne voit rien? 'Autant que rien', autant que miroir qui ne réfléchit rien; une simple possibilité de réflexion et la vision; nullement une vision réelle. L'Un est un œil qui veut voir. Mais que pourrait-il voir là où il n'y a rien? Rien, évidemment, si ce n'est soi-même. C'est donc soi-même qu'il regarde et soi-même qu'il voit, et étant ainsi sujet et objet de la vision, on peut bien dire qu'il se dédouble en se réfléchissant en lui-même. Il ne voit rien et pourtant il se voit" (Koyré 1929a, p. 332).

splitting of being, in the German idealism and, consequently, in the modern thought in general.[38]

This double (or, rather, triple) structure *vision – the eye that sees – the eye that is seen* generates the personality, the mind (*mens*, *Gemüth*) and ego (*Ichkeit*) in God. More precisely, this is just a first step towards generating consciousness and personality in the Absolute, because so long as the Absolute is not fully reflected in the mirror, it has not yet become self-consciousness and self-understanding. Indeed, says Koyré, "pure thought cannot be given to itself" (Koyré 1929a, p. 235) until it has an object, which is separate from the thought. So long as there is nothing that can be reflected in the mirror, and so long as there is no object in a strict sense, there can be no vision: indeed, a reflection in the mirror is not a genuine vision but only "a mere possibility of vision" (Koyré 1929a, b, p. 332). Light itself is not visible; it is just a medium of vision, "wherein something can become manifest, visible in itself" (Heidegger 1962, p. 51), a "horizon of visibility" (Henry 2008, p. 84) for phenomena or rather for "true objects".[39] As Henry puts it, since we cannot think of a consciousness that is not a consciousness of an object, it is precisely the (determinate) *object* that "*permits the consciousness to be what it is [...]* The transcendental can only appear under the form of the object in the internal sense" (Henry 1973, p. 112, p. 119, author's *italics*). Moreover, without "real", that is, finite, material, natural, determinate objects, the very personality of God becomes only "a scheme devoid of reality, a simulacrum, a shadow" (Koyré 1929a, p. 333). In other words, the structure of every consciousness, even of divine consciousness, is the structure of the "directedness towards [...]", that is, of intentional consciousness in Husserl's sense. Henry claims, however, that this analogy between divine and created consciousness is false; the very idea that the self-manifestation of the divine (or transcendental) consciousness somehow depends on a finite, ontic object, on "something else",[40] constitutes a "fundamental philosophical error" (Henry 2008, p. 48) that goes back to Böhme.

It is to the God of Koyré's Böhme even more than to anything else, that one can apply these words of Michel Henry: "the subjectivity of the subject" realises itself as "the objectivity of the object" (Henry 1973, p. 90), although Henry imputes this

---

[38] "Jacob Böhme's intuitions guide German Idealism and thus modern thought [...] These intuitions must be called into question again" (Henry 2008, p. 97). "*Schiedlichkeit* is the condition of consciousness. The concept of consciousness is thought of by Böhme in its solidarity with the concepts of otherness, mirror, splitting, namely in its unity with the ontological process of the internal division of Being [...]. The interpretation of the concept of consciousness which arises from 'splitting' [...] does not show up, merely through the influence of Böhme, in the *System of Transcendental Idealism*; it actually dominates all subsequent work of Schelling and notably his final philosophy" (Henry 1973, p. 79, translation modified).

[39] La clarté diffuse de l'Un s'est concentrée et est devenue lumière, mais c'est une lumière irréelle et invisible, puisqu'elle ne remplit point sa fonction d'éclairer. Il faut que quelque chose se place devant elle, que la pensée reçoive un objet véritable pour qu'elle se réalise dans le sens propre du terme et réalise ainsi son sujet (Koyré 1929a, b, p. 334–335).

[40] "The essence in its phenomenological realization is 'something else'" (Henry 1973, p. 120; this is indeed a hidden quotation from Koyré 1929a, p. 243).

identification of manifestation with objectification to the entire philosophical tradition.[41] What is striking here is the "impotence" of this objectification; the pure thought does not exist, because to "think" means to "think of something". For Henry, intentional structure of consciousness is (self-)objectification.

Why are the ontic objects necessary to achieve the manifestation (and self-manifestation) of God? This is a particular case of a general law that Koyré calls "the law of oppositions" or the "*Mysterium Magnum* of being". Manifestation is possible "only in the *other* and by the *other*": manifestation ought to be limited, because the manifested manifests itself "in the limit and in relation to the limit".[42] Koyré ascribes this proto-Hegelian thesis not only to Böhme but also to the German thought in general.[43] However, the law of oppositions is not only logical or phenomenological law; it is a cosmological one. According to this law, God produces Sophia, the Divine Wisdom. Sophia is an eye and a mirror at the same time; reflected in the Wisdom, God feels Himself capable to create a genuine object (*Objectum, Gegenwurf*) that will oppose Him and that will be "other than" Him. It is according to the law of the opposites that God is finally able to see Himself as a subject and is able to juxtapose Himself to a project of a world – a project of a separate and independent being. However, these projected ideas are not quite real; to make them fully real, God has to use imagination, "a mysterious and magical act *par excellence*" (Koyré 1929a, p. 347). God *imagines Himself* in the Wisdom, making it the

[41] "Jacob Böhme interprets divine Wisdom, as the first objectification of the divine essence, to be identical with its first manifestation" (Henry 2008, p. 97). Cf. also: "La structure phénoménologique d'une telle manifestation est clairement indiquée. Elle consiste dans une objectivation, dans cette objectivation qui est celle du monde, de telle façon que – en cette fin de Renaissance aussi bien qu'en Grèce – c'est la position de soi hors de soi qui fait surgir la manifestation. Puisqu'il s'agit en l'occurrence de la manifestation de Dieu – manifestation que Böhme appelle sa Sagesse (autre nom du Verbe) –, c'est donc comme objectivation du premier Dehors que celle-ci se produit" (Henry 2000, p. 67). Henry does not quote Koyré here, but he reproduces his formula "la Sagesse [...] est. [...] une première objectivation *ad extra* de la divinité" (Koyré 1929a, p. 297) almost literally.

[42] "La tendance à la manifestation, à la révélation est. une loi générale de l'être, une loi qui, non est seulement inhérente à l'être, mais qui, à vrai dire, le constitue et finalement l'explique [...] *L'un* ne peut arriver à s'exprimer et à se déterminer que dans *l'autre* et par *l'autre*. L'indéterminé aspire à une limite, non pour se limiter toutefois, mais pour se révéler. Il s'oppose une détermination et une limite, pour *se* révéler et *la* révéler en même temps; se révéler en elle, et par rapport à elle. C'est là le grand mystère de l'être, son *Mysterium Magnum*: les contraires s'impliques et restent unis dans leur opposition révélatrice" (Koyré 1929a, p. 245).

[43] 24 November 1924 Koyré wrote to Meyerson from London, where he worked on Paracelsus and early German mystics (and not on Böhme): "Ce gens–là font du Hegel avant la lettre et, ayant remarqué que le même *s'exprime* toujours par et dans l'autre, bien que cela ne soit nullement compréhensible, ils mettent ce fait – *Mysterium Magnum* de la Parole – à la base de tout conception et explication, disant, tout à fait comme Hegel, que le seul moyen de se débarrasser de ce *mysterium* est. de le mettre au centre même du système et d'en faire le principe du pensée vivante. Je crois que tout l'idéalisme allemand est – historiquement – parti de là" (Meyerson 2009, p. 236, italics by Koyré). See also "Or [...] si le *Mysterium* forme l'essence et base du monde, de la vie, de nous-même et de notre raison, ne faut–il pas placer résolument le *Mysterium* au centre du système et, au lieu de chercher à l'illuminer, en faire le principe explicatif par excellence?" (Koyré 1929a, b, p. 312).

first step in the transformation of infinite divine ideas into finite being. However, the ideal (and non-fictional) Cosmos, imagined by God in Sophia, is not yet a proper creation. It is "semi-real", like a plan of an artisan (Koyré 1929a, p. 348). Its existence is magical because it does not exist by itself but only in its relation to God and His imagination; it is not a genuine "real" thing but only *magia divina*. Nevertheless, the existence of this world (or, more precisely, not yet a world) enables God to begin a centrifugal movement: through it He becomes conscious of His own personality; He separates Himself from this world; and He thus begins cognition. God is expressing Himself into the imagined world in order to achieve self-knowledge. He knows Himself as His own expression in the magical world and also as something which resists this expression (Koyré 1929a, p. 348).

To know Himself, God needs nature. This is a crucial point in Koyré's conceptual scheme: neither reflection nor imagination is capable of providing true access to one's self. It is because God has a nature, that He can be a personality proper and not just a pure spirit unable to create anything new. It is because God possesses a nature, that Böhme's cosmogony becomes cosmology: the knowledge of God is linked with the knowledge of the world and the knowledge of the self.[44] According to Böhme – that is, to Koyré's Böhme – in order to know oneself, one has to know the world and its origin; the only path to one's inner self goes through the world.

In Michel Henry's interpretation, the Böhmian cosmogony serves as a model for the genesis of transcendental subjectivity: the story of God becomes a story of the transcendental subject, produced by the splitting of the empirical subject into the phenomenological onlooker and the world-constituting consciousness. He completely ignores the theosophical and, more generally, the cosmological aspects of Böhme's (and Koyré's) thought: for Henry the cosmological dimension of human experience has nothing to do with the primary mode of subjective life, that is, with pure self-affection. Actually, Henry claims that any splitting of the self – be it the splitting in order to perform phenomenological reduction or just the splitting of the self in self-reflection – already presupposes self-alienation. When we reflect upon ourselves, we look at ourselves in the same way we look at things, at worldly objects – but the self is not a thing or an object. The way I manifest to myself and the way things appear to me are radically different. So, the optical metaphor of manifestation, the metaphor of reflection, is to be subjected to the tactile metaphor, the metaphor of touching, or, better, to the feeling of pain and of passion. We feel ourselves immediately, before any reflection. I look at things and see them: they are given to me in intuitive vision, but this vision itself comes from a feeling, which is non-intentional. Henry insists that intuition is no direct access to the world and to the self, because intuition always presupposes mediation by a primary feeling and by a pathetic impression. According to Henry, the embodied human subjectivity is

[44] "La connaissance de Dieu était liée à celle du monde: la renaissance, la naissance spirituelle était pour lui quelque chose qui, permettant à l'homme de se reconquérir lui–même, lui permettait aussi de voir le monde dans sa réalité profonde, de pénétrer dans le mystère de la nature et d'y voir Dieu, non son reflet ou son symbole seulement, mais de l'y voir lui–même, en tant, bien entendu, qu'il s'y exprime" (Koyré 1929a, p. 485).

capable of "feeling itself by itself", the ability which Böhme's disincarnated God is lacking.

21.4 Conclusion

Paul Ricoeur once said that the history of phenomenology is a history of Husserlian heresies (Ricoeur 2004, p. 182). If so, Koyré, an unorthodox phenomenologist, has his own rightful place in it. He is a supporter of intentionality. The unity of being and thought, accomplished in consciousness thanks to its intentional structure, attests that the exercise of thinking is more than a purely psychological fact, that in and by an act of thought one can achieve the truth. This necessity of truth for the being of humans is a lesson of Koyré that the French "theological" phenomenology would do well to retain. There are atheistic versions of the philosophy of Levinas or of Henry; the presence of the concept of God in the phenomenological discourse is not a problem. The true problem is the problem of truth: the necessity of the most banal and most humble truth – the truth of propositions, of statements regarding the world – is lost in the post-Heideggerian and, to an even larger extent, post-Levinassian French phenomenology. The truth of being, the truth of life and the veracity of speech addressed to the Other are very powerful, essential, even fundamental concepts of contemporary phenomenology; but if phenomenology wants to be a descriptive and not merely prescriptive philosophy, it cannot *completely* renounce the truth of the world and the truth of consciousness. Separating pure philosophy from "spirituality", Michel Foucault said that pure philosophy has forgotten that in the search for truth the subject must undergo a certain transformation; however, the supporters of "spirituality" sometimes forget that the task of thinking is not just to transform the world and one's own soul but also to understand and explain them too. The task of phenomenology is twofold, practical and theoretical at the same time; philosophical conversion differs from religious conversion because the former cannot be separated from epistemological and ontological work.

A phenomenology that would not be synonymous with pure ascetics should be situated in the bordering region between the world and the beyond. It seems, therefore, that one needs to bypass the jargon of the authentic and the original, as well as the metaphysics of the ultimate foundation. If Koyré was right to speak of the "unity of human thought", then it is not possible to do phenomenological philosophy of subject without science, in particular, the science of the world. I would suggest that one should follow the way of Emmanuel Levinas and Michel Henry and go beyond them in order to regain the theoretical truth, the truth of the world – but starting from the practical truth, the truth of philosophical conversion. Indeed, the task is to reintegrate the disjointed parts of phenomenological heritage. A phenomenology that wants to be more than the philosophy of subject should stop being purely a-cosmic and should recover its scientific, epistemological, ontological dimensions and, most importantly, its cosmological dimension.

Acknowledgments I would like to express my gratitude to G. Pattison, D. Drozdova, J.-M. Salanskis, Z. Sokuler, F. Worms and G. Jean for comments and discussions.

References

Alquié F (1950) La découverte métaphysique de l'homme chez Descartes. PUF, Paris.
Baugh B (2014) French Hegel: From Surrealism to Postmodernism. Routledge, London.
Benoist J (2010) Les vestiges du donné (Apparaître, apparences, aspects). Revue canadienne de philosophie continentale 14/1:85–103.
Berdiaev N (1929) Novaya kniga o Boeme. Put' 18:116–122.
Drozdova D (2012) Interpretatsiia Nauchnoi revoliutsii v rabotakh Aleksandra Koire: dissertatsiia na soiskanie stepeni kandidata filosofskikh nauk: 09.00.03. Natsional'nyi issledovatel'skii universitet "Vysshaia Shkola Ekonomiki", Moskva.
Geroulanos S (2010) An atheism that is not humanist emerges in French thought. The Stanford University Press, Stanford.
Gilson E (1934) Sens et nature de l'Argument de Saint Anselme. Archives d'histoire doctrinale et littéraire du moyen âge 9:5–51.
Heidegger M (1962) Being and Time. Translated by John Macquarrie & Edward Robinson. SCM Press, London.
Henry M (1973) The essence of manifestation. Translated by Girard Etzkorn. Martinus Nijhoff, Den Haag.
Henry M (2008) Material phenomenology. Translated by Scott Davidson. Fordham University Press, New York.
Henry M (2000) Incarnation: une philosophie de la chair. Seuil, Paris.
Henry M (2013) Notes sur le phénomène érotique. Presentation, commentaires par G. Jean. Revue internationale Michel Henry 4:27–56.
Hering J (1926) Phénoménologie et philosophie religieuse. Étude sur la théorie de la connaissance religieuse. Alcan, Paris.
Hering J (1950) Phenomenology in France. In Farber M (ed.) Philosophic Thought in France and the United States. Essays Representing Major Trends in Contemporary French and American Philosophy. University of Buffalo Publications in Philosophy, Buffalo.
Husserl E (1960) Cartesian Meditations. An introduction to phenomenology. Translated by Dorion Cairns. Martinus Nijhoff, Den Haag.
Husserl E (1970) Logical investigations. Translated by J. N. Findlay. Edited by Dermot Moran. Volume II. Routledge, London.
Husserl E (1994) Briefwechsel. In Verbindung mit Elisabeth Schuhmann. Herausgegeben von Karl Schuhmann. Kluwer Academic Publishers, Dordrecht; London.
Janicaud D (2001) Heidegger en France. Albin Michel, Paris.
Jeangène Vilmer JB (2009) Le paradoxe de l'infini cartésien. Archives de Philosophie 72/3:497–521.
Jorland G (1981) La science dans la philosophie: les recherches épistémologiques d'Alexandre Koyré. Gallimard, Paris.
Jorland G (1994) Koyré phénoménologue ? In Vinti C (ed). Alexandre Koyré: l'avventura intellettuale. Edizioni Scientifiche Italiane, Napoli, pp. 105–126.
Kearney R, Levinas E (1986) Dialogue between E. Levinas and R. Kearney. In Face to face with Levinas. Edited by R. Cohen. State University of New York Press, Albany, pp. 13–33.
Kienzler K (1990) Die "Idee des Unendlichen" und das "ontologische Argument" bei Anselm von Canterbury und Emmanuel Levinas. Archivio di Filosofia 58:435–458.
Koyré A (1922a) Essai sur l'idée de Dieu et les preuves de son existence chez Descartes. Ernest Leroux, Paris.

Koyré A (1922b) Bemerkungen zu den Zenonischen Paradoxen. Jahrbuch für Philosophie und phänomenologische Forschung V:610–613. [French translation: Koyré 1980, pp. 9–35].
Koyré A (1923) L'Idée de Dieu dans la philosophie de St. Anselme. Ernest Leroux, Paris.
Koyré A (1929a) La philosophie de Jakob Böhme. Vrin, Paris.
Koyré A (1929b) Die Gotteslehre Jakob Böhmes: ein Fragment. Übersetzt von Hedwig Conrad-Martius. In Festschrift Edmund Husserl zum 70. Geburtstag. Max Niemeyer Verlag, Tübingen, pp. 225–281.
Koyré A (1947) La philosophie dialectique de J. Böhme. Critique 12/2:414–426.
Koyré A (1962) Introduction à la lecture de Platon. Suivi de Entretiens sur Descartes. Gallimard, Paris.
Koyré A (1957) From the closed world to the Infinite universe. The John Hopkins Press, Baltimore.
Koyré A (1971) Mystiques, spirituels, alchimistes du XVIe siècle allemand. Gallimard, Paris.
Koyré A (1980) Études d'histoire de la pensée philosophique. Gallimard, Paris.
Kozhevnikov A (Alexandre Kojève) (1929) Zaschita dissertatsii A.V. Koyré. Yevrasia 16:8.
Kojevinkoff A (Alexandre Kojève) (1934) Recension de Phénoménologie. Journées d'Etudes de la Société Thomiste. Cerf, Juvisy. 1932. In Kojevinkoff A Recherches Philosophiques 3:429.
Lescourret MA (2005) Emmanuel Levinas. Flammarion, Paris.
Levinas E (1969) Totality and infinity; an essay on exteriority. Translated by Alphonso Lingis. Duquesne University Press, Pittsburgh.
Levinas E (1989) As If Consenting to Horror. Translated by Paula Wissing. Critical Inquiry 15:485–488.
Levinas E (1996) Transcendence and intelligibility. In Peperzak AT, Critchley S, Bernasconi R (eds.) Emmanuel Levinas: Basic Philosophical Writings. The Indiana University Press, Bloomington, pp. 149–159.
Levinas E (1998) Of God who Comes to Mind. The Stanford University Press, Stanford.
Marion JL (1991) Questions cartésiennes. Méthode et métaphysique. PUF, Paris.
Marion JL (2002) Being Given: Toward a Phenomenology of Givenness. Stanford University Press, Stanford.
Meyerson É (2009) Lettres françaises. Édited par Bernadette Bensaude–Vincent et Eva Telkes-Klein. CNRS Éditions, Paris.
Patočka J (1976) Erinnerungen an Husserl. In Die Welt des Menschen — Die Welt der Philosophie: Festschrift für Jan Patočka. Herausgegeben von Walter Biemel und dem Husserl-Archiv zu Löwen. Martinus Nijhoff, Den Haag, pp. VI–XIX.
Ricoeur P (2004) À l'école de la phénoménologie. Vrin, Paris.
Schuhmann K (1977) Husserl–Chronik: Denk- und Lebensweg Edmund Husserls. Martinus Nijhoff, Den Haag.
Schuhmann K (1997) Alexandre Koyré. In Embree L et al. (eds.) Encyclopedia of Phenomenology. Kluwer Academic Publishers, Dordrecht–London, pp. 391–393.
Spiegelberg H (1981) The context of the phenomenological movement. Martinus Nijhoff, Den Haag.
Spiegelberg H (1982) The Phenomenological Movement: a historical introduction. 3rd revised and enlarged edition. With the collaboration of Karl Schuhmann. Martinus Nijhoff, Den Haag.
Zambelli P (1999) Alexandre Koyré alla scuola di Husserl a Gottinga. Giornale critico della filosofia italiana 3:303–354.

Index

R. Pisano et al. (eds.), *Hypotheses and Perspectives in the History and Philosophy of Science*, DOI 10.1007/978-3-319-61712-1

Q

R